Evgeny G. Gurvich

Metalliferous Sediments of the World Ocean

Fundamental Theory of Deep-Sea Hydrothermal Sedimentation

Evgeny G. Gurvich

Metalliferous Sediments of the World Ocean

Fundamental Theory of Deep-Sea Hydrothermal Sedimentation

With 174 Figures and 38 Tables

Springer

Dr. Evgeny G. Gurvich
Alfred Wegener Institute
for Polar and Marine Research
Bremerhaven
Germany

E-mail:
gurvich@post.com
gurvich@pochta.ru

Library of Congress Control Number: 2005930336

ISBN-10 3-540-27869-9 Springer Berlin Heidelberg New York
ISBN-13 978-3-540-27869-6 Springer Berlin Heidelberg New York

Springer is a part of Springer Science+Business Media
Springeronline.com

Printed in The Netherlands

Cover design: E. Kirchner, Heidelberg
Typesetting: camera-ready by author
Production: Christine Jacobi
Printing: Krips bv, Meppel
Binding: Stürtz AG, Würzburg

Printed on acid-free paper 32/2132/cj 5 4 3 2 1 0

CONTENTS

FOREWORD

Dramatic advances in understanding global tectonics have been made in the last half century and the information and specific data acquired on the floor of the World Ocean by the scientific community probably has exceeded that available in all previous time. With the benefit of new technology and advanced concepts in the earth sciences extensive exploration of the deep seabed became possible, and has been carried out in many parts of the world. Many features have been recognized and data recorded that are vital for understanding the fundamental processes that shape the earth's surface and control the habitable environment. The data collected to date on the ocean floor and its physical environment greatly exceeds our understanding and appreciation of their fundamental importance in the earth sciences, and our ability to apply this knowledge effectively in improving our way of life.

With his extensive scientific knowledge and unique experience from many cruises in association with scientists throughout the world, Dr. Evgeny Gurvich has made an outstanding contribution in acquiring basic data on hydrothermal and sedimentation processes in the ocean, as well as in the synthesis of data and concepts available from cruise reports and an extensive literature. This book reports his pioneer efforts and exceptional contribution to the earth sciences in demonstrating the relationship of the physical and chemical properties of hydrothermal discharge, the spreading rate, dynamics and tectonic setting of the ocean ridges through time, and the general physical and chemical features in the ocean that influence and control the deposition of metalliferous sediments and the accumulations of mineral resources of interest to world industry.

By defining and describing the specific sedimentation and genetic processes leading to mineral deposition on the ocean floor this volume is a pioneer achievement in demonstrating how fundamental scientific data on the sea floor can be used effectively in locating, exploring and evaluating the vast mineral resources that are available to the world.

The author of this exceptional volume is anxious to obtain support for continuing his scientific research and dedicated service in exploring the new frontiers of ocean geology and mineral resources.

Gordon A. Gross
Ph.D., Senior Resource Geologist, Emeritus
Geological Survey of Canada, Ottawa

PREFACE

Metalliferous sediments are the most widespread hydrothermal object in the World Ocean. Their fields are halos of dispersion of submarine hydrothermal mineral forming systems, and the metalliferous sediments carry information on these systems. I have always tried to remember it dealing with data on the metalliferous sediments.

During work on this book I have had three main goals:

- to offer in a condensed form what is, in my opinion, the most important information and data on the distribution, composition and accumulation rates of the metalliferous sediments and their components;
- to trace the process of formation of composition of the metalliferous sediments from effluence of hydrothermal metal-bearing matter to the ocean to its accumulation on the floor;
- to show the opportunities provided by studies of the metalliferous sediments for solving scientific and practical problems.

For the preparation of the book I have used original materials and the results of analyses of samples collected during many years in the expeditions of the P.P. Shirshov Institute of Oceanology, Moscow (Cruise 3 of DES *Ob*; Cruises 8, 14 and 25 of R/V *Dmitry Mendeleev*; Cruise 24 of R/V *Akademik Kurchatov*; Cruises 7, 12, 15, 21, 23, 31, 34, 35, 41 and 42 of R/V *Akademik Mstislav Keldysh*; Cruise 3 of R/V *Professor Shtokman*), of the All-Union Institute for Geology and Mineral Resources of the World Ocean, Sankt-Petersburg (Cruise 4 of R/V *Geolog Fersman*), in Legs 6 and 35 of D/S *Glomar Challenger* (collections of Alexander Lisitzin and Yury Bogdanov) and in Leg 138 of D/S *JOIDES Resolution* (collection of Mikhail Levitan). The DSDP/ODP Directory furnished many samples on our request. All samples have been analyzed in the Laboratory of Physical Geological Research of the P.P. Shirshov Institute of Oceanology or on board the ships during expeditions. Numerous published data have been used along with the results and data from these analyses.

Probably some important data and information have been overlooked or provided in insufficient detail. If this is the case the list of references will be of interest and indicate papers and publications where the desired information may be found.

The goals set in the preparation of this book will have been achieved if it provides new incentive for study of the metalliferous sediments and arouses the interest of specialists and students in the allied earth sciences and applied geology.

For more than 25 years I worked at the P.P. Shirshov Institute of Oceanology in Moscow and studied the geochemistry of river run-off and of mixing zones of rivers and seas, the geochemistry of a number of chemical elements in the ocean sedimentary cycle, the sedimentation in Arctic seas and on continental slopes, the metalliferous sediments and submarine hydrothermal activity. Since 2002 I live in Germany. In 2003 I joined the Alfred Wegener Institute for Polar- and Marine Research in Bremerhaven (AWI) as a guest scientist and worked on the composition of bottom sediments in the Russian Arctic seas. At the beginning of 2005 my contract at the AWI run out. If my knowledge and experience are of interest for some organization in any country, I ask to consider this book as an application and contact me.

Acknowledgements

I would like to express my sincere gratitude to my teachers Yury Bogdanov and Alexander Lisitzin. Not only have they introduced me to and taught me science, but also they have encouraged me to write this book. Alexander Lisitzin has always supported me and helped me in difficult situations.

I am very pleased to thank my colleagues for friendly cooperation in joint work in many ocean expeditions. In the course of my work I constantly received the support and accepted the good advice of my friends and colleagues: Genrikh Arnautov, Nikolay Bortnikov, Georgy Cherkashev, Lyudmila Demina, David Gershanovich, Yury Girin, Zinaida Gorbunova, Vladimir Gordeev, Vyacheslav Gordeev, Dmitry Grichuk, Alexandra Isaeva, Georgy Ivanov, Nikolay Katargin, Sergey Krasnov, Alexander Krasnyuk, Anatoly Kurinov, Tat'yana Kuz'mina, Lev Lebedev, Alla Lein, Mikhail Levitan, Vyacheslav Lukashin, Areg Migdisov, Alexey Mitropolsky, Lev Moskalev, Konstantin Muraviov, Inna Nemirovskaya, Alexander Novigatsky, Esper Ostroumov, Valery Peresypkin, Irina Poroshina, Anatoly Schreider, Valentina Serova, Vladimir Shevchenko, Vyacheslav Sval'nov, Evgeny Verzhbitsky, Mikhail Vinogradov, Igor' Volkov, Nina Zhabina, Valentin Zinkevich, and Lev Zonenshain.

During my studies of metalliferous sediments I benefited from the cooperation of Valery Bagin, Gleb Baturin, Galina Butuzova, Tat'yana Gendler, Yury Girin, Vesselin Dekov, Nadezhda Lisitzina, Valentin Oreshkin, Alexander Ronov, Valentin Strizhov, Vladimir Ustinov, and Victor Zaikov.

Contact and cooperative work with eminent scientists: Enrico Bonatti, Keith Crook, Harry Elderfield, Alexander Malahoff, Harald Puchelt, Peter Rona, Mark Rudnicki, and Steven Scott were especially helpful for me.

During my work in Germany Dieter Fütterer, Hannes Grobe, Gerhard Haass, Frank Schoster, Reiner Sieger, Rüdiger Stein, Tatjana Steinke, Peter Stoffers, and Jörn Thiede supported me and gave me good advice and assistance.

I would like to express my sincere gratitude to Anatoly Sagalevich for making it possible for me to participate in a series of expeditions to areas of hydrothermal activity in the oceans. Extremely successful work during these expeditions would have been impossible without the admirable experts of the Russian scientific fleet: Captain Yury Gorbach, chief mate Andrey Titov, boatswain Yury Dudinsky, and all crew members of the R/V *Akademik Mstislav Keldysh.*

I am very grateful to Dr. Gordon Gross who has been interested in my study of metalliferous sediments for many years, has kindly agreed to assist with the English text of this book, and has written Foreword.

I am much obliged to Dr. Wolfgang Engel for his support and good advice during preparation of the book for publication.

Bremerhaven, July 2005 Evgeny Gurvich

INTRODUCTION

Metalliferous sediments are unconsolidated deep-sea deposits that form in volcanically active areas on the floor of oceans and seas. An admixture of hydrothermal metal-bearing matter from the interior in the Earth's crust was and is being contributed to the ocean by submarine high-temperature springs and as a result the abiogenic part of these sediments is enriched in Fe, in a number of trace elements, in many places in Mn, and depleted in Al and Ti.

Some criteria for distinguishing metalliferous sediments have been offered. According to Boström (1973), metalliferous sediments have a ratio of Fe/(Al+Fe+Mn) greater than 0.5, a ratio of Al/(Al+Fe+Mn) less than 0.3, and a ratio of (Fe+Mn)/Al greater than 2.5. According to Strakhov (1976), metalliferous sediments have a ratio of (Fe+Mn)/Ti greater than 25. Lisitzin et al. (1976) considered that sediments that have 10% or more Fe in the abiogenic part and that are depleted in Al and Ti are metalliferous. This criterion is used in this book. Metalliferous sediments, which contain more than 30% Fe, are also referred to as "ore sediments" (Butuzova 1989; Lisitzin 1993). Nevertheless it must be remembered that all criteria mentioned here follow conventional use, and significant amounts of hydrothermal matter accumulates on the ocean floor outside the areas mapped as metalliferous sediments in sediments classified to be non-metalliferous.

Metalliferous sediments were collected for the first time on the East Pacific Rise during the expedition on board HMS *Challenger* (1873–1876) (Murray and Renard 1891). At a later time samples of metalliferous sediments were collected during the expedition on board USS *Carnegie* (Revelle 1944). However, at that time scientists paid little attention to these unusual sediments.

Skornyakova (1964) was the first scientist to presume that enrichment of Fe and Mn in bottom sediments from the East Pacific Rise was related to submarine hydrothermal activity. Also in 1964 Zelenov (1964) published a paper that reported the first visual observations of hydrothermal sedimentation on the sea floor. He filmed and studied sedimentation and the dispersal of minerals bearing Fe and Mn that entered the ocean in exhalations from an underwater volcano. Nevertheless major advances in the investigations of metalliferous sediments began after the papers of Boström and

Peterson (1966, 1969) and Bonatti (1967) were published that were devoted to metalliferous sediments of the East Pacific Rise. Almost simultaneously the results of the first comprehensive studies of metalliferous (and ore) sediments were published (Miller et al. 1966; Hot brines ... 1969; Baturin et al. 1969) that were related to hydrothermal activity and hot brines in three deeps (Atlantis II, Discovery and Chain) in the rift zone of the Red Sea. In the ensuing years marine geologists from many countries (Australia, Bulgaria, Canada, France, Germany, Great Britain, Greece, India, Italy, Japan, New Zealand, Russia, Sweden, USA, USSR and others) became aware of recent metalliferous sediments and initiated their studies on samples from the East Pacific Rise, and also from spreading ridges of the Indian, Atlantic, and Arctic Oceans, of back-arc basins, and from the vicinity of some underwater volcanoes. At present the number of known deeps with metalliferous sediments in the Red Sea rift zone exceeds twenty.

Metalliferous sediments were of special interest to many researchers before the discovery of the deep-sea high-temperature springs and the massive hydrothermal sulfide deposits. They were of special geological interest and some believed that they were most likely genetically related to submarine hydrothermal activity. However views differed on the mechanism of their origin and formation of their composition. From the time of their discovery and the initial studies of them the metalliferous sediments from the Atlantis II and Discovery Deeps of the Red Sea rift zone were considered to be genetically related to underwater hydrothermal activity.

It became apparent from the discovery of submarine high-temperature hydrothermal systems and massive sulfide deposits (Francheteau et al. 1979) that metalliferous sediments were formed by a mixture of metal-bearing matter from hydrothermal sources with background pelagic sedimentary material and that fields of metalliferous sediments were halos formed by the dispersion of the metal-bearing matter around underwater high-temperature hydrothermal systems. Subsequent discoveries of high-temperature hydrothermal fields with associated sediments bearing oxide and sulfide metallic minerals have demonstrated and confirmed that this is the case. As expected metalliferous sediments occur within and around hydrothermal fields (Haymon and Kastner 1981; Shearme et al. 1983; Marchig et al. 1986; Lisitzin et al. 1989, 1990, 1992; Bogdanov et al. 1990, 1993a,b, 1995a,c, 1997a, 2002, 2005; Lukashin et al. 1990; Cherkashev 1990, 1992; Gurvich et al. 1991, 1995a; Krasnov et al. 1992a; Gurvich 1994; Kuhn et al. 2000; Cave et al. 2002, et al.). Exceptions are in areas where due to high rate of abiogenic sedimentation there is strong dilution by lithogenic material and the content of hydrothermal components in the sediments is low. Sediments whose abiogenic part contains less than 10% Fe are not considered to be metalliferous even where there is evidence of the intensive contribution of material from hydrothermal sources. Sediments from the Guaymas Basin in the rift zone of the Gulf of California are the best example of this type. Within the Guaymas Basin, because of the high rate of abiogenic sedimentation that is approximately 1000 times higher than in rifts of the open ocean and despite very active high-temperature hydrothermal activity, metalliferous sediments do not form there (Lisitzin et al. 1990).

For the first time ancient metalliferous sediments were collected during Leg 2 of the Deep Sea Drilling Project (DSDP) in the basal layer of the sedimentary cover of the Atlantic seabed (Peterson et al. 1970). In the subsequent legs the drill ships *Glomar Challenger* and *JOIDES Resolution* metalliferous sediments were sampled repeatedly (Initial Reports … 1970–1987; Cronan 1976; Stoffers et al. 1983; Proceedings … 1988–2004; Levin et al. 1985, 1987; Lisitzin et al. 1990; Gurvich 1992, 1998; Gurvich et al. 1988, et al.). Ancient non-lithified and lithified analogues of metalliferous sediments that accumulated in paleoceans and paleoseas also occur on land on the continents and islands within ophiolite belts (Goodwin 1964, 1973; Robertson and Hudson 1973; Parrot et al. 1974; Bonatti et al. 1976; Fleet and Robertson 1980; Robertson et al. 1983, 1987; Karl 1984; Varnavas and Panagos 1984; Zaikov et al. 1984; Satian 1985; Zaikova 1985, 1991; Zaikov and Zaikova 1986; Yamamoto 1987; Gross 1984, 1988, 1991; Karpoff et al. 1988; Peters 1988; Pfeifer et al. 1988; Gurvich et al. 1989; Gurvich and Zaikov 1992, et al.).

Rona (1984) has estimated that only about 5% of the metal amount from hydrothermal sources that enters the oceans accumulates in sulfide minerals (massive sulfides) within hydrothermal fields, and about 95% outside of them. According to the estimation of Lisitzin (1993), at the early stage of the active life of hydrothermal edifices less than 5% of hydrothermal matter accumulates within them, at the mature stage about 50%, and at the ageing stage more than 90%. This suggests that at the most active early stage in the life of hydrothermal edifices metal-bearing matter entering the ocean accumulates mainly outside of the hydrothermal fields in metalliferous and transitional to metalliferous sediments. Thus the accumulation rate of hydrothermal matter in metalliferous sediments undoubtedly reflects the intensity of input of hydrothermal matter in the areas where it enters the ocean, and also the composition of the metalliferous sediments probably reflects the composition of hydrothermal fluids. All of this indicates that, with other factors being equal, higher rates of accumulation of metalliferous sediment are related to shorter distances from the sources of hydrothermal matter and that metalliferous sediments should reflect not only the presence and intensity of hydrothermal activity, but also its position.

Variations in the composition and the rate of accumulation of metalliferous sediments in time and space depend on many factors both large and small. Spreading rate of the ridges is the most important parameter that influences the large-scale factors (Boström 1973; Lisitzin et al. 1976, 1990; Lisitzin 1978; Bogdanov 1980; Levin et al. 1987; Gurvich 1998). Tectonic reorganizations of spreading ridges, migration of hydrothermal activity along the spreading axes, and also cyclicity of hydrothermal processes are the most important medium and small-scale factors. The rate of accumulation of metalliferous sediments and their composition reflect all of these factors (Lyle et al. 1986; Rea and Leinen 1986; Lyle et al. 1987; Gurvich et al. 1995; Gurvich 1998). Directions and velocities of bottom currents in areas of hydrothermal activity are also important factors that influence the transport and dispersal of hydrothermal metal-bearing matter and its accumulation in metalliferous sediments (Boström 1973; Gurvich et al. 1979; Walter and Stoffers 1985; Hauschild et al. 2003).

Many hundred reports of work devoted to metalliferous sediments have been published to date. The study of metalliferous sediments was most popular from the late sixties to the mid nineteen eighties when many tens reports of work were published annually. In later years interest in metalliferous sediments decreased and reports relating to them appeared only occasionally. Now the number of reports that appear annually worldwide is not more than 10 to 20 and they are devoted mostly to proximal metalliferous sediments from hydrothermal fields. An explanation for this appears to be that, since the discovery of modern deep-sea high-temperature hydrothermal activity, researchers have concentrated all their attention on underwater hydrothermal fields, ore-forming solutions, hydrothermal plumes and, most important of all, on massive hydrothermal mineral deposits. As a result studies of the most widespread hydrothermal object in the ocean are being neglected. The decrease in interest in distal metalliferous sediments that accumulated outside of hydrothermal fields, on the one hand, is probably because of their relative monotonous composition, and on the other hand, because of the absence of new approaches in the study of their geology. Nevertheless the potential for investigation of the information obtained during research on metalliferous sediments is far from exhausted. In the overwhelming majority of published papers and reports devoted to metalliferous sediments researchers consider them to be an "end-in-themselves" and forget that their fields are halos of dispersion of underwater hydrothermal mineral forming systems and they carry information on these systems. By recognizing this and with such an approach metalliferous sediments can be particularly useful in determining past hydrothermal activity in the oceans and the associated formation of mineral deposits in the geological past and in prospecting for hydrothermal mineral deposits. Cores of metalliferous sediments are "annals" recording not only sequences of hydrothermal events but also the intensity and temporal and spatial variations of hydrothermal activity and the conjugate deposition of minerals. The study of this aspect of metalliferous sediments is still "in swaddling-clothes".

It is necessary to emphasize that the book is devoted to and restricted to metalliferous sediments that are related to the accumulation and sedimentation of hydrothermal matter from hydrothermal plumes or from the destruction and redeposition of material from massive mineral deposits. The development of sediment hosted metalliferous sediments similar to those found at the Mound field of the Galapagos Spreading Center, in the Middle Valley of the Juan de Fuca Ridge, and in the Escanaba Trough of the Gorda Ridge have not been considered. These deposits have formed under completely different conditions and are a subject for a special major study.

CHAPTER 1 RECENT METALLIFEROUS SEDIMENTS IN THE OCEANS

Recent metalliferous sediments occur in the open parts of all oceans, on the flanks and in the axial parts of spreading ridges where submarine hydrothermal activity exists or existed in rift zones. Recent metalliferous sediments are most widely distributed in the Pacific Ocean, are much less abundant in the Indian and Atlantic Oceans and very few occurrences are known in the Arctic Ocean.

1.1. Metalliferous sediments from the Southeast Pacific

The largest areal distribution of recent metalliferous sediments in the World Ocean occurs in the Southeast Pacific. The field is almost continuous and occupies about 10 million km^2 between ~5°S and ~45°S to the west and to the east of the axis of the East Pacific Rise (EPR) (Metalliferous sediments ... 1979; Fig. 1.1[1]).

1.1.1. Conditions of formation

Metalliferous sediments are most widely distributed in the Southeast Pacific because hydrothermal activity is most intensive there and the accumulation rates of diluting abiogenic material and its proxy Al are low in the pelagic area of this southern arid zone (Lisitzin 1978; Lisitzin et al. 1980a). Within the field of metalliferous sediments the accumulation rate of Al on the bottom does not exceed 10 mg cm^{-2} ka^{-1} and the average rate is 1.75 mg cm^{-2} ka^{-1}, whereas the average rate in pelagic areas of the Pacific ocean is 25.8 mg cm^{-2} ka^{-1} (Lisitzin et al. 1980a). In the Southeast Pacific the content of hydrothermal components in bottom sediments is significant over long distances from the axis of the EPR.

---

[1] Shore line from http://www.pangaea.de/Software/PanMap/WVS/WVS.zip; plate boundaries positions from http://www.pangaea.de/Software/PanMap/Layers/PlateTectonicBoundaries.zip

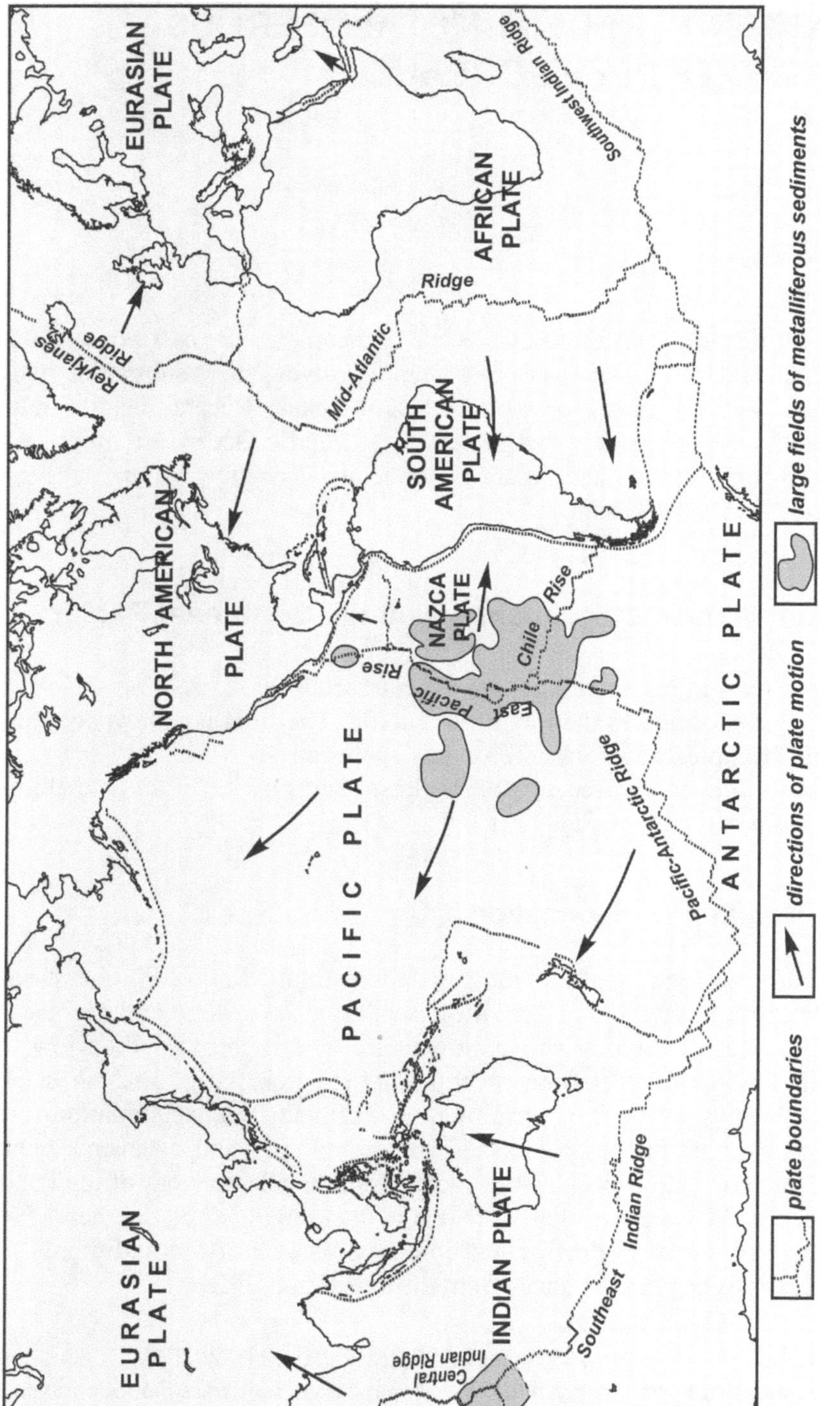

Fig. 1.1. Distribution of large fields of metalliferous sediments in the World Ocean. Directions of plate motion after Kennett (1982).

The average width of the field of metalliferous sediments in the Southeast Pacific is more than 2000 km, and the maximum width is about 3500 km. It must be noted that on the axis of the EPR and in its vicinity (within a few kilometers) the sedimentary layer is very thin or absent because of the very young or recent ocean crust here.

In the Southeast Pacific metalliferous sediments accumulate on structures related to the Nazca plate and three spreading ridges surrounding it, the East Pacific Rise in the west, the Chile Rise in the south, the Galapagos Spreading Center in the north, and also to the Pacific and Antarctic plates (Fig. 1.1). Within the Nazca plate the South Galapagos Ridge divides the Peru Basin into the western (Bauer Depression) and the eastern deeps. The South Galapagos Ridge is considered to be an extinct segment of the ancient spreading ridge. At present within the area covered by metalliferous sediments in the Southeast Pacific the full spreading rate of the EPR varies from 8 to 16–20 cm a^{-1} (according to Handschumacher (1976), up to 24 cm a^{-1}), and of the Chile Rise 3–4 cm a^{-1} (Forsyth 1972; Klitgord et al. 1973; Herron et al. 1981).

Hydrothermal solutions entering the ocean in hydrothermal fields of the EPR rift zone are sources of metal-bearing matter in the metalliferous sediments accumulating in the Southeast Pacific. Many hydrothermal fields may not have been discovered, nevertheless indirect data and the huge fields of metalliferous sediments that are known indicate that a great number of fields may exist in the EPR rift zone between 5°S and 45°S. Hydrothermal processes are most active in the EPR rift zone in the segment between 18°S and 26°S which has the maximum spreading rate (Rea 1978; Rea and Blakely 1975; Minster and Jordan 1978). Detailed studies of this geodynamically unique part of the World Ocean rift system, including the ones that used deep-sea manned submersibles (Francheteau and Ballard 1983; Bäcker et al. 1985; Renard et al. 1985; Marchig et al. 1987, et al.), have shown that the area of the greatest uplift in the axial part of the rift (2590 m) occurs at 17°26'S. Southward the basal part of the axial rift deepens gradually, and near the transform fault at 20°30'S reaches a depth greater than 2900 m. To the south of the transform fault it rises to 2800 m, and then the depth increases again near the boundary of the Easter Plate. All of the main hydrothermal fields and hydrothermal sulfide manifestations in this part of the rift have been found within the axial graben 90 m deep and some hundreds of meters wide (Bäcker et al. 1985; Renard et al. 1985; Marchig et al. 1988; Krasnov et al. 1988). These fields are located in modern spreading centers at the raised parts of the spreading axis. Hydrothermal vents occur along tectonic cracks and breaks, often in marginal parts of the axial valley of the spreading centers.

Within the area of metalliferous sediments in the Southeast Pacific sedimentation rate varies from 1 to >25 mm ka^{-1} and the distribution of sediments is not uniform but erratic. On the raised parts of the seabed where a lot of biogenic carbonaceous material accumulates the sedimentation rate is usually higher. The highest values (from 10 to >25 mm ka^{-1}) have been measured at some sites in the axial zone of the EPR between 10°S and 25°S (Bogdanov and Chekhovskikh 1979), where the highest spreading rate and the highest accumulation rate of hydrothermal matter occur. Here sedimentation rates between 5 and 15 mm ka^{-1} are typical (Dekov 1994), though much higher rates (up to 240 mm ka^{-1}) also occur (Shimmield and Price 1988). Measured accumulation rates of abiogenic matter in metalliferous sediments vary from 3 to 6000 mg cm^{-2} ka^{-1} and the average rate is about 20 mg cm^{-2} ka^{-1}. As a rule mass accumulation rates of metalliferous sediments in the axial part of the EPR are higher than on the flanks of the ridge. In the axial part of the ridge the highest mass accumulation rates of metalliferous sediment (>100 mg cm^{-2} ka^{-1}) have been reported between 10°S and 25°S (Bogdanov and Chekhovskikh 1979; Bogdanov et al. 1979a) and the average mass accumulation rate of metalliferous sediments collected between 20°30'S and 22°S is 180 mg cm^{-2} ka^{-1} (Dekov 1994).

1.1.2. Lithologic composition

Detailed studies of the lithologic composition of metalliferous sediments from the Southeast Pacific have shown that they are typical pelagic sediments enriched to some extent by hydrothermal metal-bearing matter and combined trace elements. They consist of a mixture of two main components, pelagic sedimentary material and metal-bearing matter. Pelagic sedimentary material accumulates according to circumcontinental, latitudinal, vertical and tectonic zonalities, and the accumulation of hydrothermal metal-bearing matter takes place in sediments in the vicinity of hydrothermally active zones (Lisitzin 1978; Lisitzin et al. 1976; Bogdanov et al. 1979a). The lithologic composition of metalliferous sediments depends on the location in the ocean where they accumulated. The following lithologic types of metalliferous sediments occur in the Southeast Pacific: a) carbonaceous; b) noncalcareuos; c) coarse-grained edaphogenic[2] sediments with metalliferous cement; d) turbidites (Lisitzin et al. 1976; Bogdanov et al. 1979a). Brief descriptions of these types of metalliferous sediments are as follows:

[2] Clastic sediments, which composed mainly (>70%) of products of underwater disintegration of bedrocks (Murdmaa 1979).

a) Carbonaceous metalliferous sediments are the most widespread. They occur above the critical depth of carbonate accumulation. Within the areas of metalliferous sediments in the Southeast Pacific this depth varies from 4200 to 4700m. In some places because of the redeposition of sedimentary material the carbonaceous metalliferous sediments occur at greater depths. The carbonaceous metalliferous sediments differ in appearance from the background carbonaceous pelagic ooze by their brown color and sometimes by spotty coloring that is dependent on the content of the metal-bearing matter. Biogenic material is prevalent in their composition. It consists for the main part of coccoliths that form the pelitic (<0.01 mm) grain size fraction. Significant amounts of planktonic foraminifera skeletons and organogenic detrital material are present. When uplift takes place in an area there is an increase in the development of foraminiferal carbonaceous material and a decrease in the development of coccolithic material. Colloidal clots and globules of Fe-oxyhydroxides are the main components of the abiogenic part of these metalliferous sediments. Usually pelitic coccolith ooze is more enriched in the metal-bearing matter. This is an indication of the important role of bottom water dynamics in the distribution of this material.

b) Noncalcareuos metalliferous sediments are not abundant and can be found below the critical depth of carbonate accumulation on slopes of magmatically active structures and in adjoining parts of deep-sea basins. They differ from background pelagic clays in their dark-brown or brownish-black color. They are semiliquid, colloform-caseous with a water content of up to 80% or more. The aluminosilicate part of noncalcareuos metalliferous sediments consists of fine-grained material that is indistinguishable from background material in pelagic clays. The only difference is the presence of Fe-montmorillonite in noncalcareuos metalliferous sediments.

c) Coarse-grained edaphogenic sediments with metalliferous cement occur in areas of intensive tectonic discontinuities in the vicinity of the EPR axis. They consist of edaphogenic breccia with semiliquid cement of brown metalliferous mud with an admixture of carbonaceous material. The composition of fine-grained fractions of these sediments is similar to that in types (a) and (b). Irregular fragments of basalt and dolerite occur in the coarser grain size fractions and sometimes they have traces of hydrothermal alteration. The size of fragments varies from several millimeters to several centimeters.

d) In vertical sections metalliferous turbidites may have one or more complete or incomplete sedimentation cycles. Complete cycles are up to 40 cm thick and have gradation from sandy carbonaceous material in the lower parts to pelitic colloform material in the upper parts. With changes in grain size the color of the components of the sediments changes from yellow and cream to brown and dark brown tones. The upper part of a complete cycle consists of typical metalliferous sediment. In places the upper parts of the successions have been washed out by turbid flows and the cycles are incomplete.

1.1.3. Mineral composition

In the Southeast Pacific metalliferous sediments accumulate at depths from less than 3000 m in the axial part of the EPR to depths of nearly 5000 m on distant flanks of the EPR and mainly above the critical depth of carbonate accumulation. Therefore as mentioned previously, carbonaceous metalliferous sediments are the most widespread in which biogenic calcite is the predominant mineral component. In areas of the greatest uplift of the seafloor the content of calcium carbonate in the metalliferous sediments may be 80–90% or even more and it decreases with depth (Boström 1973; Lisitzin et al. 1976; Lisitzin 1978; Bogdanov et al. 1979a). Apparent contents of biogenic opal (5–10%) occur only in the northern subequatorial part of the area of metalliferous sediments (Lisitzin et al. 1976).

As for detrital minerals, quartz, plagioclase, hornblende, monoclinic and rhombic pyroxene, volcanic glass (mainly basic), epidote, magnetite, zircon, and occasionally garnet have been identified in metalliferous sediments. Some of these minerals are terrigenous. Quartz and feldspar are mainly of eolian origin. Another part of the detrital material consists of volcanogenic and edaphogenic minerals (Lisitzin et al. 1976).

Smectite and mainly Fe-montmorillonite are the most abundant clay minerals (Sayles and Bischoff 1973; Sayles et al. 1975; Bagin et al. 1975, 1979, 1993a, 1996; Heath and Dymond 1977; Bogdanov et al. 1979a; Gorbunova 1981, 1982; Gorbunova and Lisitzin 1981; McMurtry and Yeh 1981). The crystallinity of Fe-montmorillonite is minimal in the surface layers of sediments, and increases downward (Gorbunova 1981, 1982). In areas near the flanks of the EPR the clay component in the abiogenic part of the metalliferous sediments consists almost entirely of poorly crystallized monomineralic Fe-montmorillonite and poorly crystallized goethite. Between 5°S and 30°S near to the axis of the EPR and to the east of it there is an area (province) of well to medium crystallized Fe-montmorillonite. Outside of these areas terrigenous clay minerals prevail in metalliferous sediments (Gorbunova and Lisitzin 1981).

Fe-Mn micronodules are the most abundant authigenic components of metalliferous sediments. In places on the East Pacific Rise flanks they form 10–20% of the abiogenic part (Sayles and Bischoff 1973) and up to 90% of the sandy-silt fractions of the sediments (Bogdanov et al. 1979a), near the axis they are less abundant (Dekov et al. 2003). Small Fe-Mn nodules occur in sediments in some places on the flanks of the EPR (Girin et al. 1979). Zeolite minerals have also been identified (Bogdanov et al. 1979a). Except in the axial part of the EPR, fine-grained barite occurs everywhere in metalliferous sediments (Arrhenius and Bonatti 1965; Boström et al., 1973; Gurvich et al. 1979; Walter and Stoffers 1985). The barite content in the carbonate-free part of sediments may be more than 4%.

X-rays show that amorphous colloidal and poorly crystallized Fe- and Mn-minerals are the main metal-bearing mineral components in the metalliferous sediments. Fine-grained superparamagnetic particles of Fe(III)- and Mn(IV)-oxyhydroxides prevail. Particles of crystallized ferromagnetic Fe-Mn minerals occur in very small amount (Dymond et al. 1973; Bagin et al. 1975, 1979, 1993a, 1996; Bagin and Pechersky 1977; Lisitzin et al. 1976; Heath and Dymond 1977; Bogdanov et al. 1979a; Walter and Stoffers 1985; Dekov 1994). In the vicinity of hydrothermal fields Fe-, Cu- and Zn sulfide minerals have been found in the metalliferous sediments mainly in the sandy-silt fractions. Pyrite is the most abundant, and chalcopyrite and sphalerite are less abundant and other sulfide minerals are rare (Haymon and Kastner 1981; Dekov 1994).

Metallic minerals occur in metalliferous sediments mainly as formless aggregated masses of colloform matter with rounded surfaces. The diameter of the particles is about 3 microns and they are usually yellow in color. The second colloform phase consists of globular brown particles that are found also in the nonmetalliferous pelagic clay. These particles appear to be composed of iron-silica gels (Bogdanov et al. 1979a). Mössbauer spectroscopy studies together with data from electron diffraction studies and energy dispersion analysis have shown that the metal-bearing matter of the metalliferous sediments is composed mainly of fine-grained ($d<100$Å) poorly crystallized goethite and Mn-ferroxigite (Bagin et al. 1993a,b, 1996; Gendler et al. 1993). Magnetic minerals from metalliferous sediments are mostly paramagnetic or superparamagnetic ($d<100$Å) and sometimes they are fine-grained single-domain particles. Titanomagnetite in various states of oxidation and maghemite in small amounts have been identified (Bagin et al. 1975, 1979, 1996; Bagin and Pechersky 1977). Separate crystalline phases of Mn-minerals in metalliferous sediments have been identified as δ-MnO_2 and todorokite (Dymond et al. 1973; Walter and Stoffers 1985). An increased content of Mn in the coarse grain size fractions of the metalliferous sediments has been observed (Dymond et al. 1973; Dekov and Gurvich 1991).

Native metals and alloys occur in the sandy-silt fraction of metal-bearing matter in metalliferous sediments from the Southeast Pacific, namely Cd°, Ni°, Al°, Sn°, Si°, Ti°, Al-Mg-Si, Al-Si, Al-Fe, Al-Mg, Al-Ti-Ca, Fe-Si, Fe-Cr-Ni-C, Zn-Al, Zn-Fe, Cu-Zn, Cu-Zn-Al, Ag-Cu, Sn-Cu, Sn-Zn,

Sn-Pb, Sn-Pb-Zn (Shterenberg et al. 1981; Davydov 1992; Dekov 1994). Particles of metals and alloys are mainly less than 0.1 mm in size (Dekov 1994). The widest spectrum of native metals and alloys has been found at distances up to 10–20 km from known hydrothermal fields. Among them Al°, Sn°, Cu-Zn and alloys of the Zn-, Sn- and Al groups are prominent and alloys of the Fe group, Ni°, and Cd° are much less abundant. At a distance from the hydrothermal fields the native metals and alloys are depleted and at distances of 50–100 km from the axis of the ridge particles of the Al and Fe groups are predominant (Davydov 1992).

1.1.4. Chemical composition

The average chemical composition of carbonate-free matter of metalliferous sediments from the Southeast Pacific is shown in Table 1.1. Carbonaceous metalliferous sediments are prevalent and the average chemical composition of abiogenic matter of the background pelagic carbonaceous sediments from the Pacific Ocean is also shown in this table for comparison. Enrichment in Fe, Mn and a number of trace elements and depletion in Al and Ti in comparison with their contents in background sediments are the main features apparent in the chemical composition of metalliferous sediments that are consistent with their definition. There is excessive accumulation of most elements in metalliferous sediments, for many of them excessive accumulation is predominant.

The distribution of Fe and Mn in carbonate-free matter of surface sediments from the Southeast Pacific is shown in Fig. 1.2 and 1.3. The areas with the highest contents of Fe and Mn occur in the axial zone of the EPR, and these areas are wider where the spreading rate at the axis is higher. As the distance from the EPR axis increases, Fe and Mn contents decrease and eventually are similar to the characteristic contents for background pelagic sediments. Depending on the directions of the bottom currents crossing the EPR axis, the maximum contents of Fe and Mn may be found to the west of the axis where the predominant currents flow west and to the east of the axis where the predominant currents flow east. This distribution of Fe in the sediment is clearly shown in Fig. 1.2, and also in the distribution of the (Fe+Mn)/Al ratio in metalliferous sediments from the axial zone of the EPR at 19°S–21°S, shown in Fig. 1.4, where the predominant bottom currents flow westerly (Lonsdale 1976; Reid 1981). The contents of Fe and Mn are relatively reduced in a zone of metalliferous sediments located close to the axis of the EPR, but where hydrothermal activity is greater on a ridge segment the zones are relatively wider. The Fe content reaches a maximum at 10–20 km from the axis and then decreases. The Mn content reaches a maximum at a distance of 40 km from the axis and then decreases away from it (Fig. 1.5; Davydov et al. 1997). The maximum Fe/Mn ratios are found near the axis of the EPR and higher values occur near hydrothermal fields and decrease away from them (Fig. 1.5).

Table 1.1. Average contents of chemical elements in carbonate-free matter of metalliferous sediments from the Southeast Pacific and of background sediments. Based on data from Turekian and Wedepohl (1961), Boström and Peterson (1969), Boström and Valdes (1969), Skornyakova (1970), Horowitz (1970), Boström (1973), Crocket et al. (1973), Piper and Graef (1974), Gurvich et al. (1976, 1977, 1979, 1980a,b,c), Gurvich (1977, 1987), Oreshkin (1977), Bogdanov et al. (1979a), Lisitzin (1978), Lisitzin et al. (1980b,c, 1990), Lukashin and Lisitzin (1980), Lukashin et al. (1980a,b), Cox et al. (1981), Ravizza and McMurtry (1993), Dekov (1994).

Element	Metalliferous sediments	Background sediments	Element	Metalliferous sediments	Background sediments
Fe %	16.86	5.8	U	3.8	1.3
Mn	5.07	0.75	Li	32	45
Si	16.19	23.4	Rb	33	45
Al	4.35	7.8	La	157	61.8
Ti	0.27	0.45	Ce	114	83.1
Ba	1.08	1.05	Sm	29.3	18.8
			Eu	7.7	4.3
Cu ppm	1041	136	Yb	11.0	6.8
Zn	411	159	Y	290	191
Pb	155	41	Sc	26.1	20.8
Tl	15.5	0.7	Hf	6.3	4.1
Cd	1.7	0.42	Zr	456	178
Sn	6.42	1.5	Th	5.0	7.9
Ni	826	136	Ga	10	18
Co	218	32	Ag	5.32	0.11
Cr	29	50			
V	428	91	Hg ppb	460	
As	93	13	Au	10	3.5
Sb	11.9	3.9	Pd	15	4
Mo	93	27	Ir	0.5	0.3
B	210	250	Os	0.29*	

* Content in dry bulk sediments

The content of Al that is a proxy of lithogenic matter has a reverse pattern of distribution; minimum amounts are found in the EPR axial zone. There is a gradual increase in the amount of Al with increasing distances from the axis and a gradual transition to the background contents found in pelagic sediments (Fig. 1.6). In close proximity to the EPR axis within 10 to 20 km the metalliferous sediments have higher contents of lithogenic aluminosilicate (edaphogenic) material, Al, and Ti (Dekov 1994). This results in a decrease in the content of Fe and metal-bearing matter as mentioned previously. The absence of higher contents of Al and Ti in sediments at the EPR axis at the sections shown in Fig. 1.6 and on the regional maps (Lisitzin et al. 1980a,b), and also the absence of a decrease in the contents of Fe and Mn in metalliferous sediments at the EPR axis (Fig. 1.2, 1.3), may be attributed to different levels of detail in research data or in methods of reporting it. During special work in the area at 18°30'S such phenomena were observed within 20–30 km from the EPR axis (Fig. 1.7).

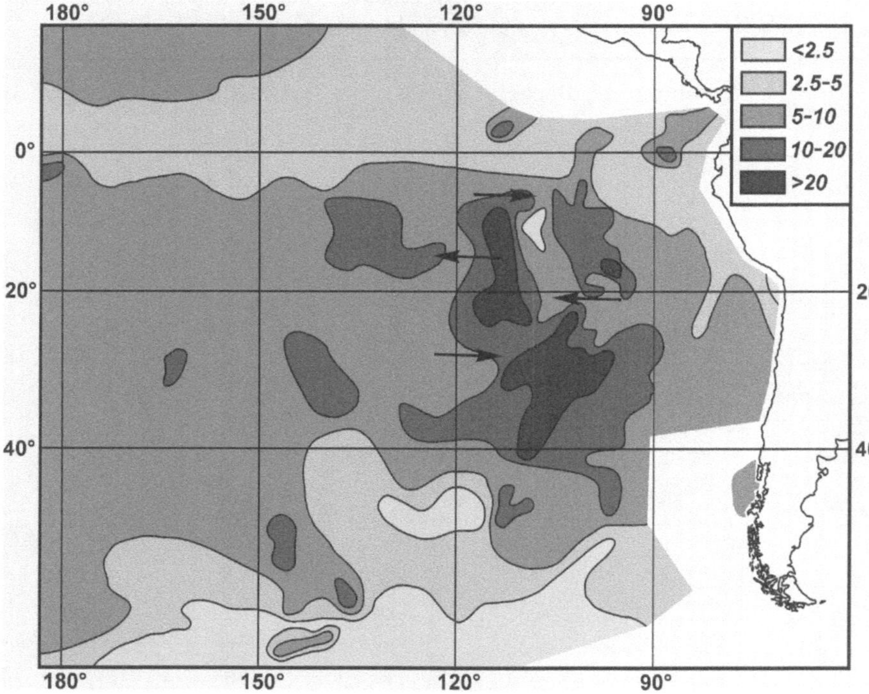

Fig. 1.2. Iron contents [%] in carbonate-free matter of surface sediments from the Southeast Pacific. After Bogdanov et al. (1979c). The arrows show the directions of the bottom currents from Lonsdale (1976), Reid (1981), and Lupton et al. (2004).

The distribution of lithogenic Si in the metalliferous sediments appears to coincide with the distribution of Al. Though metalliferous sediments in the Southeast Pacific are very poor in biogenic silica, the Si/Al ratio is not constant but increases near the axis of the EPR (Fig. 1.8). The ratio of Si/Al varies from area to area along the EPR. In near surface metalliferous sediments in the vicinity of the EPR axis at 10°S the Si/Al ratio is about 8, at 14°S – about 18, at 20°S – about 14.5, at 42°S – about 5 (Walter and Stoffers 1985), at 16°S it is about 62 (Dubinin and Volkov 1992), and between 20°30'S and 22°S it is about 6 (Dekov 1994). In all cases these values exceed 3, which is the average value for Si/Al ratios in lithogenic material of metalliferous sediments (Heath and Dymond 1977). The data show an excess of Si relative to Al, which is attributed to the accumulation of hydrothermal Si in metalliferous sediments (Dymond et al. 1973; Migdisov et al. 1979; Dubinin and Volkov 1992). However it must be noted that in the proximity of the EPR axis there is a decrease in the Si/Al ratios and in the content of hydrothermal Si in abiogenic matter of the metalliferous sediments that is related to the increase in content of lithogenic material. Evidently this is illustrated in the section across the EPR axial zone at 18°30'S (Fig. 1.7). The increase in content of hydrothermal Si in addition to the increased content of edaphogenic material already mentioned is another reason for the reduced contents of Fe and hydrothermal metal-bearing matter in metalliferous sediments in the vicinity of the EPR axis.

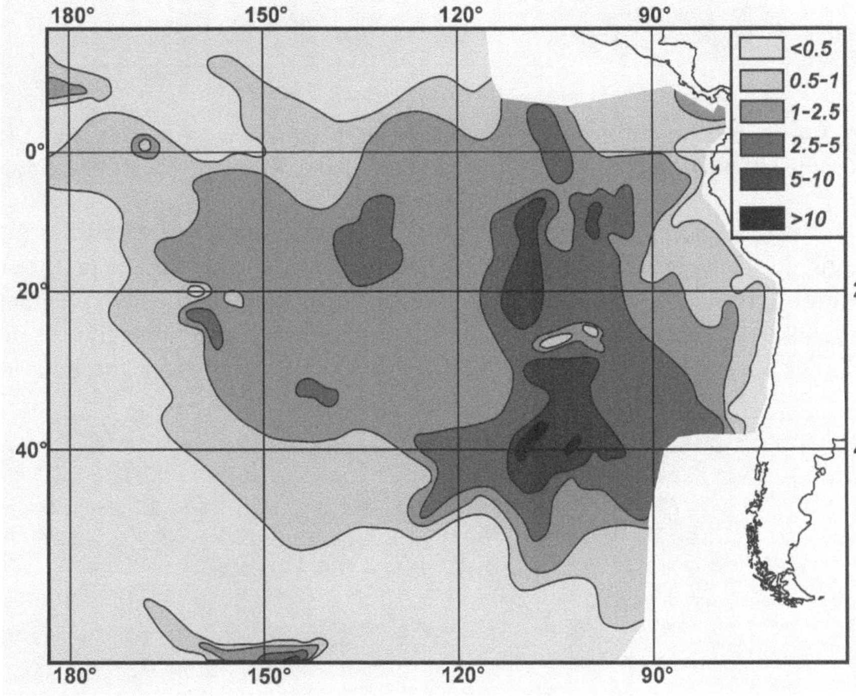

Fig. 1.3. Manganese contents [%] in carbonate-free matter of surface sediments from the Southeast Pacific. After Bogdanov et al. (1979c).

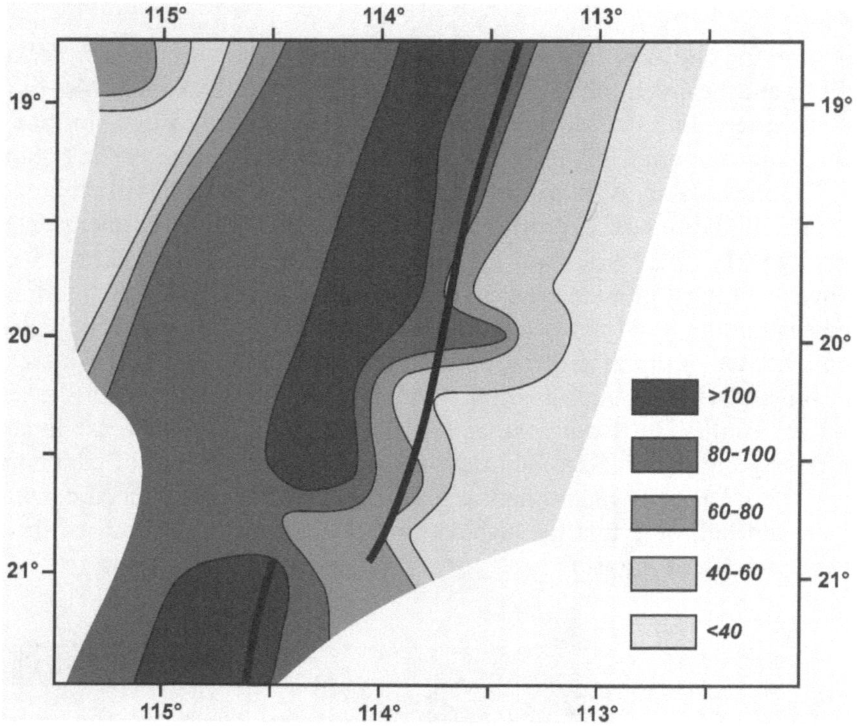

Fig. 1.4. Distribution of the (Fe+Mn)/Al ratio in metalliferous sediments from the axial zone of the East Pacific Rise at 19°S–21°S. After Walter and Stoffers (1985).

Average contents of trace elements in abiogenic matter of metalliferous sediments from the Southeast Pacific range from values higher to lower than those reported for background contents. The ratios of average element contents in abiogenic matter of metalliferous sediments to those in pelagic carbonaceous sediments of the Pacific Ocean calculated from the data in Table 1 are:

Al →	Ga →	Cr →	Ti →	Th →	Si →	Li →	Rb →	B →	Ba →
0.56	0.56	0.58	0.60	0.63	0.69	0.71	0.73	0.84	1.03
Sc →	Ce →	Y →	Hf →	Sm →	Yb →	Ir →	Eu →	U →	La →
1.25	1.37	1.52	1.54	1.56	1.61	1.7	1.79	1.90	2.54
Zr →	Zn →	Au →	Fe →	Sb →	Pb →	Pd →	Sn →	V →	Ni →
2.56	2.58	2.86	2.91	3.05	3.79	3.8	4.3	4.71	6.06
Mn →	Co →	As →	Cu →	Tl →	Ag				
6.76	6.85	7.15	7.63	22	48				

The average relative roles of the excessive accumulation of the elements in % are calculated based on the assumption that the accumulation of Al in metalliferous sediments is caused only by background process:

Al →	Ga →	Cr →	Ti →	Th →	Si →	Li →	Rb →	B →	Ba →	Sc →	Ce →
0	0	4	7	12	19	22	23	34	46	56	59
Y →	Hf →	Sm →	Yb →	Ir →	Eu →	U →	La →	Zr →	Zn →	Au →	Fe →
63	64	64	65	67	69	71	78	78	78	80	81
Sb →	Pb →	Pd →	Sn →	V →	Ni →	Mn →	Co →	As →	Cu →	Tl →	Ag
82	85	85	87	88	91	92	92	92	93	97	99

Distributions of contents of different elements in abiogenic matter of metalliferous sediments are different. The results from numerous studies have shown that there are four basic types of distribution:

Type1. The maximum contents of elements occur in metalliferous sediments in the axial zone. The contents of elements gradually decrease to background values away from the axial zone of the EPR. This type of distribution is typical for the main hydrothermal elements in metalliferous sediments – Fe and Mn (Fig. 1.2, 1.3), and for some trace elements: Zn, V, Pb, As, Cd, P, B, Hg (Fig. 1.9, 1.10; Boström 1973; Boström and Valdes 1973; Boström et al. 1973; Oreshkin 1977; Migdisov et al. 1979; Dubinin and Volkov 1992).

Type 2. The contents of elements are reduced in metalliferous sediments in the axial zone, and increase at a distance from it and have maxima on both flanks; at further distances from the axis the contents of elements decrease to background values. This type of distribution is characteristic for many trace elements, including Ni, Co, Hf, Sc, Ba, TR, Zr, Sb etc. (Fig. 1.11). Similar distributions are shown on the maps published by Boström (1973), Boström et al. (1973), Gurvich (1977), Gurvich et al. (1977, 1979, 1980a,c), Gurvich and Lisitzin (1980a,b,e), Lukashin et al. (1980b). It is of interest to note that the contents of Hf, Sc, Ba, TR and Zr have greater variation than the contents of Ni, Co and Sb. The contents of elements in the first group on the flanks of the ridge may be four or more times higher than their contents in sediments sampled in the axial zone, and the contents of elements in the second group may be 1.5 to 3 times higher.

Type 3. There are three areas where the maximum contents of elements occur in this type of distribution. The first area is in the axial zone and the other two are on the flanks, and at greater distances from the axis the element contents decrease to background values. This type of distribution is characteristic for Cu (Fig. 1.5, 1.10) and sometimes for Zn (Fig. 1.5).

Type 4. The minimum contents of elements occur in the axial zone and the element contents gradually increase to background values a distance from the axis. This type of distribution is characteristic for Al (Lisitzin et al. 1980b) and elements such as Ti, Th, and Ga (Fig. 1.12; Lisitzin et al. 1980c; Gurvich and Lisitzin 1980f; Gurvich et al. 1980b; Lukashin and Lisitzin 1980a). However at some sections across the EPR Th has Type 2 distribution (Fig. 1.12; Gurvich et al. 1980b).

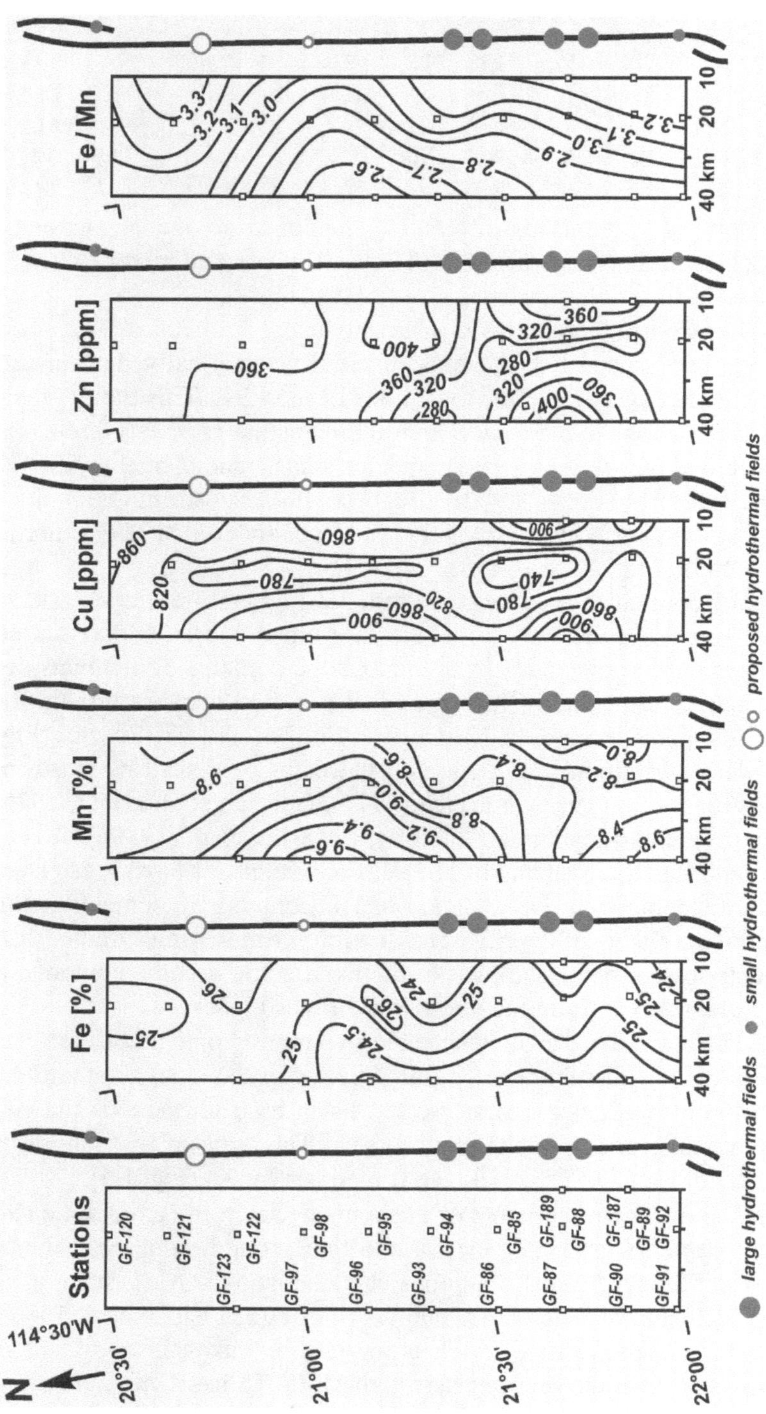

Fig. 1.5. Distributions of Fe, Mn, Cu, and Zn contents, and of Fe/Mn ratios in carbonate-free matter of surface metalliferous sediments near the axis area of the East Pacific Rise between 20°30'S and 22°S and locations of modern hydrothermal fields. Compiled from Dekov (1994) with additions. Based on data for samples collected on Cruise 4 of the R/V *Geolog Fersman*.

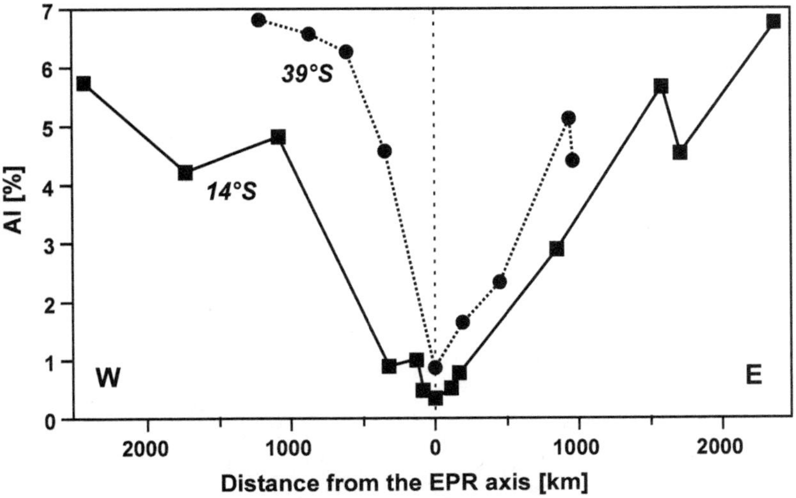

Fig. 1.6. Aluminum contents in abiogenic matter of metalliferous sediments at the sections across the East Pacific Rise. Based on data from Boström and Peterson (1969) and Migdisov et al. (1979).

Downcore variations in the contents of chemical elements in abiogenic matter of metalliferous sediments in the Southeast Pacific were investigated by a number of workers including Bender et al. (1971), Migdisov et al. (1979), Marchig and Gundlach (1982), Kunzendorf et al. (1984/1985), Varnavas (1988), Dubinin and Volkov (1992), Marienfeld and Marchig (1992), Dekov (1994), et al. Because of the low content of organic matter, the average content of organic carbon is 0.06% (Marchig and Gundlach 1982), the diagenetic vertical redistribution of chemical elements in metalliferous sediments from the Southeast Pacific is relatively insignificant. This has been proven also in studies of pore waters by Bischoff and Sayles (1972). In metalliferous sediments near the EPR axis the content of organic carbon increases and diagenetic processes may be actively developed (Dekov 1994). There are two reasons for this: activity of organisms in areas of hydrothermal activity, and shallower depths (Bogdanov et al. 1979b). The average content of organic carbon is 0.23% in metalliferous sediments from the near-axial part of the EPR at 20°30'S–22°S, but it can be as high as 0.7 to 1.1% (Dekov 1994) within small areas and in some interlayers. Diagenetic alteration in metalliferous sediments has been significant or marked in a section about 90 km long that crosses the axis of the Chile Rise (Marienfeld and Marchig 1992).

Only casual changes and major trends were found in the early studies of vertical distribution of chemical elements in metalliferous sediments. This was because samples from cores of sediments were taken only at intervals

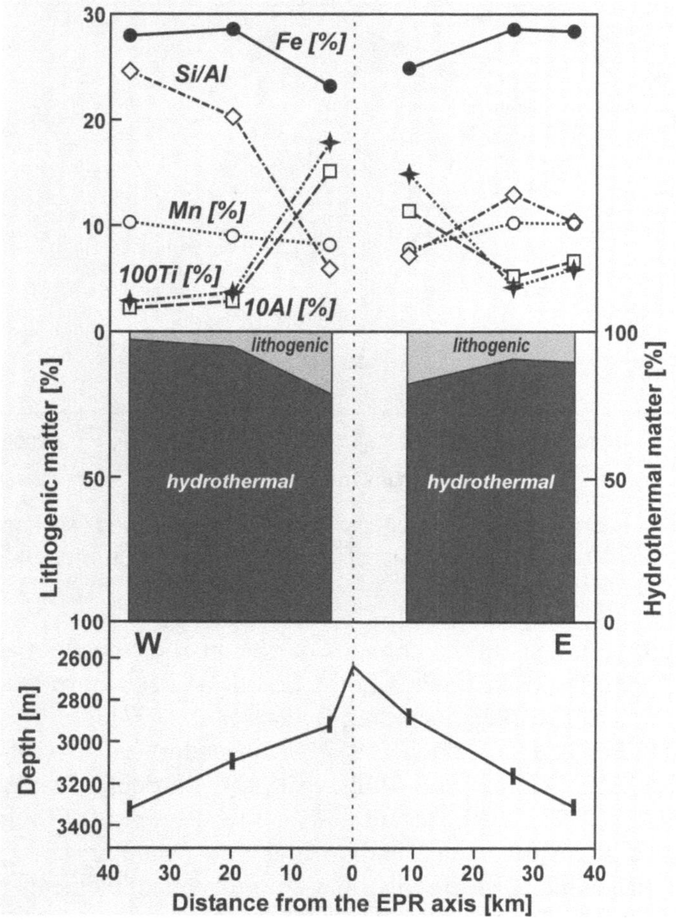

Fig. 1.7. Composition of abiogenic matter in metalliferous sediments at the section across the East Pacific Rise axial zone at 18°30'S. Based on data from Marchig et al. (1986) and Marchig and Erzinger (1986).

of 10–20 cm or greater[3]. Considering the actual sedimentation rates (see Sect. 1.1.1), such intervals sampled did not yield sufficient information for studying the short periods of time, thousands to tens of thousands of years (Fig. 1.13), in which the variations in the contents of elements developed. Consequently only the diagenetic redistribution of elements and the trends that developed over long periods of time were found.

[3] Occasionally sharp increases in the contents of some elements in metalliferous sediments were found in some horizons only (Rydell et al. 1974; Kunzendorf et al. 1984/1985).

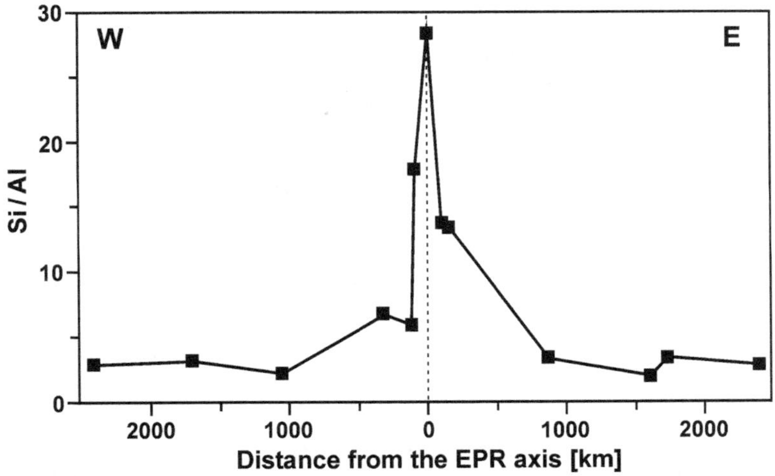

Fig. 1.8. Si/Al ratios in metalliferous sediments at the section across the East Pacific Rise at 14°S. Based on the data from Boström and Peterson (1969).

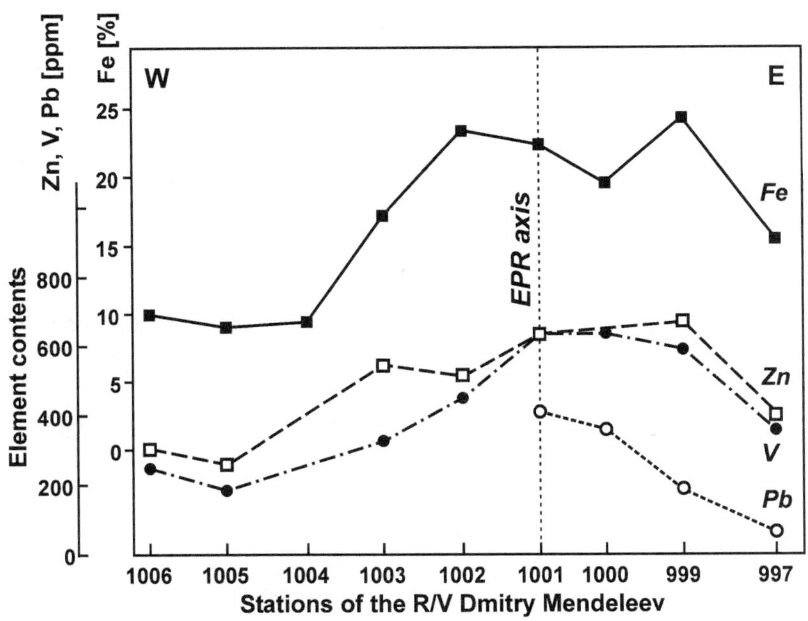

Fig. 1.9. Fe, Zn, V, and Pb contents in abiogenic matter of metalliferous sediments at the section across the East Pacific Rise at 39°S. After Migdisov et al. (1979).

In situations where variations in the contents of elements developed in short periods of time the frequency of sample intervals can result (and has resulted) in obtaining data for casual phases of variation in element contents when the intervals sampled were short or in obtaining average data for long time periods when the intervals sampled were long.

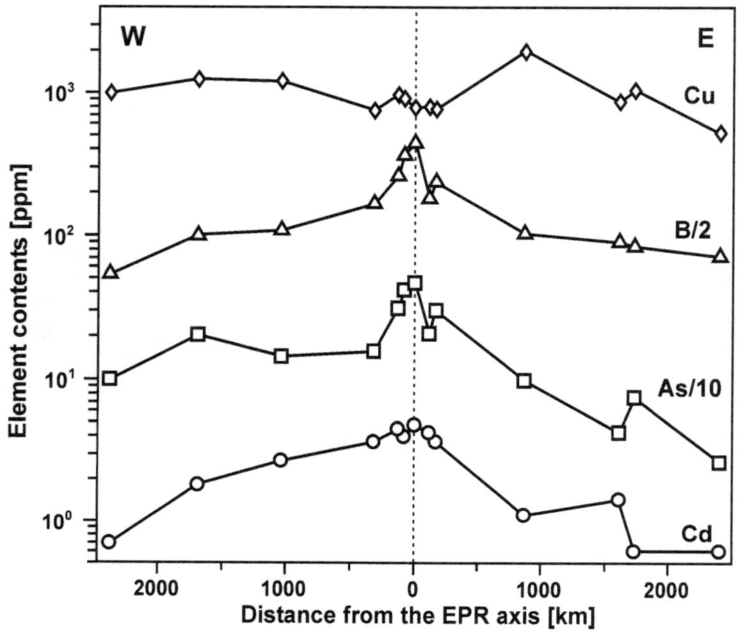

Fig. 1.10. Cu, B, As, and Cd contents in abiogenic matter of metalliferous sediments at the section across the East Pacific Rise at 14°S. Based on the data from Boström and Peterson (1969) and Boström and Valdes (1969).

Such studies can be carried out on the flanks of spreading ridges, as there the hydrothermal material in metalliferous sediments originates not from a specific hydrothermal field or vent but from hydrothermal sources in various extended parts of the rift. The sample results obtained reflect rather long periods of tens of thousands to millions of years of evolution of hydrothermal activity in these parts of the rift system.

As for studies of metalliferous sediments from the EPR axial zone that accumulated in the vicinity of specific hydrothermal fields and vents, at the sedimentation rates of 5 to 15 mm ka^{-1} (typical for the area at 21°S) sample intervals of 1 to 2 cm are required for obtaining data for rather short intervals (~1 ka) of temporal variation in hydrothermal activity (Fig. 1.13). The study of more detailed patterns in the evolution of hydrothermal activity and the distribution of elements through time would require more frequent and shorter sampled intervals. However such studies have no practical value because due to bioturbation all sediments being in the top layers (some centimeters) of the strata have been mixed and disturbed and their small structures have been broken (Kuptsov 1989). In sediments with higher sedimentation rates the rhythms of discharge and variations in the composition of hydrothermal fluids are reflected within periods of 100 to 500 years and even within 20 years (Metz et al. 1988; Lisitzin et al. 1990).

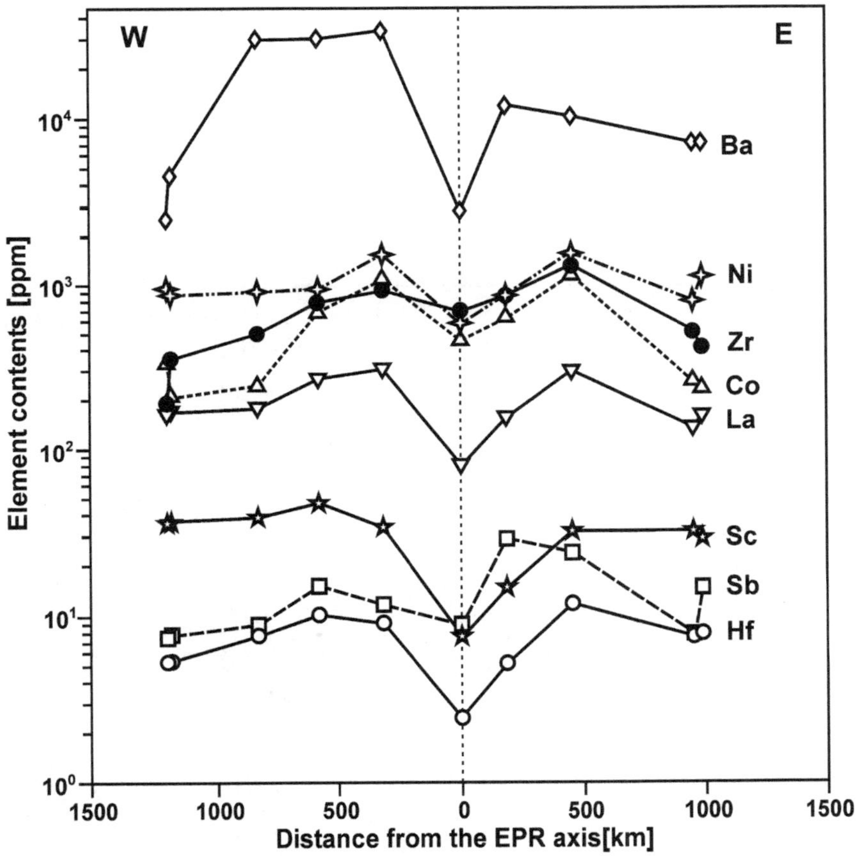

Fig. 1.11. Ba, Ni, Zr, Co, La, Sc, Sb, and Hf contents in abiogenic matter of metalliferous sediments at the section across the East Pacific Rise at 39°S.

The first study of metalliferous sediments from the Southeast Pacific with sample intervals of 2 cm was carried out by Shimmield and Price (1988). Cyclicity of sedimentation with a period of about 5 ka (Fig. 1.14) was identified in Core 154-18 located at 20°01.9'S, 113°51.4'W, at a depth of 3129 m, collected several kilometers to the west of the EPR axis.

This cyclicity was caused by cyclic accumulation of hydrothermal matter in the metalliferous sediments sampled in this core and was indicated by the in-phase variations of the Fe/Al ratio and antiphase variations of Al_{CFB}

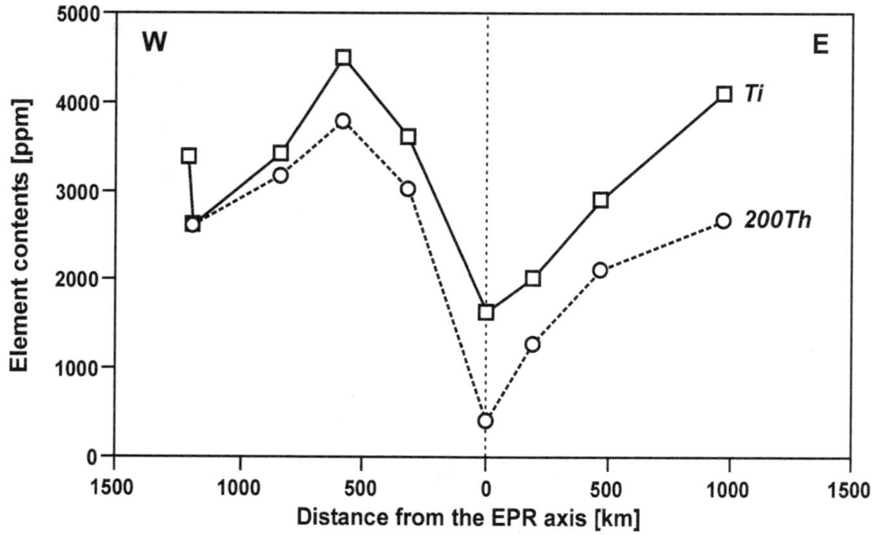

Fig. 1.12. Ti and Th content in abiogenic matter of metalliferous sediments at the section across the East Pacific Rise at 39°S.

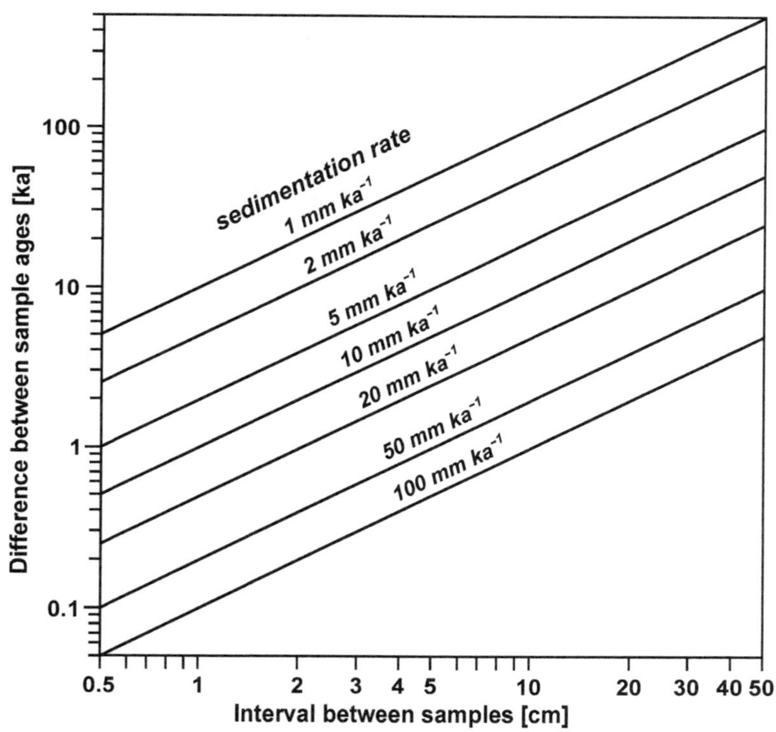

Fig. 1.13. Interrelationships between sedimentation rates, sample intervals and difference between sample ages.

(recalculated to carbonate-free base) content. Metalliferous sediments sampled in Core 154-19 located at 19°50.3'S, 116°37.8'W, at a depth of 3334 m collected 300 km to the west of the EPR axis and in Core 154-20 located at 19°39.8'S, 117°58.0'W, at a depth of 3470 m collected 440 km to the west of the EPR axis have almost unvarying sedimentation rates of 4.0 and 3.3 mm ka^{-1}; and there are no distinct trends in Fe/Al and Al_{CFB} values in metalliferous sediments in these cores.

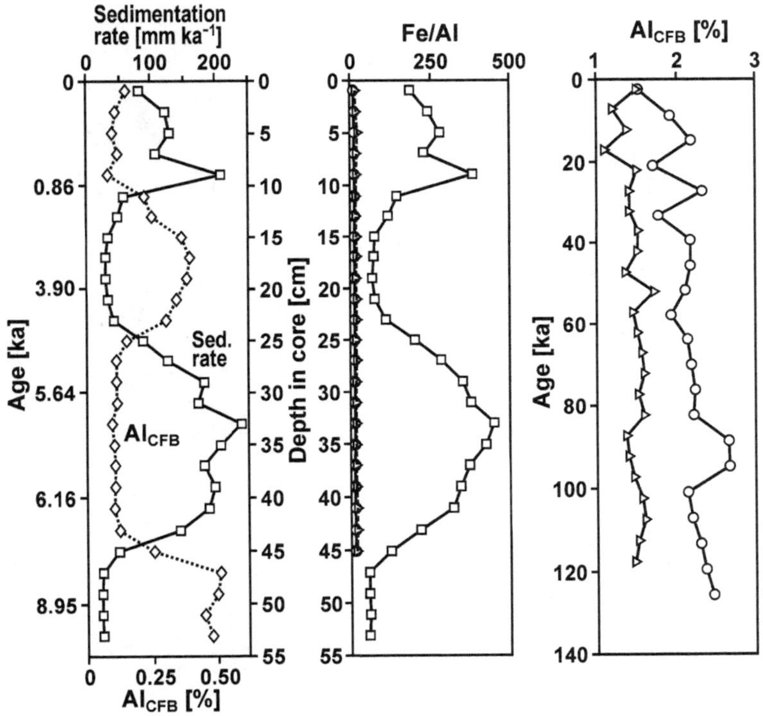

Fig. 1.14. Variations in sedimentation rate, Fe/Al ratios, and Al_{CFB} contents in Cores 154-18 (squares and diamonds), 154-19 (triangles), and 154-20 (circles). Based on data of Shimmield and Price (1988).

Twenty cores of metalliferous sediments 22 to 168 cm long (Fig. 1.5) were collected in 1987–1988 during Cruise 4 of the R/V *Geolog Fersman* within the polygon from 20°30'S to 22°S at three sections located 10, 20 and 40 km to the west of the EPR axis. A large collection of samples was obtained from these cores that were taken at intervals of 1 to 2 cm (Gurvich 1988). These samples have been studied comprehensively by Dekov (1994). The chemical composition and ages determined on these sediments showed a cyclicity in the variation in contents of Fe_{CFB} (Fig. 1.15), and also of Al_{CFB} and Ti_{CFB} within individual cores (Dekov 1994).

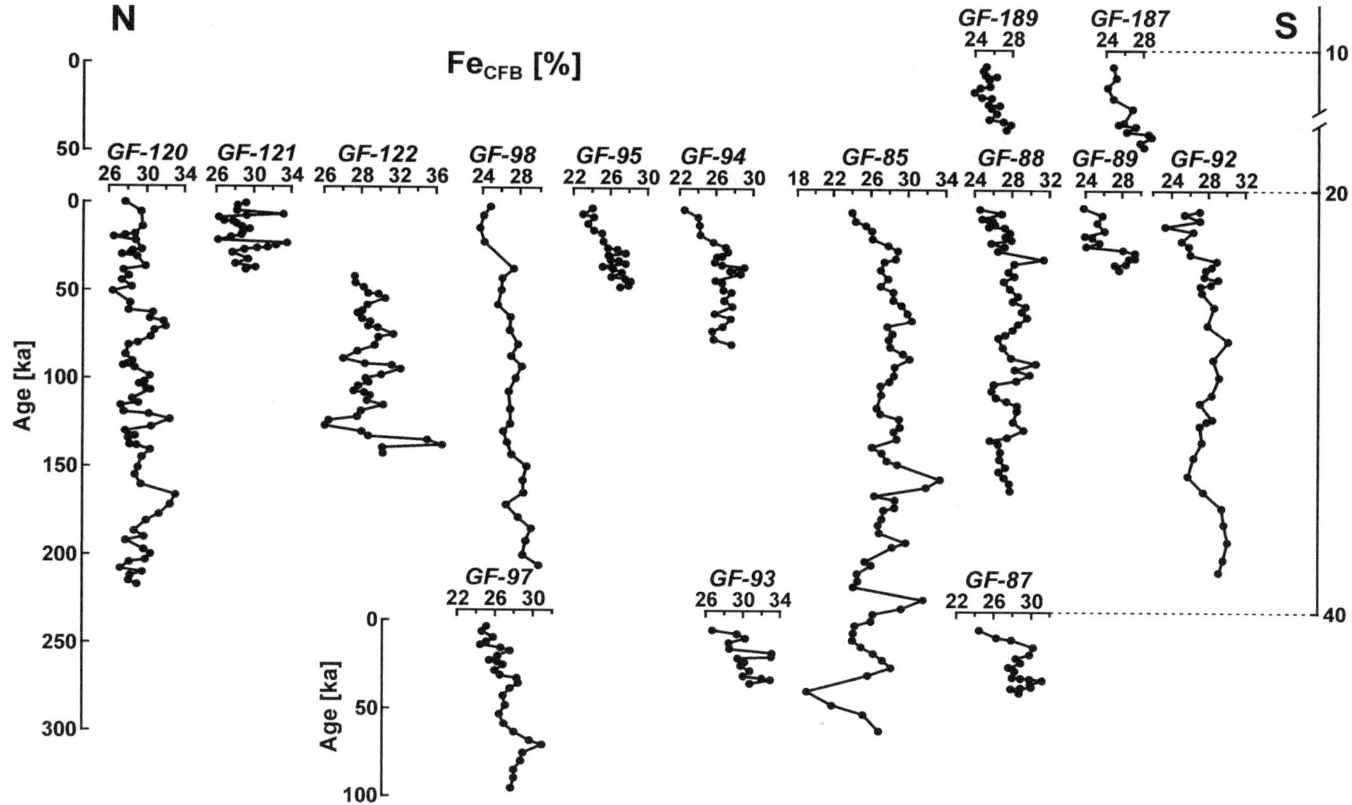

Fig. 1.15. Variation in contents of Fe_{CFB} over time in metalliferous sediments from the EPR axial zone at 20°30'S–22°S to the west of the axis. After Dekov (1994). The core positions are shown in Fig. 1.5.

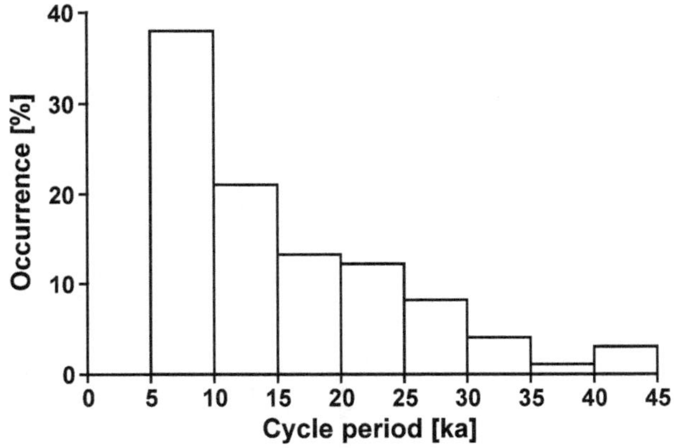

Fig. 1.16. Frequencies of occurrence of Fe_{CFB} variation cycles with different periods in metalliferous sediments from the EPR axial zone at 20°30'S–22°S to the west of the axis. After Dekov (1994).

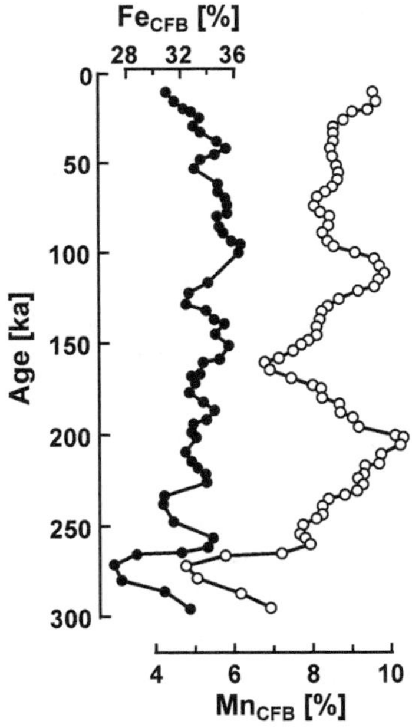

Fig. 1.17. Variations in contents of Fe_{CFB} and Mn_{CFB} in metalliferous sediments from Core GF-85 over time. After Dekov (1994). The core position is shown in Fig. 1.5.

The contents of Al_{CFB} and Ti_{CFB} vary in an antiphase manner with the variations in contents of Fe_{CFB} due to the mutual dilution of hydrothermal and lithogenic matter. The frequencies of occurrence of cycles with different periods are shown in Fig. 1.16. More than a half of the cycles have periods of 10±5 ka.

In the cores recovered for metalliferous sediment sequences that accumulated during the last 200–300 ka there are variation cycles for Mn_{CFB} with periods of about 100 ka (Fig. 1.17). Variation cycles for Fe/Mn ratios also have periods of about 100 ka (Fig. 1.18).

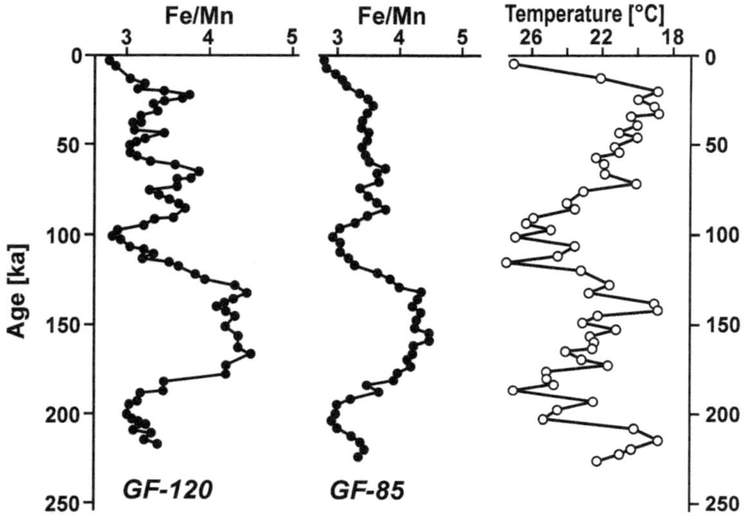

Fig. 1.18. Variations of Fe/Mn ratios over time in metalliferous sediments from Cores GF-120 and GF-85 (Dekov 1994), and variations of Equatorial Pacific surface water temperatures over time (Kennett 1982).

It is noted that the cycles with a period of ~100 ka in metalliferous sediments in Cores GF-120 and GF-85 accumulated in areas more than 100 km apart and adjacent to different hydrothermal fields (Fig. 1.5) are in-phase. Long-period cycles of Fe/Mn ratios and fragments of them in metalliferous sediments from shorter cores are in-phase with the cycles of these two cores. Cyclicity of Fe_{CFB} with the period ~10 ka is related to local cyclicity of hydrothermal activity within individual hydrothermal fields (Rona 1984). And the observed cyclicities of Fe_{CFB} in the individual sediment cores are not in-phase for the whole area from 20°30'S to 22°S. As for the cyclicities of Mn_{CFB} and Fe/Mn ratios with a period of ~100 ka they are in-phase for an area that is more than 150 km long. This cyclicity is certainly caused in a normal way. It is found in the pattern of distribution of Fe/Mn ratios in surface sediments from the western flank of the EPR at 20°30'S–22°S (Fig. 1.5) that accumulated where a westerly direction of flow of the bottom currents is predominant. If there was a decrease in the flow of currents in a westerly direction Fe/Mn ratios at the same distance west of the EPR axis would be lower and if there was an increase in flow the ratios would be higher.

The intensity of mixing of ocean waters and the velocities of bottom currents are connected to climate changes. In Pleistocene and Holocene time they were especially strong. Variations in the temperature of surface water are an indication of these changes. The intensity of mixing of ocean waters increases with lower temperatures and decreases with higher temperatures. Comparison of variations of Fe/Mn ratio in metalliferous sediments from Cores GF-120 and GF-85 over time, and variation of the temperature of surface waters from the Equatorial Pacific over time show distinct synchronism (Fig. 1.18). This synchronism confirms the assumptions that were made about the common causes of cyclic variations of Fe/Mn ratios in metalliferous sediments from the polygon area that was explored.

1.1.5. Mineral carriers of chemical elements

The contents and distribution of chemical elements in metalliferous sediments are influenced to a great extent by the minerals that carry them. Special attention was given to this aspect of the study of the geochemistry of metalliferous sediments in the seventies and eighties of the last century by Dymond et al. (1973), Sayles and Bischoff (1973), Sayles et al. (1975), Bagin et al. (1975), Lisitzin et al. (1976), Heath and Dymond (1977), Gurvich et al. (1977, 1979, 1980a-c), Migdisov et al. (1979), Cronan (1980), Varnavas (1988), Dekov (1994), and others.

In the metalliferous sediments accumulated 10 km to the west of the EPR axis the main carriers of most chemical elements and especially of the heavy metals are the Fe- and Mn-oxyhydroxide and the Fe-smectite minerals. They carry more than 90 to 95% of the total contents of these elements. About 50% of the Si is bound chemically to hydrothermal silica and Fe-smectite minerals (Dekov 1994).

The relative roles of different mineral carriers of some chemical elements in metalliferous sediments of the EPR axial zone between latitudes 10°S and 25°S at distances up to 150 km from the EPR axis, in the Bauer Depression and in the Central Basin (the southern part of the Peru Basin) are shown in Fig. 1.19 based on the data of Heath and Dymond (1977).

Fe- and Mn-oxyhydroxides and Fe-smectites are the prevalent mineral carriers of all of the elements studied except Si and their role as mineral carriers decreases with distances away from the EPR axis. About 80% of the non-lithogenic Fe is carried in Fe-oxyhydroxides and about 20% in the Fe-smectites in metalliferous sediments from the EPR axial zone. At a distance from the axis the proportion of the Fe in the respective minerals changes and its relative content carried in the Fe-smectites increases and in the Fe-oxyhydroxides decreases. In metalliferous sediments from the Bauer Depression and Central Basin the relative contents of the Fe carried in these minerals are about equal. Most (≥90%) of the Mn and Ni and 50 to 80% of the Cu and Zn are carried in the Fe- and Mn-oxyhydroxides. At a distance from the EPR axis the relative contents of elements carried in these minerals decrease. Only in metalliferous sediments from the EPR axial zone Si is carried mainly in the non-lithogenic minerals, Fe-smectites and opal, and this is reflected by the high Si/Al ratio (Fig. 1.8). There is a sharp increase in the amount of Si carried in lithogenic minerals at a distance from the axis.

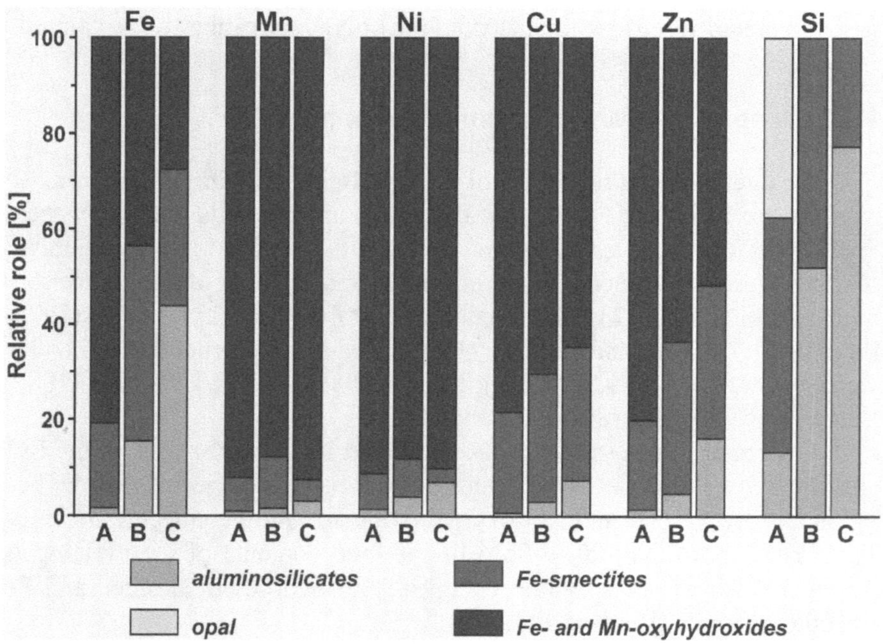

Fig. 1.19. Relative roles of different mineral carriers of chemical elements in metalliferous sediments from the EPR axial zone between 10°S and 25°S (A), Bauer Depression (B) and Central Basin (C). Based on data from Heath and Dymond (1977).

It has been possible by using partition chemical analysis to identify changes in the role of different mineral carriers for twelve chemical elements in transitions of metalliferous sediments deposited from 300 km to the west of the EPR axis to those accumulated 1200 km to the west of it and in the deep-sea red clays in the South Basin (Fig. 1.20).

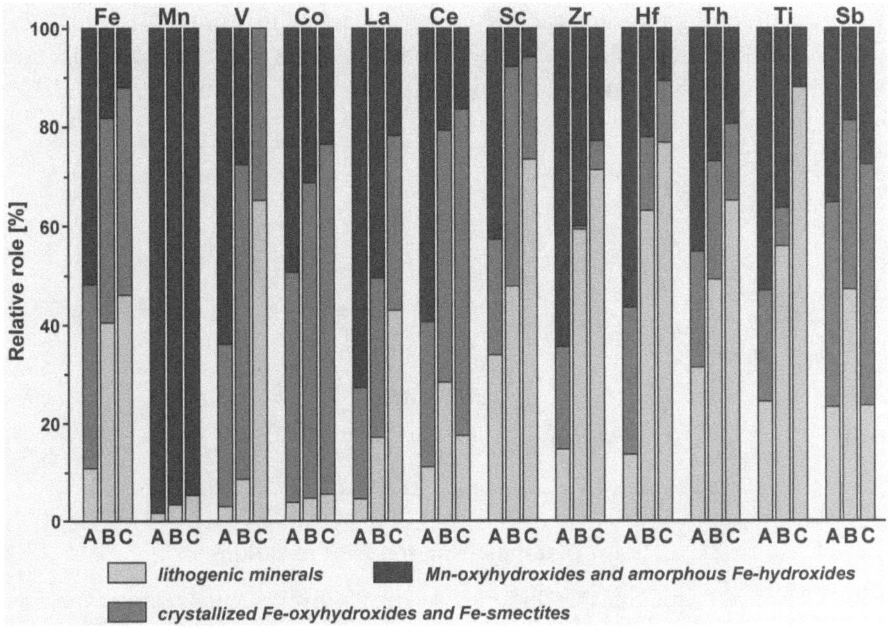

Fig. 1.20. Relative roles of different mineral carriers of chemical elements in metalliferous sediments accumulated 300 km (A) and 1200 km (B) to the west of the EPR axis as well as in deep-sea red clays (C). Based on the results of partition chemical analysis. The data of Gurvich (1977), Gurvich et al. (1977, 1980a-c), and Migdisov et al. (1979)

Non-lithogenic minerals are the main carriers for all the elements of the group in metalliferous sediments accumulated 300 km to the west of the EPR axis. In sediments 1200 km west of the axis the role of these non-lithogenic mineral carriers is lower and becomes secondary (<50%) for Ti, Th, Zr and Hf, but it remains significantly higher than it is in the red clays. At some distance from the EPR axis the role of Mn-oxyhydroxide minerals and of amorphous Fe-hydroxides as element carriers decreases and the role of crystallized Fe-oxyhydroxides and Fe-smectite minerals increases. A similar trend has been observed with an increase in the age of metalliferous sediments (Marchig and Gundlach 1982; Varnavas 1988, et al.) and the role of leached Fe is also reduced with an increase in the age of metalliferous sediments (Dubinin and Volkov 1992).

The data for the distribution and occurrence of Ti in metalliferous sediments may appear strange or inconsistent when first examined. However they are confirmed in the very high Ti/Al ratios in metalliferous sediments from the EPR axial zone (Fig. 1.21) and by the study of leached forms of this element. Practically all of the Ti in metalliferous sediments from the

EPR axial zone has been leached, about a half of it in metalliferous sediments on the EPR flanks and about 20% of its bulk content in metalliferous sediments in the Bauer Depression (Dubinin and Volkov 1992).

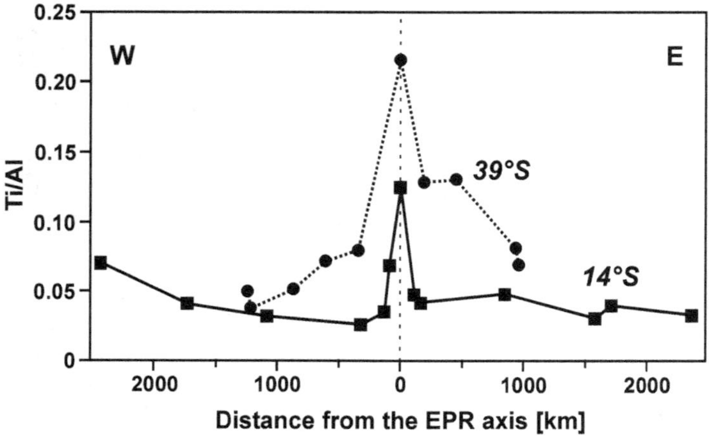

Fig. 1.21. Ti/Al ratios in metalliferous sediments from sections across the East Pacific Rise at 14°S and at 39°S. Based on the data from Boström and Peterson (1969) and Migdisov et al. (1979).

Fe- and Mn-oxyhydroxides and Fe-smectite minerals are the prevalent carriers of the rare-earth elements (REE) in metalliferous sediments in most of the section across the EPR at 16°S (Fig. 1.22). Despite significant fluctuations in the contents of the REE 55% to 85% of their bulk contents are carried in these minerals in almost all sediments of the section. Lithogenic minerals in sediments from the EPR axial zone carry minimal part of the REE content but their role increases at a distance from the axis. Some part of the REE content is bound in biogenic carbonate in the carbonaceous and low-carbonaceous metalliferous sediments.

Transitions in the role of different mineral carriers of barium in the generalized section through metalliferous sediments from the EPR axial zone to deep-sea red clays in the Southern Basin are shown in Fig. 1.23. Fe- and Mn-oxyhydroxides are major carriers of Ba in metalliferous sediments of the EPR axial zone where the Ba content is minimal (Boström et al. 1973; Gurvich et al. 1979). Metalliferous sediments are enriched in Ba at a distance from the EPR axis where fine-grained authigenic barite becomes the main carrier of the element (Gurvich et al. 1979). This barite has formed during decomposition and mineralization of organic matter (Boström at al. 1978; Gurvich et al. 1978).

According to the above data and that of Lukashin (1983), the average roles of Fe- and Mn-oxyhydroxide and Fe-smectite minerals as carriers of

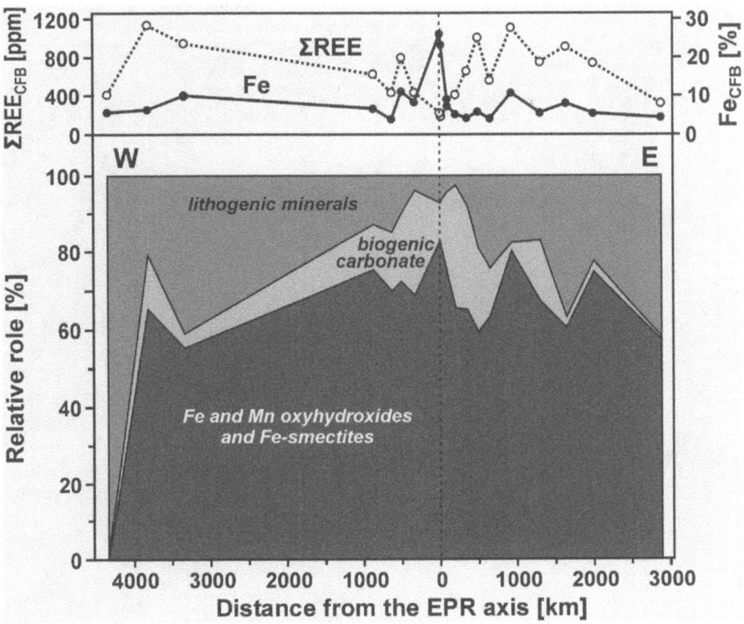

Fig. 1.22. Relative roles of different carriers of the REE in bottom sediments from the section across the EPR at 16°S. After Dubinin and Volkov (1986).

different chemical elements in metalliferous sediments from the Southeast Pacific are reduced in the following sequence in the order of magnitude shown in %:

Mn →	(Ni, Co) →	(V, Cu) →	Zn →	(Fe, REE) →	(Sb, Zr, Hf, Sc) →	(Ti, Th) →	Ba →	Si
>98	95–98	90–95	80–90	70–80	60–70	50–60	20–40	10–20

This sequence is eloquent evidence indicating that most of the chemical elements investigated, including the elements-hydrolysates, have been contributed to the metalliferous sediments from solution and are bound within the hydrothermal and authigenic mineral carriers. If we compare this sequence with the data showing excessive accumulation of chemical elements in metalliferous sediments (Sect. 1.1.4), we can conclude that most of the Ni, Co, V, Cu, Zn, REE, Sb, Zr, Hf and Sc was contributed from solutions and bound in hydrothermal mineral carriers. Si is also a part of this sequence of elements because it also has entered metalliferous sediments from solution and is bound not only with Fe-smectite minerals but also with opal. According to Cronan (1980), Ni and Co are bound mainly in the manganese minerals, but Cu and Zn are bound mainly in the iron minerals. As for Ti and Th that were contributed to metalliferous sediments from solutions, most of their contents are bound in authigenic mineral carriers.

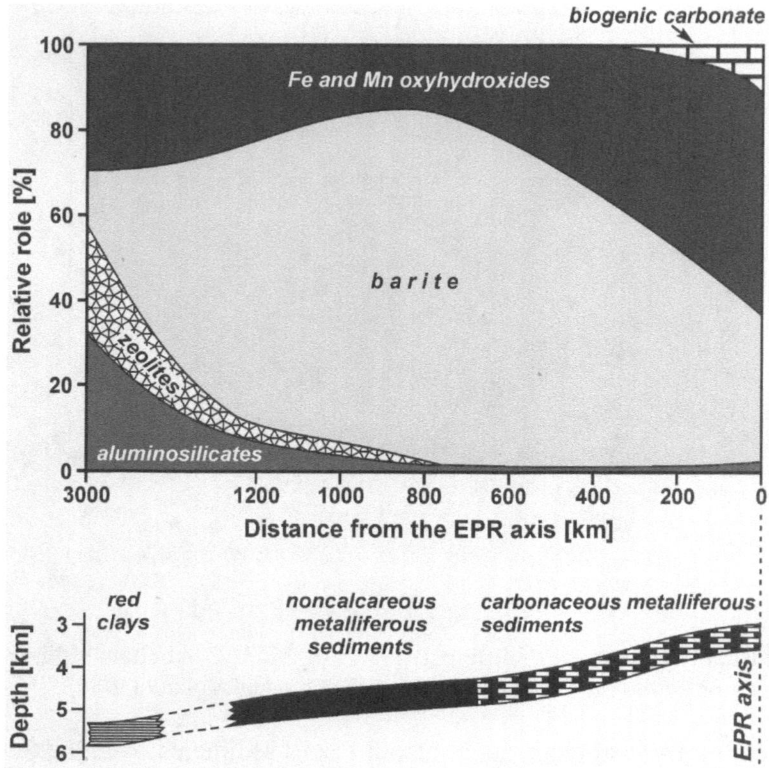

Fig. 1.23. Transitions in the relative roles of different mineral carriers of barium in the generalized section in the Southeast Pacific from metalliferous sediments at the EPR axial zone to deep-sea red clays in the Southern Basin.

It is important to consider the carriers of aluminum in metalliferous sediments. According to the results of the partition chemical analysis, Al is not bound with amorphous Fe-hydroxides and Mn-oxyhydroxides. The only forms leached in the partition analysis are the poorly crystallized aluminosilicate minerals that occur in metalliferous sediments (Lisitzin et al. 1980b). According to Varnavas (1988), some leaching of Al from metalliferous sediments during the sequential partition analysis results from the partial dissolution of newly formed low temperature authigenic smectite minerals (Hein et al. 1979; Cole 1985).

Dymond and Heath (1977) attempted to estimate the ratio of hydrothermal Al and Fe in metalliferous sediments by using the minimum Al/Fe ratio. At that time the minimum Al/Fe ratio measured was 0.00625. Later Shimmield and Price (1988) measured a ratio of 0.0022. In buoyant and nonbuoyant hydrothermal plumes the minimum Al/Fe ratios are about 0.001–0.002 and 0.0014–0.0027 respectively. If 0.002 is accepted as a minimum ratio, the average content of hydrothermal Al in metalliferous sediments is 0.027%, which is only about 0.6% of its average bulk content (Table 1.1). This permits consideration of Al as a proxy for lithogenic material in metalliferous sediments, and in the calculations of other elements by assuming an absence of excessive accumulation of Al.

1.1.6. Accumulation rates of chemical elements

Sedimentation of hydrothermal matter on the ocean floor results not only in an increase in the contents of Fe and Mn in sediments but also in an increase in their accumulation rates (Fig. 1.24 and 1.25).

Accumulation rates of Fe and Mn in background pelagic sediments from the southern arid zone are respectively 1 to 5 mg cm^{-2} ka^{-1} and less than 0.5 mg cm^{-2} ka^{-1} (in some places 0.5 to 1 mg cm^{-2} ka^{-1}). And in metalliferous sediments from the EPR axial zone the accumulation rates of Fe and Mn exceeding 25 mg cm^{-2} ka^{-1} and 10 mg cm^{-2} ka^{-1} are reported (Boström 1973; Boström et al. 1973; Froelich et al. 1977; Lisitzin 1978; Bogdanov et al. 1979c; Heinze 1985; Lisitzin et al. 1990; Dekov 1994, et al.). The highest accumulation rates of Fe and Mn in the EPR axial zone occur between 10°S and 25°S in the part of the EPR that has the greatest spreading rate and where the highest rates of abiogenic sedimentation for the EPR axial zone are reported (Bogdanov and Chekhovskikh 1979). Fe and Mn accumulation rates decrease at a distance from the axis and gradually attain values that are characteristic for background pelagic sediments. The greatest decrease in the accumulation rates occurs in the axial zone and becomes lower at a distance from the axis. Annually $680 \cdot 10^3$ tons of Fe, $200 \cdot 10^3$ tons of Mn, and $175 \cdot 10^3$ tons of Al accumulate within the area of metalliferous sediments in the Southeast Pacific where the average accumulation rates are respectively 6.8, 2.0 and 1.75 mg cm^{-2} ka^{-1}. The accumulation of Fe in excess of that in the background pelagic sediments is $550 \cdot 10^3$ t a^{-1}, and the excess of Mn accumulation over the background is $180 \cdot 10^3$ t a^{-1} (Table 1.2).

Patterns of the distribution of accumulation rates of trace elements in recent bottom sediments from the Southeast Pacific are dependant in many respects on the mode of distribution of their contents.

The areal distribution of accumulation rates of Zn, V, Pb, As, Cd, P and B, elements with Type 1 distribution of their contents (Sect. 1.1.4), can be described only in general terms as maps of their distribution have not yet been drawn. Comparing the distribution of contents of these elements with those of Fe and Mn it is apparent that the distribution of accumulation rates of these elements and of Fe and Mn are similar.

Table 1.2. Comparison of the annual accumulation of chemical elements in metalliferous sediments and the estimated annual hydrothermal contribution in the Southeast Pacific based on the assumption that hydrothermal contribution and excessive accumulation of Fe are related.

Element	Average accumulation rate $\mu g\ cm^{-2}\ ka^{-1}$	Total accumulation $t\ a^{-1}$	Background accumulation (B) $t\ a^{-1}$
Fe	6800	680000	130000
Al	1750	175000	175000
Ga	0.4	40	40
Cr	1.2	120	110
Th	0.2	20	18
Ti	150	15000	13000
Si	6500	650000	520000
Li	1.3	130	100
Rb	1.3	130	100
B	8.5	850	560
Ba	435	43500	24000
Sc	1.05	105	47
Ce	4.6	460	190
Hf	0.25	25	9
Y	12	1200	430
Sm	1.18	118	42
Yb	0.44	44	15
Eu	0.31	31	10
Zn	16	1600	360
La	6.3	630	140
Zr	18	1800	400
U	0.15	15	3
Sb	0.48	48	9
Pb	6.2	620	90
Cd	0.069	6.9	0.9
V	17	1700	200
Mn	2000	200000	20000
Ni	33	3300	300
Co	8.8	880	70
Cu	42	4200	300

Distribution of accumulation rates of elements with Type 2 distribution of their contents may have three somewhat different patterns.

Accumulation rates of Hf, Sc, Ba, REE, and Zr have minimum values in the axial zone. And accumulation rates of Ba and Zr usually have maximum values on the flanks of the EPR, which gradually decrease to background

Table 1.2. (continued):

Element	Excessive accumulation (E) t a^{-1}	Hydrothermal input (H) t a^{-1}	E/B	H/E
Fe	550000	550000	4.2	1.0
Al		860		
Ga	0	?	0	
Cr	10	?	0.09	
Th	2	?	0.11	
Ti	2000	?	0.15	
Si	130000	1300000	0.25	10
Li	30	9000	0.30	300
Rb	30	1900	0.30	63
B	290	16000	0.52	55
Ba	19500	5600	0.81	0.29
Sc	58	?	1.2	
Ce	270	3.1	1.4	0.011
Hf	16	?	1.8	
Y	770	?	1.8	
Sm	76	0.37	1.8	0.0049
Yb	29	0.056	1.9	0.0019
Eu	21	1.5	2.1	0.071
Zn	1240	20000	3.4	16
La	490	1.8	3.5	0.0037
Zr	1400	?	3.5	
U	12	0	4.0	0
Sb	39	~4	4.3	~0.1
Pb	530	190	5.9	0.36
Cd	6.0	24	6.7	4.0
V	1500	22	7.5	0.015
Mn	180000	180000	9.0	1.0
Ni	3000	220	10.0	0.073
Co	810	24	11.6	0.030
Cu	3900	4500	13.0	1.15

values away from it (Gurvich et al. 1979; Lukashin et al. 1980), but accumulation rates of Hf, Sc, and REE have maximum values on the flanks at some sites only and the values for them gradually increase from those in the axial zone to those in background pelagic sediments (Gurvich 1977; Gurvich and Lisitzin 1980c,d).

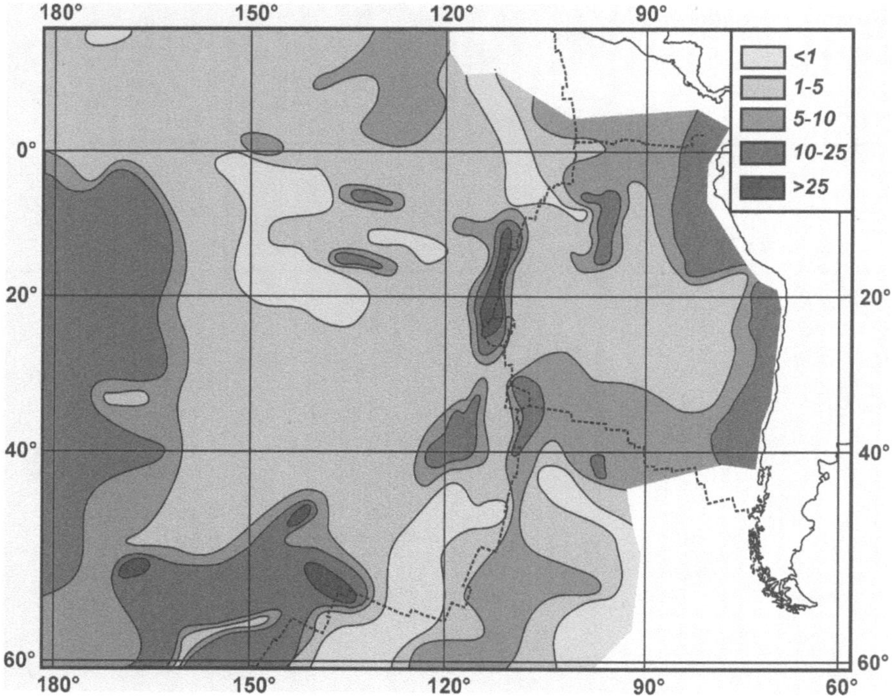

Fig. 1.24. Accumulation rates of Fe [mg cm^{-2} ka^{-1}] in recent bottom sediments from the Southeast Pacific. After Bogdanov et al. (1979c).

Accumulation rates of Ni, Co, and Sb have no minimum values in the axial zone of the part of the EPR that has the highest spreading rate but instead they may have maximum values in this area. Some examples of this kind of distribution are: the axial maximum for the accumulation rate of Ni at 17°S (Dymond and Veeh 1975), both high and low accumulation rates of Ni in the EPR axial zone were measured by Boström (1973) and Boström et al. (1973) as well as the distribution of accumulation rates of Sb (Fig. 1.26). The distribution of the accumulation rate of Cu probably has a similar pattern. According to Boström (1973) and Boström et al. (1973), metalliferous sediments of the EPR axial zone have both high and low accumulation rates of Cu. The distribution of accumulation rate of Th is characterized by gradual increase from the EPR axial zone to backgro-und sediments (Gurvich et al. 1980b).

The mentioned minima of accumulation rates of many chemical ele-ments result from partial redeposition of sedimentary material by bottom currents from the EPR axial zone to the flanks where its accumulation rates increase. This phenomenon is characteristic to all underwater rises (Lisi-tzin 1978). On the example of Ba it has been shown that in the areas of bottom currents crossing the EPR axis on the average 50–70% of Ba (of the amount, which would accumulate without the currents) have been redeposited from the axial zone to the flanks. In the flank metalliferous sediments, on the contrary, the amount of accumulated Ba on the average is 50% more than it would be without this redeposition (Gurvich et al. 1979). Similar results have been obtained during $CaCO_3$ studies with the only difference: $CaCO_3$ redeposited from the EPR axial zone has been accumulated on the flanks not completely due to its partial dissolution below the lysocline (Bogdanov and Gurvich, unpublished data).

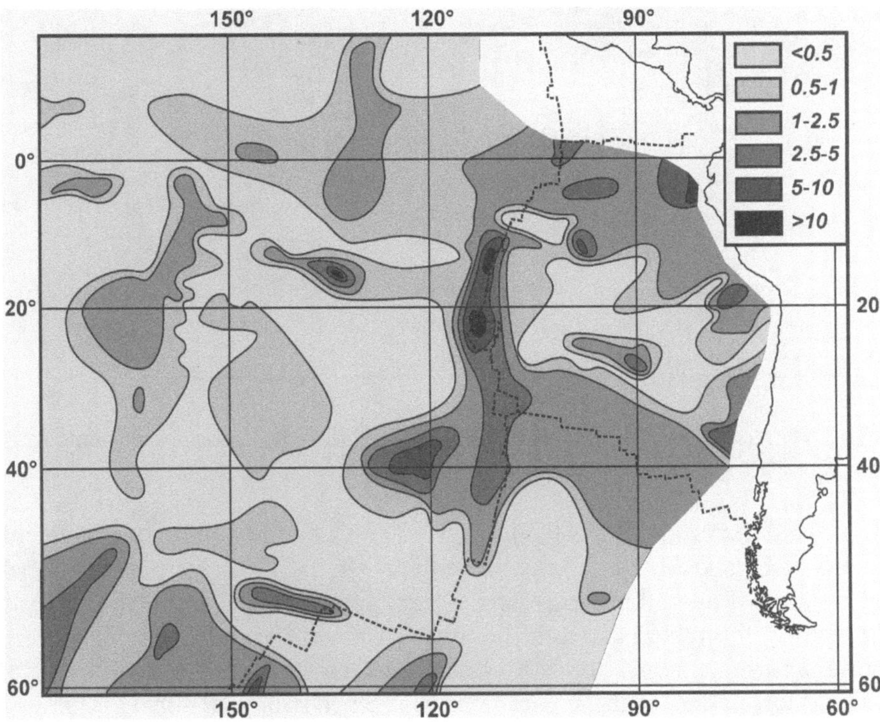

Fig. 1.25. Accumulation rates of Mn [mg cm^{-2} ka^{-1}] in recent bottom sediments from the Southeast Pacific. After Bogdanov et al. (1979c).

It is necessary to note that such "not catastrophic" (occurring mainly in particulate matter) redeposition of sedimentary material almost does not result to differentiation of its abiogenic part (as in turbidites), but changes accumulation rate.

If to assume, that Al is a proxy of background abiogenic sedimentary material, at no redeposition of this material by bottom currents within the Southeast Pacific field of metalliferous sediments located in the southern arid climatic zone accumulation rate of Al within the field should not vary

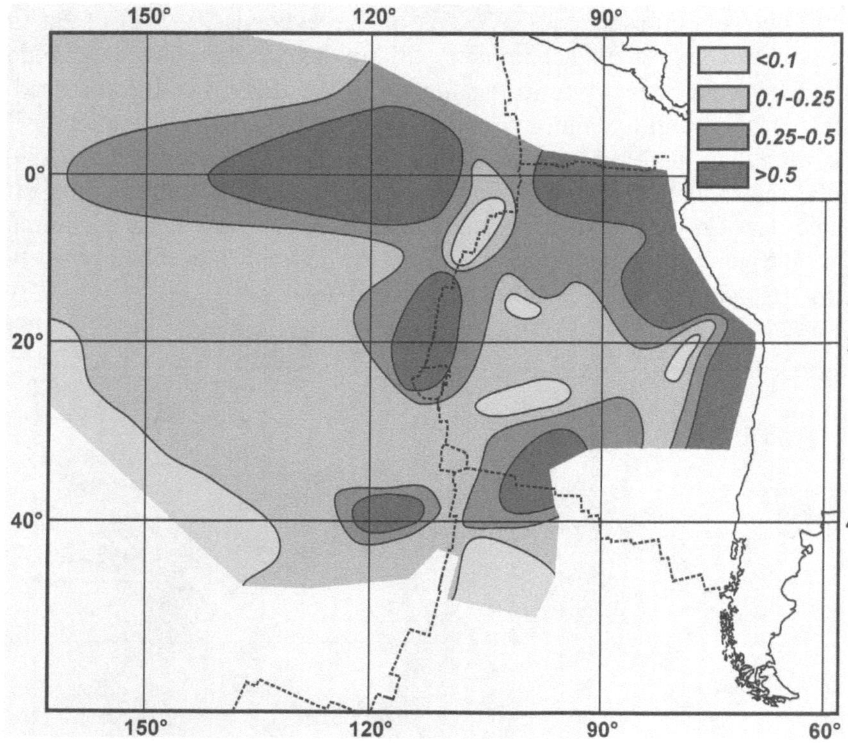

Fig. 1.26. Accumulation rate of Sb [μg cm^{-2} ka^{-1}] in recent bottom sediments from the Southeast Pacific.

a lot. And the ratio (Fe_{ex}+Mn_{ex})/Al (where Fe_{ex} and Mn_{ex} are excessive contents of Fe and Mn, and Al is Al content) should be directly proportional to the sum of Fe and Mn excessive accumulation rates (Fe_{ex}*+Mn_{ex}*). It is proved by the materials from the areas of metalliferous sediments where bottom currents across the EPR axis are absent or very weak (Fig. 1.27). Linear correlation of these parameters is rather strong (R=0.94). The regression equation is:

$$Fe_{ex}^* + Mn_{ex}^* = 1.78(Fe_{ex}+Mn_{ex})/Al, \quad (1.1)$$

where element contents are in %, and element accumulation rates are in mg cm^{-2} ka^{-1}.

The equation:

$$(Fe_{ex}+Mn_{ex})/Al = (Fe_{ex}^*+Mn_{ex}^*)/Al^*$$

is also true (Al* is accumulation rate of Al in mg cm^{-2} ka^{-1}). Hence

$$Fe_{ex}^* + Mn_{ex}^* = 1.78(Fe_{ex}^*+Mn_{ex}^*)/Al^*.$$

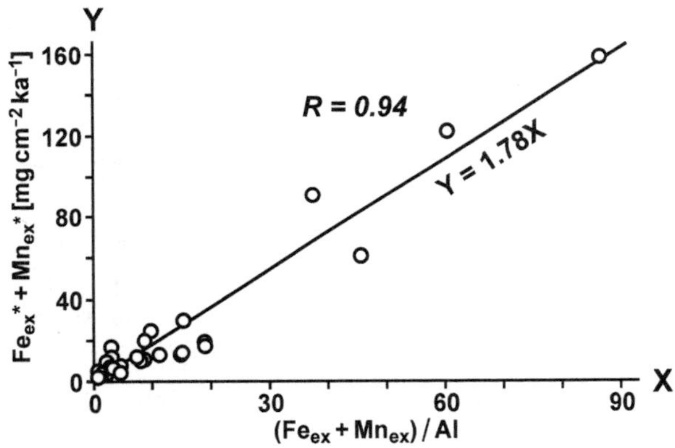

Fig. 1.27. Connection of $(Fe_{ex}+Mn_{ex})/Al$ and $Fe_{ex}{*}+Mn_{ex}{*}$ in metalliferous sediments from the areas where bottom currents across the EPR axis are absent or very weak. Based on the data of Boström (1973) and Migdisov et al. (1979).

It follows from this equation that Al* = 1.78 mg cm^{-2} ka^{-1}. This average Al accumulation rate in the sediments from the sampling used for the calculation of the regression equation is very close to the average Al accumulation rate (1.75 mg cm^{-2} ka^{-1}, Table 1.2) in metalliferous sediments from the Southeast Pacific.

If these calculations are correct, the ratio $(Fe_{ex}+Mn_{ex})/Al$ is a rather exact reflection of the "undistorted" sum $(Fe_{ex}{*}+Mn_{ex}{*})$ in concrete metalliferous sediments, which would be without the redeposition of sedimentary material from the EPR axial zone to the flanks by bottom currents. As the background ratio $(Fe_b+Mn_b)/Al_b = 0.84$ (where Fe_b=5.8%, Mn_b=0.75%, Al_b=7.8%; Table 1.1),

$$(Fe_{ex}+Mn_{ex})/Al = (Fe-AlFe_b/Al_b+Mn-AlMn_b/Al_b)/Al =$$

$$= (Fe+Mn)/Al - (Fe_b+Mn_b)/Al_b = (Fe+Mn)/Al - 0.84,$$

where Fe, Mn, and Al are contents of Fe, Mn, and Al in these concrete sediments.

Thus the "undistorted" accumulation rate of the excessive Fe+Mn in metalliferous sediments can be calculated from the equation:

$$Fe_{ex}{*} + Mn_{ex}{*} = 1.78((Fe+Mn)/Al-0.84) = 1.78(Fe+Mn)/Al - 1.5 \qquad (1.2)$$

Estimations of total and excessive accumulation of chemical elements in metalliferous sediments from the Southeast Pacific are shown in Table 1.2. They have been made at the assumption that there is no excessive Al accumulation in metalliferous sediments (see above). For most elements – heavy metals excessive accumulation exceeds background accumulation. Relative roles of excessive accumulation of the elements increase in the same row as ratios of element contents in metalliferous sediments to ones in background sediments.

1.1.7. Sources of excessive accumulation of chemical elements

The excessive accumulation of Fe in metalliferous sediments is derived entirely from hydrothermal sources. As for other chemical elements, it is interesting to compare their excessive accumulation with the maximum amounts that could possibly be contributed to metalliferous sediments from hydrothermal sources, based on the assumption that the other elements associated with Fe enter metalliferous sediments in the same proportions as they occur in primary hydrothermal fluids. The average concentrations of the elements in hydrothermal fluids from the EPR[4] have been used for calculating the input of elements from hydrothermal sources. The results of these calculations are shown in Table 1.2. It is seen from this table that hydrothermal input could provide excessive accumulation not only of Fe but also of Si, Li, Rb, B, Zn, Cd, Mn, and Cu. It is interesting to note that the estimated amount of Al (860 t a^{-1}) is 0.66% of its bulk accumulation in metalliferous sediments from the Southeast Pacific and is close to the estimated content of hydrothermal Al in metalliferous sediments of its average bulk content (~0.6%, Sect. 1.1.5). For most trace elements given in Table 1.2 even the maximum possible hydrothermal input does not account for their excessive accumulation in metalliferous sediments ($H/E < 1$).

Ocean water as a source of chemical elements also contributes to their excessive accumulation in metalliferous sediments. Elements from ocean water may be contributed to the metalliferous sediments due to scavenging (co-precipitation and/or sorption) by Fe- and Mn-oxyhydroxides, and Fe-smectites.

If hydrothermal solutions are the main sources of trace elements that leads to their excessive accumulation in metalliferous sediments, any increase in the content of hydrothermal metal-bearing matter (including the content of its proxy $Fe_{ex}+Mn_{ex}$) in abiogenic matter of metalliferous sediments may cause an increase in the excessive contents of trace elements (El_{ex}). This being the case one should expect a direct covariance relationship of El_{ex} and the sum $Fe_{ex}+Mn_{ex}$.

If ocean water is the main source of trace elements and their excessive accumulation, two extreme variants can be considered depending on the concentration of the trace elements in ocean water:

a) Concentrations of trace elements or of those forms of trace elements that are accessible for scavenging are high enough in ocean water and con-

[4] Element concentrations in hydrothermal fluids are shown in Sect. 4.2.

ditions for their scavenging are favorable; and ocean water could be an inexhaustible source of the trace elements, and their scavenging from ocean water and transfer by hydrothermal metal-bearing matter to metalliferous sediments would be defined only by properties of this hydrothermal metal-bearing matter and by the concentration of trace elements in ocean water. In this case an increase in the contents of the metal-bearing matter in metalliferous sediments would result in the increase in contents of the trace elements bound with them, and one would expect a direct covariance relationship of El_{ex} and the sum of Fe_{ex}+Mn_{ex}.

b) Concentrations of trace elements or of those forms of trace elements that are accessible for scavenging are low in ocean water, and ocean water is a limited source of these trace elements. In this case scavenging of the trace elements from ocean water does not depend on the amount of the scavenger – particulate metal-bearing hydrothermal matter migrating through ocean water from hydrothermal vents to the ocean bottom or on the amount of ocean water migrating through the sediments, and all of the amount of trace elements that are accessible for scavenging entry into the metalliferous sediments. Theoretically under these conditions it is assumed that there is a constant rate of input of the trace elements to the metal-bearing matter of the metalliferous sediments regardless of their accumulation rate. In this situation an increase in the accumulation rate of metal-bearing matter, where it is undistorted by the redeposition, results in a decrease in the content of the trace elements in it. It is concluded that the content of trace elements in the metal-bearing matter varies inversely to the accumulation rate of this metal-bearing matter. One can expect a direct covariance relationship of the ratio of $El_{ex}/(Fe_{ex}+Mn_{ex})$ and the inverse accumulation rate of the metal-bearing matter. As shown above, the ratio $(Fe_{ex}+Mn_{ex})/Al$ is a direct indication of the actual accumulation rate of $Fe_{ex}^{*}+Mn_{ex}^{*}$ in specific metalliferous sediments in which the redeposition of sedimentary material by bottom currents is absent. Therefore one can expect a direct proportional relationship to be shown by $El_{ex}/(Fe_{ex}+Mn_{ex})$ to $Al/(Fe_{ex}+Mn_{ex})$.

The ideal conditions in the examples considered above do not exist because they assume a constant contact time for hydrothermal metal-bearing matter with ocean water. Actually this does not happen because the time required for sedimentation of this matter on the ocean floor and hence the length of time for its contact with ocean water increases with the distance from its source. By increasing this distance the sum $Fe_{ex}^{*}+Mn_{ex}^{*}$ decreases, and the ratio $Al/(Fe_{ex}+Mn_{ex})$ increases.

If scavenging of a trace element by the metal-bearing matter from ocean water comes to an end quickly, considered as Variant 1, at longer contact of the metal-bearing matter with ocean water or at longer distance from the hydrothermal source, the contents of Fe_{ex}+Mn_{ex} and El_{ex} decrease, the

$Al/(Fe_{ex}+Mn_{ex})$ increases, and the $El_{ex}/(Fe_{ex}+Mn_{ex})$ ratio becomes constant. A direct proportional relationship of $El_{ex}/(Fe_{ex}+Mn_{ex})$ to $Al/(Fe_{ex}+Mn_{ex})$ does not exist in this case because it is inconsistent with the constant $El_{ex}/(Fe_{ex}+Mn_{ex})$ ratio.

If scavenging of trace elements by the metal-bearing matter from ocean water is a lengthy process because of their low concentration in ocean water and/or because of their properties, considered as Variant 2, a longer distance from the hydrothermal source provides a longer contact of the metal-bearing matter with ocean water resulting in a decrease in the contents of $Fe_{ex}+Mn_{ex}$, an increase in the $Al/(Fe_{ex}+Mn_{ex})$ ratio, and an increase in the $El_{ex}/(Fe_{ex}+Mn_{ex})$ ratio. The content of El_{ex} can increase, remain constant or decrease depending on the gradients of $Fe_{ex}+Mn_{ex}$ and of $El_{ex}/(Fe_{ex}+Mn_{ex})$. Where there is an equal time of contact the sum $Fe_{ex}+Mn_{ex}$ is lower, the $Al/(Fe_{ex}+Mn_{ex})$ ratio is higher, and the $El_{ex}/(Fe_{ex}+Mn_{ex})$ ratio is higher.

Also a mixed situation can occur in which the $El_{ex}/(Fe_{ex}+Mn_{ex})$ ratio increases initially and then as the scavenging ceases conditions gradually become stabile. This situation is universal for trace elements scavenged from ocean water. With very fast or fast stabilization of the $El_{ex}/(Fe_{ex}+Mn_{ex})$ ratio the maximum contents of El_{ex} are in the vicinity of and within a first few kilometers of the hydrothermal source; and at longer distances from the source the distribution patterns correspond to those in Variant 1. With slow stabilization of the $El_{ex}/(Fe_{ex}+Mn_{ex})$ ratio, within a few hundreds of kilometers from hydrothermal sources, the maximum contents of El_{ex} occur in the vicinity of the areas of stabilization. In this case the pattern of the $El_{ex}/(Fe_{ex}+Mn_{ex})$ ratio before stabilization corresponds to that developed in the conditions of Variant 2, and after stabilization to that of Variant 1. In the absence of the stabilization of the $El_{ex}/(Fe_{ex}+Mn_{ex})$ ratio within areas of metalliferous sediments the patterns correspond to those under the conditions of Variant 2. In this case El_{ex} contents reach maxima, but they are not as evident as in the previous cases.

As the bulk element content in metalliferous sediments (El) is the sum of the background content (El_b) and the excessive content (El_{ex}), and where the El_{ex} content is prevalent the pattern of an El distribution is similar to the patterns of El_{ex} distribution, and where the El_b content is prevalent the pattern of distribution is similar to the pattern of El_b distribution. Intermediate patterns in the distribution of the El contents are also possible.

Resulting patterns in the variation of El distribution with increasing distances from the EPR axis with different values and characters of variation of the $El_{ex}/(Fe_{ex}+Mn_{ex})$ ratio are shown in Fig. 1.28, where data derived from the mathematical modeling of the described hypothetical situations have been plotted.

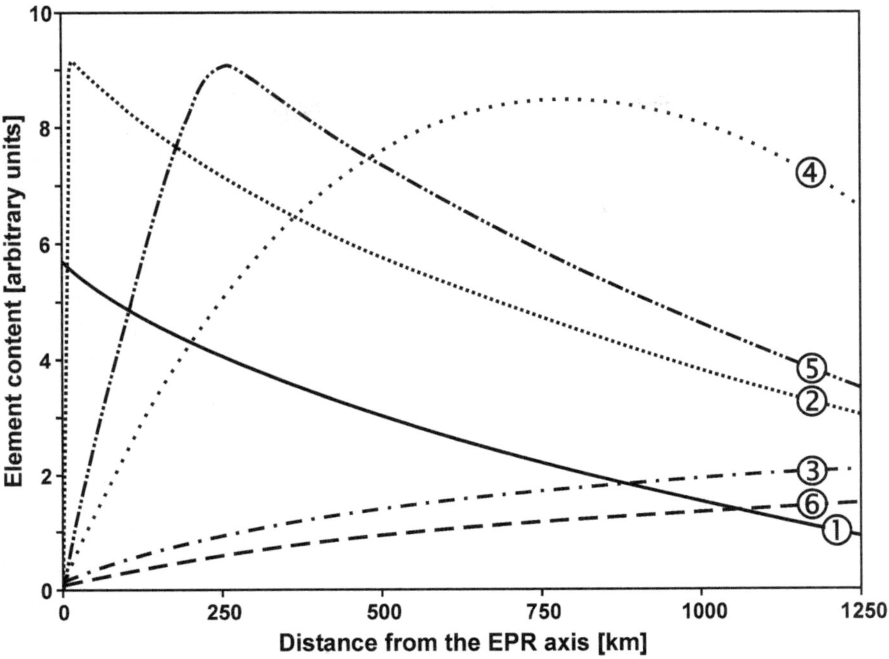

Fig. 1.28. Hypothetical variations of chemical element contents (El) in abiogenic matter of metalliferous sediments at increasing distances from the EPR axis and with different types of $El_{ex}/(Fe_{ex}+Mn_{ex})$ variation. ① $El_{ex}/(Fe_{ex}+Mn_{ex})$ = const (hydrothermal elements); ②–⑤ elements scavenged from ocean water: ② from the axis to 1 km distance $El_{ex}/(Fe_{ex}+Mn_{ex})$ = 2 arbitrary units km^{-1} and then decreases to nil; ③ $El_{ex}/(Fe_{ex}+Mn_{ex})$ = 0.002 arbitrary units km^{-1}; ④ $El_{ex}/(Fe_{ex}+Mn_{ex})$ = 0.025 arbitrary units km^{-1}; ⑤ from the axis to 200 km distance $El_{ex}/(Fe_{ex}+Mn_{ex})$ = 0.05 arbitrary units km^{-1} and then decreases to nil; ⑥ $El_{ex}/(Fe_{ex}+Mn_{ex})$ = 0 (elements without excessive accumulation). ① and ⑥ are also Fe/5 and Al/5 contents in %.

Curve 1 shows the situation when hydrothermal solutions are the main source of excessive accumulation of a trace element.

Curve 2 illustrates the situation for Variant 1 in which case scavenging of a trace element by the metal-bearing matter from ocean water comes to an end quickly.

Curves 3 and 4 show the situations for Variant 2 in which the scavenging of a trace element by the metal-bearing matter from ocean water is over a lengthy time, but the gradient of the $El_{ex}/(Fe_{ex}+Mn_{ex})$ ratio for Curve 3 is lower than for Curve 4.

Curve 5 shows the mixed situation with slow stabilization of the $El_{ex}/(Fe_{ex}+Mn_{ex})$ ratio.

Curve 6 shows the situation when element scavenging from solution is absent.

Curves 1 and 2 in Fig. 1.28 also show Type 1 element distribution with maximum values in the axial zone (see Sect. 1.1.4), Curves 3 and 6 show Type 4 element distribution with minimum values in the axial zone and no maximum values on the flanks of ridges, Curves 4 and 5 show Type 2 element distribution with minimum values in the axial zone and maximum values on the flanks, and mixtures of Curves 1 and 5 or Curves 1 and 4 show Type 3 element distribution with maximum values in the axial zone and maximum values on the flanks. Type 3 element distribution can be the result of well defined dualism in which an element enters the metal-bearing matter of metalliferous sediments from both hydrothermal solutions and ocean water.

The congruence of a real pattern with one in the proposed situations can be estimated by considering correlation coefficients of data pairs of the two kinds. In the first kind of the pairs data for El_{ex} and (Fe_{ex}+Mn_{ex}), and in the second kind of the pairs data for the El_{ex}/(Fe_{ex}+Mn_{ex}) ratio and for the Al/(Fe_{ex}+Mn_{ex}) ratio are considered.

It is important to note that Fe- or Mn-minerals or both of them are carriers of trace elements and the correlation coefficient of the pair Fe_{ex} and Mn_{ex} in metalliferous sediments from the Southeast Pacific is 0.95.

Correlation coefficients for data pairs of elements of the two kinds have been calculated for 15 trace elements. They are shown in Fig. 1.29.

V, Zn, P, Pb, As, and Cd have high positive correlation coefficients ($R \geq +0.80$) for the first kind of pairs and low positive or negative correlation coefficients for the second kind of pairs. Sc, Th, Ni, Zr, Hf, La, and Co have high positive correlation coefficients ($R > +0.85$) for the second kind of pairs and low positive or negative correlation coefficients for the first kind of pairs. Sb and Cu have correlation coefficients about +0.7 for the pairs of both the first, and the second kinds[5].

High correlation coefficients for the second kind pairs for Sc, Th, Ni, Zr, Hf, La, and Co can be unequivocally interpreted as a result of their scavenging by the metal-bearing matter from ocean water. Concentrations of these trace elements, or their forms accessible for sorption, are low in ocean water, and ocean water is a limited source of these elements.

High correlation coefficients also indicate the variations in contents of Sc_{ex}, Th_{ex}, Ni_{ex}, Zr_{ex}, Hf_{ex}, La_{ex}, and Co_{ex} with distance from the EPR axis and are close to those in Variant 2. This kind of variations coincides with measured (real) variations in the contents of Sc, Th, Ni, Zr, Hf, La and Co

[5] Fe and Mn have correlation coefficients for the first kind of pairs +0.996 and +0.97, and for the second kind of pairs –0.21 and +0.21.

in metalliferous sediments at a distance from the EPR axis. Sc, Ni, Zr, Hf, La and Co contents increase with distance from the EPR axis, reach maxima, and then decrease. Variations in the contents of Th are similar or they may gradually increase with distance from the EPR axis (see Sect. 1.4). The prevalent role of ocean water as a source of Sc_{ex}, Th_{ex}, Ni_{ex}, Zr_{ex}, Hf_{ex}, La_{ex} and Co_{ex} is also confirmed by the data in Table 1.2. Even with hydrothermal sources for the elements having the greatest possible role, they could supply only an insignificant part in the excessive accumulation of these elements in metalliferous sediments.

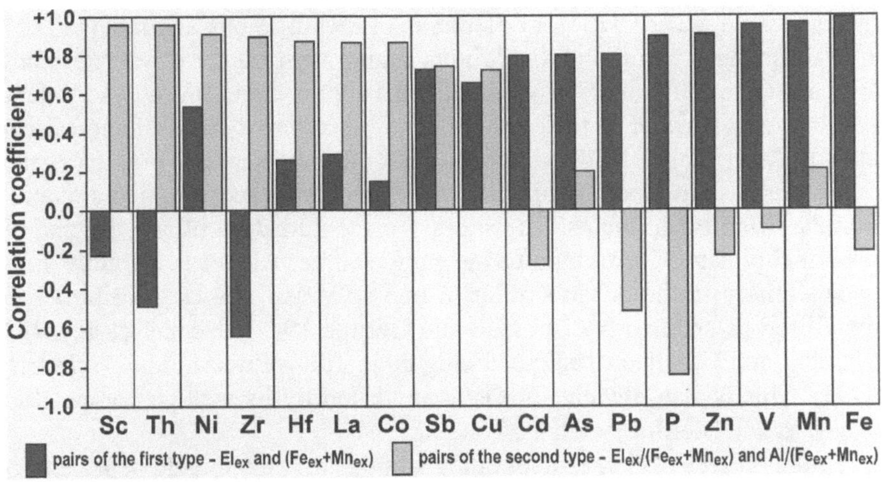

Fig. 1.29. Correlation coefficients for the data pairs of the two kinds calculated for abiogenic contents of chemical elements bound with the metal-bearing matter of metalliferous sediments from the Southeast Pacific.

The REE patterns also indicate that ocean water is the main source contributing to the excessive accumulation of these elements in metalliferous sediments from the Southeast Pacific. Since the publication of the paper of Bender et al. (1971) all studies of distribution of the REE in metalliferous sediments from the Southeast Pacific (Dymond et al. 1973, 1976; Piper 1974; Piper and Graef 1974; Balashov et al. 1974; Balashov 1976; Gurvich et al. 1980c; Dubinin and Volkov 1986, 1988; Ruhlin and Owen 1986; Barrett and Jarvis 1988; Kunzendorf et al. 1988; Owen and Olivarez 1988; Olivarez and Owen 1989; Lisitzin et al. 1990; Halliday et al. 1992; Dekov and Gurvich 1993; Strekopytov 1997; Dubinin 2004, et al.) show similarities of their patterns to the REE pattern of ocean water.

High correlation coefficients for the first kind of pairs for V, Zn, P, Pb, As and Cd indicate either that hydrothermal solutions are the main source for their excessive accumulation in metalliferous sediments, or that the excessive accumulation of these elements could be derived from ocean water where element concentrations are high enough, and conditions are favorable for their rapid scavenging by the metal-bearing matter. It is necessary to note that, unlike many other elements, V, P and As enter metal-bearing matter mainly by their co-precipitation from ocean water during the formation of hydrothermal Fe-oxyhydroxides (Trefry and Metz 1989; Feely et al. 1991a,b, 1994, 1998; Rudnicki and Elderfield, 1993; Ludford et al. 1996; Savenko 2000; Edmonds and German 2004, et al.). A solution for this dilemma is impossible in the framework of the approach discussed here. The data for the greatest possible role of hydrothermal solutions contributing to the excessive accumulation of elements in metalliferous sediments (Table 1.2) show that hydrothermal solutions cannot provide the observed excessive amounts of V and Pb accumulated in these metalliferous sediments and that ocean water is their source. The behavior of these elements, and also of P, As, Zn, and Cd in hydrothermal plumes (Sect. 4.3) indicates that ocean water is the main source that leads to their excessive accumulation in metalliferous sediments. The excessive accumulation of Mo, Pd and Os presumably can be attributed to the same source. All these elements have geochemistry in the ocean similar to one of V and similar distribution in metalliferous sediments (Turekian and Bertine 1971; Crocket et al. 1973; Marchig and Gundlach 1982; Marchig et al. 1986; Ravizza and McMurtry 1993). This is also evident for Os as indicated by its isotope composition (Ravizza and McMurtry 1993).

Similar correlation coefficient data, about +0,7, for the element pairs of the first and second kinds for Cu and Sb indicate a dual nature for their excessive accumulation in metalliferous sediments in the Southeast Pacific. The question of the source of the excessive accumulation of Sb is unequivocally decided in favor of ocean water, because the greatest possible role that hydrothermal solutions could make in its accumulation is only 11% (Table 1.2). The dual nature of Sb accumulation probably is caused by its presence in ocean water in two oxidation states: Sb(III) and Sb(V) (Vinogradov 1967; Onishi 1969). In the oxidizing conditions of ocean water Sb(V) dominates whereas Sb(III) makes up only 1–5% of the total concentration of dissolved Sb (Bertine and Lee 1983). Therefore it can be assumed that during the scavenging Sb(III) behaves similar to Ni, Co and other elements that have high correlation coefficients of pairs of the second kind, as indicated by low Sb contents in metalliferous sediments from the EPR axial zone (Gurvich 1977), and Sb(V) behaves similar to V, P, and As that have high correlation coefficients of the first kind of pairs. The double nature of the excessive accumulation of Cu may be caused by the prevalence of hydrothermal Cu in metalliferous sediments from the EPR axial zone as shown by the maximum of contents in axial areas, and by Cu derived from ocean water in flank metalliferous sediments as shown by the maximum contents of Cu in these areas.

Boström (1973) showed a strong correlation coefficient between the ratio (Fe+Mn)/Al in metalliferous sediments and the spreading rate of the segments of the EPR where these sediments accumulated. The predominant parts of the Fe and Mn contents in metalliferous sediments are composed of Fe_{ex} and Mn_{ex} (Sect. 1.1.4) and there is a strong correlation between the ratio $(Fe_{ex}+Mn_{ex})/Al$ in metalliferous sediments and the spreading rate. It appears that metalliferous sediments accumulated at equal distances from the EPR segments with faster spreading rates have lower ratios of $Al/(Fe_{ex}+Mn_{ex})$ and concentrations of trace elements scavenged over a long period of time than in sediments accumulated near the segments that have slower spreading rates.

1.2. Metalliferous sediments from the northern part of the East Pacific Rise

In the mid nineteen eighties a field of metalliferous sediments in an area of $\sim 200 \cdot 10^3$ km^2 in the northern part of the EPR between 9°N and 14°N was discovered and studied during expeditions of the Sevmorgeologiya Association (Sankt-Petersburg). The field occurs on both sides of the EPR axis near areas where high-temperature hydrothermal activity and massive sulfide deposits are present (Fig. 1.1; Cherkashev 1990; Krasnov et al. 1992a; Davydov et al. 2002). The field is asymmetric, its east boundary is 300 km, and west boundary is 200 km from the EPR axis. About two thirds of the field is located to the east and one third to the west of the EPR axis. It is developed by predominant easterly flowing bottom currents (Lonsdale 1976).

The full spreading rate of this EPR segment is 11 cm a^{-1} (Herron 1972; Hekinian et al. 1985). Between 9°N and 14°N the axial part of the EPR represents an echelon chain of shield volcanoes with non-transform displacement at intervals of 30 to 120 km. The axial uplift area located at a depth of about 2600 m is 2 to 10 km wide and is raised 200 to 400 m above the base of the axial graben. Cross-sections with various forms, triangular, dome-shaped or trapezoidal, are found in different parts of the uplift area (Hekinian et al. 1983; Francheteau and Ballard 1983; Choukroune et al. 1984; Gente et al. 1986; Staritsyna et al. 1989; Cherkashev 1990). Hydrothermal zones that occur within structural segments that have the greatest uplift along longitudinal sections have trapezoidal or dome-shaped cross-sections and are complicated in the uppermost parts by grabens 200 to >600 m wide and 20 to 50 m deep (Staritsyna et al. 1989).

Most of the information on the lithology and geochemistry of recent metalliferous sediments from the northern part of the EPR has been obtained by Cherkashev (1990, 1992) and further description of these sediments will be based mainly on his data.

Background non-hydrothermal sedimentation in the northern part of the EPR is related to the boundary zone between the northern arid and equatorial zones and vicinity to the North American continent. Sedimentation rates in this area are 3 to 10 mm ka^{-1} (Lisitzin 1974). In the axial part of the EPR the sedimentation rates are higher because of the additional input of edaphogenic and hydrothermal material. Low carbonaceous and noncalcareuos silty-clay and clay sediments occur within the field. Clay minerals, ferruginous clay aggregates, coccolith-foraminiferal material, ferruginous silica gel, Fe-smectite minerals and quartz are the main components of these sediments. The sediments in the axial part of the ridge are enriched in basalt detrital material, the content of biogenic silica is negligible to absent and the sediments are oxidized at the surface. The subsurface sediments east of the EPR axis are reduced and are oxidized to the west of it.

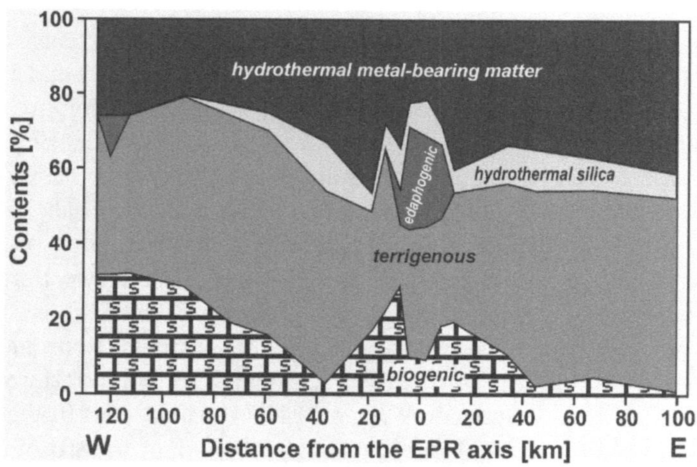

Fig. 1.30. Proportions of the main components in surface metalliferous sediments at the section across the EPR at 12°50'N. After Cherkashev (1992) with modification.

Variations in the proportions of the main components of the surface metalliferous sediments at the section across the EPR at 12°50'N are shown in Fig. 1.30. The contents of biogenic components in the sediments that consist mainly of $CaCO_3$ are higher west of the EPR axis and at uplifted sites on the sea bottom than to the east of it, except in the axial zone. Because of the significant contribution of edaphogenic material in sediments in the axial zone the lithogenic components are predominant. Some edaphogenic material is also present in sediments in the west part of the section because of the exposure of bedrock. In flank areas near the axis where edaphogenic material is not apparent the amount of lithogenic components is minimal and the amount of hydrothermal material is maximal.

Table. 1.3. Average contents of chemical elements in abiogenic matter of metalliferous sediments from the northern part of the East Pacific Rise. Based on data from Turekian and Wedepohl (1961), Skornyakova (1970), Crocket et al. (1973), Piper and Graef (1974), Gurvich et al. (1976, 1977, 1979, 1980a,b,c), Gurvich (1977), Oreshkin (1977), Lisitzin et al. (1980b,c), Lukashin and Lisitzin (1980), Lukashin et al. (1980), Cox and McMurtry (1981), Cherkashev (1990, 1992), Davydov et al. (2002).

Element	Metalliferous sediments	Background sediments	Element	Metalliferous sediments	Background sediments
Fe %	12.79	5.06	Rb ppm	34	45
Mn	2.65	0.41	Mo	146	27
Si	16.80	27.8	La	21.7	24.5
Al	4.77	8.4	Ce	49.9	45.6
Ti	0.29	0.37	Sm	6.0	5.1
Ba	0.40	0.51	Eu	2.0	1.2
			Tb	1.58	
Cu ppm	335	136	Yb	2.66	3.6
Zn	371	159	Lu	1.41	
Pb	52	34	Y	98.1	38.6
Cd	1.73	0.7	Sc	24.7	18.9
Sn	3.0	1.5	Zr	306	112
Ni	245	136	Th	5.68	7.4
Co	110	32	Ga	20.3	18
Cr	89	50	Ag	0.41	0.11
V	288	91	Hg	2.12	
As	49	13			
Sb	5.9	1.7	Au ppb	20.3	3.5
U	1.8	1.3	Pd	8.9	4

The content of hydrothermal material is higher in sediments on the eastern flank at a distance from the axis. The predominance of hydrothermal material in sediments to the east of the EPR axis is attributed to bottom currents flowing in an easterly direction (Lonsdale, 1976) and also to the enrichment of surface sediments in manganese that has been redistributed from the reduced subsurface strata. In the opinion of Davydov et al. (2002) the enrichment of metalliferous sediments in hydrothermal matter on the eastern flank of the ridge is the result of off-axial pneumatolytic-hydrothermal activity.

Average contents of chemical elements in abiogenic matter of metalliferous sediments from the northern part of the EPR and background sediments are shown in Table 1.3. One can see from these data that the contents of Fe, Mn, and most trace elements bound in the metal-bearing matter in metalliferous sediments from the northern part of the EPR are considerably lower and the contents of Al, Ti, and Si are higher than in metalliferous sediments from the Southeast Pacific (Table 1.1). This is because of the greater amount of background sedimentary material and its high accumulation rate (200–500 mg cm^{-2} ka^{-1}) (Lisitzin 1978) that is greater than that in the Southeast Pacific (av. 20 mg cm^{-2} ka^{-1}).

The average contents of the majority of trace elements in abiogenic matter of metalliferous sediments from the northern part of the EPR are higher than in background sediments. This is shown in Table 1.3 and in ratios of the average contents of elements in metalliferous sediments to those in background sediments:

Al →	Si →	Rb →	Th →	Ba →	Ti →	La →	Ce →	Sm →	Sc →	U →
0.57	0.66	0.76	0.77	0.78	0.78	0.89	1.09	1.18	1.31	1.50
Pb →	Eu →	Cr →	Ni →	Sn →	Pd →	Zn →	Cu →	Cd →	Fe →	Y →
1.53	1.67	1.78	1.80	2.00	2.23	2.33	2.46	2.47	2.53	2.54
Zr →	V →	Co →	Sb →	Ag →	As →	Mo →	Au →	Mn		
2.73	3.16	3.44	3.47	3.73	3.77	5.41	5.80	6.46		

Based on the assumption that the accumulation of Al in metalliferous sediments results only from background process the average relative roles of the excessive accumulation of the elements in % are as follows:

Al →	Si →	Rb →	Th →	Ba →	Ti →	La →	Ce →	Sm →	Sc →	U →
0	14	25	26	28	28	36	48	52	56	61
Pb →	Eu →	Cr →	Ni →	Sn →	Pd →	Zn →	Cu →	Cd →	Fe →	Y →
63	66	68	68	72	74	76	77	77	78	78
Zr →	V →	Co →	Sb →	Ag →	As →	Mo →	Au →	Mn		
79	82	84	84	85	85	89	90	91		

In general this sequence is similar to one for metalliferous sediments from the Southeast Pacific. However the excessive accumulation of chemical elements in metalliferous sediments from the northern part of the EPR is lower, than in metalliferous sediments from the Southeast Pacific.

A large part of the excessive accumulation reported for Ti is caused by the significant number of samples collected within the axial part of the EPR where because of the admixture of basic edaphogenic material the contents of Ti in metalliferous sediments are much higher than in sediments outside the axial part of the EPR (Fig. 1.31).

At times the contents of this admixture are significant and can show ambiguities in the interpretation of the contents of elements. In order to avoid this ambiguity, calculation of the contents of elements not only in abiogenic but also in non-edaphogenic matter has been suggested (Cherkashev

1990). The advantages of such calculation can be seen in Fig. 1.32, where the "corrected" contents of Fe distributed in metalliferous sediments at the section across the axial zone of the EPR are shown. However, some part of the excessive accumulation of Ti may be derived from hydrogenic sources and related to metal-bearing material that is similar to that in metalliferous sediments from the Southeast Pacific (Sect. 1.5, Fig. 1.20).

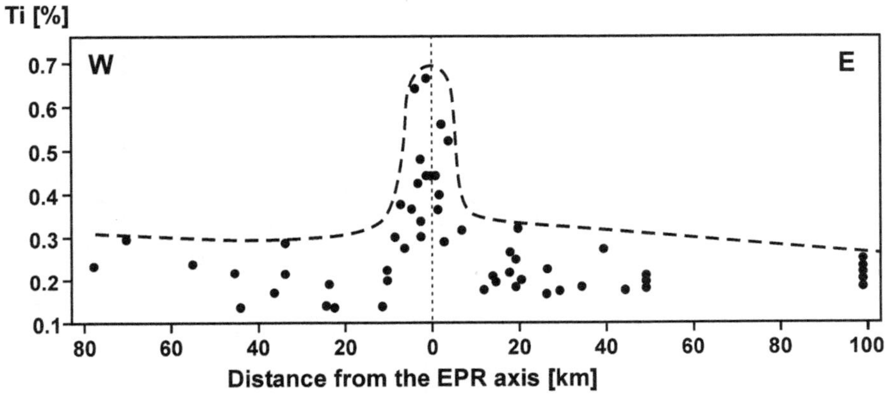

Fig. 1.31. Titanium contents in carbonate-free matter of metalliferous sediments at various distances from the EPR axis in the area of 11°N–13.5°N. After Cherkashev (1992).

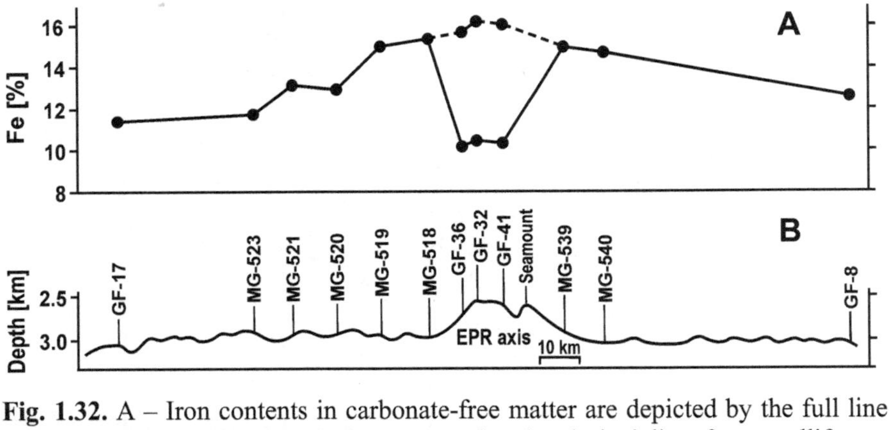

Fig. 1.32. A – Iron contents in carbonate-free matter are depicted by the full line and in carbonate-edaphogenic-free matter by the dashed line for metalliferous sediments at the section across the EPR at 12°50'N. B – Bottom relief and station numbers. After Cherkashev (1992).

As to the excessive accumulation of Cr, this can no longer be explained by the presence of edaphogenic material or on the assumption that all excessive Ti is bound with this material. The excess of Cr is likely attributable to the oxidation-reduction state of this element. In the reduced conditions of subsurface strata of metalliferous sediments to the east of the EPR axis reduction of Cr(VI) to Cr(III) can occur (Krauskopf 1963; Lukashin 1981). Cr(III) unlike to Cr(VI) behaves similar to Co and is easily sorbed by metal-bearing matter from the ocean and pore water.

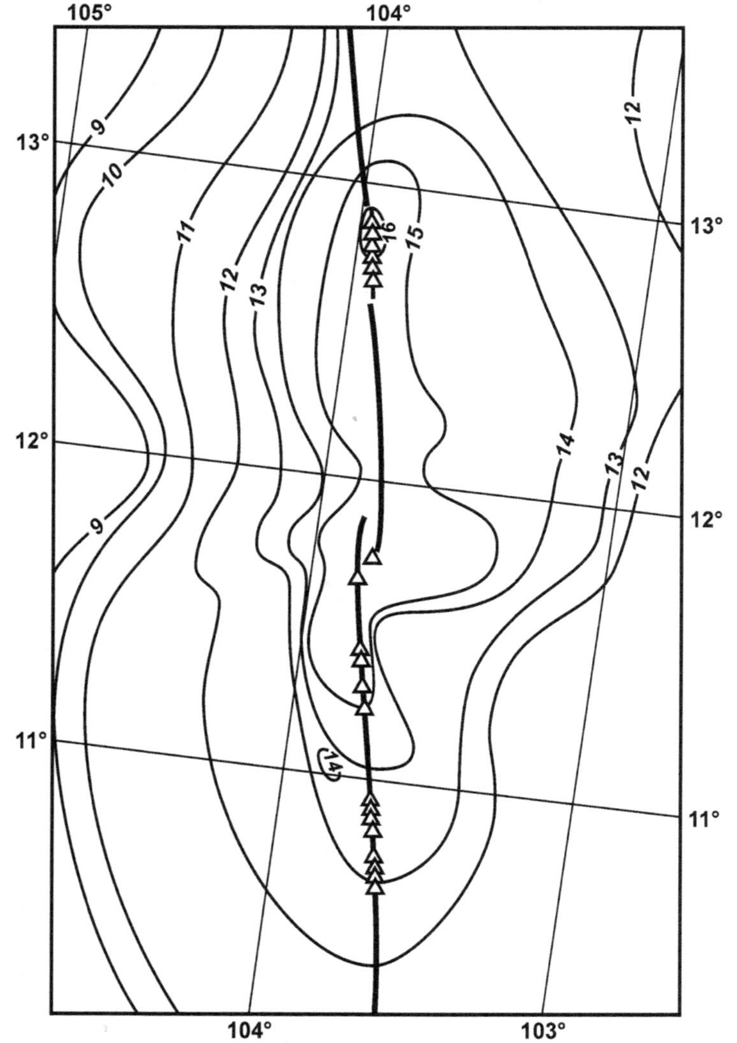

Fig. 1.33. Iron contents [%] in carbonate-edaphogenic-free matter of surface sediments from the northern part of the EPR. Compiled from Cherkashev (1990) and Sudarikov at al. (1992) with simplification. Triangles show sites of accumulation of massive sulfides on the bottom.

The distribution of Fe in surface metalliferous sediments from the northern part of the EPR (Fig. 1.32, 1.33) shows maximum content values in the axial zone of the EPR and a gradual decrease at distances away from the axis and from sources of hydrothermal material. There is small asymmetry in the distribution of Fe because of the predominance of bottom currents flowing in an easterly direction.

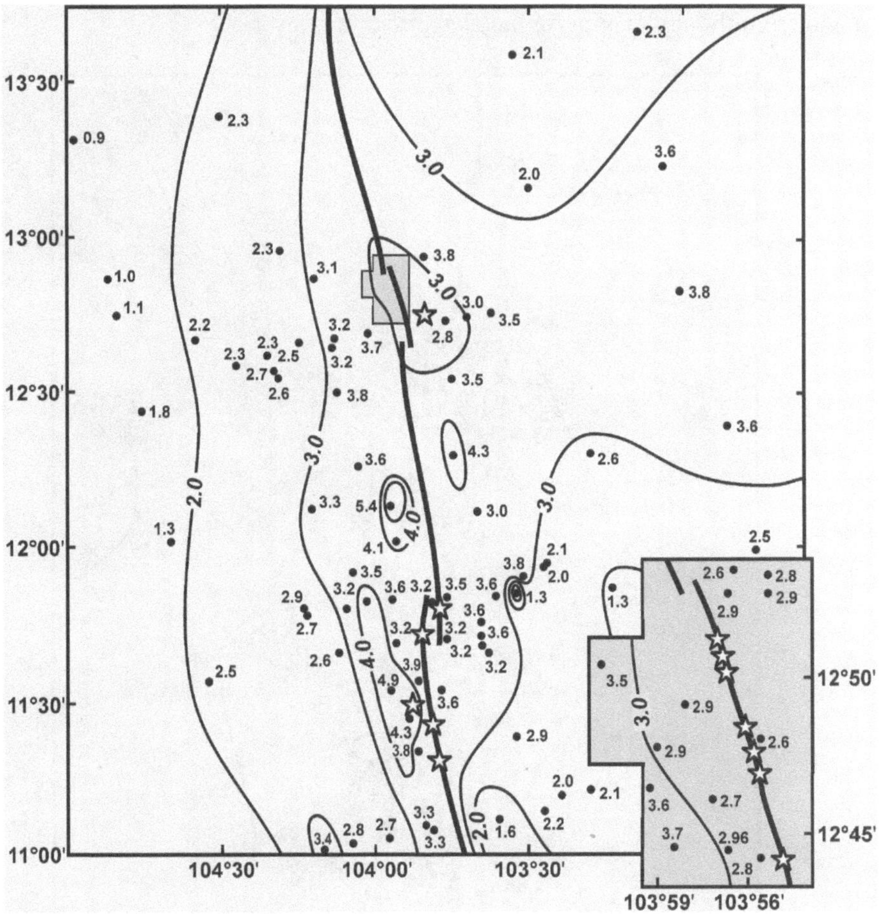

Fig. 1.34. Manganese contents [%] in carbonate-edaphogenic-free matter of surface sediments from the northern part of the EPR. After Cherkashev (1990). Stars – hydrothermal fields (active and inactive).

The distribution of Mn in surface metalliferous sediments from the northern part of the EPR (Fig. 1.34) is more complicated. As a rule the maximum contents of Mn occur in off-axial sediments and are displaced to the west of the EPR axis in the vicinity of hydrothermal fields. In the southern half of the area the maximum contents of Mn gradually decrease at distances east and west from the axis. In the northern half of the area gradual decreases in the content of Mn occur only to the west of the sites with maximum contents. In the vicinity of hydrothermal fields the Mn contents decrease at distances east of the sites of maximum content and then increase within the large fields where they remain high. This enrichment of surface sediments to the east of the EPR axis is likely caused by the redistribution of Mn from the underlying reduced sedimentary strata.

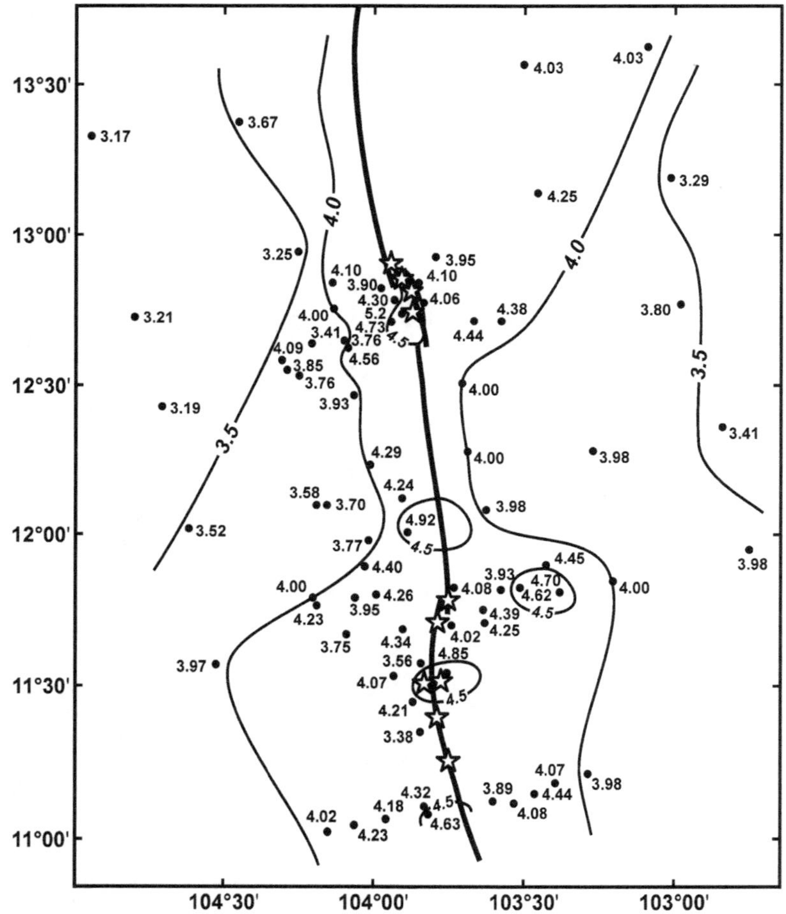

Fig. 1.35. Distribution of the SiO_2/Al_2O_3 ratio in surface sediments from the northern part of the EPR. After Cherkashev (1992). Stars – hydrothermal fields (active and inactive).

Excessive Si accumulation occurs in metalliferous sediments in the northern part of the EPR where the Si appears to be from hydrothermal sources. The pattern of its distribution is similar to the distribution of the SiO_2/Al_2O_3 ratio (Fig. 1.35). The higher ratios, >4.0, occur in sediments from the near-axial and axial areas and maximum ratios, >4.5, occur in the vicinity of the hydrothermal fields. Where the ratios of SiO_2/Al_2O_3 are 4.0, about 20% of the total Si content is derived from hydrothermal sources and where SiO_2/Al_2O_3 ratios are 4.5, about 30 to 35% appears to be from hydrothermal sources.

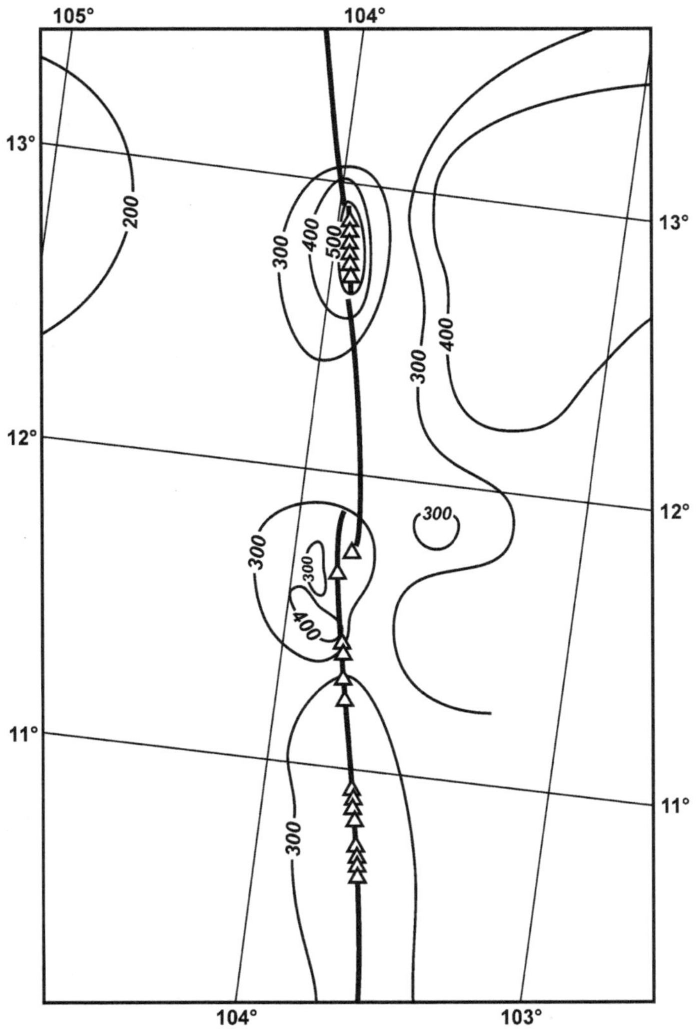

Fig. 1.36. Copper contents [ppm] in carbonate-edaphogenic-free matter of surface sediments from the northern part of the EPR. Compiled from Cherkashev (1990) and Sudarikov at al. (1992) with simplification. Triangles show sites of accumulation of massive sulfides on the bottom.

The higher contents of Cu in sediments from the EPR axial zone in the vicinity of hydrothermal sources are characteristic (fig. 1.36). At distances east of the axis the contents of Cu first decrease and then increase further to the east. Moving west from the axis the contents of Cu gradually decrease and an increase, like that on the eastern flank, has not been observed in the area studied.

In general the distribution of Zn contents in sediments (Fig. 1.37) has a similar pattern. Similar distribution patterns with maximum content values at the EPR axis and at some distance away from it also occur in the Southeast Pacific. The increased contents in sediments in the vicinity of hydrothermal vents in the axial zone are caused by the settling of Cu and Zn to the sea bottom in sulfide particles formed in buoyant hydrothermal plumes (see Sect. 4.3.1) and the off-axial maximum contents are developed by the sorption of Cu and Zn from ocean water on to metal-bearing suspended particles.

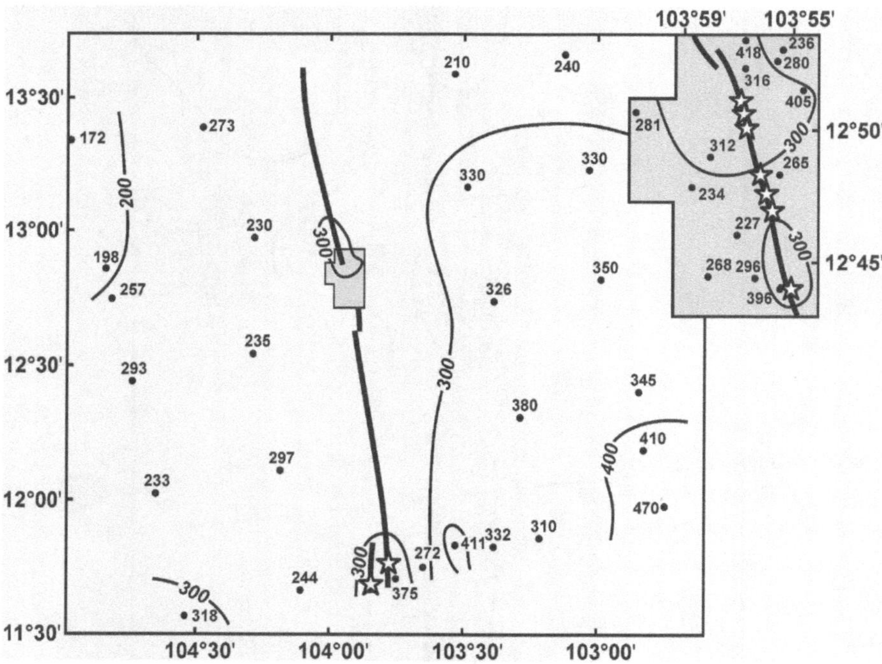

Fig. 1.37. Zinc contents [ppm] in carbonate-edaphogenic-free matter of surface sediments from the northern part of the EPR. After Cherkashev (1990). Stars – hydrothermal fields (active and inactive).

Ni, unlike Cu and Zn, is not concentrated in sulfide particles, and its excessive accumulation in metalliferous sediments results from the sorption

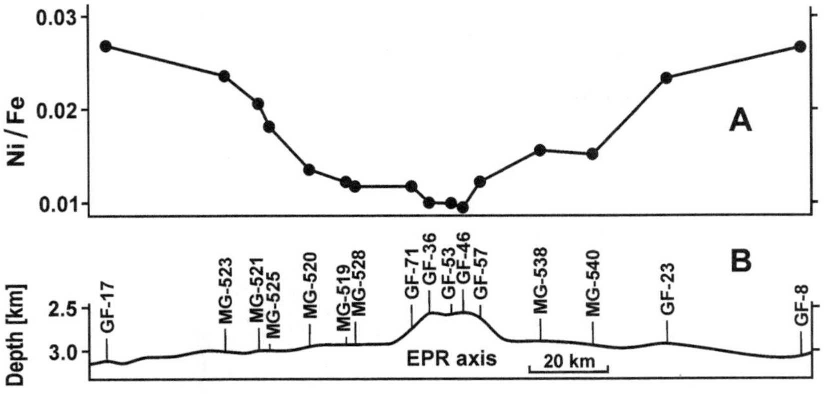

Fig. 1.38. A – Ni/Fe ratio in metalliferous sediments at the section across the EPR at 12°50'N. B – Bottom relief and station numbers. After Cherkashev (1992).

onto metal-bearing particles from ocean water. Therefore Ni contents in sediments from the axial zone and especially in the vicinity of hydrothermal fields are low and they increase at distances from the EPR axis to have maximum contents on the flanks and then decrease further away on the flanks of the ridge (Davydov et al. 2002). With increasing distance from the axis the Ni/Fe ratio increases (Fig. 1.38) which indicates the increase in the amount of sorbed Ni in metal-bearing matter of metalliferous sediments.

1.3. Metalliferous sediments from the Juan de Fuca Ridge

Juan de Fuca Ridge is the northern prolongation of the EPR. On this ridge metalliferous sediments occur only in the vicinity of the sites of recent high-temperature hydrothermal activity. There is hydrothermal activity in the northern, southern and central parts of the Juan de Fuca Ridge (Fig. 1.39), however it occurs much less frequently here than in the fast-spreading segments of the EPR (Baker and Hammond 1992).

Metalliferous sediments from the Axial Seamount and at least up to 50–60 km to the west of the Juan de Fuca Ridge have been investigated (Malahoff et al. 1984; McMurtry et al. 1984; Gross and McLeod 1987; Bogdanov et al. 1989, 1990). There is information on metalliferous sediments from the southern part of the Juan de Fuca Ridge, 16 km from its axis (Massoth et al. 1984; Olivarez and Owen 1989). Within the Juan de Fuca Ridge, and also in the Explorer Ridge, its northern prolongation, as well, metalliferous sediments in sporadic samples and some non-metalliferous sediments contain higher than normal amounts of trace elements (Bornhold et al. 1981; Grill et al. 1981; Gross 1987; Gross and McLeod 1987; Blaise and Bornhold 1987; German et al. 1997). Information on sediments from the intersection area of the Blanco Fracture Zone and the Juan de Fuca Ridge indicates that sediments in this area are enriched in Mn and some trace elements that are derived from hydrothermal activity (Selk 1982; Embley et al. 1982).

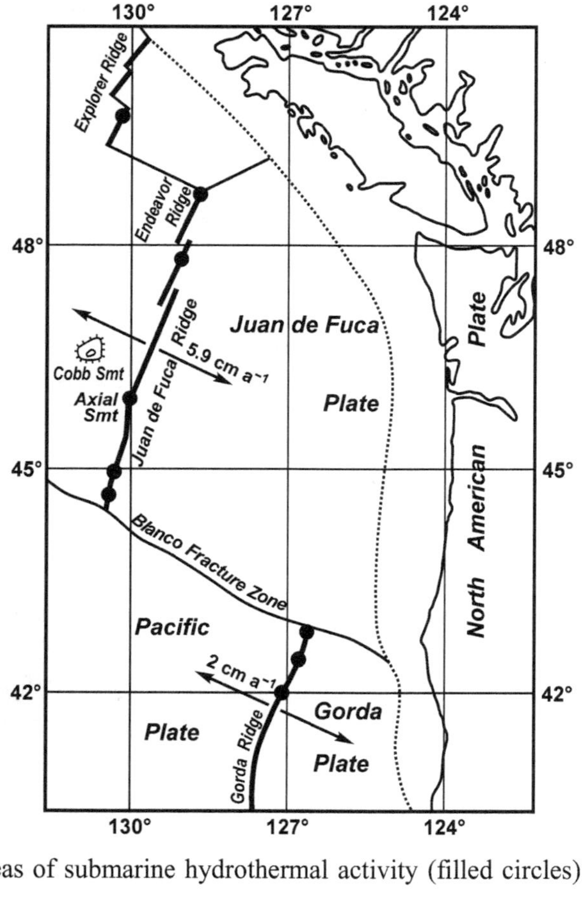

Fig. 1.39. Areas of submarine hydrothermal activity (filled circles) in the Northeast Pacific.

Juan de Fuca Ridge is the spreading zone between the Pacific and Juan de Fuca plates. The recent full spreading rate on the ridge is about 6 cm a^{-1} with movement in the directions 294° and 114° (Wilson et al. 1984) and the ridge is oriented perpendicular to the spreading directions (Fig. 1.39). The ocean depth at the axial part of the ridge is 2000 to 2500 m, which is almost 1000 m shallower than the average depth of the World Ocean rift system.

The full spreading rate of the Explorer Ridge is about 4 cm a^{-1} (Riddihough 1977). The depth in its axial part is less than 1900 m. Throughout the whole area of the Juan de Fuca Ridge the morphology of its axial zone is typical for intermediate- and fast-spreading ridges and is characterized by the absence of deep rift valleys. Only in some places there are axial troughs (~100 m deep and 1 km wide) in the axial zone. The Juan de Fuca Ridge spreading center has existed for about 18 Ma and was derived from the former boundary between the Pacific and Farallon plates. During this time spreading direction has varied from latitudinal in the period from 17 to 8.5 Ma BP to WNW–ESE in later time. The spreading rate has also varied from 6.5 to 7.5 cm a^{-1} in the period from 17 to 3.5 Ma BP and then decreased to 5.9 cm a^{-1}.

The Axial Seamount (or Axial Volcano) is located in the central part of the Juan de Fuca Ridge at the point of its intersection with the chain of volcanic mounts of the Cobb system (Desonie and Duncan 1990). This chain of volcanic mounts has been developed by the movement of the lithosphere plate over the Cobb hot spot. The Axial Seamount is the most easterly and the youngest mountain of this chain. It has been uplifted above the adjacent part of the ridge by about 1000 m to depths of 1400 to 1600 m. The mount has the form of a regular truncated cone with the apex cut off. Its dimensions viewed from the top are 10 by 15 km. The summit is opened to the south to form a horseshoe-shaped caldera 7 km long and 2.5 km wide.

The recent sedimentation rate in the Axial Seamount area varies from 10 to 30 mm ka^{-1} (Lisitzin 1974), and the accumulation rate of abiogenic matter is 500 to 2000 mg cm^{-2} ka^{-1} (Lisitzin 1978). Less intensive hydrothermal activity and higher accumulation rates of abiogenic matter in comparison with those at the East Pacific Rise result in the less frequent occurrence of metalliferous sediments on the Juan de Fuca Ridge.

Most information on regularities of distribution and composition of the most studied metalliferous sediments from the area of the Axial Seamount has been obtained during manned submersible dives and coring (Bogdanov at al. 1989, 1990). The sediments have been sampled along a sublatitudinal section crossing the caldera, summit, and slopes of the Axial Seamount as well as the adjacent basins (Fig. 1.40). The northern part of the caldera floor has almost no sedimentary cover. Sediments have been found only in the southern part and also on the small hill in the northwestern part.

There is a thin layer of Holocene sediment, tens of cm thick, on the summit and on the upper part of the slopes. The section sampled consists of two strata in which hydrothermal and volcanogenic material are prevalent. The upper one is enriched in hydrothermal material accumulated during the last 7.5 ka that consists of brown ferruginous mud with an admixture of biogenic tests. Its contents vary and include fresh hyaloclasts and a mixture of grains of basic plagioclase, olivine, pyroxene and basalt fragments in some places. The lower stratum is enriched in volcanogenic material that accumulated 10 to 7.5 ka BP. It consists of gray mud containing up to 40% in some parts of an admixture of volcanic mictic silt and fine sand and minerals from greenstone altered volcanic rocks as well as edaphogenic hydrothermal material consisting of fragments of Fe-Mn crusts, barite and other minerals. Insignificant amounts of biogenic components are present in sediments from the summit. X-ray amorphous phases are prevalent in some of the dispersed minerals. They consist mainly of Fe-oxyhydroxide minerals and opal. Some of the oxyhydroxide material is partially crystallized and forms goethite (Gorbunova 1990).

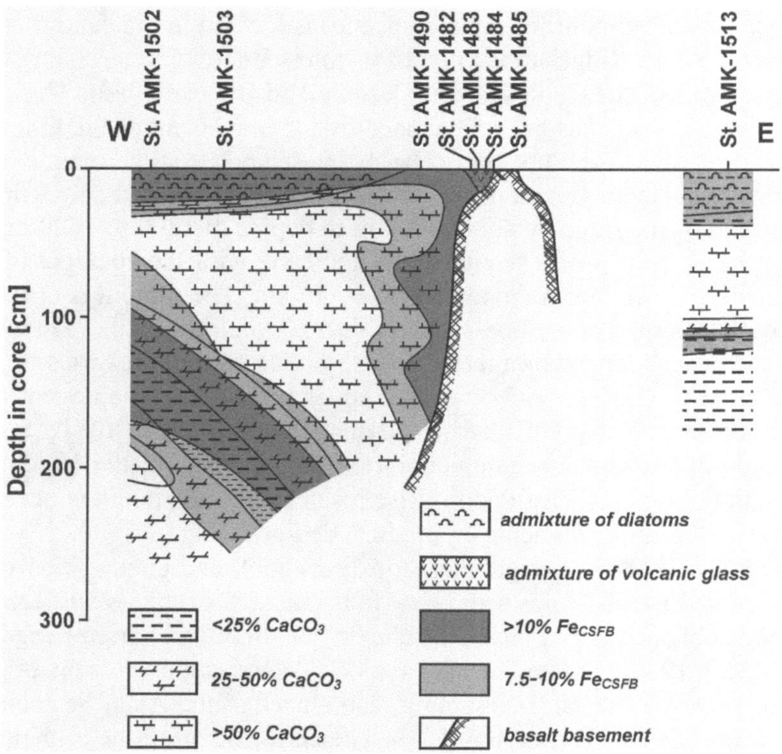

Fig. 1.40. Distribution of bottom sediments along the sublatitudinal cross-section through the Axial Seamount. After Bogdanov et al. (1990). Fe_{CSFB} – iron contents in abiogenic (carbonate- and silica free) matter.

Sediments on the lower parts of the slopes are somewhat different. The sediments from the upper stratum are brown and from the lower one are greenish-gray and some parts are in a reduced state. Fe-oxyhydroxide minerals are prevalent in fine-grained dispersed material. The silica content decreases downcore from 30–34% to 13% and is lower than in the sediments at the summit where it is 30–40%. The content of $CaCO_3$ is significant and increases downcore up to 72%. The content of admixture of volcanogenic and edaphogenic material is insignificant.

The composition of sediments is different in the adjacent Juan de Fuca basin to the west and the Cascadia basin to the east of the Axial Seamount. Sediments in the Juan de Fuca Basin 40 to 50 km west of the Axial Seamount are composed of brown clayey mud with an admixture of carbonaceous and siliceous material in the thin upper part. There are three sediment layers in the section sampled, a carbonaceous clay ooze, mud with a low content of carbonaceous material and carbonaceous ooze (Fig. 1.40). Two strata of metalliferous sediments are present. The upper stratum accumulated during the last 7.5 ka at the same time that the upper sediment layer on the summit formed. No strata analogues to the lower stratum that has been dated 25 to 35 ka have been found on the summit and slopes of the Axial Seamount. The sediments in the upper stratum are enriched in Fe and also in Mn. Sediments in the lower stratum are depleted in Mn because of its diagenetic migration upward. Both sedimentary strata are enriched in X-ray amorphous material. The clay consists of hydromica (40–50%), montmorillonite (25–30%), and illite (25–30%) minerals and Fe-montmorillonite is abundant. In general, features in the section of sediments studied from the Cascadia Basin, 30 km east of the Axial Seamount, are similar to those described at the seamount. The difference is that isochronal sediments from east of the seamount are depleted in biogenic components because of the dilution by terrigenous material, they have less hydrothermal material and metalliferous sediments are absent. However, the sedimentary strata to the east contain 7.5 to 10% Fe in the abiogenic matter (Fig. 1.40) and are synchronous to the metalliferous sedimentary strata accumulated to the west of the Axial Seamount.

When analyzing metalliferous sediments in the study of the hydrothermal activity at the Axial Seamount in the past, special attention was given to the identification and distribution of components derived from volcanism and hydrothermal activity (Bogdanov et al. 1990; Lisitzin et al. 1990).

Biogenic $CaCO_3$ and SiO_2 as well as terrigenous detrital and clay minerals are the main components in the sediments of the area.

A large part of the local sedimentary material consists of volcanogenic and edaphogenic clastic material that includes altered greenstone fragments that were produced in the decomposition of the hydrothermal and edaphogenic material and fine grained particles that precipitated during the mixing of hydrothermal fluids and ocean water. The fine-grained hydrothermal particles that accumulated in sediments in the vicinity of hydrothermal fields are composed mainly of Fe- and Mn-oxyhydroxide minerals.

Volcanogenic material produced by the decomposition of the friable glassy crusts on the basalt rocks is widely distributed in the sediments in the Axial Seamount area. Decomposition of volcanic glass is taking place at the present time and the presence of glass dispersed in the sediments is an indication that volcanic rocks are exposed on the seafloor in the area. Layers composed almost completely of volcanic glass are an indication of intensive decomposition of the glassy crusts either by tectonic movements or volcanic activity[6].

Two groups of detrital minerals that are absent in the background sediments are present in sediments from the Axial Seamount area. Epidote and chlorite minerals belong to the first group. In the Axial Seamount area these minerals are associated only with greenstone alteration of volcanic rocks. They may have been formed by the alteration of basalt by high-temperature hydrothermal jets. The second mineral group includes fragments of Mn-crusts, barite, opal, and Fe-oxyhydroxide minerals. The fragments of Mn-crusts, barite, and opal may have been deposited in the sediments during destruction of hydrothermal material. Coarse-grained Fe-oxyhydroxide minerals have formed during the oxidation of Fe-bearing sulfide minerals of the hydrothermal bodies and fine-grained Fe-oxyhydroxide minerals have precipitated during the mixing of hydrothermal fluids and ocean water.

Table 1.4. Average contents of chemical elements in abiogenic matter of metalliferous sediments from the area of the Axial Seamount. After Lukashin et al. (1990).

Element	Metalliferous sediments	Background sediments	Element	Metalliferous sediments	Background sediments
Fe %	11.09	5.8	Zn ppm	364	159
Mn	3.13	0.75	Ag	3.4	0.11
Si	19.8	27.8	Ni	272	136
Al	5.25	8.4	Co	63	32
Ti	0.56	0.57	Cr	104	50
C_{org}	0.96		V	208	91
Ba ppm	8450	5109	Yb	4.4	3.6
P	1250	850	Y	55	39
Cu	204	136	Zr	165	112

The average contents of chemical elements in abiogenic matter of the metalliferous and background sediments from the area of the Axial Seamo-

[6] In January 1998 a dike injection and an eruption in the volcano's caldera were observed; the dike injection produced more than 8000 earthquakes over several days (Embley et al. 1999; Dziak and Fox 1999; Baker et al. 2004a).

unt are shown in Table 1.4. The contents of Fe, Mn, and of the trace elements, except for Cr, bound within metal-bearing matter in metalliferous sediments from the area of the Axial Seamount, are much lower than those in sediments from the Southeast Pacific. They have higher contents of Al, Ti and Si that are derived from the greater amount of lithogenic material in metalliferous sediments in the area of the Axial Seamount where its accumulation rate (average Al*=20÷180 mg cm^{-2} ka^{-1}) (Bogdanov 1990a) is much higher in comparison with that in metalliferous sediments from the Southeast Pacific (average Al*=1.75 mg cm^{-2} ka^{-1}).

The contents of most trace elements as found in the study of metalliferous sediments from the area of the Axial Seamount area are higher than in background sediments (Table 1.4). The ratios are as follows:

Al →	Si →	Ti →	Yb →	Y →	Zr →	P →	Cu →	Ba →	Fe →	Co →
0.62	0.71	0.98	1.22	1.41	1.47	1.47	1.50	1.65	1.91	1.97
Ni →	Cr →	V →	Zn →	Mn						
2.00	2.08	2.28	2.29	4.17						

Based on the assumption that the accumulation of Al in metalliferous sediments is caused only by background process, it is possible to calculate the role of excessive accumulation of the elements in %:

Al→	Si→	Ti→	Yb→	Y→	Zr→	P→	Cu→	Ba→	Fe→	Co→	Ni→	Cr→	V→	Zn→	Mn
0	12	36	49	56	58	58	58	62	67	68	69	70	73	73	85

In general these data are similar to the data for metalliferous sediments from the Southeast Pacific but there are essential differences. The role of excessive accumulation of all the elements, except for Ti and Cr, in metalliferous sediments from the area of the Axial Seamount is lower due to the higher content of background abiogenic material. The excessive accumulation of Ti and part of the Cr is related to the significant amount of basic volcanogenic material that is present on the summit and the slopes and especially in the caldera of the seamount. Enrichment of metalliferous sediments in the Axial Seamount area and particularly of the subsurface sediments on its slopes and in the adjoining Juan de Fuca Basin[7] in chromium is caused by its sorption in the form of Cr(III) by metal-bearing matter from ocean water and pore water in reducing conditions (see Sect. 1.2).

Another feature of metalliferous sediments from the Axial Seamount area is the content of Zn that is greater than that of Cu (Table 1.4), which probably is because of the relatively low concentration of Cu in hydrothermal fluids in the whole Axial Seamount and Juan de Fuca Ridge area (Von Damm 1995).

---

[7] These sediments are low in Mn.

The distribution of the content of some elements in abiogenic matter of surface metalliferous sediments along the generalized section from the summit of the Axial Seamount to the Juan de Fuca Basin is shown in Fig. 1.41.

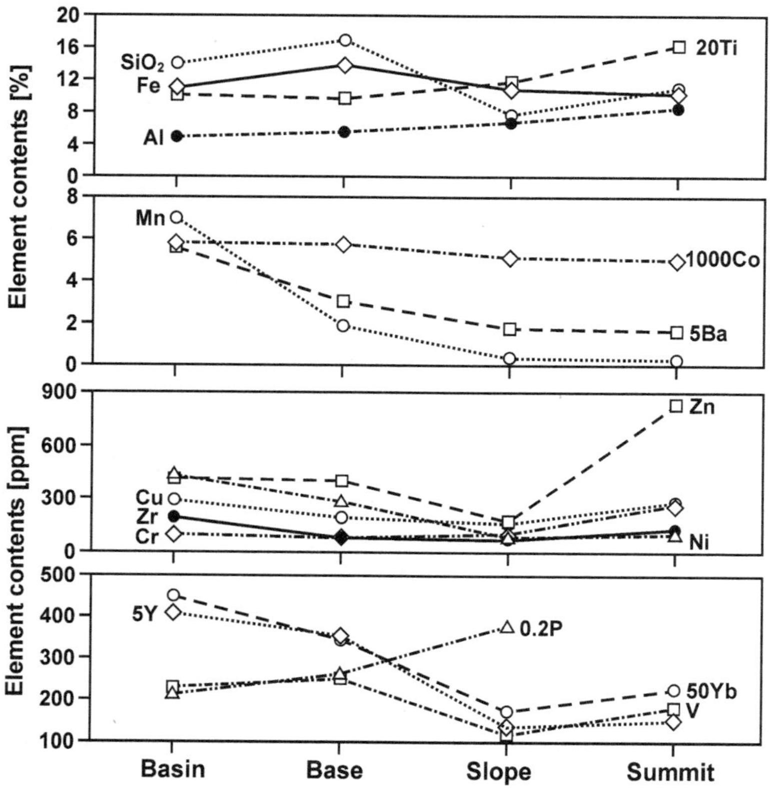

Fig. 1.41. Chemical element contents in abiogenic matter of surface metalliferous sediments along the generalized section from the summit of the Axial Seamount to the Juan de Fuca Basin. Based on data of Lukashin et al. (1990).

The content of Fe is higher in sediments from the base of the seamount than in sediments in the vicinity of hydrothermal vents. This is attributed to the presence of significant amounts of basic volcanogenic material that is enriched in Al, Ti, and Cr in sediments of the summit area. The content of Mn increases gradually with the distance from the summit[8]. Three types of distribution of the trace elements in the sediments are recognized. The first type is characterized by a gradual increase in the contents of elements from the summit to the basin and is shown in the distribution of the con-

[8] Transfer of Mn to metalliferous sediments depends highly on the activity of microbiological processes. This question will be discussed in Sect. 4.3.

tents of Mn, Ba and Co. The second type is characterized by increased or maximum contents in summit sediments and minimum contents in sediments on the slope, and increased contents in sediments in the Juan de Fuca Basin; the contents of Cu, Cr, Ni, Zr, Yb, and Y exhibit such distribution. The maximum content of Zn also occurs in sediments in the summit area but its contents in sediments at the base and in the basin are equal. The third type of element distribution is characterized by the maximum contents in sediments on the slope or at the base of the Axial Seamount; V and P contents have this kind of distribution.

There is probably a common reason for an increase in the contents of elements from that in the slope sediments to that in the basin sediments, which is observed for the elements having the first and the second types of distribution (except for Ba). It can result from an increase in the role of scavenging of the elements from ocean water by hydrothermal metal-bearing matter with the distance from its source. Similar results have been obtained in the study of the contents of REE in sediments from the flanks of Juan de Fuca Ridge that contain hydrothermal components. Continuous scavenging from ocean water indicates that REE_{ex}/Fe_{ex} ratios increase with increasing distances from the ridge axis (German et al. 1997). The scavenging of Zn, probably, ceases closer to its source than the scavenging of other elements that have the first and second types of distribution.

As for the increased or maximum contents in the summit sediments of the elements of the second type of distribution, the high contents of Cu and Zn most likely result from their deposition in the vicinity of hydrothermal sources, and of Cr, Ni, Zr, Yb, and Y from the high content of volcanogenic material. Barium content has the first type of distribution and Ba is one of the main elements in the low-temperature hydrothermal deposits at the Axial Seamount and is present mainly in barite (Lebedev and Tsepin 1990; Bogdanov and Lisitzin 1990). Barite is the main carrier of Ba in metalliferous sediments, and it may be assumed that in this area barite was formed in the zone of discharge of hydrothermal fluids and was transported in particulate matter to places of deposition. However, barite particles formed at the discharge sites of hydrothermal fluids on the Juan de Fuca Ridge are very unstable in ocean water at depth and are dissolved rapidly, and apparently do not participate in forming the sediments in the vicinity of the hydrothermal fields (Feely et al. 1987; Lukashin et al. 1990). Therefore it is reasonable to suggest that barite accumulation is due mainly to biological processes in the area of the Axial Seamount as it is in metalliferous sediments from the Southeast Pacific (Gurvich et al. 1979).

The third type of distribution may indicate rapid co-precipitation and sorption of elements from ocean water by hydrothermal metal-bearing matter. Maximum contents of P occur closer to the hydrothermal sources than the maximum contents of V, which indicates that the scavenging of P is completed before that of V.

These assumptions have been confirmed by the correlation analysis of data pairs of two kinds in metalliferous and transitional sediments from the area of the Axial Seamount[9] (Fig. 1.42). The first kind of data pairs are for El_{ex} and $(Fe_{ex}+Mn_{ex})$ and the second kind of data pairs are for $El_{ex}/(Fe_{ex}+Mn_{ex})$ and $Al/(Fe_{ex}+Mn_{ex})$ (see Sect. 1.1.7).

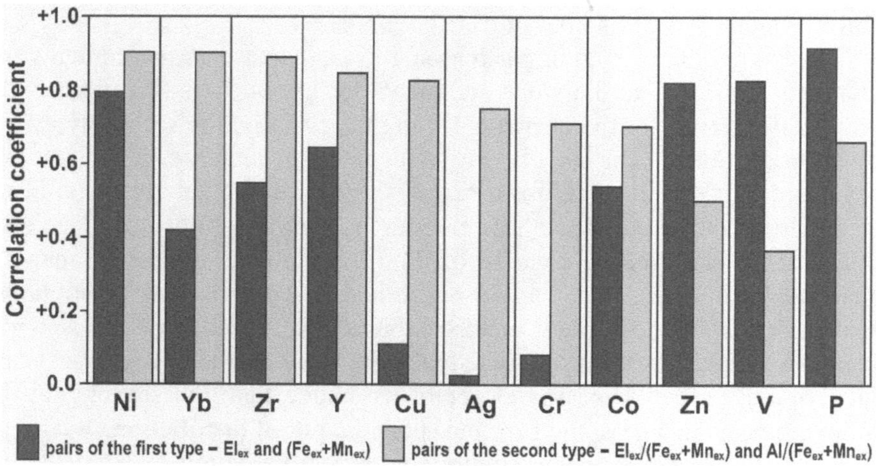

Fig. 1.42. Correlation coefficient data for the pairs of two kinds calculated for abiogenic contents of chemical elements bound with the metal-bearing matter in metalliferous and transitional sediments from the Axial Seamount area. Calculations based on the data of Lukashin et al. (1990).

High positive correlation coefficients (R>+0.70) for the second kind of pairs for Ni, Yb, Zr, Y, Cu, Ag, Cr, and Co can be unequivocally interpreted as the result of their scavenging by the metal-bearing matter from ocean water. Concentrations of these trace elements, or their forms accessible for sorption, are low in ocean water, and ocean water is a limited source for these elements. These results as well as the data for the distribution of the element contents indicate that hydrothermal fluids play an insignificant role as a source of these trace elements in metalliferous sediments from the area of the Axial Seamount and that they are the only main source for Cu in metalliferous sediments in the summit area.

High positive correlation coefficients (R>+0.8) for the first kind of pairs for Zn, V, and P together with the data for the distribution of their contents indicate that the excessive contents of these elements in metalliferous sedi-

[9] The data from the summit sediments have not been used in these calculations.

ments from the area of the Axial Seamount, at least outside of the summit area, are derived from ocean water. Concentrations of these elements in ocean water are high enough, and conditions are favorable for rapid scavenging of them from ocean water and their transfer to metalliferous sediments in hydrothermal metal-bearing matter. The low concentrations of V and P in the hydrothermal fluids[10] do not indicate that they are significant sources of V and P in metalliferous sediments from the Axial Seamount area or from the summit. The distribution of Zn (Fig. 1.41) does not indicate that hydrothermal fluids are an essential source of this element in metalliferous sediments outside of the summit area, but indicates that Zn is mainly of hydrothermal origin in the sediments in the summit area.

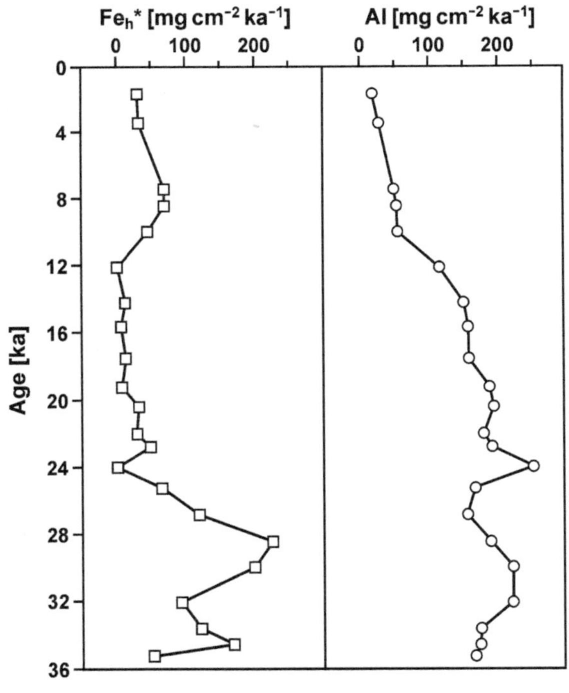

Fig. 1.43 Temporal variations in the accumulation rates of aluminum and hydrothermal iron in metalliferous sediments from Core AMK-1503. Based on data of Bogdanov (1990a).

Information on accumulation rates of chemical elements in sediments from the Axial Seamount area is given by Bogdanov (1990a). During the last 35 ka accumulation rates of background and hydrothermal matter in the se-

[10] Concentration of V in hydrothermal fluids is only ~5 times higher than in ocean water, and concentration of P is ~3 times lower than in ocean water.

diments from Core AMK-1503 varied considerably (Fig. 1.43). In the upper stratum of metalliferous sediments accumulated from 7.5 ka BP to the present the accumulation rate of hydrothermal Fe was variable: from 3.5 to 1.7 ka BP it was about 30 mg cm^{-2} ka^{-1} and 10 to 3.5 ka BP it was 45–68 mg cm^{-2} ka^{-1}. These accumulation rates are comparable to the rates observed in recent metalliferous sediments in the axial zone of the EPR (Sect. 1.1.6). In subsurface metalliferous sediments with ages from 35 to 25 ka the accumulation rate of hydrothermal Fe was higher, 50–220 mg cm^{-2} ka^{-1}, which indicates that there was considerably more hydrothermal activity in the Axial Seamount area at that time.

1.4. Metalliferous sediments from the Indian ocean

Recent metalliferous sediments occur in the Indian Ocean in the area of the Rodriguez Triple Junction (Fig. 1.1). Metalliferous sediments also occur along the axes of the Central Indian Ridge, Southeast Indian Ridge, and Southwest Indian Ridge (Boström 1973; Metalliferous sediments ... 1987; Herzig and Plüger 1988; Kuhn et al. 2000; German 2003) and in some sites in the Maria Celesta and Argo fault zones (Gordeev 1986a,b).

Typical mid-ocean spreading centers marked by well developed axial valleys are developed in the Rodriguez Triple Junction area at the intersection of the Central Indian Ridge, Southeast Indian Ridge, and Southwest Indian Ridge (Price et al. 1986; Zhivago 1987). During the last 4 Ma the average spreading rate of the Central Indian Ridge in this area was 5.4 cm a^{-1} (Tapscott et al. 1980, Schlich et al. 1986). Its axial valley is from 5 to 8 km wide and on approaching the triple junction from 15°S to 22°S the ocean depth in its central part decreases from 3500–4000 m to less than 3000 m (Price et al. 1986). The spreading rate of the Central Indian Ridge near the triple junction is 5–6 cm a^{-1} and the width of the rift valley in this area is about 14 km (Schlich et al. 1986; Kuhn et al. 2000). The spreading rate the Southeast Indian Ridge near the triple junction is 7.0 cm a^{-1} (Kuhn et al. 2000). The Southwest Indian Ridge has a triangular form with a sharp top and has a slow spreading rate of 1.6 cm a^{-1} (Tapscott et al. 1980). The Rodriguez Triple Junction area has played an important role in the structural evolution of the Indian ocean since Cretaceous time (Schlich 1982; Krishna et al. 1995). From Eocene time to the present the Southwest Indian Ridge has been developed by the rapid eastward movement of the triple junction area (Sclater et al. 1981; Krishna et al. 1995). Detached magmatic chambers are believed to exist in the area of the triple junction under each of the three converging ridges (Schlich et al. 1986).

Before the high-temperature hydrothermal vents were discovered in the Indian Ocean (Hashimoto et al. 2001; Gamo et al. 2001) increased concentrations of Mn, Fe, CH_4, and He as well as increased values of δ^3He in bottom water were measured and hydrothermal plumes were mapped in the areas of the Rodriguez Triple Junction, Central Indian Ridge, and Southwest Indian Ridge (Tikhomirov 1987; Herzig and Plüger 1988; Plüger et al. 1990; Jean-Baptiste et al. 1992; Gamo et al. 1996; Scheirer et al. 1998; German et al. 1998, 2001; Sohrin et al. 1999; Rudnicki and German 2002).

Enrichment in Fe and Mn and depletion in Al and Ti of the sediments in the area of the Rodriguez Triple Junction and along the axis of the Central Indian Ridge and the Southeast Indian Ridge were first mentioned by Boström (1973). The distribution of the ratios of Al/(Al+Fe+Mn) (Fig. 1.44) in the sediments of the area and ratios lower than 0.30 are evident and significant on his chart.

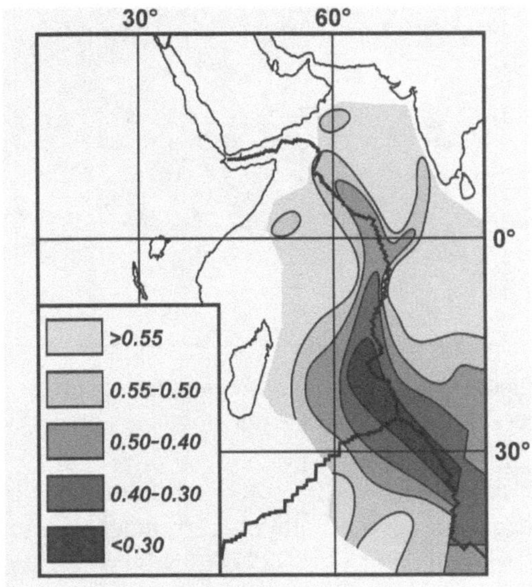

Fig. 1.44. Distribution of the Al/(Al+Fe+Mn) ratio in bottom sediments from the northwest Indian Ocean. After Boström (1973) with changes.

The first large-scale multidisciplinary studies of the ~$2 \cdot 10^6$ km^2 metalliferous sediment field in the area of the Rodriguez Triple Junction were carried out in 1980 during Cruise 25 of R/V *Dmitry Mendeleev* (Metalliferous sediments ... 1987; Fig. 1.1, 1.45). In 1983 and 1986 multidisciplinary studies, including sediment studies, within two polygons in the axial and near-axial parts of the Central Indian Ridge close to 23°S and 21.5°S were carried out during expeditions of R/V *Sonne* on GEMINO Project (Walter et al. 1986; Herzig and Plüger 1988). In 1993 and 1995 sediment cores near the Rodriguez Triple Junction were collected during Cruise 92 of R/V *Sonne* and Cruise 33/2 of R/V *Meteor* (Kuhn et al. 2000).

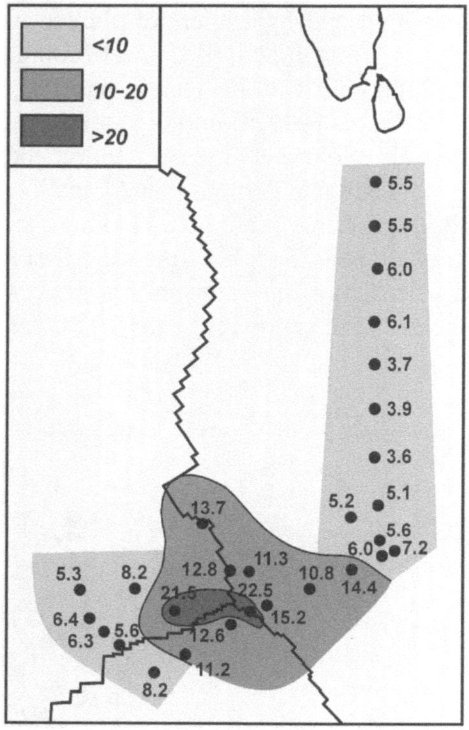

Fig. 1.45. Iron contents [%] in carbonate-free matter of surface sediments from the area of the Rodriguez Triple Junction. After Bogdanov et al. (1987) with changes.

The following types of metalliferous sediment have been found within the field studied: carbonaceous, clayey-carbonaceous, carbonaceous-clayey, and clayey (Bogdanov et al. 1987). Carbonaceous metalliferous sediments occur at depths from 3050 to 3770 m. The contents of $CaCO_3$ are usually greater than 90% and up to 96%. Carbonaceous-clayey and clayey-carbonaceous metalliferous sediments occur at depths of 3660 to 4210 m. $CaCO_3$ contents in them range from 13 to 20% and from 81 to 87% respectively. Clayey metalliferous sediments occur at depths greater than 4320 m. The content of $CaCO_3$ in them is usually insignificant, <1%, but may be as high as 8–10% in some places. The carbonaceous part of the sediments is composed of coccoliths with an admixture of foraminiferal material and the content of biogenic silica is low (Bogdanov et al. 1987).

In the non-carbonate part of the metalliferous sediments the Fe-hydroxides occur in very small grains and aggregates, ferruginous globules, and clots of silt size ferruginous clay and Fe-Mn micronodules are present in places. Fragments of edaphogenic material composed of basalt up to 1 cm in size commonly occur, fragments of Fe-Mn crusts and fragments of basalt glass are less common and the amounts of pyroxene, basic plagioclase, biotite, quartz, and zeolite minerals are insignificant (Bogdanov et al. 1987; Herzig and Plüger 1988). There are clay and feldspar minerals, quartz and an abundance of X-ray amorphous matter in the <1 μm size fraction. Goethite and Fe-montmorillonite minerals, which are typically present in metalliferous sediments in the Southeast Pacific, are rare (Gorbunova 1987).

Average contents of chemical elements in abiogenic matter of metalliferous and background sediments from the area of the Rodriguez Triple Junction are shown in Table 1.5. The contents of Fe, Mn and most of the trace elements in metalliferous sediments from the Indian Ocean are considerably lower than their contents in metalliferous sediments from the Southeast Pacific, and the contents of Al, Ga, Ti, and Cr are higher. The lower average rate of excessive accumulation of Fe, 1.2 mg cm^{-2} ka^{-1}, in metalliferous sediments from the Indian Ocean than in metalliferous sediments from the Southeast Pacific, 5.5 mg cm^{-2} ka^{-1}, indicates considerably less intensive accumulation of hydrothermal matter in metalliferous sediments from the Indian Ocean.

Table 1.5. Average contents of chemical elements in abiogenic matter of metalliferous sediments from the Indian Ocean. Compiled from data of Gurvich et al. (1980c), Lukashin and Lisitzin (1980), Lukashin (1981), Lisitzin et al. (1987), Herzig and Plüger (1988), Kuhn et al. (2000), Sirocko et al (2000).

Element	Metalliferous sediments	Background sediments	Element	Metalliferous sediments	Background sediments
Fe %	12.7	5.0	Pb ppm	167	56
Al	5.27	7.3	As	88	6.2
Mg	4.34		Mo	105	11.5
Mn	2.01	0.70	Ni	307	106
K	0.60		Co	155	37
Ti	0.31	0.42	V	307	119
P	0.27	0.19	Cr	104	128
Ba	0.22	0.47	Yb	22.3	10.3
			Y	353	131
Cu ppm	564	260	Zr	291	160
Zn	257	110	Ga	12.9	17.0

The ratios of average element contents in metalliferous and background sediments from the Indian Ocean are as follows:

Al → Ti → Ga → Cr → P → Zr → Yb → Cu → Zn → Fe → V →
0.72 0.74 0.76 0.81 1.42 1.82 2.17 2.17 2.34 2.54 2.58
Y → Mn → Ni → Pb → Co → Mo → As
2.69 2.87 2.90 2.98 4.19 9.13 14.2

Based on the assumption that the accumulation of Al in metalliferous sediments is related only to background process, it is possible to calculate the roles of excessive accumulation of the elements in %:

Al → Ti → Ga → Cr → P → Zr → Yb → Cu → Zn → Fe → V → Y → Mn → Ni →
0 2 5 11 49 60 67 67 69 72 72 73 75 75
Pb → Co → Mo → As
76 83 92 95

For elements listed from Zr to As the role of excessive accumulation is more than 50%. Nevertheless, this role in metalliferous sediments from the Indian Ocean is less than that in metalliferous sediments from the South-east Pacific.

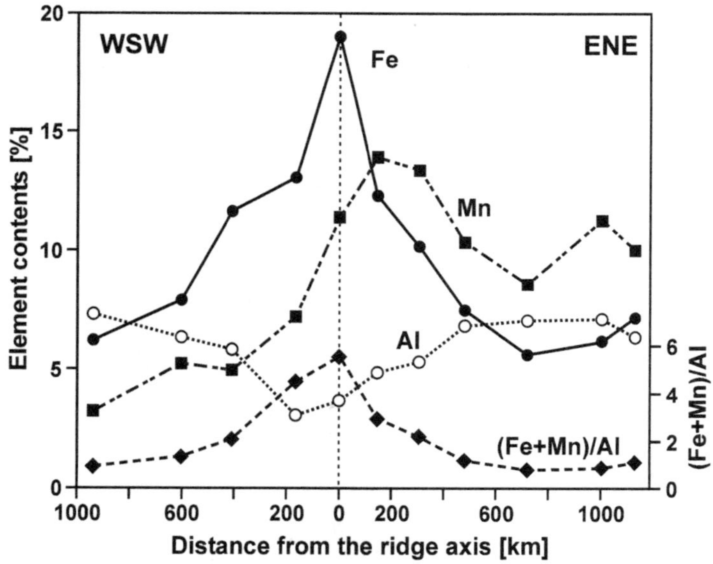

Fig. 1.46. Fe, Mn, and Al contents in abiogenic matter of surface sediments from the section across the Central Indian Ridge near the Rodriguez Triple Junction. Based on the data of Lisitzin et al. (1987).

Variations in the contents of elements in surface sediments from the section across the Central Indian Ridge near the Rodriguez Triple Junction are shown in Fig. 1.46 and 1.47. The maximum contents of Fe, Zn, and V occur in the axial zone of the ridge. As distances increase from the ridge axis toward the abyssal basins the contents of elements gradually decrease to those characteristic for background eupelagic sediments of the Indian Ocean (Sval'nov 1991). But the content of Al increases with distances from the ridge axis. The maximum contents of Mn and Cu occur in the axial zone, and are shifted somewhat to the east. The distribution of the contents of Ni and Co have a bimodal character, and their maximal contents occur in sediments on the eastern flank of the ridge. The shift to the east obviously is caused mainly by bottom currents flowing in an easterly direction (Lisitzin 1984; Lukashin 1987).

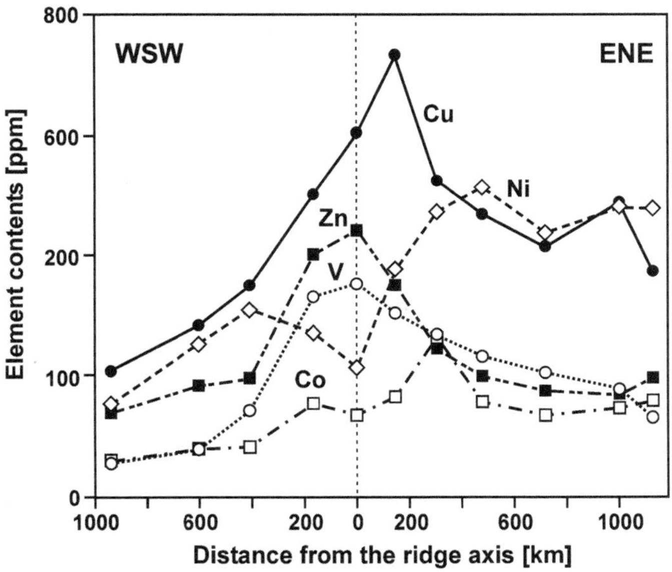

Fig. 1.47. Cu, Zn, Ni, Co, and V contents in abiogenic matter of surface sediments from the section across the Central Indian Ridge near the Rodriguez Triple Junction. Based on the data of Lisitzin et al. (1987).

The study of sediment cores has shown that in the area of the Rodriguez Triple Junction the Quaternary metalliferous sediments are less distributed than the Late Pliocene metalliferous sediments in which the excessive accumulation of elements is greater (Bogdanov et al. 1987; Lisitzin et al. 1987). Accumulation rates of lithogenic material in the sediments of both ages are similar and it is evident that accumulation rate of hydrothermal metal-bearing matter in the studied Late Pliocene metalliferous sediments is higher.

The relative roles of hydrothermal carriers of Fe, Mn, Cu, Zn, and V in the Quaternary metalliferous sediments are considerably lower than in the Late Pliocene metalliferous sediments (Lukashin 1987). The relative roles of hydrothermal carriers of Ni, Co, and Zr are similar in the Quaternary and Late Pliocene metalliferous sediments (Fig. 1.48). The role of hydrothermal carriers of the elements in the Quaternary metalliferous sediments to the east of the Rodriguez Triple Junction is higher than in the sediments to the west of it. This obviously results from the easterly flowing bottom currents (Lukashin 1987).

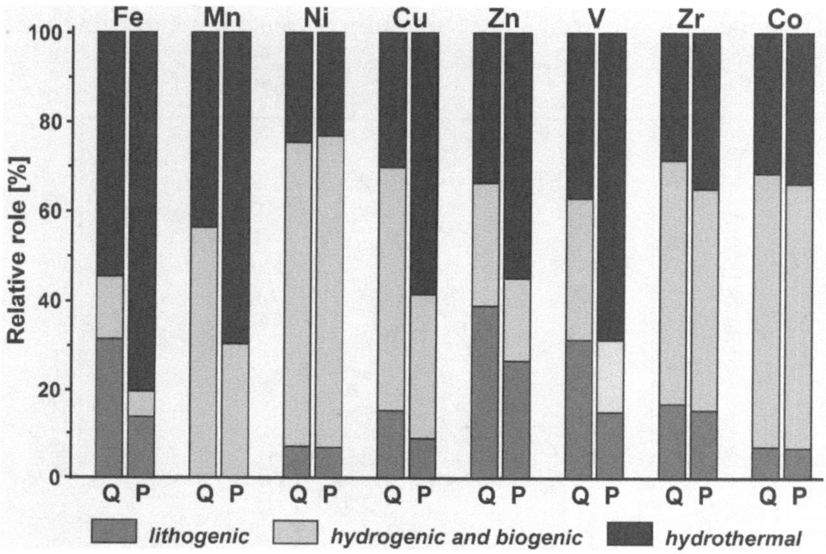

1.48. Relative roles of different carriers of chemical elements in Quaternary (Q) and Late Pliocene (P) metalliferous sediments from the area of the Rodriguez Triple Junction. Based on data of Lukashin (1987).

According to the results of partition chemical analysis (Lukashin 1983, 1987), Fe- and Mn-oxyhydroxides play a greater role as element carriers in the Late Pliocene metalliferous sediments than in the Quaternary metalliferous sediments and much less than in metalliferous sediments from the Southeast Pacific (Fig. 1.49).

Unlike the metalliferous sediments from the Southeast Pacific, crystallized Fe-oxyhydroxide minerals are prevalent minerals of Fe in the metalliferous sediments from the Indian Ocean and are more abundant in the Late Pliocene than in the Quaternary metalliferous sediments. About a half of the leachable Cu, Zn, V, and Zr in the Quaternary metalliferous sediments and most of the contents of these elements and about a half of the leachable Y in the Late Pliocene metalliferous sediments are bound in crystallized Fe-oxyhydroxide minerals. A predominant part of the leachable Ni, Co, and Y is bound with amorphous Fe-hydroxides and Mn-oxyhydroxides in the Quaternary metalliferous sediments and a predominant part of the Ni and Co is bound in these carriers in the Late Pliocene metalliferous sediments.

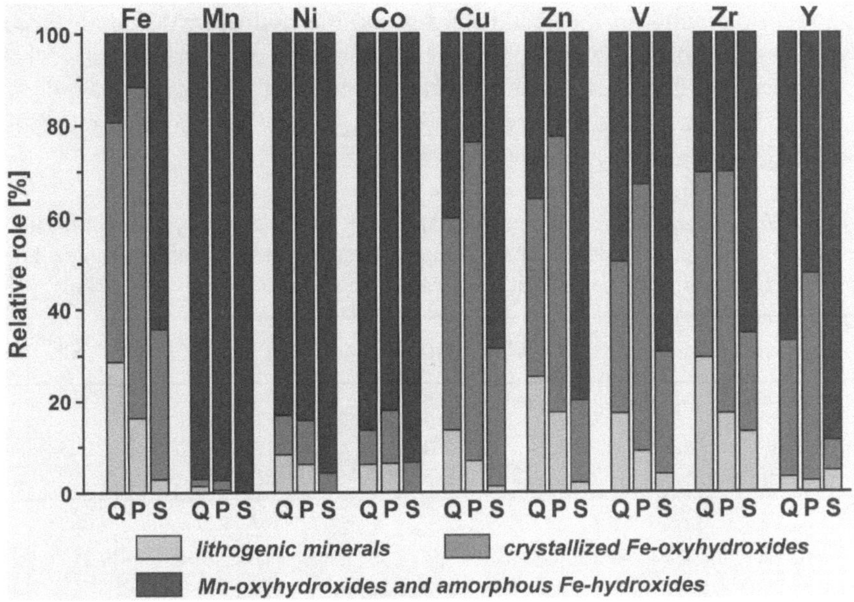

Fig. 1.49. Relative roles of different carriers of chemical elements in Quaternary (Q) (average for 10 samples) and in Late Pliocene (P) (average for 8 samples) metalliferous sediments from the area of the Rodriguez Triple Junction and in metalliferous sediments from the Southeast Pacific (S) (average for 18 samples). Based on data of Lukashin (1983, 1987).

Because of the lower sedimentation rate the role of the crystallized Fe-oxyhydroxide minerals as carriers of the trace elements is higher in metalliferous sediments from the Indian Ocean than in metalliferous sediments from the Southeast Pacific (Lukashin 1987). The average sedimentation rate for metalliferous sediments in the Indian Ocean is about 1 mm ka^{-1} (Lisitzin et al. 1987), and is 3–5 mm ka^{-1} for metalliferous sediments in the Southeast Pacific (Bogdanov and Chekhovskikh 1979).

The average bulk accumulation rates of Fe and Mn and the accumulation rates of the excessive Fe and Mn in metalliferous sediments from the Indian ocean (Fe*=1.7 mg cm^{-2} ka^{-1}, Fe_{ex}*=1.2 mg cm^{-2} ka^{-1}, Mn*=0.27 mg cm^{-2} ka^{-1}, Mn_{ex}*=0.20 mg cm^{-2} ka^{-1}) are much lower in comparison to the rates for metalliferous sediments from the Southeast Pacific (Fe*=6.8 mg cm^{-2} ka^{-1}, Fe_{ex}*=5.5 mg cm^{-2} ka^{-1}, Mn*=2.0 mg cm^{-2} ka^{-1}, Mn_{ex}*=1.8 mg cm^{-2} ka^{-1}). Metalliferous sediments from the Indian Ocean also have much lower total excessive accumulation not only of Fe and Mn and but trace elements (Table 1.6).

The following can be used for estimating the roles of different sources of chemical elements in the metalliferous sediments from the Indian Ocean: the types of spatial distribution of the contents of elements (Fig. 1.46, 1.47), and the correlation coefficients for pairs of data of two kinds, the first for the data pairs El_{ex} and (Fe_{ex}+Mn_{ex}) and the second for the data pairs El_{ex}/(Fe_{ex}+Mn_{ex}) and Al/(Fe_{ex}+Mn_{ex}) (Fig. 1.50).

Table 1.6. Average accumulation rates and rates of the excessive accumulation of chemical element in metalliferous sediments from the area of the Rodriguez Triple Junction (RTJ) and from the Southeast Pacific (SEP).

Area (km^2)	Fe	Mn	Cu	Zn	Co	Ni	Pb	V	Zr	Mo	As
	$mg\ cm^{-2}\ ka^{-1}$		$\mu g\ cm^{-2}\ ka^{-1}$								
RTJ	1.2	0.20	5.0	2.4	1.7	3.1	1.7	3.0	2.4	1.3	1.1
SEP	5.5	1.8	39	12.5	8.0	30	5.3	15	14		4.4
	$10^3\ t\ a^{-1}$		$t\ a^{-1}$								
RTJ (>$2 \cdot 10^6$)	>24	>4.0	>100	>48	>34	>62	>34	>60	>48	>26	>22
SEP ($10 \cdot 10^6$)	550	180	3900	1240	810	3000	530	1500	1400		440

The high positive correlation coefficients (R>+0.6) for the second kind of data pairs for Y, Co, Zr, Ni, and Yb indicate that these trace elements have been sorbed from ocean water by metal-bearing matter. Concentrations of these elements or concentrations of their forms available for sorption in ocean water are low, and ocean water is a limited source of these elements. For them ocean water is the prevalent source. For Co and Ni it is also confirmed by the bimodal distribution of their contents in abiogenic matter of metalliferous sediments of the section across the Central Indian Ridge near the Rodriguez Triple Junction (Fig. 1.47). Hydrothermal contribution is insignificant in the accumulation of these elements in metalliferous sediments.

The first kind of data pairs for the main hydrothermal elements, Fe and Mn, and for V, Mo, and Zn have high positive correlation coefficients (R>+0.6) and these elements have maximum contents in metalliferous sediments from the axial zone (Fig. 1.46, 1.47). This indicates either that hydrothermal fluids are the main source for the excessive accumulation of V, Mo, and Zn in metalliferous sediments, or that they were derived from ocean water and their concentrations in ocean water are high enough, and conditions are favorable for rapid scavenging of them by metal-bearing matter and their transfer to metalliferous sediments. Considering the low concentration of V and very low concentration of Mo in hydrothermal fluids and the behavior of Zn in hydrothermal plumes (see Chap. 4), it is concluded that ocean water is the main source for excessive accumulation of these elements in metalliferous sediments.

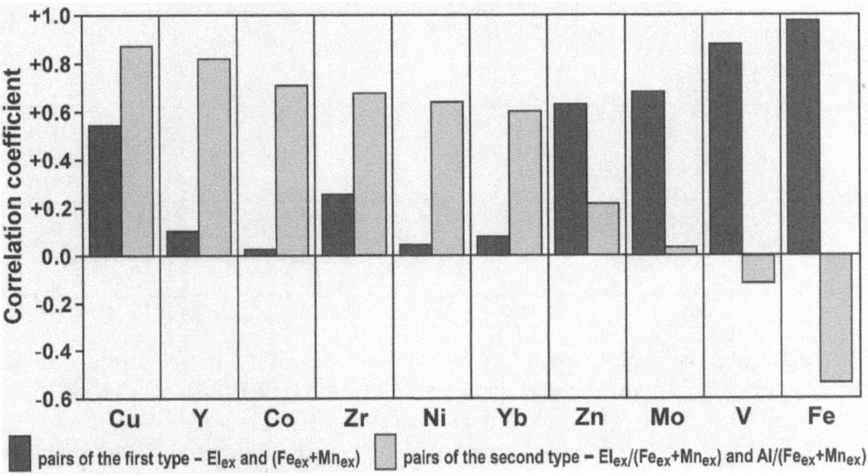

Fig. 1.50. Correlation coefficients for the pairs of two kinds calculated for abiogenic contents of chemical elements bound with the metal-bearing matter of metalliferous sediments from the area of the Rodriguez Triple Junction. Calculations based on the data of Lisitzin et al. (1987).

The correlation coefficient for Cu for the pairs of the second kind is much higher than that for the pairs of the first kind. Obviously this indicates that ocean water is the main source for the excessive accumulation of Cu, and hydrothermal fluids are less important, but for Cu they are of greater importance than for Y, Co, Zr, Ni, and Yb.

1.5. Sediments in the areas of hydrothermal activity in the Southwest Pacific

Bottom sediments containing hydrothermal matter occur in the Southwest Pacific in the hydrothermally active parts of the back-arc spreading areas in the Manus Basin of the Bismarck Sea, in the North Fiji Basin and Lau Basin, and also in the intercontinental rift zone of the Western Woodlark Basin (Fig. 1.51) where metalliferous sediments are rare and have been found mainly within hydrothermal fields and in thin layers of the sedimentary cover.

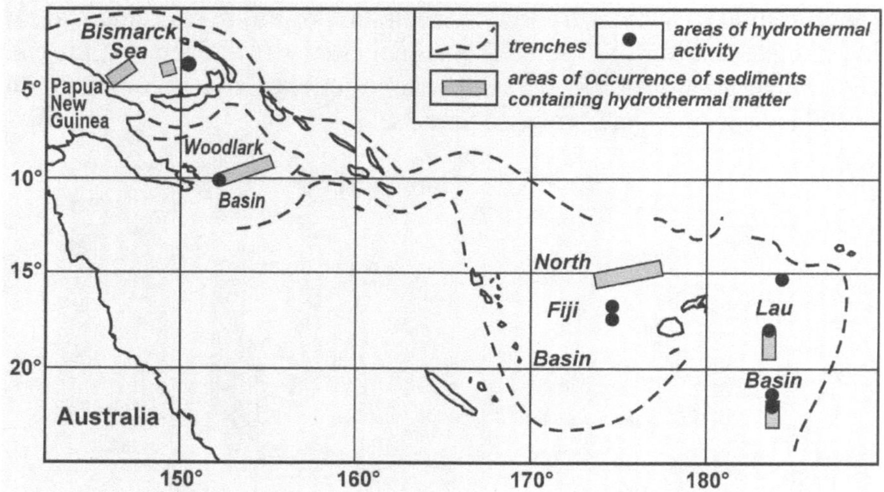

Fig. 1.51. Areas of high-temperature hydrothermal activity and occurrence of sediments containing hydrothermal matter in the Southwest Pacific. Based on materials from Cronan (1983), Moorby et al. (1986), Bogdanov (1990b), Gurvich (1990), Walter et al. (1990), McMurtry et al. (1991), Lisitzin et al. (1992), Bogdanov et al. (1992, 1997), Rona and Scott (1993), Binns and Scott (1993), Bendel et al. (1993), Fouquet et al. (1993), Daesslé et al. (2000), Daesslé and Cronan (2002).

The most comprehensive study of bottom sediments in the Southwest Pacific has been carried out by Cronan (1983). He has analyzed sediment samples from more than 300 stations. Factor analysis of the chemical content data for sediment samples from 261 stations shows that the factor of hydrothermal Fe- and Mn-oxyhydroxides is of secondary importance, and that the factor for lithogenic and biogenic materials is of first importance.

Both the low-carbonaceous and high-carbonaceous sediments in the Western Woodlark Basin contain hydrothermal matter. The content of $CaCO_3$ varies from 8 to 79% and the average content is 44% (Gurvich 1990; Bogdanov et al. 1997). The carbonaceous material consists mainly of calcite skeletons of planktonic (rarely – benthic) foraminifera and coccoliths. Aragonitic pteropod shells are abundant at depths shallower than 2000 m. Content of biogenic silica is insignificant. The abiogenic part of the sediments consists of highly variable mixtures of terrigenous material from the neighboring islands, volcanogenic material and its alteration products, as well as material of hydrothermal origin (Bogdanov 1990b; Bogdanov et al. 1997). The average content of hydrothermal matter in the abiogenic part of the sediments studied is about 5%, the maximum content is 12–13% (Fig. 1.52).

Noncalcareuos and low-calcareous sediments containing hydrothermal matter are prevalent in the Manus Basin and the content of $CaCO_3$ in these sediments varies from 0 to 40%. Volcanogenic matter, fragments of rhyolite pumice and globules of Fe- and Mn-oxyhydroxides are abundant in the abiogenic matter associated with the terrigenous clayey material (Bogdanov 1990b). The average content of hydrothermal matter in the abiogenic part of the sediments studied is about 7%, the maximum content is about 17% (Fig. 1.52).

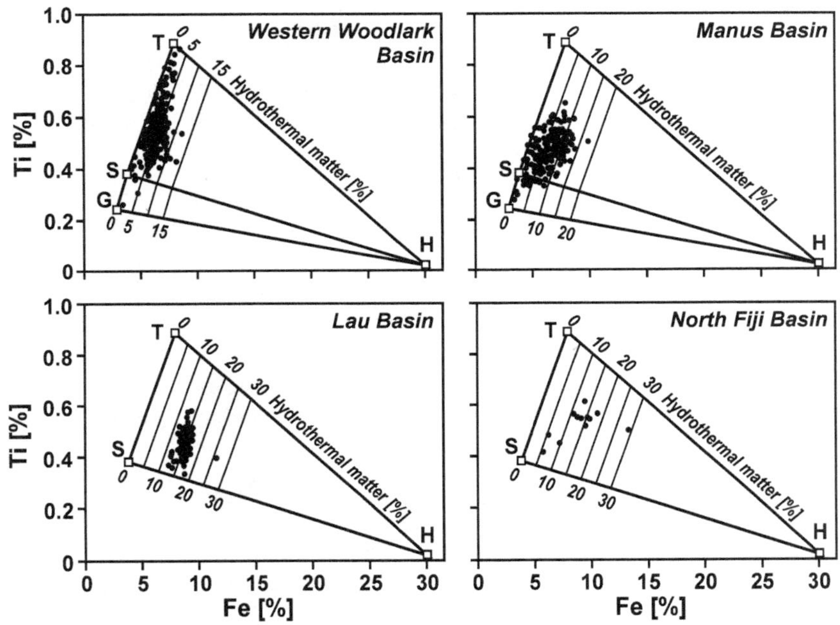

Fig. 1.52. Ti versus Fe contents and estimation of contents of hydrothermal matter in the abiogenic part of bottom sediments from the basins of the Southwest Pacific. Based on data from Moorby et al. (1986) – Lau Basin, Gurvich (1990) – Western Woodlark, Manus, and Lau Basins, and McMurtry et al. (1991) – North Fiji Basin. H – average for hydrothermal matter, from Marchig et al. (1986) and Lisitzin et al. (1990); T – average for oceanic tholeiites, from Taylor (1968); S – average for continental sedimentary rocks, from Ronov et al. (1990); G – average for granites and granodiorites, from Vinogradov (1962).

In the sediments containing hydrothermal matter in the Lau Basin the content of $CaCO_3$ varies from 1 to 80% (Moorby et al. 1986; Gurvich 1990; Riech 1990; Walter et al. 1990; Daesslé et al. 2000; Daesslé and Cronan 2002). Coccoliths are prevalent in the carbonate matter and the content of foraminifera skeletons does not exceed 25% (Moorby et al. 1986; Bogdanov 1990b; Riech 1990). The content of biogenic silica is insignificant and does not exceed 8–10%. Volcanoclastic material consisting mainly of volcanic glass is prevalent in the coarse grain size fractions of the abiogenic matter and terrigenous material and volcanic ash occur in the fine-grained fractions (Cronan et al. 1984; Moorby et al. 1986; Bogdanov 1990b; Riech 1990; Riech et al. 1990; Walter et al. 1990; Daesslé et al. 2000; Daesslé and Cronan 2002). The content of hydrothermal matter in the abiogenic part of the sediments studied from the northern Lau Basin varies from 11 to 26%, the average content is about 13%[11] (Fig. 1.52). In the carbonate-free matter of the sediments from the southern Lau Basin, the Valu Fa Ridge, the highest content of hydrothermal matter calculated is 35% (Walter et al. 1990). The hydrothermal matter consists mainly of films and thin coverings of Fe- and Mn-oxyhydroxides on particles. The content of this oxyhydroxide material in dry samples of sediments is usually 5 to 15% (Moorby et al. 1986; Riech et al. 1990).

In the North Fiji Basin the sediments that contain hydrothermal matter are mainly carbonaceous and the content of $CaCO_3$ usually varies from 70 to 95%. Volcanic detrital material is prevalent in the >2 μm size fraction of the non-carbonate part of these sediments and in the <2 μm size fraction clay minerals consisting mainly of smectites, amorphous Fe-oxyhydroxides, and volcanic ash are prevalent (Riech 1990; McMurtry et al. 1991). The content of hydrothermal matter in the abiogenic part of the sediment samples studied varies from 5 to 30% and the average content is about 15% (Fig. 1.52).

The content of hydrothermal matter in the abiogenic part of the sediments from the areas of hydrothermal activity in the Southwest Pacific are considerably lower than in the abiogenic part of the metalliferous sediments from the Southeast Pacific and Rodriguez Triple Junction areas (Fig. 1.53).

The low contents of the hydrothermal matter and the rare occurrences of metalliferous sediments in the areas of hydrothermal activity in the Southwest Pacific result from the high accumulation rate of the diluting abiogenic material. Here the accumulation rate of Al, considered to be a proxy of the lithogenic material, is 20 to 250 mg cm^{-2} ka^{-1} or even higher (Lisitzin 1978; Walter et al. 1990; Daesslé et al. 2000; Daesslé and Cronan 2002). This accumulation rate of Al is comparable only to that for the metalliferous sediments in the area of the Axial Seamount (where it varies from 20 to 180 mg cm^{-2} ka^{-1}) and is considerably higher than that for the metalliferous sediments from the northern part of the East Pacific Rise (where it is 25 to 50 mg cm^{-2} ka^{-1}), and it is especially higher than that for metalliferous sediments from the Southeast Pacific where the average rate is 1.75 mg cm^{-2} ka^{-1}, and from the area of the Rodriguez Triple Junction in the Indian ocean where the average rate is 0.7 mg cm^{-2} ka^{-1}.

[11] After Walter et al. (1990), the average content of hydrothermal precipitates including hydrogenous precipitates in the non-carbonate fraction of the surface sediments from the northern and southern Lau Basin are 11.1% and 11.2%.

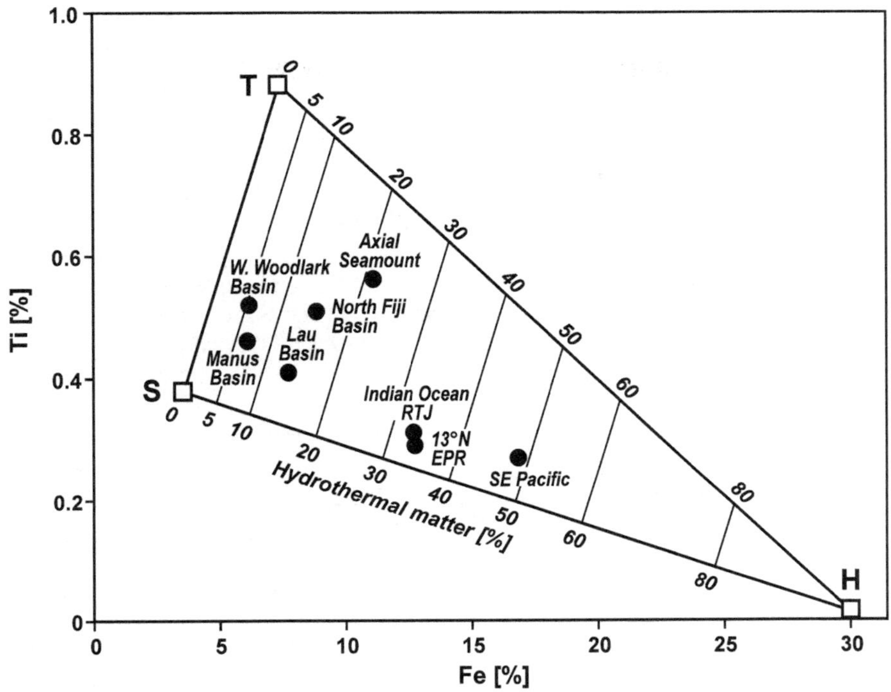

Fig. 1.53. Ti versus Fe contents and the estimated average contents of hydrothermal matter in the abiogenic part of bottom sediments from the areas of hydrothermal activity in the Southwest Pacific and in metalliferous sediments from the Pacific and Indian oceans. H, T, and S abbreviations are the same as for Fig. 1.52.

The average chemical composition of the carbonate-free matter of the sediments from the areas of hydrothermal activity in the Southwest Pacific is shown in Table 1.7. The contents of Fe, Mn, and other elements that usually are combined with the hydrothermal metal-bearing matter in the abiogenic matter of these sediments are considerably lower than those in metalliferous sediments from the East Pacific Rise, and from the area of the Rodriguez Triple Junction (Tables 1.1, 1.3, 1.5).

In the areas of hydrothermal activity in the Southwest Pacific the sediments from the Western Woodlark Basin have the lowest excessive contents of elements and the lowest content of the hydrothermal metal-bearing matter. The sediments from the Lau and North Fiji Basins have the highest excessive contents of the elements and the highest enrichment in the hydrothermal metal-bearing matter (Fig. 1.54; Table 1.8).

The average relative roles in % of excessive accumulation of the elements in the sediments from the areas of hydrothermal activity in the Southwest Pacific are as follows:

Ni →	Co →	Zn →	Ba →	Fe →	V →	Cu →	Pb →	Mn →	As
28	29	32	35	36.5	37	38	47	59	≤86

Only the roles of the excessive accumulation of Mn and As exceed 50%. Unlike the metalliferous sediments in the mid-ocean ridge areas of the Pacific and Indian Oceans, the excessive accumulation of most of the other chemical elements is of secondary importance in the sediments from the areas of hydrothermal activity in the Southwest Pacific.

Table 1.7. Average contents of chemical elements in carbonate-free matter of sediments from the areas of hydrothermal activity in the Southwest Pacific. Based on the data from Cronan (1983), Moorby et al. (1986), Gurvich (1990), Walter et al. (1990), McMurtry et al. (1991), Daesslé et al. (2000), Daesslé and Cronan (2002).

Element		Western Woodlark Basin	Lau Basin	North Fiji Basin			Manus Basin
				bulk	<2 μm	>2 μm	
Si	%	22.7	25.4	24.1	20.5	25.2	
Al		7.37	6.5	7.8	7.4	7.9	9.8
Ti		0.55	0.40	0.51	0.61	0.48	
Fe		6.00	7.7	8.9	13.6	7.4	7.5
Mn		0.31	1.4	0.93	1.88	0.61	1.4
P		0.125	0.17				
Cu	ppm	134	230	300	491	240	153
Zn		183	140	179	356	128	192
Pb		73	51	69	126	51	55
Cd				0.58	1.4	0.32	
As			40	19	55	9.3	
Sb			1				
Ni		112	96	129	192	111	116
Co		25.6	35	75	136	55	
Cr		111	42	74	106	65	
V		168	250				
La			23				
Mo			8.4				
Bi			0.43				
Be			0.6				
Ba		860	580				
Li			11	20	24	19	

Studies of sediments from the North Fiji Basin carried out by McMurtry et al. (1991) have shown that the fine-grained material is enriched in hydrothermal metal-bearing matter. The carbonate-free part of the material in the <2 μm grain size fraction, unlike the material of the >2 μm fraction and that in the bulk sediments, contains more than 10% Fe and is enriched in Mn and many trace elements (Table 1.7).

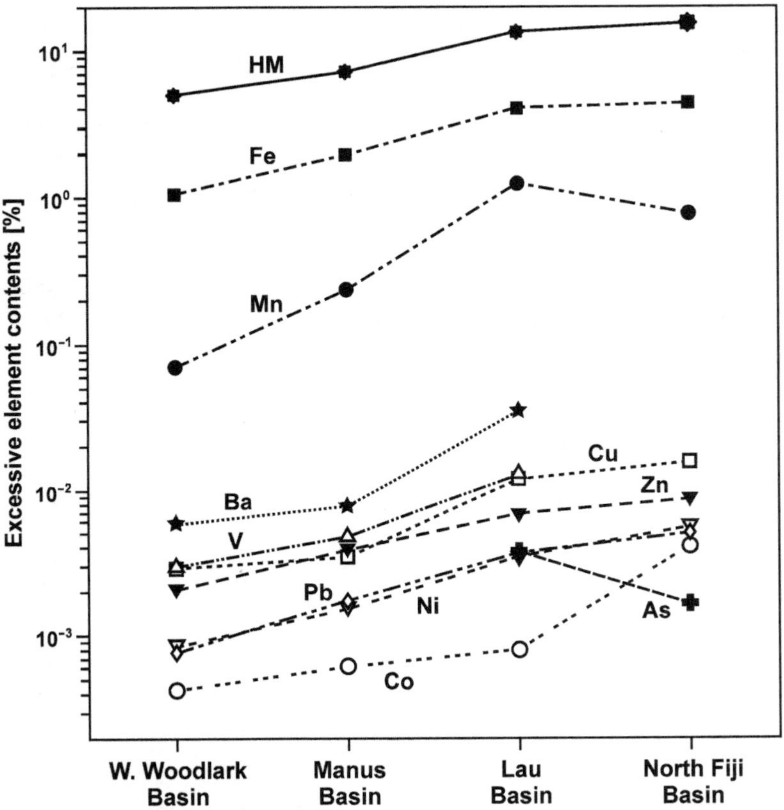

Fig. 1.54. Average contents of hydrothermal matter (HM) and average excessive contents of chemical elements in the abiogenic matter of bottom sediments from the areas of hydrothermal activity of the Southwest Pacific.

Factor analysis of data for the chemical composition indicates that for the carbonate-free matter of the <2 μm grain size fraction the hydrothermal factor is major and accounts for 36% of the variance while for the carbonate-free matter of the >2 μm grain size fraction the hydrothermal factor accounts for only 8% of the variance (McMurtry et al. 1991).

On the average, about a quarter of the carbonate-free matter of the sediments from the North Fiji Basin is composed of the <2 μm size fraction which contains from 60 to 85% of the bulk carbonate-free contents of Ni, Fe, Cu, Co, Pb, Zn, Cd, Mn, and As (McMurtry et al. 1991).

Results of the factor analysis of bulk chemical composition of sediments from the Lau Basin (Walter et al. 1990) indicate that the hydrothermal factor is of secondary importance and accounts for 28.9% of the variance.

The results from the partition chemical analysis carried out by McMurtry et al. (1991) show that more than a half of the contents of Fe, Co, Cd, Mn, and As in the carbonate-free matter of the <2 μm grain size fraction and a half of the Ni content are carried in the Fe- and Mn-oxyhydroxide minerals. About a half of the contents of Cu, Zn, and Pb are carried by Fe-smectite minerals and 20 to 40% are carried by Fe- and Mn-oxyhydroxide minerals (Fig. 1.55).

Table 1.8. The average roles [%] of excessive accumulation of chemical elements in abiogenic matter of bottom sediments from the areas of hydrothermal activity in the Southwest Pacific.

Element	Western Woodlark Basin	Manus Basin	Lau Basin	North Fiji Basin
Fe	17	31	50	48
Mn	39	23	91	83
Cu	21	31	48	52
Zn	11	24	46	48
Pb	10	33	70	73
As			92	80
Ni	7	21	38	44
Co	17	23	21	55
V	17	31	50	
Ba	9	13	59	

Partition chemical analysis of 63 sediment samples from the Lau Basin, carried out by Daesslé et al. (2000), and Daesslé and Cronan (2002), shows that on the average 95% of the total content of Mn and >30% of the total content of Fe are bound in Fe- and Mn-oxyhydroxide minerals. These minerals also carry more than a half of the total contents of Ni, Mo, Pb, Co, Cu, and REE and about a half of the contents of Zn and V.

30 km west of the axis of the Valu Fa Ridge, 85% of the total Fe and about 99% of the total Mn are bound with oxyhydroxide minerals. These minerals carry 73% of the total Zn, 84% of the total Cu, and almost all of the Ni. The role of crystallized Fe-oxyhydroxides as carriers of all mentioned elements increases downcore.

The average rates of excessive accumulation of some chemical elements in the Lau and North Fiji Basin bottom sediments that contain hydrothermal matter are shown in Table 1.9. The average rates of excessive accumulation of the elements in the Lau Basin sediments are considerably higher than the average rates of the excessive accumulation of these elements in metalliferous sediments from the Southeast Pacific and from the area of the Rodriguez Triple Junction (Table 1.6).

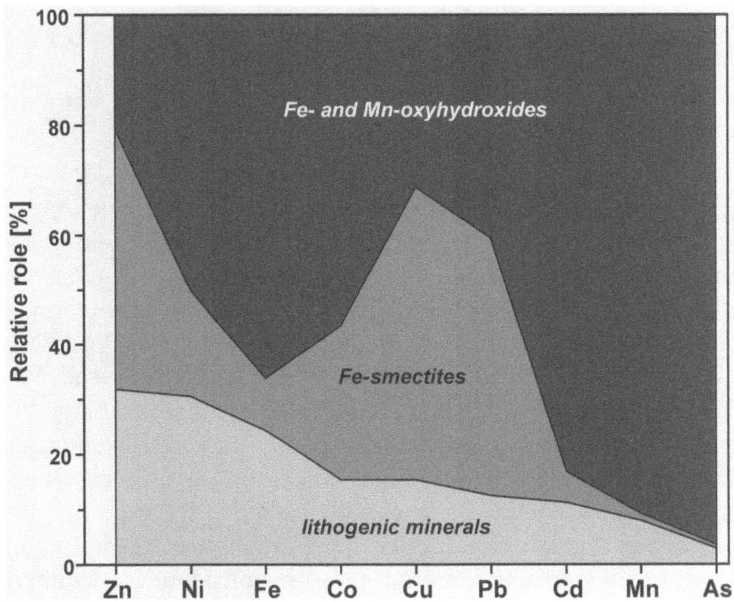

Fig. 1.55. Relative roles of different carriers of chemical elements in carbonate-free matter of the <2 mm grain size fraction of bottom sediments from the North Fiji Basin. Based on the data from McMurtry et al. (1991).

Table 1.9. Average rates of excessive accumulation of chemical elements in the Lau and North Fiji Basin bottom sediments that contain hydrothermal matter. Based on the data from Cronan et al. (1986), Riech (1990), Walter et al. (1990), McMurtry et al. (1991), Daesslé et al. (2000), Daesslé and Cronan (2002).

Area	Fe	Mn	Cu	Zn	Co	Ni	As
	mg cm^{-2} ka^{-1}		µg cm^{-2} ka^{-1}				
Lau Basin	26	11	115	56	3.1	29	22
North Fiji Basin	11.5	1.6	49		14	18	

The reason for this may be that most of the sediment cores from the Lau Basin were collected in the vicinity of the spreading axis and in the Southeast Pacific and in the area of the Rodriguez Triple Junction the average values were calculated using data from all of the fields of metalliferous sediments. In the metalliferous sediments from the axial zone of the East Pacific Rise the rates of the excessive accumulation of the elements are comparable and higher than those in the sediments from the Lau Basin. Comparisons of this kind should be made for sediments that are equidistant from hydrothermal sources, and the influence of bottom currents must be considered.

1.6. Metalliferous sediments in the rift zone of the Mid-Atlantic Ridge

There are no large fields of recent metalliferous sediments in the Atlantic Ocean and small fields occur mainly within the rift zone of the slow-spreading Mid-Atlantic Ridge. Metalliferous sediments have been found within the TAG (26°08'N) and MARK (23°22'N) hydrothermal fields, within and near the Logachev (14°45'N) and Rainbow (36°14'N) hydrothermal fields, in the areas of 45°30'N (Cronan 1972) and FAMOUS (ARCYANA 1975, Hekinian and Fevrier 1979), in several sites found by Emel'yanov (1975, 1982), and some others. The metalliferous sediments from the areas where the modern high-temperature hydrothermal activity is known, the TAG, MARK, Logachev and Rainbow hydrothermal fields, are the most interesting. Conditions for the formation of these metalliferous sediments are different and therefore they are considered separately.

1.6.1. Metalliferous sediments from the TAG hydrothermal field

The Mid-Atlantic Ridge rift zone in the area of the TAG hydrothermal field is asymmetric. To the west the rift valley has a system of faults where the depth is about 1000 m shallower than the depth of the valley floor. To the east the rift valley adjoins a large uplifted structure; here the depth is about 1000 m shallower than it is along the western boundary of the rift valley. This structure was developed due to the uplift of a large block of the ocean crust. Serpentinization of ultrabasic rocks in the lower crust and upper mantle zone may have caused this uplift (Zonenshain et al. 1989, 1992). The rift valley is 10 to 15 km wide in the TAG hydrothermal field area and 3650 to 3700 m deep and is 1000 m deeper than in typical ocean rift zones. Nevertheless there is an elevation of the axial zone here and the depth is about 400 m shallower than in the areas to the north and to the south (Lisitzin et al. 1990).

Both modern and relict high-temperature hydrothermal activity and mineral formation within the TAG hydrothermal field are spatially confined to the eastern part of the rift valley (Fig. 1.56). Formation of the hydrothermal mound that is active at the present time began about 50 thousand years ago and has continued to be active with some interruptions to the present time (Lalou et al. 1990). The inactive high-temperature hydrothermal Mir mound that is located to the east-northeast of the Active hydrothermal mound is the largest within the TAG field and it is almost not covered by sediments (Rona et al. 1993a; Bogdanov et al. 1994). The formation of its high-temperature parts began about 100 ka BP and continued with interruptions until ~700 years ago (Lalou et al. 1995). The chain of relict inactive high-temperature hydrothermal mounds in the Alvin Zone is located to the north-northwest of the Mir mound. (Fig. 1.56; Rona et al. 1993b) and is covered extensively by sediments. The northern edifices were formed at an earlier time than the southern ones as indicated by the increase in thickness of the sediment layer to the north (Lisitzin et al. 1989; Bogdanov et al. 1995b).

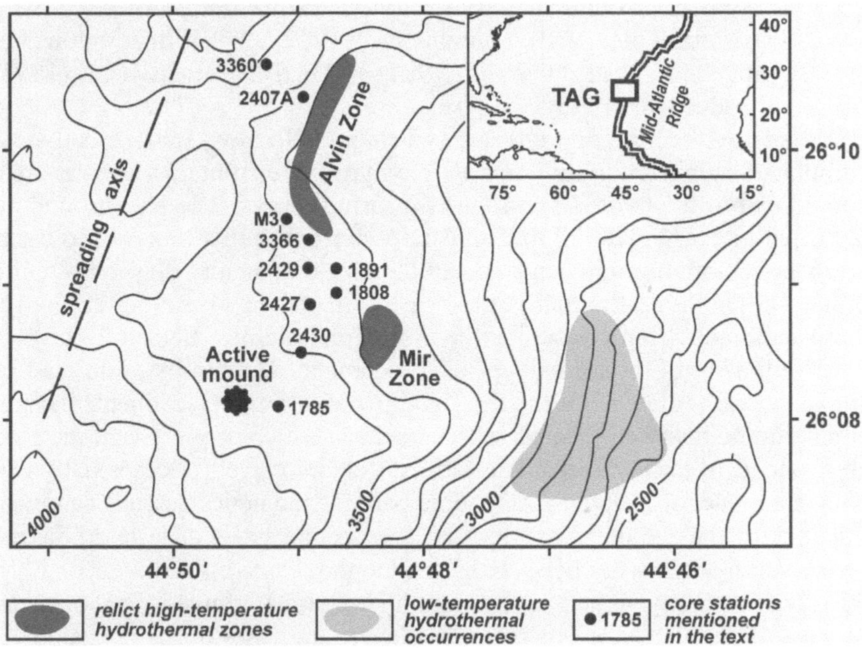

Fig. 1.56. Bathymetric chart of the TAG hydrothermal field, hydrothermal occurrences within the field, and core stations mentioned in the text.

Information was published in 1978 by Scott et al. (1978) on sediments collected 10 to 15 km north of the TAG hydrothermal field, before the high-temperature hydrothermal occurrences were known, that have excessive contents of Fe, Mn, and some of the trace elements. Metalliferous sediments from the TAG hydrothermal field were described for the first time in 1983 (Shearme et al. 1983) and now there is information on metalliferous sediments in the area from more than 30 cores (Metz et al. 1988; Gurvich and Bogdanov 1991; Hydrothermal... 1993; German et al. 1993; Gurvich 1994; Bogdanov et al. 1995a,d; Cherkashev 1995). Most of the cores, including the longest one, were collected during Cruises 15, 23, and 34 of the R/V *Akademik Mstislav Keldysh.*

The metalliferous sediments in the TAG hydrothermal field occur between 26°08'N and 26°11'N within a band 1 to 2 miles wide along the eastern slope of the rift valley. They are distributed in a small area of about 15–20 km^2. They have accumulated in the vicinity of the high-temperature hydrothermal zones mentioned, the Active mound, and the relict Mir and Alvin Zones. Apparently these sediments were deposited during the hydrothermal activity that contributed to the formation of these accumulations, and also during their disintegration.

The metalliferous sediments from the TAG hydrothermal field have been subdivided into four main types following the studies of Shearme et al. (1983), Metz et al. (1988), Lisitzin et al. (1989, 1990), Hannington et al. (1990), Bogdanov et al. (1993a,b, 1994, 1995a,d), German et al. (1993), Mills and Elderfield (1995).

a) Sediments of the first type are typical metalliferous sediments that accumulated "grain-by-grain" and Fe-oxyhydroxide minerals are the main component of the hydrothermal metal-bearing matter. The background pelagic sedimentary material that is usually present in this type of sediments consists of foraminiferal and coccolithophorid remains, clay and zeolite minerals. The hydrothermal metal-bearing matter has been contributed to these sediments from nonbuoyant hydrothermal plumes (Sect. 4.3.2).

b) Sediments of the second type also accumulated "grain-by-grain" and are similar to the sediments of the first type. However, these sediments have an admixture of fine well faceted sulfide crystals, which in places are the main components of the sediments and form sulfide interlayers. Occasionally edaphogenic material is abundant in some parts of the noncalcareuos sandy-silt fractions of the sediments of the first and second types. The hydrothermal metal-bearing matter has been contributed to the sediments of the second type from both the buoyant and nonbuoyant hydrothermal plumes. The formation of the sulfide interlayers is attributed to sharp increases in the deposition of material from the buoyant plumes (Sect. 4.3.1).

c) Sediments of the third type are composed mainly of coarse-grained material derived from the disintegration of the hydrothermal edifices and often are referred to as debris-flow deposits (Nelsen et al. 1986; Trefry et al. 1986; Lisitzin et al. 1989, 1990; Bogdanov et al. 1994). Clastic material in sediments of this type is composed of Fe- and Mn-oxyhydroxides, nontronite, and sulfide minerals. Their relative amounts vary throughout a vertical section. The clasts "are cemented" by silty-clay, silty and sandy material composed mainly of Fe-oxyhydroxides, well faceted crystals of sulfide minerals and in places amorphous silica.

d) Sediments of the fourth type are thin-laminated and composed of a great variety of sandy material and other components similar to those in the third type of sediments. However, each lamina is usually enriched in one of the components mentioned. The sediments of the third and the fourth types were formed by the horizontal transportation of the products of disintegration of the hydrothermal edifices and have undergone partial gravitational sorting during their deposition (the fourth type).

Edaphogenic material, the products of the disintegration of basalt and dolerite rocks exposed on the seafloor, is the main component of the non-calcareuos part of the sandy-silt fraction of nonmetalliferous matter in the bottom sediments. Within zones where volcanic rocks have been extruded the edaphogenic material is composed almost completely of volcanic glass of basaltic composition with only minor amounts of terrigenous material. Hydrothermal activity in the sediments in the vicinity of the hydrothermal mounds is indicated by the Mn-crusts on the basalt and glass fragments and the occurrence of Fe-hydroxides, hematite, talc, zoned quartz and sulfide minerals. The bulk of the highly dispersed fractions of the sediments is composed of hydrothermal material consisting of an abundance of X-ray amorphous components along with goethite, hematite, and nontronite mineral particles and the minor part of the sediments consists of illite, kaolinite, and palygorskite clay minerals (Bogdanov et al. 1993a).

Panning analyses show that the abiogenic part of the metalliferous sediments from the TAG hydrothermal field has a high content of Fe-hydroxide material, and considerable amounts of Fe-, Cu-, and Zn-sulfide minerals and particles of native metals and alloys. The sediments in the vicinity of the extinct hydrothermal mounds, unlike the sediments near the Active mound, have a higher ratio of oxidized to non-oxidized sulfide minerals, as well as considerably higher contents of atacamite, Fe-Mn micronodules and incrustations of manganite and Fe-oxide material. At a distance of 2 to 2.5 km from the hydrothermal mounds the products of their disintegration are almost not found in the sediments (Andreeva and Stepanova 1992). Typically the prevalence of sulfide minerals and products of their oxidation are found in the metal-bearing part of the sediments that accumulated in the close proximity, within one kilometer, of the hydrothermal mounds during intensive hydrothermal activity and the partial disintegration of the mounds. At greater distances from the mounds there is a transition from metalliferous sediments bearing primary sulfide minerals to typical metalliferous sediments containing metal-bearing matter that settled from the hydrothermal plume (Bogdanov et al. 1993b).

The sediments accumulated near hydrothermal mounds reflect different stages in their formation within one or several hydrothermal cycles, which lasted for about 10 thousand years (Bogdanov et al. 1992b, 1994). During the first initial stage of formation of a mound hydrothermal metal-bearing fluids were discharged through the broken basement of the ocean floor. At this time non-metalliferous sediments or low-metalliferous sediments accumulated in the vicinity of the hydrothermal orifices. After cementation of the fractures and the circulation canals by the precipitated metal-bearing matter the fluids were discharged only in the central parts of the mound. When temperature and flow rates of the hydrothermal fluids increased a considerable amount of sulfide matter, rounded grains of minerals from greenstone rocks as well as 2–3 mm or larger size fragments of altered volcanic rocks were removed. All of this is an indication of the high flow rates of the hydrothermal fluids, which reached several meters per second. During this second stage of hydrothermal activity the output of material of hydrothermal origin was more intensive and resulted in more extensive accumulation of it in the sediments, in the enrichment of Fe and Cu in the sediments in the adjacent areas. During the second stage the partial disintegration of the hydrothermal mound took place. As the fragmental material moved down the slopes it formed turbidite sediments composed of hydrothermal material and debris-flow deposits under the mounds that are composed of Fe- and Mn-oxyhydroxide, nontronite and sulfide minerals. There is a marked increase in the contents of secondary metamorphic minerals in the sediments surrounding the orifices. During the third stage of mound development the rate of discharge and temperature of the fluids decreased gradually. Crusts of Fe- and Mn-oxyhydroxide material could form on the surface of the hydrothermal mounds. It protected the sulfide minerals from oxidation. Globules and crust fragments accumulated in the surrounding sediments. During the third stage metalliferous sediments with low contents of Fe and Cu and non-metalliferous sediments accumulated before new stages of increases and decreases in the amount of hydrothermal activity took place.

The sedimentation rates of the metalliferous sediments vary within the TAG hydrothermal field. In the first type of sediments they depend mainly on variations in the rate of accumulation of hydrothermal matter that is related to the intensity of hydrothermal activity and their distance from the hydrothermal mounds. The sedimentation rates for the third and fourth types of sediments depend on the deposition and horizontal transport of the products of disintegration of the hydrothermal mounds. According to the data from biostratigraphic and radiocarbon dating, the sedimentation rates of the undisturbed metalliferous sediments of the first and second types vary from <1 to 6 cm ka^{-1} (Metz et al. 1988; Kuptsov 1993; Lukashina 1993; Bogdanov et al. 1994). Minimum values were found in sediments of the first type and maximum values in sediments of the second type. The average sedimentation rate of the undisturbed metalliferous sediments is about 2 cm ka^{-1}. From a geological point of view, layers of sediments of the third and fourth types were formed almost instantaneously and accumulated material deposited in their vicinity during a much longer period, probably in several thousand years (Lisitzin et al. 1990). The measured thicknesses of layers of these sediments range up to several tens of centimeters (Metz et al. 1988; Lisitzin et al. 1990; Bogdanov et al. 1994, 1995a,b).

Table 1.10. Average chemical composition of carbonate-free matter of the sediments from the TAG hydrothermal field in comparison with the average compositions of carbonate-free matter of the metalliferous sediments from the Southeast Pacific and sediments from the other areas of the Atlantic. Based on data from Scott et al. (1978), Shearme et al. (1983), Gordeev (1988), Metz et al. (1988), Gurvich and Bogdanov (1991), Mills (1992), German et al. (1993), Dekov and Gurvich (1993), Mills et al. (1993, 1994), Godfrey et al. (1994), Gurvich (1994), Krasnov et al. (1995), and Table 1.1.

Sediments	Si	Al	Ti	Fe	Si_{ex}	Mn	Cu	Zn	P
					%				
					metalliferous				
TAG (Fe 10÷50%)	12.00	2.90	0.34	23.84	4.2	0.33	1.106	0.488	0.32
SE Pacific	16.19	4.35	0.27	16.86	3.1	5.07	0.104	0.041	
					non-metalliferous				
TAG (Fe 8÷10%)	18.54	6.87	0.51	8.86	0	0.47	0.162	0.057	0.28
TAG (Fe 6÷8%)	19.83	7.34	0.52	6.97	0	0.36	0.072	0.037	0.18
TAG (Fe <6%)	20.71	7.67	0.48	5.63	0	0.32	0.052	0.037	0.13
10–15 km to the north of TAG		6.2		5.0	0.3	0.021			
North Atlantic, pelagic clays		9.0		5.0	0.4	0.012			

Sediments	Pb	Ni	Co	Cr	V	Ba	La	U	Th
					ppm				
					metalliferous				
TAG (Fe 10÷50%)	117	63	120	82	270	630	10.5	10	4.3
SE Pacific	155	826	218	29	428	10800	157	3.8	5.0
					non-metalliferous				
TAG (Fe 8÷10%)	27	140	72	150	230	790	47.6	1.6	7.3
TAG (Fe 6÷8%)		130	62	155	190	640	37.9		
TAG (Fe <6%)		110	60	120	120	510			
10–15 km to the north of TAG		240	120	200					
North Atlantic, pelagic clays		79	39	80					

The average chemical composition of metalliferous and transitional sediments from the TAG hydrothermal field based on the calculation of data obtained from the study of more than 1000 samples is shown in Table 1.10. For comparison the average compositions of the metalliferous sediments from the Southeast Pacific and of sediments from other regions of the Atlantic Ocean are shown in this table.

In comparison with non-metalliferous sediments, the metalliferous sediments from the TAG hydrothermal field are enriched in Fe, Cu, Zn, Pb, Co, P, V, U and depleted in Al, Si, Ti, Th, Cr, Ni, REE. The average contents of Mn and Ba in the metalliferous and non-metalliferous sediments are similar. The Si/Al ratio in the metalliferous sediments (4.1) is higher than in the non-metalliferous sediments (2.7) which indicates that about one third of the Si is of hydrothermal origin.

Comparison with the metalliferous sediments from the Southeast Pacific shows that the metalliferous sediments from the TAG hydrothermal field have higher average contents of Fe, U, and especially of Cu and Zn, but lower contents of Si and Al, and metalliferous sediments from the TAG field are enriched in Ti and Cr. The average contents of Mn, Co, V, Pb, Ni, Ba, and REE in the TAG metalliferous sediments are lower than in metalliferous sediments from the Southeast Pacific (Table 1.10). On the average about 3/4 of the metal-bearing matter in the metalliferous sediments from Southeast Pacific is composed of iron minerals and in the TAG metalliferous sediments more than 9/10 of the metal-bearing matter is composed of iron minerals. The average content of Fe in the metalliferous sediments studied from the TAG hydrothermal field is higher than the sum of the average contents of Fe and Mn in metalliferous sediments from the Southeast Pacific. The reason for this is that the most part the metalliferous sediment samples studied from the TAG field were proximal to hydrothermal sources and metalliferous sediment samples studied from the Southeast Pacific were distal from them. On the other hand, the area of the TAG hydrothermal field where the hydrothermal matter accumulates on the seafloor is within a narrow and deep rift valley and the hydrothermal matter is not dispersed outside of it. In the Southeast Pacific the hydrothermal matter is dispersed and accumulates on the ocean floor many hundreds and even thousand of kilometers away from the East Pacific Rise rift zone.

The ratios of the average element contents in the carbonate-free matter of metalliferous sediments to those in non-metalliferous sediments containing <6% Fe (Table 1.10) are as follows:

Al	→	Ni	→	Cr	→	Mn	→	Ba	→	V	→	Co	→	P	→	Fe	→	Zn	→	Cu
0.37		0.57		0.68		1.0		1.2		1.8		2.0		2.5		4.2		13		21

Based on the assumption that the accumulation of Al in metalliferous sediments is caused only by background process the average relative roles of the excessive accumulation of the elements in % are as follows:

Al	→	Ni	→	Cr	→	Mn	→	Ba	→	V	→	Co	→	P	→	Fe	→	Zn	→	Cu
0		34		45		63		69		79		81		85		91		97		98

It must be noted that these results are underestimated slightly because in the sediments containing <6% Fe, that were considered as background sediments, there is a small amount of hydrothermal metal-bearing matter that contains significant quantities of Cu, Zn and some other elements.

It is interesting to consider changes in the contents of the elements with increases in the contents of the metal-bearing matter in the abiogenic part of the sediments from the TAG hydrothermal field. Other factors being equal the content of the metal-bearing matter in the sediments is a reflection of the distance from a hydrothermal source and of the intensity of hydrothermal activity. As Fe minerals constitute more than 9/10 of the hydrothermal metal-bearing matter in the sediments from the TAG hydrothermal field it is convenient to use Fe content as a proxy of hydrothermal matter content and the average contents of chemical elements have been calculated for sediments with different contents of Fe (Fig. 1.57–1.60).

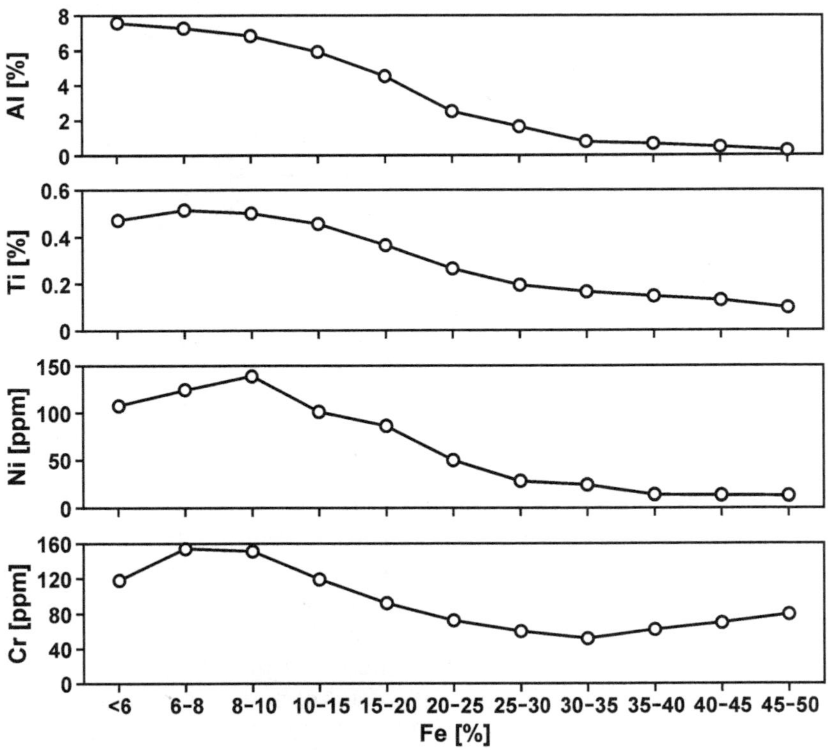

Fig. 1.57. Changes in the contents of Al, Ti, Ni, and Cr with increase in Fe content in the carbonate-free part of the sediments from the TAG hydrothermal field.

The contents of lithogenic Al and Ti in metalliferous and transitional sediments increase as the contents of Fe decrease. Where transition to non-metalliferous sediments takes place the content of Ti decreases slightly with increase in the Al content (Fig. 1.57) and this indicates a decrease in the amount of basic material in the lithogenic matter.

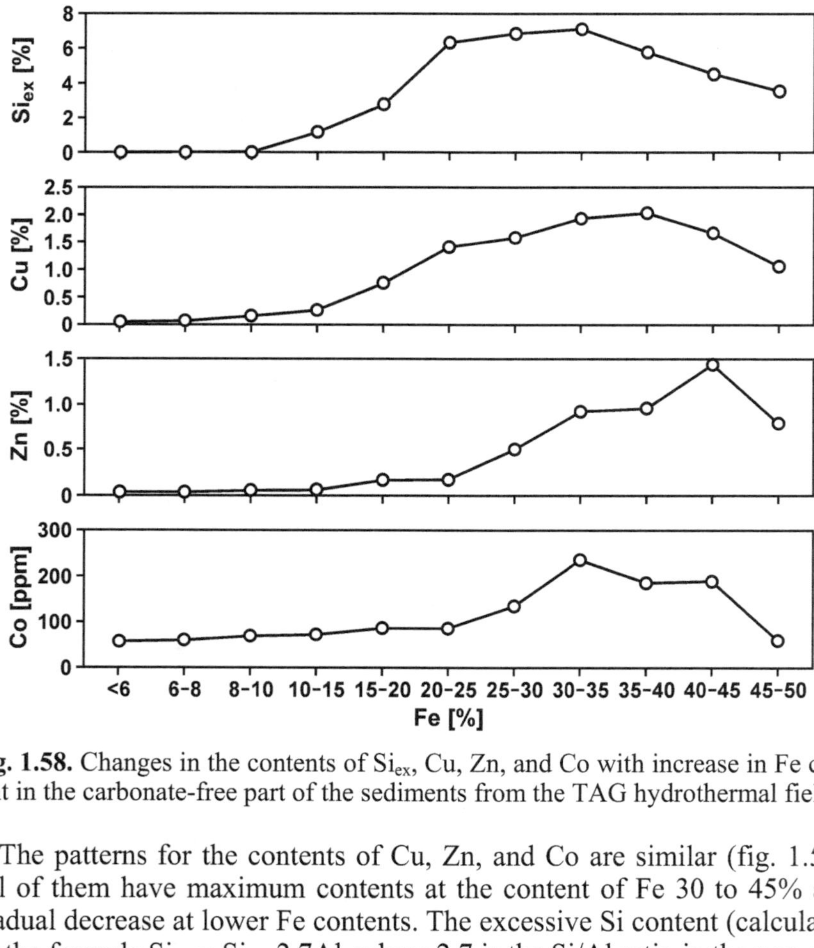

Fig. 1.58. Changes in the contents of Si_{ex}, Cu, Zn, and Co with increase in Fe content in the carbonate-free part of the sediments from the TAG hydrothermal field.

The patterns for the contents of Cu, Zn, and Co are similar (fig. 1.58). All of them have maximum contents at the content of Fe 30 to 45% and gradual decrease at lower Fe contents. The excessive Si content (calculated by the formula $Si_{ex} = Si - 2.7Al$, where 2.7 is the Si/Al ratio in the non-metalliferous sediments) has a similar pattern, but it is at a maximum when the content of Fe is from 20 to 35%. Patterns for the changes in contents of Pb and U (Fig. 1.59) are similar to the patterns for the contents of Cu, Zn, and Co, but there are no data for the Pb and U contents in the metalliferous sediments containing >41.3% Fe.

The gradual increase to maximum contents when the content of Fe is 8 to 20%, and decrease in content when the Fe content is higher, are common features in the patterns for the contents of Mn, Ba, La, and Th.

The highest contents of P and V occur in sediments with the maximum contents of Fe and they decrease to the point where the Fe content is 30 to 35%, increase slightly and show minor changes when the contents of Fe are 15–25%, and then decrease gradually in content at lower contents of Fe (Fig. 1.60).

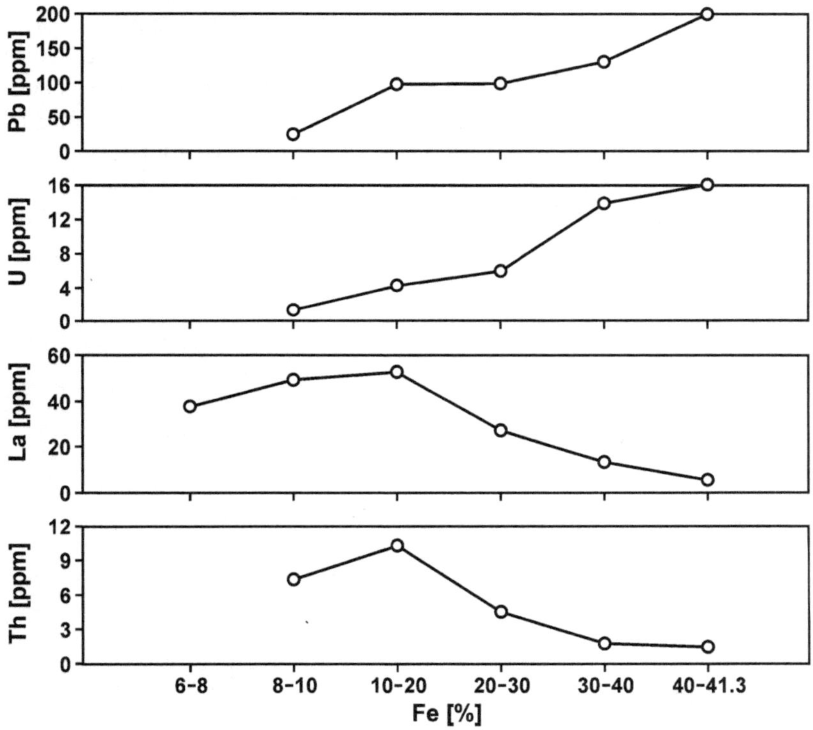

Fig. 1.59. Changes of the contents of Pb, U, La, and Th with increase in Fe content in the carbonate-free part of the sediments from the TAG hydrothermal field.

The patterns for changes in the contents of Ni and Cr are similar to those for the contents of Al and Ti (fig. 1.57). But there are distinctions. The Cr contents increase at the maximum Fe contents, like the P and V contents, but less pronounced. And the contents of Ni increase as the contents of Fe decrease from 25 to 8% and are more pronounced than the changes for Ti and Cr contents.

The reasons for variations in the chemical composition of the carbonate-free part of the sediments from the TAG hydrothermal field are shown clearly by factor analysis. The results from the study of 1058 samples of both metalliferous and non-metalliferous sediments have been used in the factor analysis that revealed four reasonable factors, which account for 95.1% of the variance (Fig. 1.61). The data for Pb, La, U, and Th were not used because of the considerably smaller number of data points available.

The first factor accounts for 79.0% of the variance in chemical composition. The most abundant element of the hydrothermal metal-bearing matter, Fe, and the excessive silicon (Si_{ex}) have the highest positive loadings. The positive loadings of two other main hydrothermal elements, Cu and Zn, are also significant. The main lithogenic elements, Al and Ti, and the metals Ni and Cr have the greatest negative loadings. This distribution of the loadings indicates that the variation in chemical composition of the carbonate-free matter of the TAG sediments depends mainly on the mutual dilution of the hydrothermal and lithogenic matter and that basic material enriched in Ti, Ni, and Cr is a predominant component in the lithogenic material. These results are in good agreement with the results from lithologic studies (see above). The prevalence of basic material also accounts for the higher average contents of Ti and Cr in the metalliferous sediments from the TAG hydrothermal field in comparison with metalliferous sediments from the Southeast Pacific despite the higher content of the metal-bearing matter in the TAG sediments.

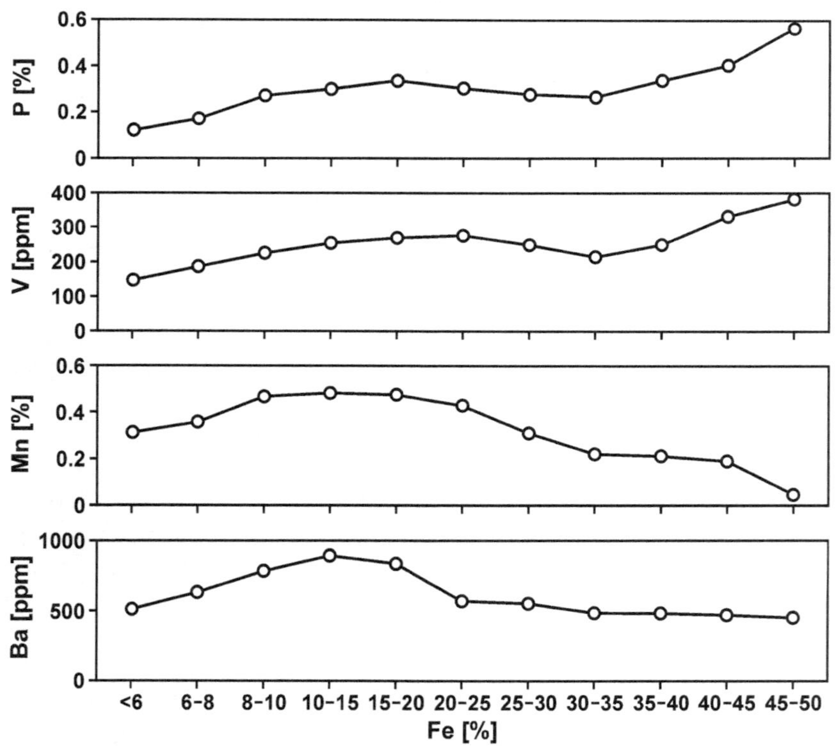

Fig. 1.60. Changes in the contents of P, V, Mn, and Ba with increase of Fe content in the carbonate-free part of the sediments from the TAG hydrothermal field.

The second factor accounts for 8.6% of the variance. V and P have high positive loadings. The positive loadings of Fe and Cr are also evident. Such association of the elements is typical for the matter formed in buoyant hydrothermal plumes where there is co-precipitation of V, P, and Cr with Fe-hydroxides (Rudnicki and Elderfield 1993; Feely et al. 1994; Sect. 4.3). The rather low positive loading of Cr apparently is because it is mainly present in the lithogenic material (see the first factor).

The third factor accounts for 4.5% of the variance. Cu, Zn, and Co have the highest positive loadings and the positive loading of Fe is also evident. The lithogenic elements Al and Ti have negative loadings. This association of the elements indicates the presence of hydrothermal sulfide material and/or products of its alteration in the sediments.

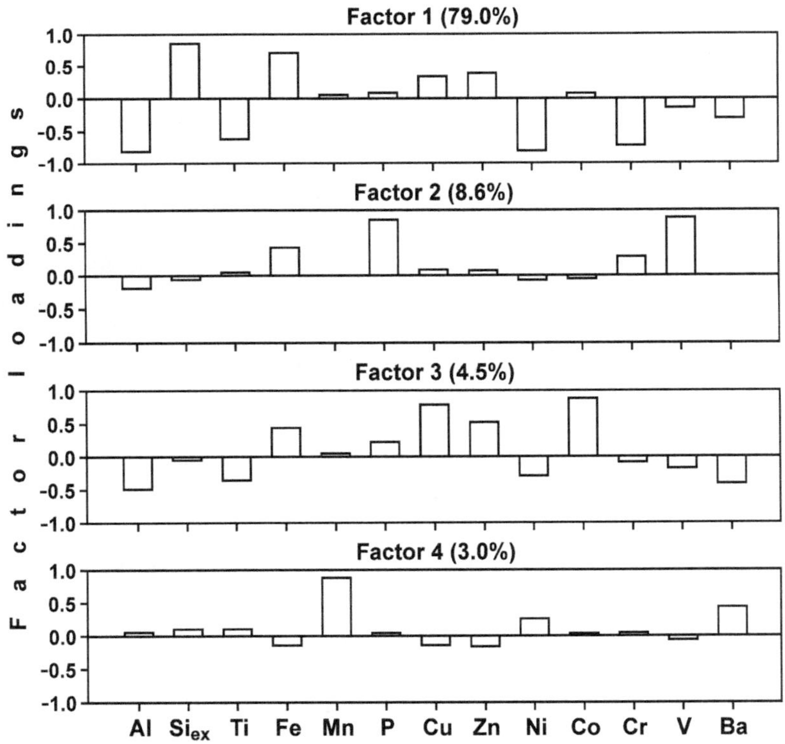

Fig. 1.61. Factor loadings of four varimax rotated factors for the carbonate-free element contents in sediments from the TAG hydrothermal field.

The fourth factor accounts for 3.0% of the variance. Mn and Ba have high positive loadings; and Mn and Ba have positive loadings and that of Ni is evident. This association of the elements indicates the presence of material that settled from nonbuoyant hydrothermal plumes (see Sect.

4.3.2). Ba has geochemical affinity with Mn especially in hydrothermal sedimentary formations but not with Fe (Fisher and Puchelt 1972; Gurvich et al. 1979). The rather low positive loading of Ni apparently is because of its presence in the lithogenic material (see the first factor).

The changes in the patterns of the element contents related to increases in the contents of Fe in the carbonate-free part of the metalliferous sediments from the TAG hydrothermal field (Fig. 1.57–1.60) and the results of the factor analysis indicate that the matter formed in the buoyant hydrothermal plume has the greatest influence on the composition of the metalliferous sediments that contain 40 to 50% Fe. The hydrothermal sulfide material and/or the products of its alteration have significant influence on the composition of the sediments that contain 30–45% Fe. The content of uranium in the hydrothermal sulfide material is insignificant but it enriches the products of oxidation of the sulfide material in the sediments after their deposition entering from ocean water due to bacterial activity (German et al. 1993; Mills et al. 1993, 1994). The matter formed in the nonbuoyant plume has the greatest influence on the composition of the sediments containing 8 to 25% of Fe.

In the metalliferous sediments that have different contents of Fe the contents of REE differ and there are also different patterns of REE (Fig. 1.62). In the non-metalliferous sediments and in the metalliferous sediments containing 10–20% of Fe, which are most enriched in REE, there is a small negative Ce anomaly that is a typical feature of the REE patterns in the ocean water and in particulate matter of the buoyant and nonbuoyant hydrothermal plumes, and a small enrichment of the middle REE is characteristic. With an increase in the content of Fe the contents of REE decrease, the negative cerium anomaly persists and a positive Eu anomaly appears. With a further increase in the content of Fe, at least up to 41.3%, the positive Eu anomaly also increases. This is a typical feature in the REE patterns of high-temperature hydrothermal fluids, sulfide and Fe oxide materials from the active hydrothermal mound, and also of particulate matter from the buoyant hydrothermal plume (Fig. 1.62).

It is important to note, that even the lowest contents of REE in the most metalliferous (containing >40% of Fe) sediments are much higher than in the massive sulfides and products of their oxidation from the active hydrothermal mound. These data and ratios of the REE contents to the contents of Fe (Fig. 1.63) show that the hydrothermal sulfides and products of their oxidation are only of minor importance as a source of the REE not only in the low metalliferous sediments, the most enriched in REE, but also in the most metalliferous sediments. For the most metalliferous sediments the particulate matter from the buoyant hydrothermal plume apparently plays an important role in REE supply. In this matter the ratios of the REE contents to the Fe content as well as the positive Eu anomaly are comparable with ones in the most metalliferous sediments (Fig. 1.63). The role of the particulate matter from the buoyant plume is also significant in the REE supply to the metalliferous sediments containing 30–40% of Fe. As to less metalliferous sediments and non-metalliferous sediments having 7–10% of Fe, the REE/Fe ratios as well as the Eu anomaly observed in these sediments (Fig. 1.63) indicate that the particulate matter from the nonbuoyant plume plays the main role in the REE supply to these sediments.

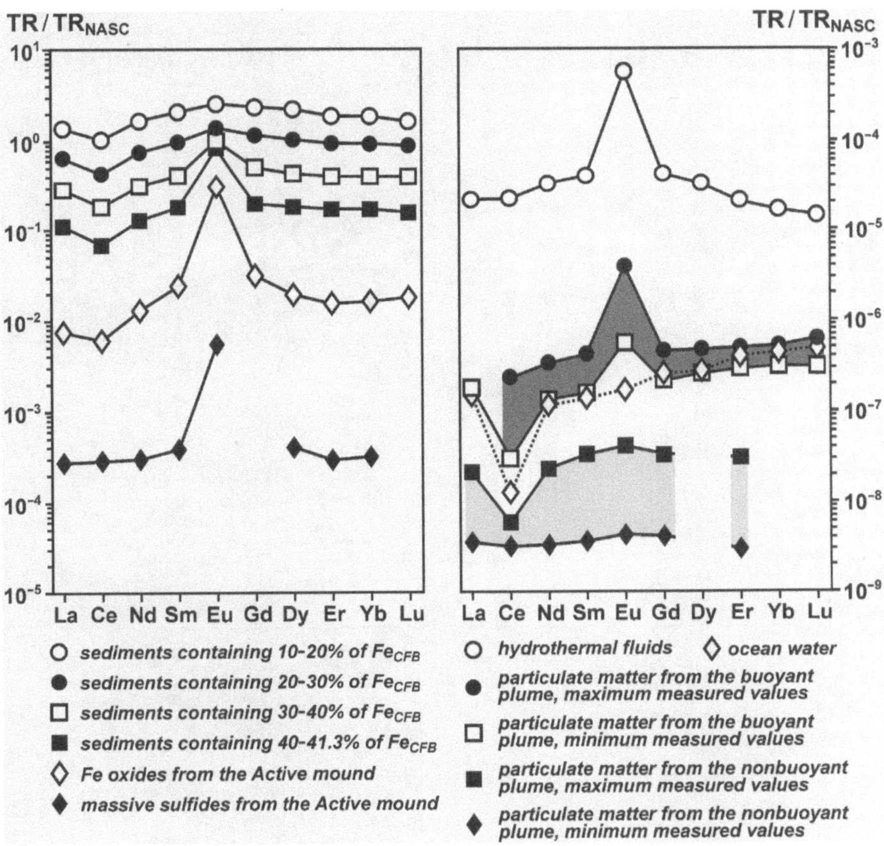

Fig. 1.62 Shale-normalized REE patterns of the metalliferous sediments (carbonate-free matter), massive sulfides and products of their oxidation (from the Active hydrothermal mound), primary hydrothermal fluids, and particulate matter of the hydrothermal plumes from the TAG hydrothermal field, as well as in the background ocean water. Based on data from German et al. (1990, 1993), Mills (1992), Dekov and Gurvich (1993), and Mitra et al. (1994).

Study of cores of the metalliferous sediment from the TAG hydrothermal field has shown considerable downcore variations, which mainly reflect variations in accumulation rate of the hydrothermal metal-bearing matter, and also a history of evolution of hydrothermal mounds if the cores have been collected nearby (Shearme et al. 1983; Metz et al. 1988; Lisitzin et al. 1990; Bogdanov et al. 1994, 1995a,d). As to variations in the accumulation rate of the background sedimentary material in the absence of redeposition during the time period recorded in the cores (up to 30 ka BP) it changed twice (Lukashina 1993).

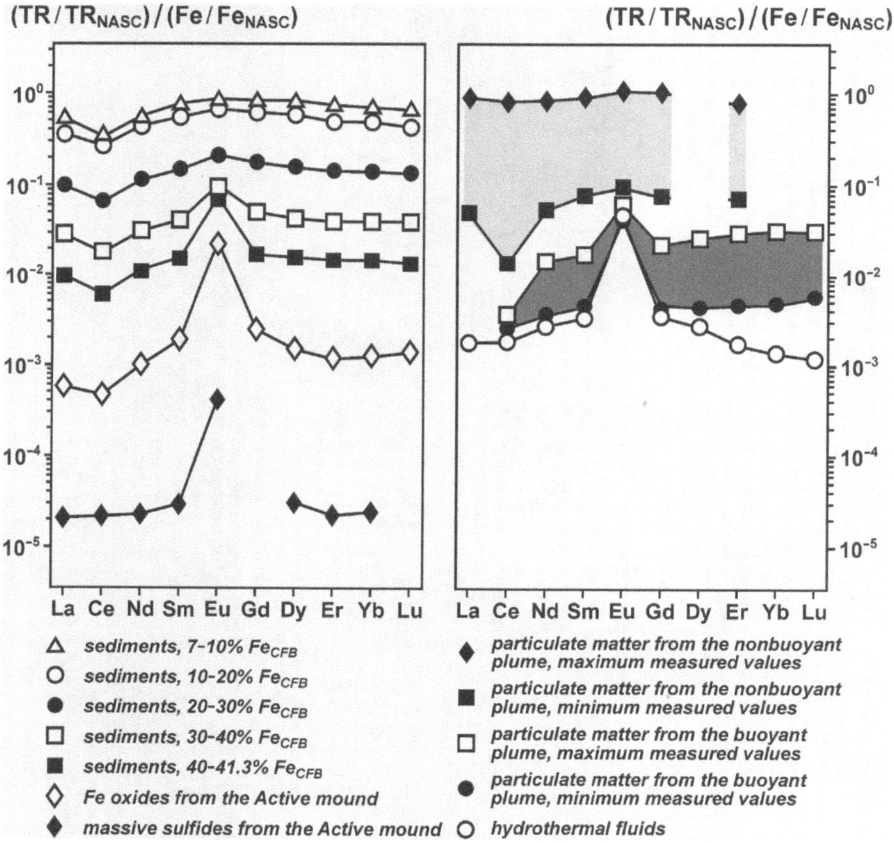

Fig. 1.63. Ratios of the shale-normalized REE to Fe concentrations in the metalliferous sediments, massive sulfides and products of their oxidation, primary hydrothermal fluids, and particulate matter of the hydrothermal plumes from the TAG hydrothermal field.

As an example the downcore variations in compositions of sediments from Core 3366 collected near the southern part of the chain of relict hydrothermal mounds of the Alvin Zone, and from Core 3360 collected in the vicinity of the southern part of the Alvin Zone (Fig. 1.56) can be shown (Fig. 1.64). The downcore variations in the chemical composition of the sediments distinctly reflect ones in their lithologic composition.

The lithologic composition of sediments from Core 3366 correlates with one from Core M3 (Fig. 1.65) studied by Metz et al. (1988). The core M3 has been collected at a greater distance from the southern part of the Alvin Zone (Fig. 1.56), which has been a source of the metal-bearing matter for sediments from the both cores.

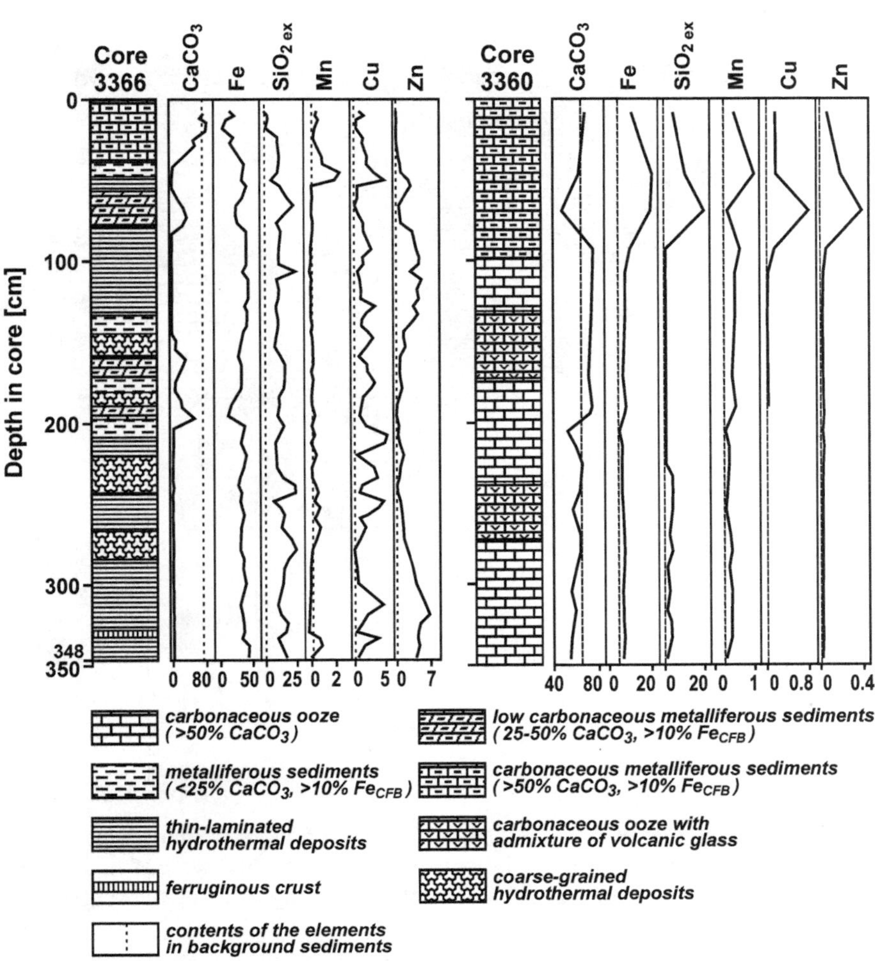

Fig. 1.64. Lithologic and chemical compositions of sediments from Cores 3366 and 3360. The core positions see at Fig. 1.56. Contents in %. Contents of $CaCO_3$ – in dry sediments, of other components – in carbonate-free matter.

Due to closer position of Core 3366 to the source it has greater thickness of fine-grained thin-laminated hydrothermal deposits, which have accumulated as a result of horizontal transport of the hydrothermal material. Besides, in Core 3366 there are coarse-grained hydrothermal deposits accumulated at "hydrothermal explosion". Such deposits are absent in Core M3, but it also has redeposited material (it is indicated by an inversion in sediment ages in the lower part of the core). And thicknesses of layers composed mainly of non-hydrothermal material, which accumulation depends in a lesser degree on distance from the hydrothermal source, in both cores are very similar.

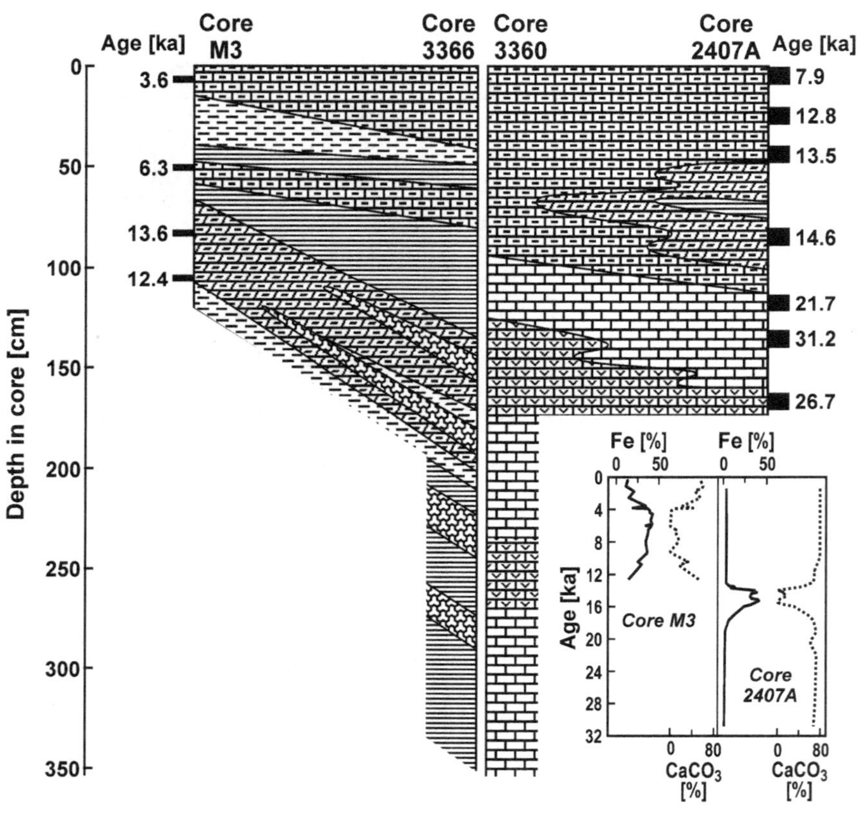

Fig. 1.65. Correlation of lithologic compositions of bottom sediments from Cores M3 and 3366 and from Cores 2407A and 3360. Legend is the same as in Fig. 1.64. Contents of $CaCO_3$ and Fe – in dry sediments. After Bogdanov et al. (1995a).

Lithologic composition of Core 3360 correlates with one of Core 2407A (Fig. 1.65) collected much closer to the source of the hydrothermal material – the northern part of the Alvin Zone (Fig. 1.56).

Owing to greater distance from the source of the hydrothermal material there are only carbonaceous metalliferous sediments in Core 3360, whereas in Core 2407A there are also low-carbonaceous metalliferous sediments and thin-laminated hydrothermal deposits. Due to higher content of the hydrothermal matter, thickness of the metalliferous sediment layer in Core 2407A is rather more than in Core 3360.

Average rates of the excessive accumulation of a number of chemical elements in the metalliferous sediments from the TAG hydrothermal field formed "grain-by-grain" are shown in Table 1.11. For comparison in this table also the data on the metalliferous sediments from the Southeast Pacific and from the ERP axial zone at 20°30'S–22°S are given.

Table 1.11. Average rates of the excessive accumulation of chemical elements in metalliferous sediments from the TAG hydrothermal field, Southeast Pacific, and 10–20 km to the east of the East Pacific Rise axis at 20°30'S–22°S. Calculated from the data of Bogdanov et al. (1979c), Dekov (1994), and Gurvich (1998).

Area	Fe	Mn	Cu	Zn	Co	Ni	V	Ba
	mg cm^{-2} ka^{-1}		μg cm^{-2} ka^{-1}					
TAG	50	1.5	2400	370	39	18	90	240
Southeast Pacific	5.5	1.8	39	12.5	8	30	15	200
EPR 20°30'S–22°S	50	15	130	50	13	100	130	220

The average rate of the excessive accumulation of Fe in the metalliferous sediments from the TAG hydrothermal field is an order higher than one in the metalliferous sediments from the Southeast Pacific; however it is similar to the average value for the metalliferous sediments from the EPR axial zone at 20°30'S–22°S. The rates of excessive accumulation of Cu, Zn, and Co entering the TAG metalliferous sediments mainly with the hydrothermal sulfides and products of their oxidation are considerably higher than in the metalliferous sediments from two other areas. The rates of excessive accumulation of Ba in all three areas are almost equal. This, probably, indicates the constant rate of removal of this element from the ocean water by means of scavenging by Mn-oxyhydroxides. The rate of excessive accumulation of Ni (entering metalliferous sediments mainly due to scavenging from the ocean water by particles of Fe-oxyhydroxides) in the TAG metalliferous sediments is similar to one in the metalliferous sediments from the Southeast Pacific, but is lower than in the metalliferous sediments from the EPR axial zone at 20°30'S–22°S. The ratios Ni_{ex}/Fe_{ex} are 0.00036, 0.0055, and 0.002 respectively (Table 1.11). It indicates that time of contact of the Fe-oxyhydroxide particles with ocean water within the TAG hydrothermal field is the shortest among three areas in question.

The rate of the excessive accumulation of V (entering metalliferous sediments mainly due to co-precipitation with Fe-hydroxides) in the TAG metalliferous sediments is higher, than in the metalliferous sediments from the Southeast Pacific, but a bit lower than in the metalliferous sediments from the EPR axial zone at 20°30'S–22°S. The V_{ex}/Fe_{ex} ratios in the metalliferous sediments from the Southeast Pacific and from the EPR axial zone at 20°30'S–22°S are practically identical 0.0027 and 0.0026, and in the TAG metalliferous sediments the ratio is 0.0018. If it is assumed that within the TAG hydrothermal field in the hydrothermal sulfides and Fe-hydroxides (formed at mixing of the hydrothermal fluids with the ocean water) the average V_{ex}/Fe_{ex} ratios are 0.0001 (Mottl and McConachy 1990; Feely et al. 1994) and 0.0026 respectively, and also if one neglects the sorption of V from the ocean water by the Fe-hydroxides and by the products of the sulfide oxidation then in the metalliferous sediments formed "grain-by-grain" about two thirds of the excessive Fe is bound with particles of "primary" Fe-hydroxides, and about one third with the sulfide particles and products of their oxidation. At distances from hydrothermal orifices the role of the heavy sulfide minerals decreases, due to their sinking, and that of Fe-oxyhydroxides increases. The prevalence of the Fe-oxyhydroxides is typical for the sediments.

1.6.2. Metalliferous sediments from the MARK hydrothermal field

The MARK hydrothermal field[12] is located 30 km south of the Kein transform fault within a large dome in a neovolcanic zone in the axial part of the Mid-Atlantic Ridge rift valley. In the field area the valley, 10 to 17 km wide, extends close to the meridian direction (~10°NE) (Detrick et al. 1988). The neovolcanic zone represents a ridge structure that rises approximately 500 m on the east side and 900 m on the west side above the valley floor. The youngest volcanic rocks at the top of the ridge are almost free of sedimentary material and the pillow-lavas of the lower part of the slope have a thin cover of pelagic sediments. It appears that in the area of the MARK field the volcanic dome is in an active phase of growth over a magmatic cell located under the ridge (Karson et al. 1987; Zonenshain et al. 1992). There is a graben in the axial part of the ridge that extends parallel to the meridian (Mevel et al. 1989; Gente et al. 1991). Manifestations of metallic mineral deposition have been found in the MARK hydrothermal field within this graben or in the vicinity of it in the zone of tectonic dislocation. The field is 400 m long and 125 to 250 m wide and it is elongated east-northeasterly.

The information on metalliferous sediments from the MARK hydrothermal field is inadequate. Because the field is very young, about 4–5 ka, many attempts to obtain samples have been unsatisfactory or sediment cores

[12] The field is called Snake Pit in some reports.

have been very short and only centimeters in length (Lalou et al. 1990, 1993). The most successful sampling of the field was carried out in Cruise 3 of R/V *Professor Logachev* that provided data for outlining the distribution of the metalliferous sediment in the central part of the field as shown in Fig. 1.66.

The areal distribution of the metalliferous sediments and massive sulfides within the MARK field is only about 0.06 km^2, and approximately 300 times less than that of the TAG hydrothermal field.

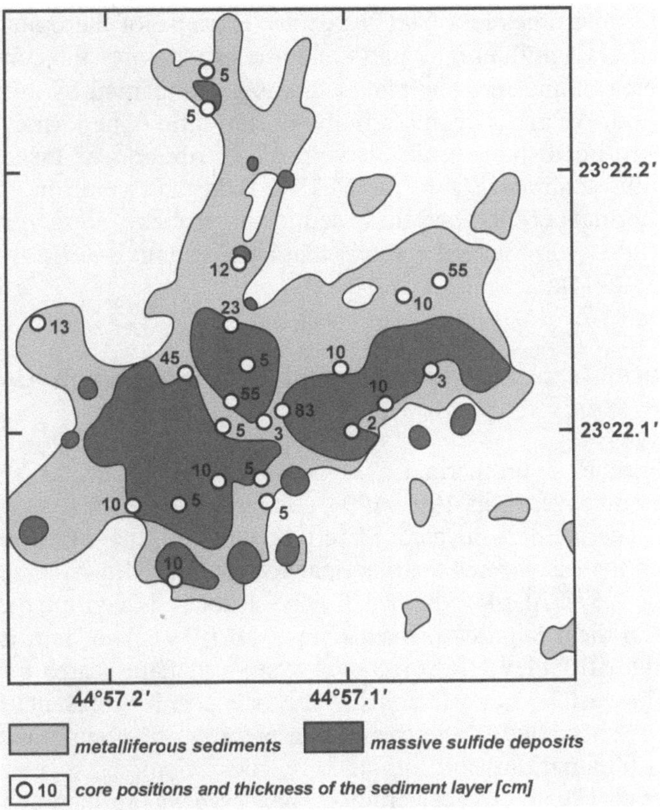

Fig. 1.66. The location of the metalliferous sediments in the central part of the MARK hydrothermal field. After Cherkashev (1995).

The thickness of the metalliferous sediment layer in the MARK field is relatively small and 19 of 24 core samples are less than 15 cm long; and among them cores less than 10 cm in length are prevalent. Of the 5 cores that are more than 15 cm long, four of them including the longest one, 83 cm in length, were collected at the base of the hydrothermal edifices. The cores collected from the edifices have not exceeded 10 cm in length (Fig.

1.66). This indicates redeposition of the metalliferous sediments from the slopes of the edifices at their base.

The metalliferous sediments in the MARK hydrothermal field are mainly proximal and composed of a significant amount of sulfide material. Upper parts of some cores are composed of oxidized sulfide material. The sediments are rich in Fe, 22–47%. The average contents of chemical elements are as follows: Fe 38.3%, Cu 6.6%, Zn 1.3%, Au 0.6 ppm, Ag 4.3 ppm (Cherkashev 1995). The metalliferous sediments from the TAG hydrothermal field that have a similar content of Fe also have a similar content of Zn, but three times less Cu (Fig. 1.58). Because of the oxidation of the sulfide materials in the upper parts of some of the cores the contents of Fe and S decrease; and in some places this is accompanied by enrichment in the content of Au and a decrease in the Eu/Sm ratio (Cherkashev 1995).

There are no distinct peaks in vertical distribution of the elements in metalliferous sediments from the MARK field that would indicate events of hydrothermal activity, and these sediments probably formed in the course of a single hydrothermal episode that began about 4–5 ka ago (Cherkashev 1995, 2004).

1.6.3. Metalliferous sediments from the Logachev hydrothermal field

The Logachev hydrothermal field located at 14°45'N on the Mid-Atlantic Ridge was discovered in 1993–1994 (Batuyev et al. 1994; Krasnov et al. 1995a,b). The field is located 35 miles south of the transform fault at 15°20'N on the east side of the marginal scarp of the Mid-Atlantic Ridge rift valley (Fig. 1.67). At 14°45'N the rift zone structure is asymmetric and similar to one in the TAG area (Zonenshain et al. 1989). The "normal" western slope of the rift valley is formed by a system of fault scarps composed of basalts. The eastern slope is considerably steeper. It represents a marginal scarp of the rift valley where there is an outcrop of serpentinite rocks, serpentinized ultramafic rocks, and gabbroic rock (Geologic research ... 1991, Krasnov et al. 1995a). The eastern marginal scarp was formed by the vertical uplift of a large block of the ocean crust developed by the serpentinization of ultramafic rocks of the lower crust and upper mantle zone and the intrusion of serpentinite rocks. Gentle slopes are covered by a layer of carbonaceous sediments that is up to 2.5 m thick.

The Logachev hydrothermal field is located within a zone of serpentinized ultramafic and gabbroic rocks (Bogdanov et al. 1995c; Krasnov et al. 1995). This essentially distinguishes the field from the majority of known hydrothermal fields connected with axial basalt hosted circulation systems, in which hydrothermal mineral-forming solutions are formed by the interaction of ocean water with basalt and dolerite rocks in the roof zones of magma chambers at depths of 1.5 to 3 km below the ocean floor.

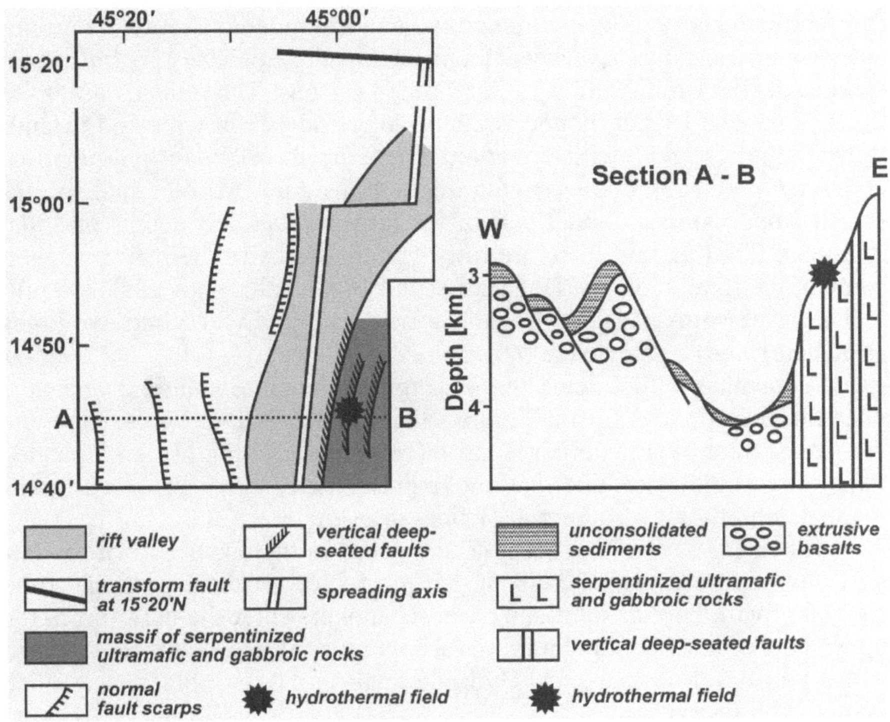

Fig. 1.67. Axial part of the Mid-Atlantic Ridge in the area of the Logachev hydrothermal field. After Bogdanov et al. (1995c).

The Logachev hydrothermal field is connected with a deeply circulation system (Bogdanov et al. 1995c, 1997a; Bogdanov and Sagalevich 2002). The existence of such hydrothermal circulation systems in rift zones of slow-spreading ridges was considered in theory previously (Rona et al. 1987; Lisitzin et al. 1990). In such circulation systems the ocean water penetrates much deeper than in the axial basalt hosted systems, probably to the lower part of the ocean crust und to the upper mantle zone, and is involved in the serpentinization of the ultramafic rocks. The hydrothermal solutions that deposit metals are developed in the deeply circulation hydrothermal system that differs from those in the axial basalt hosted systems (Charlou et al. 2002; Douville et al. 2002). Significant changes in hydrostatic pressure during the migration of the solutions through the earth's crust to the surface of the ocean floor result in the repeated phase separation of the solutions and in the boiling of the solutions at various levels near the surface of the crust (Bogdanov et al. 1995c, 1997a).

According to Lalou et al. (1996), hydrothermal activity within the Logachev field has existed at least during 60 ka, according to later information of Cherkashev (2004), during 66.5 ka. At the present time the major part of the field is inactive. Twelve mounds of sulfide minerals and two zones of sulfide mineral talus have been found within the Logachev hydrothermal field at depths from 2900 m to 3060 m (Fig. 1.68). The mounds are up to 20 m high. The largest mound is 200 m long and 125 m wide and extends down-slope in a northwest direction. There are three active hydrothermal zones up to 10 m in diameter with high-temperature hydrothermal springs (black smokers) in the axial part of the largest mound at depths of 3005, 2960 and 2940 m, and there are low-temperature hydrothermal springs in some of the small mounds (Bogdanov et al. 1995c,e; Krasnov et al. 1995a).

The black smokers in the Logachev field can be divided into two kinds (Bogdanov and Sagalevich 2002):

a) At depths of 3005 and 2960 m, high-temperature solutions enter the ocean from chimneys in small craters that are up to 20 to 50 cm high and composed of sulfide material. Hydrothermal plumes with black suspended matter are not buoyant and they move horizontally in the bottom currents at the depth of the hydrothermal orifices or even downhill.

b) Black smokers of the second kind occur in the upper part of the largest mound at a depth of 2940 m that are typical of those in ocean rifts. Buoyant hydrothermal solutions of these smokers enter the ocean from the upper part of a vertical column of sulfide material that is about 3 m high and 0.5 to 0.8 m in diameter. Hydrothermal plume with black suspended matter is buoyant.

The existence of two kinds of black smokers is an indication that the last boiling of the hydrothermal solutions during their migration through the ocean crust takes place in a subsurface zone near the ocean floor (Bogdanov et al. 1995c).

The accumulations of massive sulfide material in the Logachev field are enriched mainly in copper. They are enriched also in Co and As in comparison with those from other hydrothermal fields in the Mid-Atlantic Ridge rift zone (Bogdanov et al. 1997; Cherkashev 2004).

There are few published data on metalliferous sediments from the area of the Logachev hydrothermal field[13]. There is a small field, about 0.5 km^2,

---

[13] In 1995 Alexander Ashadze showed me numerous data and charts of element distributions in sediments within and around the Logachev field. Unfortunately these data were never published and no one has worked with them since his death. Some unpublished data on the Logachev field sediments were kindly made available to me during preparation of this book by G.A. Cherkashev and V.V. Shilov from the All-Russia Institute for Geology and Mineral Resources of the World Ocean (VNIIOkeangeologiya), Sankt-Petersburg.

of metalliferous sediments around the hydrothermal mounds (Fig. 1.68). In the proximity, within tens of meters of the mounds, sediments composed of sulfide material have accumulated (Krasnov et al. 1995a; Sudarikov and Roumiantsev 2000). Metalliferous sediments with a predominance of Fe oxyhydroxide material in their metal-bearing part occur at distances up to 200 to 300 m from the mounds and at greater distances from them the sediments are non-metalliferous. But at a distance of about 5 km to the north of the Logachev field sediments have been sampled that contain from 9.0 to 9.6% Fe_{CFB} (Cherkashev 1995).

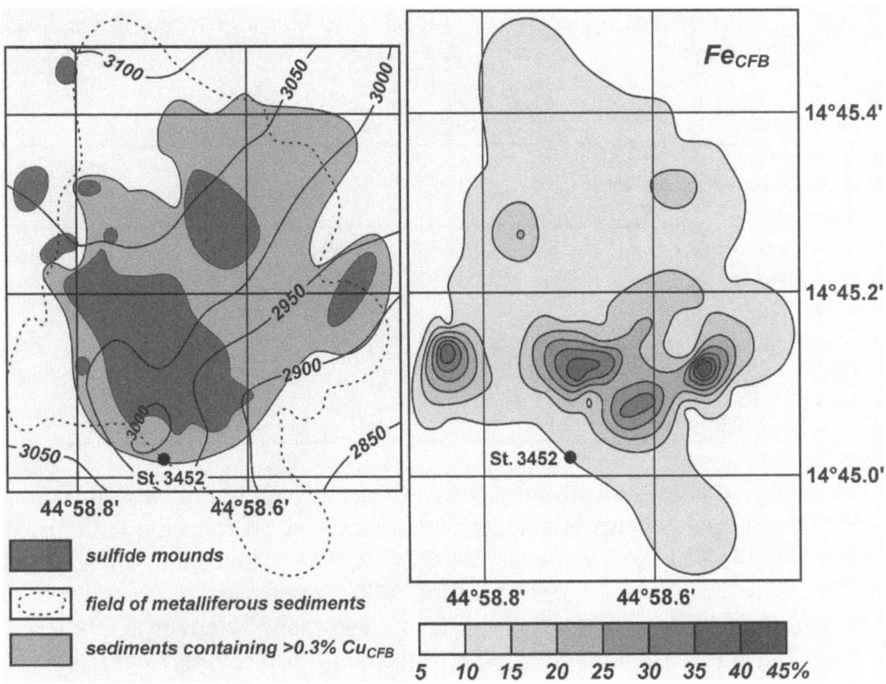

Fig. 1.68. Mounds of sulfide minerals and metalliferous sediments in the Logachev hydrothermal field. Compiled from data of Krasnov et al. (1995), Sudarikov and Roumiantsev (2000), and recent data of VNIIOkeangeologiya. St. 3452 – position of Station 3452 of R/V *Akademik Mstislav Keldysh.*

According to results obtained during studies of VNIIOkeangeologiya, the main factors controlling sedimentation process within the Logachev hydrothermal field are: hydrothermal activity, accumulation of biogenic carbonate material, bottom currents, and slumps.

The content of biogenic carbonate material in sediments of the field varies from 0 to 78.7%, the average is 33.1%. The average carbonate-free contents of the main elements of the metal-bearing matter in sediments from the Logachev field vary with the different content of Fe (Table 1.12). The metalliferous sediments are enriched in Cu and its contents are higher compared to those in sediments from the TAG hydrothermal field with similar contents of Fe (Fig. 1.58). The contents of Zn are much lower than the contents of Cu. Apparently this is because of the predominance of Cu over Zn in the massive sulfide material in the Logachev field (Cherkashev 2004). The contents of both elements in metalliferous and transitional sediments increase with increase in the content of Fe. The content of Mn in metalliferous sediments decreases as the content of Fe increases.

Table 1.12. Average carbonate-free contents of chemical elements in the Logachev hydrothermal field sediments with different contents of Fe and in background sediments. Based on data from Gurvich et al. (1995a), Cave et al. (2002), and on the unpublished data of VNIIOkeangeologiya.

Sediments / Fe content	Fe	Cu	Mn	Zn
		%		
metalliferous	19.4	1.6	0.40	0.051
Fe 30÷34.2%	32.2	5.2	0.22	0.078
Fe 20÷30%	25.5	3.3	0.35	0.061
Fe 10÷20%	15.1	0.72	0.47	0.040
Fe 6÷10%	7.3	0.20	0.31	0.038
background	4.9	0.022	0.28	0.0116

A sediment core, 241 cm long, was collected close to the base of the large hydrothermal mound of sulfide material, about 80 m to the south of it, at Station 3452 of R/V *Akademik Mstislav Keldysh*, located at 14°45.02'N, 44°58.70'W, at a depth of 2980 m (Fig. 1.68). Apparently the hole penetrated, and the core is a sample of, the whole sedimentary cover and where it stopped in the hard substratum. The following description of this core is based on the reports of Bogdanov et al. (1998) and Bogdanov and Sagalevich (2002).

There are two distinct layers in the core (fig. 1.69). The upper layer is 46 cm thick and is composed of high-carbonaceous (69.7 to 78.7% $CaCO_3$) sandy metalliferous sediments that contain 10.1 to 10.2% Fe_{CFB}. Shells of planktonic foraminifera are predominant in the carbonaceous material. The metal-bearing matter is composed mainly of Fe-oxyhydroxide and Fe-montmorillonite minerals and there are lower contents of nontronite. Aggregates and individual crystals of sulfide minerals have also been found. Sediments in the lower layer, 46–241 cm, are much finer grained than the sandy sediments in the upper layer (Fig. 1.69). They are non-metalliferous and carbonaceous with 53.2 to 61.2% $CaCO_3$. The carbonaceous part is composed of coccoliths mixed with a significant amount of foraminifera shells. The non-metalliferous sediments of the lower layer also contain hydrothermal metal-bearing material but the contents of it are lower than in the metalliferous sediments in the upper layer.

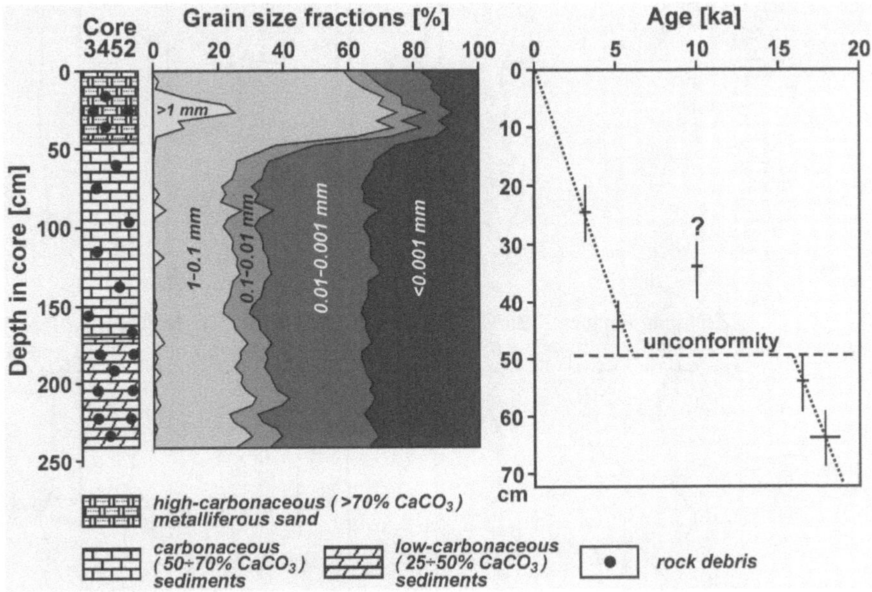

Fig. 1.69. Composition and age of sediments from Core 3452. After Bogdanov et al. (1998). Position of the core station is shown in Fig. 1.68.

Numerous angular and subrounded fragments of crystalline rocks up to 3 cm in size are distributed throughout the whole core. They are most abundant, up to 25.8%, within the 15 to 20 cm part of the core. Serpentine is the main lithogenic mineral in the coarse-grained fraction of the sediments because of proximity to outcroppings of serpentinite rock or the presence of serpentinized deep-seated rocks. The content of serpentine material in the core increases with depth, apparently because of the gradual burial of the basement rocks by the sediments. Products of the destruction of gabbroic rocks occur throughout the whole core as well. Products of the destruction of basalt rocks are rare.

The lithologic boundary between the upper and lower layers corresponds to the age boundary between Würmian and Holocene time (Fig. 1.69). There is an unconformity between Würmian and Holocene sediments caused either by washout, or a hiatus in sedimentation that lasted for about 10 ka. The age of the sediments at depth 70 cm is about 20 ka. The average sedimentation rate for the Würmian sediments is about 7 cm ka^{-1}, and for the Holocene sediments, about 9 cm ka^{-1}.

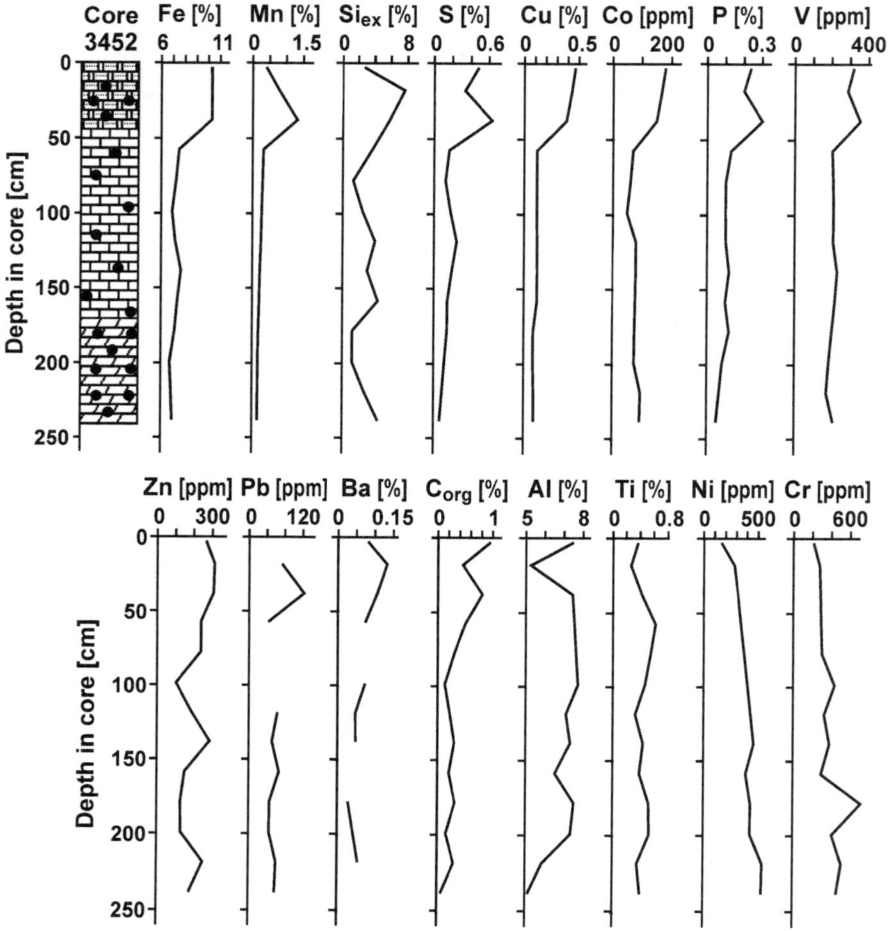

Fig. 1.70. Chemical composition of carbonate-free matter in sediments from Core 3452.

Distribution of carbonate-free chemical element contents in the sediments along Core 3452 is shown in Fig. 1.70. The metalliferous sediments in the upper layer of the core compared with those in the lower layer are enriched not only in Fe but also in Mn, S, P, C_{org}, Ba, Cu, Zn, Pb, Co, V, and Si_{ex} and depleted in Al, Ti, Cr, and Ni. The contents of Co, Ni, and Cr, the elements enriched in ultramafic rocks (Turekian and Wedepohl 1961; Vinogradov 1962), in the metalliferous sediments of the upper layer, are higher than in Atlantic pelagic clays and metalliferous sediments from the TAG hydrothermal field (Table 1.10). The high contents of Cr and Ni in non-metalliferous sediments in the lower layer are the result of significant intermixing of edaphogenic material from depth.

There is also a small amount of hydrothermal metal-bearing matter intermixed in the non-metalliferous sediments. Relative to the Atlantic pelagic clays they are enriched in S, Cu, and Fe, elements that are typically found in the massive hydrothermal deposits of the Logachev field.

The data for the chemical composition of the carbonate-free matter of the sediments from Core 3452 have been used for factor analysis with the varimax rotation of the factor matrix. Four reasonable factors account for 91.9% of the variance (Fig. 1.71).

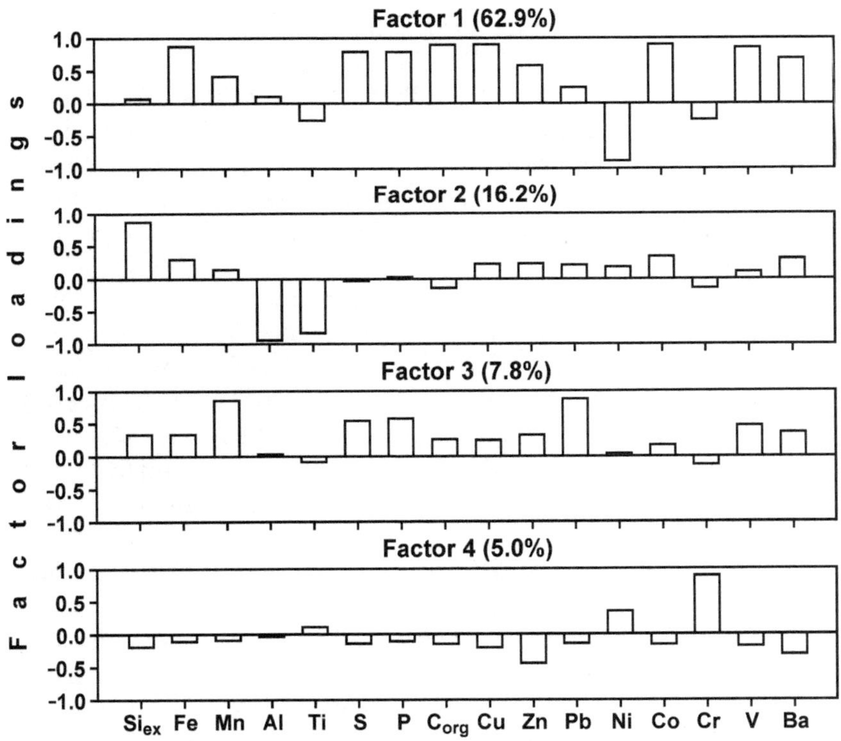

Fig. 1.71. Factor loadings of four varimax rotated factors for the carbonate-free element contents in the sediments from Core 3452.

The first factor accounts for 62.9% of the variance. Fe, S, P, C_{org}, Cu, Zn, Co, V, Ba, and Mn have significant positive loadings, Ni has a significant negative loading and Ti and Cr have marked negative loadings. The loadings of Al and Si_{ex} are close to zero. This factor shows that mutual dilution of lithogenic ultramafic matter and of components accumulated under the influence of hydrothermal process is the main reason for variations in the chemical composition. The influence of the hydrothermal process is reflected in: accumulation of hydrothermal components, components scavenged by the hydrothermal components from ocean water, and components combined with organic matter (sediments accumulated near hydrothermal vents and fields are enriched in organic matter).

The second factor accounts for 16.2% of the variance. Si_{ex} has a significant positive loading and Fe, Cu, Zn, Pb, Co, Ni, and Ba have marked positive loadings. Al and Ti have significant negative loadings. This factor, probably, reflects some influence of the mutual dilution of hydrothermal matter and lithogenic matter (background pelagic material and edaphogenic basic material).

The third factor accounts for 7.8% of the variance. The positive loadings of Mn, Pb, Si_{ex}, Fe, S, P, C_{org}, Cu, Zn, Co, V, and Ba, probably, indicate some role of low-temperature hydrothermal minerals as carriers of these elements. Low-temperature hydrothermal deposits from the Logachev field have notable contents of these elements (Bogdanov et al. 1997b, 2004).

The fourth factor accounts for 5.0% of the variance. A high positive loading of Cr and a notable loading of Ni apparently reflect the presence of ultramafic material.

The analysis of mineral and chemical compositions shows that hydrothermal components occur in the sediments throughout the whole core 3452. It indicates that hydrothermal activity at the Logachev field has existed much longer than 20 ka. According to data of Lalou et al. (1996), the oldest dated hydrothermal deposits from the field formed at least 60 ka BP. According to results obtained during studies of VNIIOkeangeologiya Institute, hydrothermal material has been diagnosed in sediments that are 128 ka to 170 ka of age.

It is important to note that no cyclicity in temporal evolution of the chemical composition of the core sediments and in the accumulation of the sulfide deposits at the Logachev field have been documented. This is not a typical feature in the development of hydrothermal process in the axial parts of slow-spreading ridges where cyclicity often exists with a period of 5 to 10 ka. During these cycles phases of hydrothermal activity alternate with phases when hydrothermal activity completely ceases (Bogdanov 1992; Bogdanov and Sagalevich 2002).

According to the results of the studies of Core 3452 (Fig. 1.69 and 1.70), a strong increase in hydrothermal activity took place about 10 ka ago. This increase was accompanied by or was simultaneous with considerable intensification of the water dynamics near the seafloor. A strong influence by the bottom currents on the Logachev field sediments, aged 0–11 ka, has been shown also during the studies of VNIIOkeangeologiya Institute. Most of the fine-grained sedimentary material including dispersed hydrothermal metal-bearing matter was removed by bottom currents and sandy material that was depleted in fine hydrothermal components accumulated on the floor. If it is assumed that, without the removal of the fine-grained sedimentary material by the bottom currents, the grain size composition of the uppermost sediments of the core would be similar to that in sediments from the lower layer of the core and that the existing difference is caused only by the partial removal of the <0.1 mm fraction, one can estimate the amount that has been removed. The calculation for the data for the uppermost sediments of the core (60% of the fraction >0.1 mm, 40% of the fraction <0.1 mm) and the data for the lower layer of the core (25% of the >0.1 mm fraction and 75% of the <0.1 mm fraction) shows that the amount of material removed is 3.5 times greater than the part remaining and about 78% of the fine grained material (<0.1 mm) has been removed and about 22% of it has accumulated in the sediment on the seafloor. Apparently this accounts for the low contents of elements combined with the hydrothermal matter in the metalliferous sediments from Core 3452 that accumulated in the vicinity of the large mound composed of hydrothermal sulfide material (Fig. 1.68).

1.6.3. Metalliferous sediments near the Rainbow hydrothermal field

The Rainbow hydrothermal field (Fig. 1.72) was discovered in the axial part of the Mid-Atlantic Ridge on the western slope of the axial nonvolcanic Rainbow ridge (German et al. 1996a,b; Fouquet et al. 1997, 1998). The center of the field is located at 36°13.8'N, and 33°54.12'W, at depths between 2270 m and 2320 m. In the area of the field the internal rift of the Mid-Atlantic Ridge has structural elements that are typical of slow-spreading ridges.

The Rainbow ridge is not in a constructive volcanic edifice and it is composed mainly of serpentinite rocks. Fragments of rocks from all of the main layers of the ocean crust including gabbro, diabase, and basalt have been found on the surface of the ridge. These rock fragments are highly altered and deformed. The ridge has been formed by the protrusion and uplift of serpentinite and associated rocks to the bottom of the ocean (Barriga et al. 1997; Bogdanov et al. 1999, 2000, 2002). Modern volcanic activity and a subsurface magmatic chamber are absent (German et al. 1996b; Parson et al. 2000). This part of the rift appears to have been in a tectonic phase of evolution during the last 200 ka or longer (Bogdanov et al. 2002). The axial rift fracture formed by the spreading of the lithosphere plates has been filled by serpentinite rock and materials squeezed out from the lower part of the crust and upper part of the mantle and not by new volcanic material (Bogdanov et al. 2005).

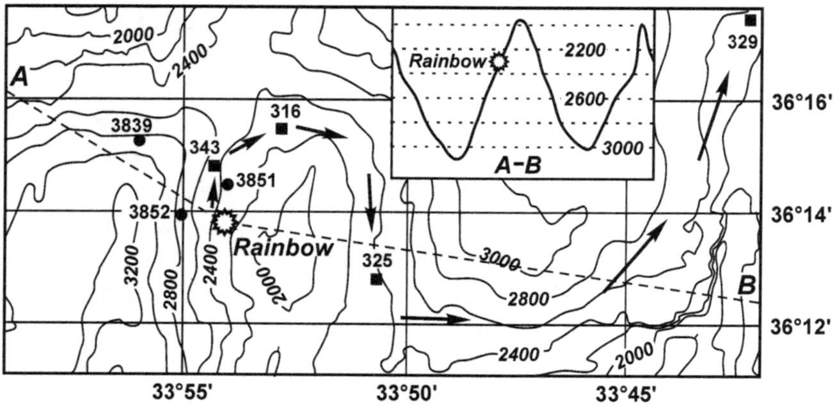

Fig. 1.72. Location of the Rainbow hydrothermal field and core stations. Compiled from Bogdanov et al. (2002) and Cave et al. (2002). Filled circles – stations of R/V *Akademik Mstislav Keldysh*, filled squares – stations of FS *Poseidon*. Arrows show the path of the hydrothermal plume (Thurnherr and Richards 2001).

Part of the ocean water that penetrates the crust through open fractures is taken up by the process of serpentinization of the ultramafic rocks at temperature from 350° to 500°C and part of the ocean water is transformed to high-temperature hydrothermal fluid (Barriga et al. 1997; Bogdanov 1997).

The Rainbow hydrothermal field is 250 m long and up to 100 m wide. It is the most intensive hydrothermal field known within the Mid-Atlantic Ridge (Parson et al. 1997; Fouquet et al. 1998). At least 10 groups of high-temperature black smokers occur within the field. At the Rainbow field, unlike the other known hydrothermal fields in the Mid-Atlantic Ridge except for the Logachev field, hydrothermal activity exists continually for more than 100 ka and did not cease at intervals of 5 to 10 ka (Bogdanov et al. 2002).

The Rainbow hydrothermal field like the Logachev field is connected to a deeply circulation system (Bogdanov et al. 1995c, 1997a, 2002, 2005; Barriga et al. 1997). Hydrothermal fluids and hydrothermal accumulations in the Rainbow field characteristically have high contents of Ni and Co that apparently were leached from ultramafic rocks. Hydrothermal fluids have very high concentrations of Fe (Charlou et al. 2002; Douville et al. 2002), H_2, CO, CH_4 and other hydrocarbons (Lein et al. 2000; Bortnikov et al. 2001; Holm and Charlou 2001; Bogdanov et al. 2002, 2005; Charlou et al. 2002; Douville et al. 2002).

In the area near the Rainbow hydrothermal field seven sediment cores up to 345 cm long have been collected (Fig. 1.72). Five cores have been collected directly below and along the path of the hydrothermal plume at distances of 1.2 to 25 km from the hydrothermal field. Sedimentation rates varied from 2.65 to 6.2 cm ka^{-1} throughout the Holocene period of time and between 1.7 and 10.1 cm ka^{-1} during the pre-Holocene period (Bogdanov et al. 2002; Cave et al. 2002). The sediments in the cores are oxidized and highly carbonaceous. The content of C_{org} varies from 0.08 to 0.46% and the content of $CaCO_3$ varies from 75.8 to 92.8% and averages 85.0% (Gurvich et al. 1998; Bogdanov et al. 2002; Cave et al. 2002). Lithogenic components consist mainly of eolian continental material and serpentinite fragments. Hydrothermal matter was found in all the cores, but it is not abundant. Holocene sediments from the upper parts of all the cores are metalliferous (Gurvich et al. 1998; Bogdanov et al. 2002; Cave et al. 2002). Also sediments from Core 3851 in the part from 195 to 250 cm and from Core 3852 in the part from 185 to 235 cm are metalliferous. According to data from Bogdanov et al. (2002), the age of the part from 185 to 235 cm from Core 3852 is estimated to be from 85 to 110 ka.

The average chemical compositions of carbonate-free matter of metalliferous sediments containing >10% Fe, and non-metalliferous sediments that have different contents of Fe, and that of background sediments and serpentinite fragments from Core 343 are shown in Table 1.13. Hydrothermal matter is present not only in the metalliferous but also in the non-metalliferous sediments.

Table 1.13. Average chemical composition of carbonate-free matter of sediments with different Fe contents from the area near the Rainbow hydrothermal field as well as those for background sediments and serpentinite in the sediments. Based on data from Bogdanov et al. (2002) and Cave et al. (2002, 2003).

Sediments / Fe contents	Fe	Al	Ti	Mn	Cu	Zn	P	V
	%				ppm			
Fe >10%	11.5	3.9	0.37	0.54	1560	194	5460	671
Fe 8÷10%	9.1	4.3	0.35	0.43	1820	239	4170	498
Fe 6÷8%	7.3	5.0	0.37	0.41	940	221	3100	261
background	4.9	7.2	0.68	0.28	220	116	1790	171
serpentinite	8.8	0.33	0.016	0.137	116	45	77	54

Sediments / Fe contents	Ni	Co	As	Cr	Pd	Ir	Pt	Os
	ppm				ppb			
Fe >10%	133	75	213	116	12.4	0.27	6.9	0.97
Fe 8÷10%	217	85	140	199	11.0	0.78	6.5	1.40
Fe 6÷8%	274	83	69	239	8.7	0.64	6.7	0.72
background	72	39	6.2	127	~4.8	~0.13	~3.9	~0.23
serpentinite	1370		13	977	7.45	1.75	2.69	1.27

If it is assumed that carbonate-free matter of the metalliferous sediments is composed of background material, serpentinite fragments, and hydrothermal matter and that the El/Al ratios in the background material and the El/Ni ratios in the serpentinite fragments are equal to those calculated from Table 1.13, one can estimate the average roles of hydrothermal matter as a carrier of the elements in %:

Al	= Ni	= Cr	= Ti	→ Ir	→ Co	→ Zn	→ Pt	→ Mn	→ Fe	→ Pd	→ Os	→ P	→ V	→ Cu	→ As
0	0	0	7	29	~60	66	67	70	72	75	78	82	86	92	98

For most of the elements studied the roles are significant in spite of underestimations related to the assumption that all excessive Ni is bound only with the serpentinite material. Of course, some of the excessive Ni is also bound in the hydrothermal matter as it is in metalliferous sediments from other areas of the World Ocean.

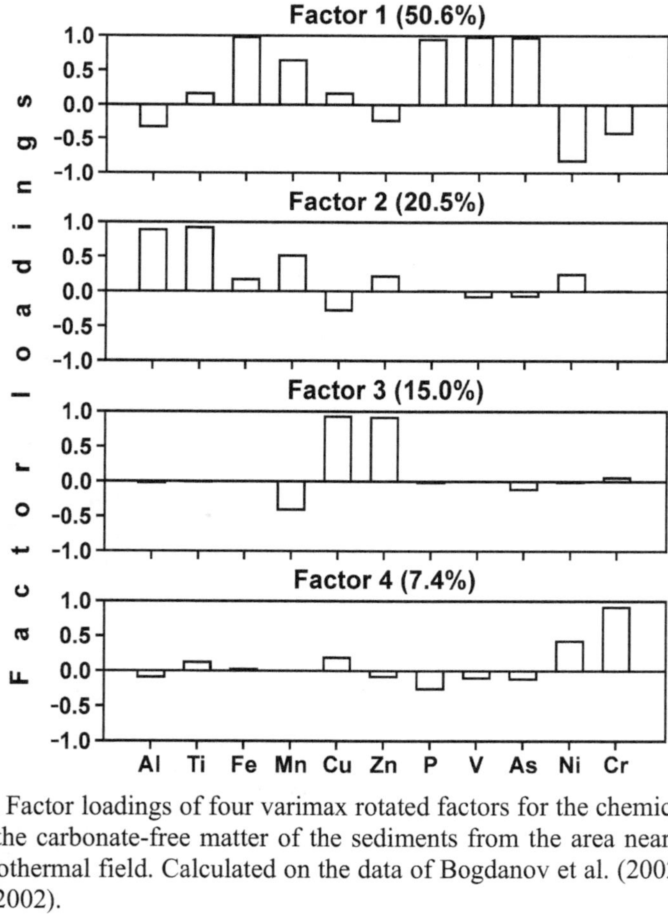

Fig. 1.73. Factor loadings of four varimax rotated factors for the chemical composition of the carbonate-free matter of the sediments from the area near the Rainbow hydrothermal field. Calculated on the data of Bogdanov et al. (2002) and Cave et al. (2002).

Factor analysis of the carbonate-free contents of the elements with varimax rotation of the factor matrix has been done for the sediments from the area of the Rainbow hydrothermal field. Four reasonable factors account for 93.5% of the variance (Fig. 1.73).

The first factor accounts for 50.6% of the variance. Fe, P, V, and As have high positive loadings and Mn has a significant positive loading. Al and Cr have marked negative loadings and Ni has a significant negative loading. The factor apparently shows mutual dilution of hydrothermal Fe-oxyhydroxide and ultramafic serpentinite material. The association of P, V, and As with the Fe-oxyhydroxide material indicates that the material has formed in a buoyant hydrothermal plume (Rudnicki and Elderfield 1993; Feely et al. 1994; Sect. 4.3).

The background and mafic materials contribute less to the variation in chemical composition of the sediments. This is evident from the second factor that accounts for 20.5% of the variance. Al and Ti have high positive loadings and Mn has a significant positive loading.

The third factor accounts for 15.0% of the variance. The high positive loadings of Cu and Zn are evidence of the presence of hydrothermal sulfide material or products of its alteration. This material is a dilutant for Mn that has a notable negative loading.

The fourth factor that accounts for 7.4% of the variance has a high positive loading of Cr and a notable positive loading of Ni. This apparently is further evidence of the presence of ultramafic material.

Variations in the average contents of the elements in carbonate-free matter of the Holocene sediments in the 25 km long section along the path of the hydrothermal plume (Fig. 1.72) are shown in Fig. 1.74. Sediments that accumulated 1.2 km to the north of the field show the highest contents of Fe and Zn and apparently of Cu and the lowest content of Mn. Further distances away from the field the contents of Cu and Zn decrease and the content of Mn increases. The content of Fe gradually increases away from the minimum found at 2 km north of the field and the contents of P, V and As have a similar distribution. The maximum contents of Al are found 1.2 km and 25 km from of the field. The content of Ti increases gradually from 2 km to 25 km from the field and the content of Ni decreases gradually from 1.2 km to 10 km from the field and then remain constant. The gradual decrease in the content of Ni apparently is evidence of the decrease in content of the serpentinite material.

The highest contents of Fe, Cu and Zn in the sediments accumulated in the vicinity of the Rainbow field coincide with the highest content of hydrothermal matter. The contents of Cu and Zn reflect the presence of sulfide material or products of its alteration and they decrease with distance from the field. Apparently the decrease in content of Fe in the area 1.2 to 2 km away from the field is attributable to the same reason. There is not enough data to account for the minimum contents of Fe and associated elements, P, V, and As, in the sediments accumulated from 2 to 5 km from the field and their increase in content beyond this distance. The presence of hydrothermal silica (see Sect. 1.1.4) and serpentinite material accumulated in the proximity of the hydrothermal field could account for this.

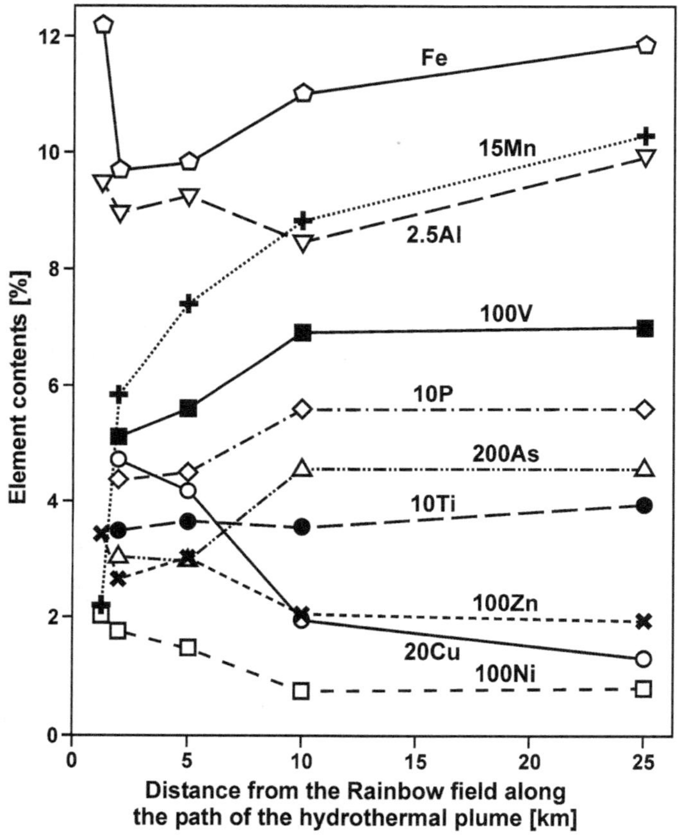

Fig. 1.74. Average content of elements in carbonate-free matter of Holocene metalliferous sediments in the cores from the section along the path of the Rainbow hydrothermal plume. The core stations are shown in Fig. 1.72. Calculations are based on the data from Bogdanov et al. (2002) and Cave et al. (2002).

The gradual increase in the content of Mn may be related to the properties of hydrothermal Mn in nonbuoyant plumes. This question will be considered in Sect. 4.3.2.

Accumulation rates of some chemical elements bound with the hydrothermal matter in bottom sediments of the cores collected in the area of the

Rainbow hydrothermal field are shown in Table 1.14. Accumulation rates of all the elements in Holocene sediments are much higher than in pre-Holocene sediments. This is evidence for the intensification of hydrothermal activity at the Rainbow field in Holocene time.

Table 1.14. Accumulation rates of chemical elements bound with hydrothermal matter in bottom sediments from the area of the Rainbow hydrothermal field. The core stations are shown in Fig. 1.72.

Core	Distance from the field along the section km	Age	Fe	Mn	P	Cu	V	As
			mg cm^{-2} ka^{-1}			μg cm^{-2} ka^{-1}		
343	2	Holocene	19	0.81	1.4	660	130	45
		pre-Holocene	3.4	0.34	0.24	74	16	7
316	5	Holocene	19	1.1	1.0	550	130	41
325	10	Holocene	24	1.4	1.4	240	180	64
329	25	Holocene	41	2.6	2.1	230	280	100
3852		Holocene	42	2.0				
		pre-Holocene	6.1	0.15				
3839		Holocene	47	5.3				
		pre-Holocene	7.6	1.6				

The accumulation rates of hydrothermal Fe and combined elements P, V, and As in the Holocene sediments do not show gradual decrease with distance from the Rainbow field as could be expected. This may be related to the accumulation of sedimentary material on slopes of 10–15°. The sedimentation rate patterns on such slopes in rift zones are usually irregular (Kuptsov and Barteneva 1989; Lisitzin et al. 1989). As for the accumulation rates of Cu and Mn that are bound with the hydrothermal matter, that for Cu decreases gradually and that for Mn increases gradually with distance from the Rainbow field.

The values for Fe in the Holocene sediments are comparable with those in metalliferous sediments from the TAG hydrothermal field and from the East Pacific Rise at 20°30'–22°S, 10 to 20 km to the west of the axis. These data confirm the very high intensity of the hydrothermal activity at the Rainbow field.

The accumulation rates of Mn in the Holocene sediments are higher or comparable to those for the TAG metalliferous sediments and lower than those for the metalliferous sediments from the East Pacific Rise at 20°30'–22°S, 10 to 20 km to the west of the axis. In contrast, the accumulation rates of Cu in the Holocene sediments are lower than in the TAG metalliferous sediments and higher than in the metalliferous sediments from the East Pacific Rise at 20°30'–22°'S, 10 to 20 km to the west of the axis (Table 1.11). It appears evident that the Holocene sediments from the area of the Rainbow hydrothermal field that were studied are more distal from a hydrothermal field than the TAG sediments that were studied and less distal than the sediments studied from the East Pacific Rise.

CHAPTER 2 METALLIFEROUS SEDIMENTS OF THE RED SEA

2.1. Conditions of formation

The Red Sea rift zone is a part of the global rift system. Volcanic and hydrothermal processes within the Red Sea rift are similar to those in other ocean rifts. The Red Sea rift was formed by the spreading apart of the Arabian and African lithosphere plates (Fig. 2.1). Apparently in the first stage of spreading the tension and thinning-out of the continental crust was accompanied by numerous intrusions of volcanic material. The splitting of the continental lithosphere began only 3 to 5 Ma BP and an active rift and the typical ocean crust began to form in its axial part. At the present time the boundary between the thinned continental crust and the ocean crust extends through the central part of the Red Sea basin. The plains slope gently from the coastal area and pass step-by-step to the axial rift zone that is some tens of kilometers wide and extends submeridionally along most of the length of the Red Sea. In the southern part of the sea the maximum depth of the rift is about 2900 m and in the northern part it is about 1500 m. The morphology of the rift does not vary much along its strike length and only a brief description of the rift structure with an example from the part investigated in detail at 18°N is given here[14].

The Red Sea rift zone has a distinct linear orientation in the area at 18°N and it consists of an internal rift 4 to 5 km wide and a system of fault scarps. The internal rift consists of a central uplifted extrusive volcanic zone of highly dissected relief and the uppermost part of the uplifted area is at a depth of 1300 to 1500 m and the base of the depressions that frame the uplift are at depths from 1700 to 1900 m. The extrusive zone consists of a disordered pile of separate volcanic mounds that are up to 300 m high. The slopes of the volcanic mounds are composed of chaotic piles of broken lava pipes (Bogdanov et al. 1983b). The lava rocks are composed of low-potassium primitive tholeiites similar to the basalt rocks of the mid-ocean

[14] Detailed description is given in the publications of Monin et al. (1980b, 1985), Sborshchikov et al. (1981), Bogdanov et al. (1983b, 1986).

ridges (Almukhamedov et al. 1981, 1983, 1985). Considering the morphology and volcanic activity in the Red Sea rift zone it can be classified as a typical slow-spreading rift zone.

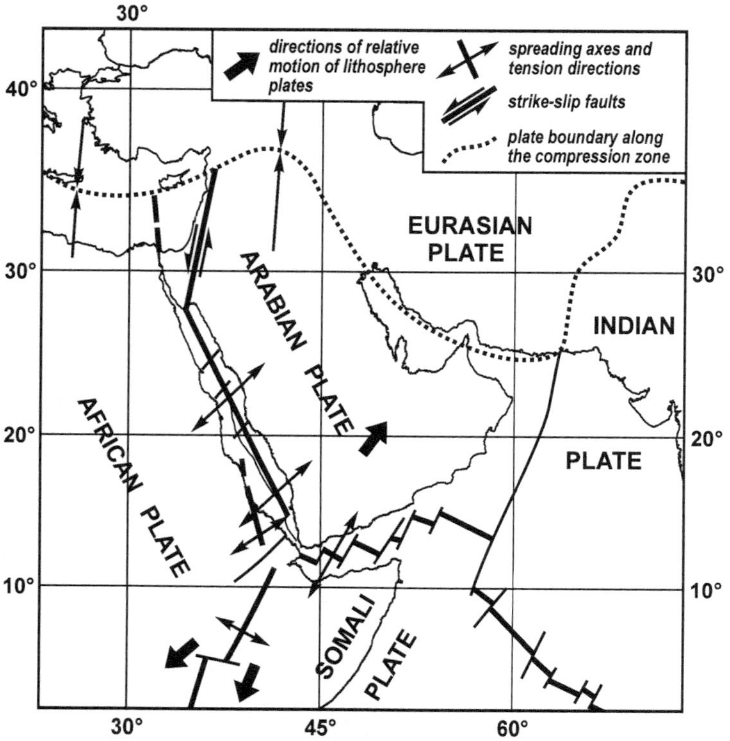

Fig. 2.1. Geodynamic features and situation in the Red Sea region. After Monin et al. (1985) with changes.

The average spreading rate of the Red Sea rift throughout the last 3 Ma was 1.6 cm a^{-1}. Present spreading rate at 18°N is 1.04 cm a^{-1} (Zonenshain et al. 1981). At 17°N spreading rate is 1.49 cm a^{-1} (Girdler 1985), at 23°N 1.06 cm a^{-1}, at 25°N 1.0 cm a^{-1} (Cochran and Martinez 1988), at 26–28°N 0.5 cm a^{-1} (Wong and Degens 1984).

The following features in the Red Sea rift zone in the area around 18°N differ from those in typical slow spreading rift zones: the rift is "entrenched" in the surrounding area of the seabed and no median ridge is found, and most of the rift along its total length is shallower than 2500–2700 m.

The thickness of the sedimentary layer covering the basalt basement rocks and the age of the basal rocks, as in others ocean rift zones, increase from the axis of the rift toward the border scarps. Very little sediments are present in the extrusive zone in the central part of the rift and young volcanic rocks are exposed on the surface of the seabed.

In a number of deeps of the axial or internal rift metalliferous sediments have been found (Fig. 2.2). They are mixture of variable amounts of hydrothermal sedimentary material and background Red Sea sedimentary material. Composition of the background sedimentary material reflects the peculiarities of the sedimentation process within this semi-enclosed basin. It has a high content, 40 to 50%, of terrigenous material whereas biogenic components prevail, from 80 to 90%, in sediments from the rift zones of the open ocean areas. The terrigenous material in the sediments of the Red Sea rift zone is mainly of eolian origin.

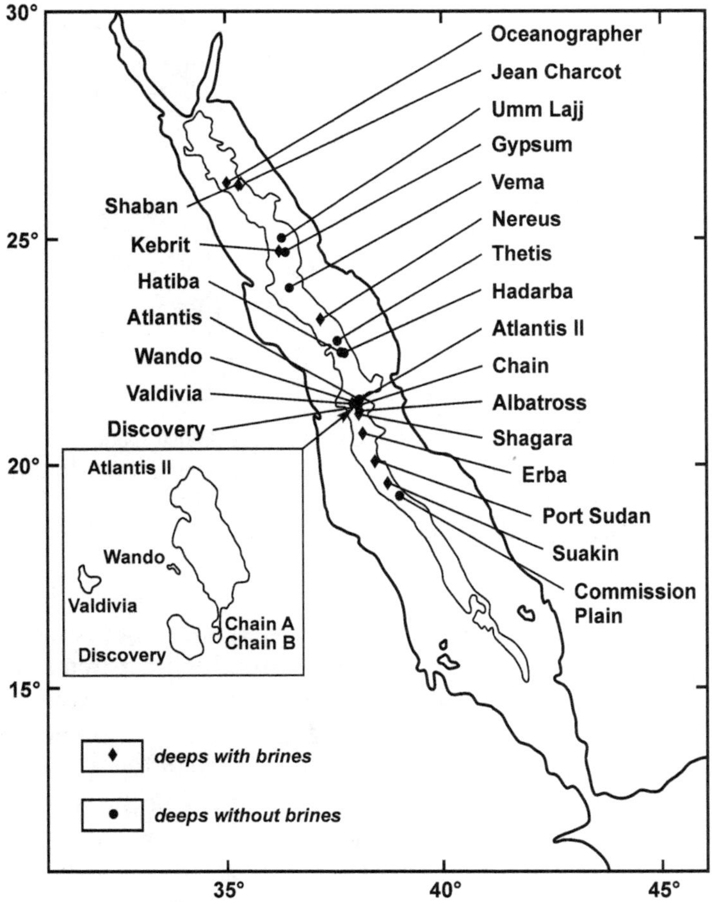

Fig. 2.2. Deeps of the Red Sea rift zone where metalliferous sediments or sediments influenced by hydrothermal action have been found. Compiled from Pautot et al. (1984), Bogdanov et al. (1986), Abu-Gharah (1997), Scholten et al. (2000).

The structures of the Red Sea sedimentary strata have an important distinctive feature that is essential for understanding the specific manner in which the mineral deposits in this area were formed. During Late Miocene time evaporite strata up to 7.5 km thick accumulated before the rifting and splitting of the lithosphere and the new ocean crust began to form (Hutchinson and Engels 1970). Because of the high plasticity of the evaporite strata in the early stages of the accretion of the ocean crust the eruption of basalt rocks took place under the salt strata. Similar conditions have existed in the northern part of the Red Sea rift zone to the present time (Fig. 2.3).

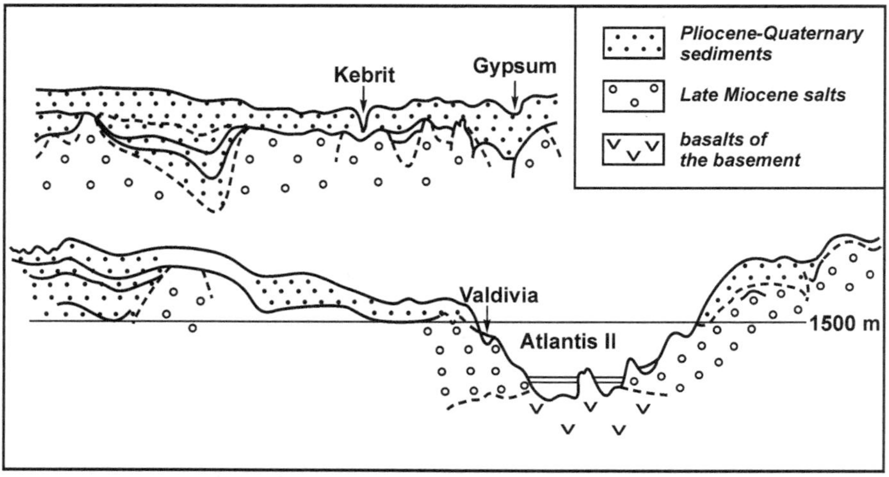

Fig. 2.3. Geological sections across the Gypsum and Kebrit Deeps from the northern part of the Red Sea rift zone and across the Atlantis II and Valdivia Deeps from the middle part of the Red Sea rift zone, based on data from the seismic profiling. After Bogdanov et al. (1986).

During the early stages of accretion of the ocean crust, and in the northern part of the rift zone at the present time, metal-bearing hydrothermal solutions flow out of the crust below the salt strata and apparently discharge the bulk of their load of mineral matter. The salt strata are exposed in border scarps in parts of the rift zone where fault structures were developed by the spreading action (Fig. 2.3). In both cases the dissolving of the salts results in the formation of highly mineralized brines on the seafloor. They act as traps for heat and matter that are carried into the basin deeps along with the high-temperature hydrothermal solutions.

Characteristics of the bottom brines from the deeps of the Red Sea rift zone that are associated with metalliferous sediments are shown in Tables 2.1 and 2.2. These tables are based on data from Brewer and Spenser (1969), Brooks at al. (1969), Craig (1969), Bäcker and Schoell (1972), Baumann et al.

Table 2.1. Locations of the deeps with metalliferous sediments or sediments influenced by hydrothermal action in the Red Sea rift zone, occurrence of the bottom brines in these deeps and their characteristics.

Deep	Coordinates of the deep center		Maximal depth	Maximal brine thickness	Brine surface	Maximal temperature
	Lat., N	Long., E	m	m	km^2	°C
Suakin SW	19°36.7'	38°43.6'	2850	74	2.5	23.2÷23.9
Suakin NE	19°38'	38°46.3'	2830	54	10	24.6
Port Sudan	20°03.8'	38°30.8'	2800	286	5	35.9÷36.2
Erba	20°43.8'	38°11.0'	2395	19	7	27.9
Shagara	21°07.8'	38°05.3'	2496	8	1	
Albatross	21°11.9'	38°07.0'	2133	72	1.5	24.4
Chain A	21°18'	38°04.9'	2072	83	0.7	52.1÷53.2
Chain B	21°17.17'	38°04.95'	2130	140	0.8	45.3÷46.7
Discovery	21°17'	38°03.2'	2237			
transition zone				63	11.5	23.6÷44.6
brine layer				179		44.7÷50.8
Wando W	21°21.45'	38°01.8'	2013	28	0.2	24.1
Wando E	21°21.20'	38°02.4'	2007	22	0.5	29.3
Valdivia	21°20.5'	37°57'	1673	123	4	29.5÷29.8
Atlantis II	21°22.5'	38°04.5'	2170			
transition zone				27	55	
upper brine layer				29÷31		41.2÷56.3
lower brine layer				122	43.5	55.9÷67.2
Atlantis	21°26'	38°05.5'	1960	no brines		
Hatiba	22°30'	37°40'				
Hadarba	22°27.5'	37°46'	2200	no brines		
Thetis NE	22°47.5'	37°35.5'	1800	no brines		22.6
Nereus E	23°11.5'	37°15'	2458	39	3	29.9÷30.2
Nereus W	23°11'	37°12'	2432	11	1	
Vema	23°52'	36°30.5'	1611	no brines		
Gypsum	24°42.1'	36°24.8'	1196	no brines		
Kebrit	24°43.35'	36°16.6'	1573	84	2.5	23.34
Shaban	26°21'	35°14'	1590	220	2–12	25.38
Red Sea deep water						21.8÷22.0

(1973), Hartmann (1973, 1980, 1985), Bäcker (1976, 1982), Cronan (1980), Danielsson et al. (1980), Puchelt and Laschek (1984), Bogdanov et al. (1986), Oudin and Cocherie (1988), Anschutz and Blanc (1996), Puchelt and Stoffers (1997), Hartmann et al. (1998), Anschutz et al. (1999), Scholten et al. (2000), Schmidt et al. (2003).

Table 2.2. Chemical composition of bottom brines in the deeps of the Red Sea rift zone, as well as of bottom seawater in the deeps without brines (shown in italics).

Deep	Cl^-	Na^+	K^+	Ca^{2+}	Mg^{2+}	SO_4^{2-}	Si	Fe	Mn
	g kg^{-1}						mg kg^{-1}		
Suakin SW	85.8	50.6	1.13	2.08	1.44	3.18		0.05	29
Suakin NE	85.9	50.6	1.13	2.19	1.42	3.18		0.04	18
Port Sudan	125			1.1	1.5	4.0			
Erba	86.5			1.0	1.5	4.1			
Shagara	113								
Albatross	143.3								
Chain A	153.8								
Chain B	156.0								
Chain C	154.6								
Discovery	153.8	93.0	2.14	5.12	0.81	0.7	4.5	0.27	54.6÷64.9
Wando E	73.5								
Valdivia	141	82		0.8	1.9	6.8			
Atlantis II									
upper layer	82	46.9	1.07	2.47	1.19	2.26	21	<0.1÷8	71.5÷101
lower layer	156.5÷158.7	92.6	2.4	5.3	0.78	0.84		75÷90	81÷117
Thetis	*22.9*								
Nereus E	129.5	76		8.0	1.5	0.9		0.77	47
Vema	*22.7*								
Gypsum	*23.7*								
Kebrit	154	90		1.7	2.4	2.2		0.08÷<1.3	7.8÷9.5
Red Sea deep water	21.13	11.8	0.42	0.45	1.41	2.96	2.4	<0.02	0.004

The thickness of the brine layers varies from several meters to many tens of meters. Their salinity may be up to 7.4 times greater than that of seawater. Frequently the temperature of the brines is much higher than that of the bottom seawater. The temperature of the lower brine layer in the Atlantis II Deep in the year 1998 was 67.2°C (Scholten et al. 2000). Brines are absent in some deeps where there is only a small excess of salinity and where temperatures of the bottom water are slightly higher than in the deep water in other parts of the Red Sea.

The brines differ considerably from seawater not only in salinity, but also in the salt composition (Fig. 2.4). The most typical difference is the high relative depletion of Mg^{2+} and SO_4^{2-} ions in the brines. This is most pronounced in the brines in the Atlantis II and Discovery Deeps.

A distinctive feature of the brines, especially of the high-temperature ones, are the high concentrations of dissolved Fe, Mn, Zn, Cu, Pb, Co, Ba, Li, Si in comparison to the Red Sea deep water and the low concentration of

Table 2.2. (continued)

Deep	Zn	Cu	Pb	Co	Ni	Ba	Li	O_2
	$\mu g\ kg^{-1}$							$ml\ l^{-1}$
Suakin SW	50	5	20					
Suakin NE	40	7	10					0.07
Port Sudan	170							
Erba	360							
Shagara								
Albatross								
Chain A								
Chain B								
Chain C								
Discovery	770	75	165	129	3.42		400	0.1
Wando E								
Valdivia	400							
Atlantis II								
upper layer	1100÷8000	0.5÷100	8.8	0.8	1.2		1700	0.05
lower layer	2000÷7000	0.5÷600	100	95		900	4400	0
Thetis								
Nereus E	490	20	130					
Vema								
Gypsum								
Kebrit	2500	300	20					
Red Sea deep water	5	5.5	0.03	0.07	0.5÷2	16.7	207	1.7÷2.1

dissolved oxygen. The latter is absent in the lower brine layer of the Atlantis II Deep, which has the highest temperature. The concentrations of Ni in the brines and in seawater outside of the deeps are similar (Table 2.2).

Studies of the lithology and mineralogy of the bottom sediments from the deeps of the Red Sea rift zone are reported in many publications (Miller et al. 1966; Baturin et al. 1969; Bischoff 1969; Baturin 1971; Baumann et al. 1973; Bäcker and Richter 1973; Hackett and Bischoff 1973; Bäcker 1976, 1982; Bignell and Ali 1976; Bignell et al. 1976a; Brockamp et al. 1978; Mossman and Hefferson 1978; Al-Karghuli 1979; Butuzova et al. 1979, 1980, 1983, 1988a; Shanks and Bischoff 1980; Zierenberg and Shanks 1983; Puchelt and Laschek 1984; Bogdanov et al. 1986; Missack 1988; Scholten et al. 1991, 2000; Anschutz and Blanc 1995; Butuzova 1998, et al.).

These studies have shown that in some deeps all of the sediments are metalliferous and in other deeps metalliferous sediments occur only in single

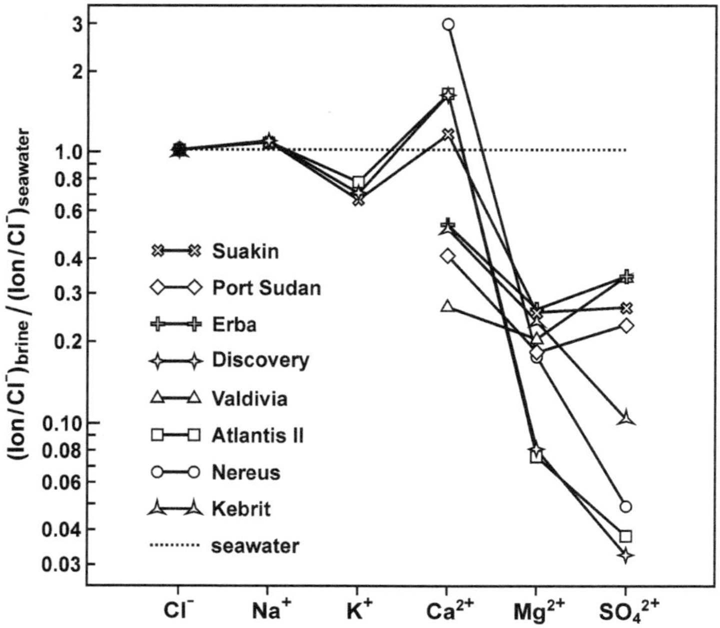

Fig. 2.4. Salt composition of the bottom brines from the deeps of the Red Sea rift zone normalized to Cl^- and to the salt composition of the Red Sea deep water.

thin layers. The sediments in the Atlantis II Deep have the highest content of metals of all of the metalliferous sediments in the deeps that are filled with brines, and in the metalliferous sediments in the deeps without brines those from the Northeast Thetis Deep have the highest content of metals. The metalliferous sediments from these two deeps will be considered in detail.

2.2. Atlantis II Deep

The Atlantis II Deep represents a typical depression of the Red Sea rift zone that is elongated along the spreading axis. It is 14 km long and 5 km wide. Its maximum depth is 2170 m. There are three elevated areas in the central part of the deep that are elongated along the spreading axis and that could be considered as separate parts of the extrusive zone in the rift structure. A seismic profile across the Atlantis II Deep (Fig. 2.3) shows a reflection horizon C in the upper parts of the slopes. This horizon corresponds to the boundary between Upper Miocene salt strata and overlying Pliocene-Pleistocene sediments. In the upper tectonic step the top of the salt stratum is located 200 to 250 m below the seafloor. It appears to be exposed on the slopes of the deep where the cover of sediments is very thin or irregular. This reflection boundary observed at the base of the slopes is considered to be the boundary between the sediment layer and the underlying basalt rocks of the basement. The thickness of the sedimentary layer in the Atlantis II Deep is 0–20 m or more (Hackett and Bischoff 1973; Mustafa et al. 1984). The surface of the brines that fill the deep is clearly shown on the seismic profiles. The level of this surface coincides with the depth of the surface mark of the reflection horizon C on the slopes (Whitmarsh et al. 1974).

The brines of the Atlantis II Deep are clearly stratified. There are two main layers, a lower and an upper, that differ in temperature and salinity (Tables 2.1, 2.2). The temperature of the lower layer that has a density of about 1.18–1.19 g cm^{-3} (Emery et al. 1974; Anschutz et al. 1998) is steadily rising and from 1965 to 1998 it increased from 55.9° to 67.2°C[15] (Scholten et al. 2000). The mean heat flux to the deep in the period from 1965 to 1992 was about 540 megawatts. This heat was held almost completely within the brines and did not dissipate into the overlying seawater (Anschutz and Blanc 1996). According to the estimates of Hartmann (1980), the flux of hydrothermal solutions to the deep exceeds 1000 m^3 h^{-1}. The later estimates of Anschutz and Blanc (1996) for the period from 1965 to 1992 gave 2000–3000 m^3 h^{-1}. The temperature of the hydrothermal solutions that enter the deep, according to the estimates of Sakai et al. (1970), is 250°C, of Truesdell (1975) 261°C, of Hartmann (1980) 280°C, of Monin et al. (1980a) >300°C, of Schoell (1980) >210°C, of Shanks and Bischoff (1980) 200°–250°C, of Grinenko and Ustinov (1982) ~400°C, of Thisse et al. (1983) 350°–400°C, of Ramboz et al. (1988) 390°–403°C, of Missak et al. (1989) <450°C, of Ramboz and Danis (1990) 353°–432°C, of Anschutz and Blanc (1996) 195°–310°C, and according to different estimates of Pottorf and Barnes (1983), from 210° to >334°C.

It has been mentioned that high concentrations of dissolved Fe, Mn, Zn, Cu, Pb, Co, Ba, Li, and Si have been observed in the brines of the Atlantis II Deep and that the brines of the lower layer are considerably richer in all of these elements, except for Mn, in comparison to their concentrations in the brines of the upper layer. The concentrations of Mn in the brines of both layers are similar. Dissolved oxygen is absent in the brines of the lower layer, and there is a small amount of oxygen in the brines of the upper layer where its concentration is about 35–40 times lower than in the Red Sea deep water (Table 2.2). There is a small amount of hydrogen sulfide in the brines of the lower layer in the southwest basin (Fig. 2.5) of the deep.

---

[15] According to the information of Hartmann et al. (1998), the estimated temperature of the lower brine layer was 71.6°–71.7°C in 1995 in the SW part of the Atlantis II Deep.

It has not been found in other parts of the deep or in the brines of the upper layer (Miller et al. 1966; Brewer and Spencer 1969; Bäcker and Schoell 1972). Concentrations of dissolved elements in the brines are not uniform. In the brines of the lower layer the highest concentrations of Fe, Mn, and Zn occur in the central part of the southwest basin and the lowest concentrations occur in the north basin (Brewer and Spencer 1969; Bäcker 1976).

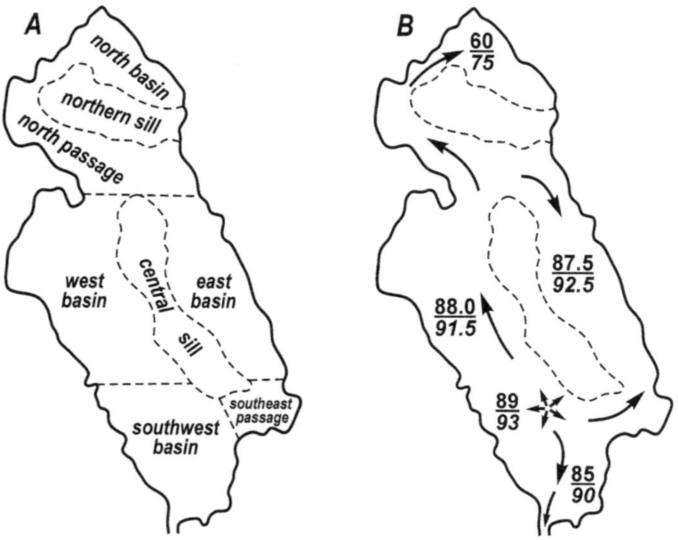

Fig. 2.5. A – parts of the Atlantis II Deep. B – distribution of concentrations of dissolved Fe (numerator) and Mn (denominator) in g m^{-3}, and current directions in the brines of the lower layer. Compiled from Bäcker (1976) with simplification.

Hartmann (1973, 1985) has studied particulate matter and concentrations of particulate Fe, Mn, Zn, and Cu in the lower and upper brine layers and in the transition zone from the upper brine layer to seawater in two phases of hydrothermal activity in the Atlantis II Deep: in 1971/1972 in the phase of relatively high activity, and in 1977 in the phase of relatively low activity (Fig. 2.6).

The concentration of the particulate Fe (Fe_{part}) shows maximum values in the upper brine layer and especially in the boundary zone between the lower and upper brine layers. The increase in 1977 in the concentration of the Fe_{part} near the bottom in the lower brine layer was probably caused by the resuspension of freshly deposited material (Hartmann 1985). The highest concentrations of the Mn_{part} occur in the transition zone.

In 1971/1972 the Zn_{part} and the Cu_{part} showed the highest concentrations in the lower brine layer and gradual decrease upward. In 1977 the Zn_{part} and the Cu_{part} showed some enrichment in the upper brine layer.

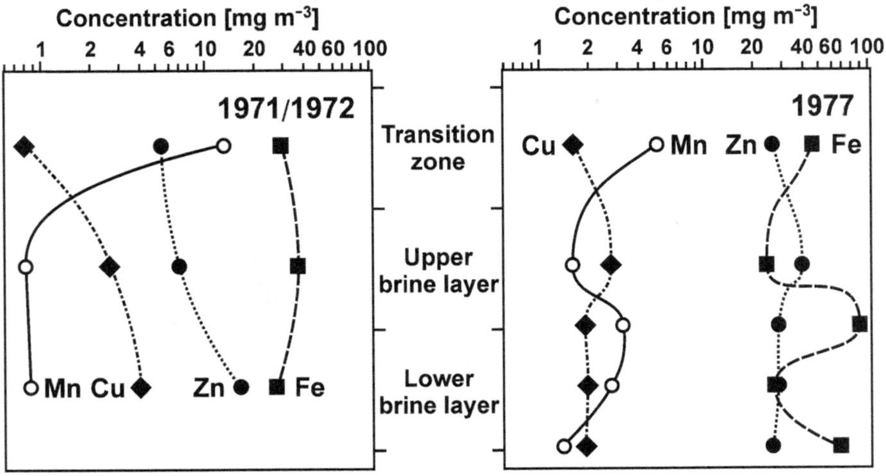

Fig. 2.6. Geometric mean concentrations of the particulate Fe, Mn, Zn, and Cu in the brines and in the transition zone of the Atlantis II Deep in the years 1971/1972 and 1977. Calculated and based on data from Hartmann (1973, 1985).

In 1971/1972 within the lower brine layer the highest concentration of the Fe_{part} occurred in the north basin of the deep, within the upper brine layer in the southwest basin and in the northern part of the west basin, and within the transition zone in the southwest basin. The highest concentration of the Mn_{part} within the transition zone occurred in the southwest and north basins. The highest concentrations of the Zn_{part} and the Cu_{part} within the lower brine layer occurred in the southwest basin, within the upper brine layer in the southwest basin and in the southern part of the east basin, and within the transition zone in the southwest basin.

Table 2.3. Specific roles (%) of particulate forms of the elements in their total concentrations (particulate + dissolved) in the brines of the Atlantis II Deep in the years 1971/1972 and 1977. Calculated on the data from Hartmann (1973, 1985).

Brine layer	Cu	Zn	Fe	Mn	Cu	Zn	Fe	Mn
	1971/1972				1977			
upper	6.0	0.42	0.44	0.00088	79	2.3	0.31	0.0020
lower	5.2	0.43	0.026	0.00085	78	0.86	0.078	0.0030

In 1971/1972 in the southwest basin in the lower brine layer the particulate Fe, Zn, and Cu were present in sulfide particles, and their concentrations were similar; in other basins the particulate Fe was present in sulfide and Fe-hydroxide particles, the concentrations of Zn- and Cu-sulfide particles were much lower. In the upper brine layer the particulate Fe was present in the Fe-hydroxide particles, which contained some Zn and Cu. In the transition zone the particulate Mn and Fe were present in the Mn- and Fe-hydroxides, which contained a little Zn and Cu.

In 1977 particulate sulfides containing Fe, Zn, and Cu were not found in the lower brine layer, even in the southwest basin. In the uppermost part of the lower brine layer, where the concentration of the Fe_{part} was maximal, it was presented in the Fe-hydroxides. In the transition zone the particulate Mn and Fe were presented in the Mn(IV)- and Fe(III)-hydroxides.

Specific roles of particulate forms of the elements in their total concentrations have been estimated for observations of Hartmann (1973, 1985) in the years 1971/1972 and 1977 (Table 2.3). These estimates show that for the Zn, Fe and particularly for the Mn they are of minor importance. For the Cu the role of the particulate form is much higher, and in 1977 it was dominant.

2.2.1. Spatial-temporal evolution of the lithology and mineral composition and accumulation rates of the bottom sediments

A complex of metalliferous sediments has been deposited in the Atlantis II Deep that contains quantities of metals of economic significance and their formation is continuing at the present time. The sediments from the deep have been studied intensively because of their economic importance. During expeditions of Preussag AG from 1969 to 1981 sediments were studied within an area of about 60 km^2 and more than 600 sediment cores were collected (Bäcker 1982; Mustafa et al. 1984), which is more than has been collected in any area of the same size anywhere else in the rift systems of the World Ocean. Most of the material from these cores has been used for estimating the mineral resources in the deep and for technological research but the original and published results of lithological, mineralogical and geochemical studies of about 100 cores collected during British, French, German, USA, and USSR expeditions are integrated here.

The first general description of the sediments from the Atlantis II Deep was given by Miller et al. (1966). Bischoff (1969) carried out more detailed study and described the following facies in these sediments: detrital, Fe-montmorillonite, goethite and amorphous Fe-hydroxide, sulfide, manganosiderite, anhydrite, and manganite. In his opinion these facies reflect the temporal sequence of sedimentation and are present within the whole deep and each facies shows a uniformity of conditions for the formation of the mineral components. Bäcker and Richter (1973) have shown that the composition of the sediments is more variable; and that as a rule the facies are interrupted, alternate in a section and are seldom undiluted. Despite the extreme variability in the composition of the sedimentary layers within the Atlantis II Deep, Bäcker and Richter have created a generalized scheme for their distribution (Fig. 2.7).

According to Bäcker and Richter (1973), there are five main lithostratigraphic units within the vertical section listed from the oldest to the youngest: Detrital-Oxidic-Pyritic Zone (DOP), Lower Sulfidic Zone (SU_1), Central Oxidic Zone (CO), Upper Sulfidic Zone (SU_2) and Amorphous-Silicatic Zone (AM). The characteristics of the sedimentary strata in the southwest basin (Fig. 2.5) as well as their mineral and chemical compositions are essentially different to those in other parts of the deep. Bäcker and Richter have also distinguished a Sulfidic-Oxidic-Anhydritic Zone (SOAN) that in part is overlain by an Oxidic-Anhydritic Zone (OAN), and a Sulfidic-Amorphous-Silicatic Zone (SAM) in the southwest basin. The SOAN Zone and the OAN Zone are of the same age as the SU_2 Zone, and the SAM Zone is the same age as the AM Zone.

It is noted that X-ray amorphous Fe-hydroxides and silica (Fe-Si gel) are the main components of the hydrothermal matter in almost all of the metalliferous sediments and especially those sediments of interest as resources in the Atlantis II Deep, as well as in the sediments from most other deeps of the Red Sea, except for those in the SOAN and OAN Zones (Butuzova and Lisitzina 1986a, b). As a rule, all variations in the composition of the sediments that are distinguished as lithofacies zones of the sedimentary strata have a prevalent background content of these components.

The accumulation of sediments of the oldest **Detrital-Oxidic-Pyritic (DOP) Zone** began about 25 ka BP. The upper boundary of the DOP Zone or the lower boundary of the SU_1 Zone is from 9.5 to 13.5 ka old and the average age of the boundary is about 11.7 ka (Fig. 2.7). Lithified carbonate sediments are present in this zone (Fig. 2.7; Bäcker and Richter 1973). Their time of formation extended from the end of the Würmian to the beginning of the Holocene (Herman 1965; Sval'nov 1986), which proves to be the correct dating of the boundary. The DOP Zone is composed mainly of detrital-biogenic material with an admixture of sulfide, anhydrite and Fe-hydroxide minerals in minor amounts that are not distributed uniformly and fragments of basalt are rare (Bäcker and Richter 1973).

The age of the upper boundary of the **Lower Sulfidic (SU_1) Zone** and the lower boundary of the CO Zone is from 7 to 10 ka and the average is about 8.6 ka (Fig. 2.7). The zone formed during a period of about 3.1 ka and the sediments in this zone are enriched by intercalated sulfide minerals and as a result have higher contents of S, Zn, Pb, and Cu.

Pyrite is the main sulfide mineral in the sediments of the SU_1 Zone in the western, eastern, and northern parts of the deep and sphalerite and chalcopyrite are less abundant. Sphalerite is the most abundant sulfide mineral in

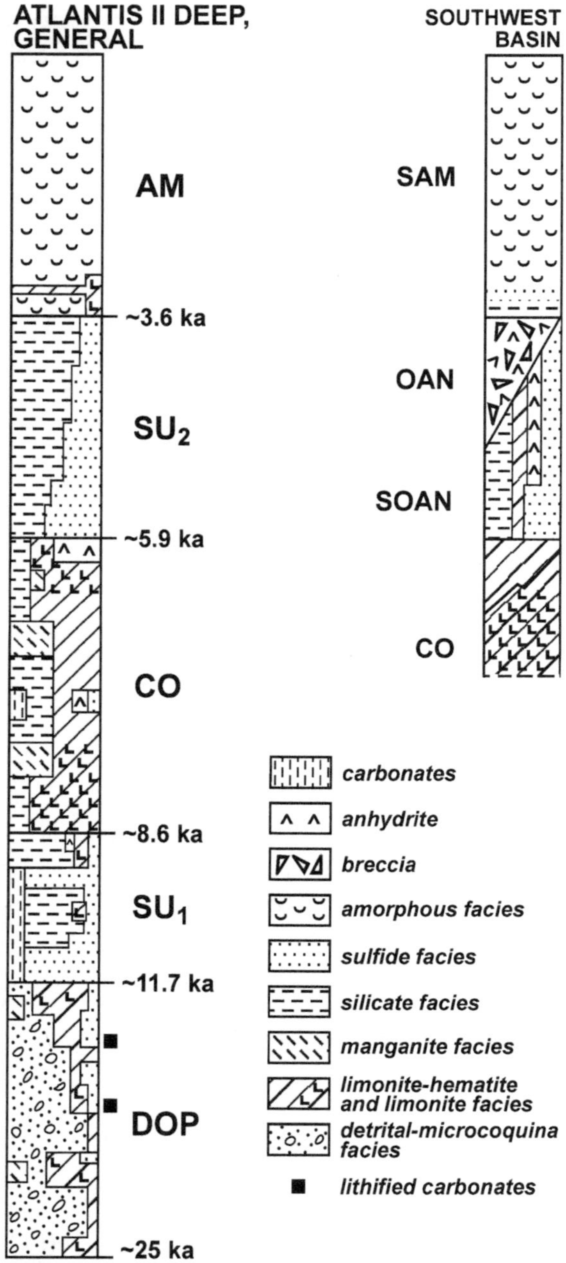

Fig. 2.7. Lithostratigraphic units of the sediment sequence in the Atlantis II Deep after Bäcker and Richter (1973) with simplification and additions. Estimations of the ages of the sediments have been made on materials from Ku (1969), Ku et al. (1969), Geyh and Höhndorf (1976), Shanks and Bischoff (1980), Kuptsov and Palkina (1986).

the sediments in the southern part of the deep and chalcopyrite occurs in small amounts (Shanks and Bischoff 1980; Butuzova and Lisitzina 1986b). Analyses of the sulfur isotope composition of the sulfide minerals provide evidence for their hydrothermal origin (Shanks and Bischoff 1980). Among other crystal phases present in the Fe-Si gel manganosiderite minerals are the most abundant in association with hematite, anhydrite, authigenic silicate minerals and barite along with the sporadic distribution of atacamite.

The distribution of some of the minerals other than the Fe-Si gel and sulfide minerals in the SU_1 Zone sediments and the thickness of the zone are shown in Fig. 2.8. The thickness of the zone varies from <1 m to >3 m and the maximum thickness is about 4 m. The sedimentation rates within the zone have been calculated after accepting that the SU_1 Zone formed during a period of about 3.1 ka and usually they range from 32 to 96 cm ka^{-1}. The maximum sedimentation rate was about 130 cm ka^{-1}. The minimum thickness of sediments in the zone, less than 1 m, occurs on the periphery of the deep where the sedimentation rate was <32 cm ka^{-1}. Lower thickness of the zone and sedimentation rates were found also in the northern part of the deep and on elevated surfaces underwater. In most parts of the deep the SU_1 Zone is 1to 2 m thick and the sedimentation rate was 32 to 64 cm ka^{-1}. In the central parts of the east and west basins the thickness of the zone increases to 2 to 3 m and the sedimentation rate was 64 to 96 cm ka^{-1}. The greatest thickness of sediments in the zone, more than 3 m, occurs in the northern part of the west basin where the sedimentation rate exceeds 96 cm ka^{-1}. The areal patterns showing the distribution of thickness and sedimentation rate of the sediments in the zone also indicate the distribution of accumulation rates of dry salt-free material in the sediments (Fig. 2.8). The total amount of this material within the SU_1 Zone is about 24 million tons and the average annual accumulation within the Atlantis II Deep during the time interval from 11.7 to 8.6 ka BP was about 5800 tons.

Sediments of the **Central Oxidic (CO) Zone,** overlying the SU_1 Zone, accumulated during a period of 2 to 3 ka (on the average 2.7 ka) from 8.6 to 5.9 ka BP (Fig. 2.7). The metal-bearing matter in the CO Zone sediments, unlike the sediments in the underlying zones, consists mainly of Fe-oxyhydroxide minerals, and intercalations are preset in these sediments that are highly enriched in Mn-oxyhydroxide minerals, manganite, todorokite, and X-ray amorphous Mn-hydroxide. Fe-smectite and manganosiderite minerals are distributed irregularly. In some parts there is an admixture of minerals consisting essentially of anhydrite and small amounts of gypsum. The content of detrital-biogenic material is somewhat greater in the sediments of the CO Zone (Butuzova and Lisitzina 1986b).

The Fe-oxyhydroxide material in most parts of the deep is mainly X-ray amorphous. The goethite and hematite occur in the central and southern parts

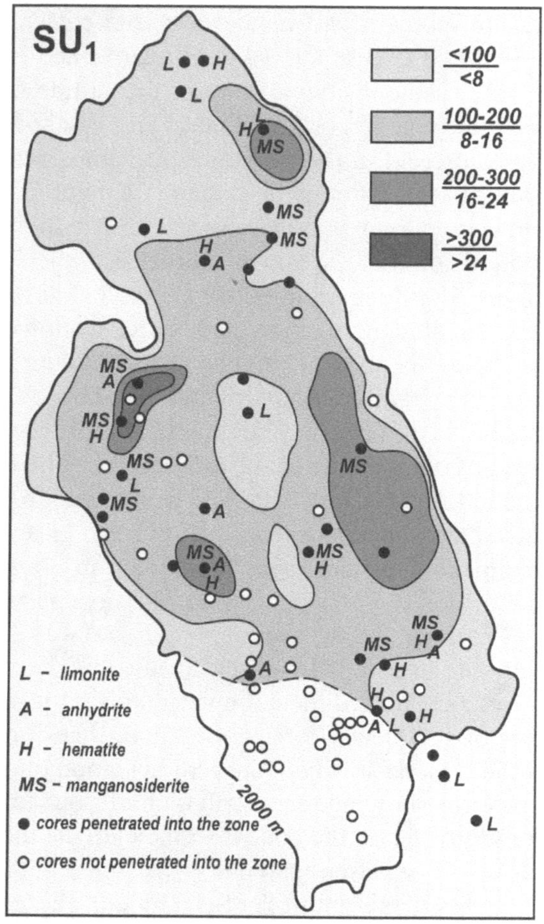

Fig. 2.8. Thickness of sediments of the SU_1 Zone shown as nominators [cm], their accumulation rates shown as denominators [g cm^{-2} ka^{-1}], and the occurrence of some minerals. After Bogdanov and Gurvich (1986b).

of the deep (Fig. 2.9). Usually they are present only in small amounts and are poorly crystallized. Their contents and degree of crystallinity are higher in the southern and southwestern parts of the deep. In cores collected in the north of the southwest basin of the deep (here probably the hydrothermal orifices are located) the hematite is highly crystallized and present in significant amounts. According to Bäcker and Richter (1973) and Hackett and Bischoff (1973), magnetite is also present in this area along with anhydrite that is typically present in the sediments in this and the adjacent areas (Fig. 2.9).

Silicate minerals are prevalent in some parts of the Atlantis II Deep in sediments that correspond in their stratigraphic position to the CO Zone. The

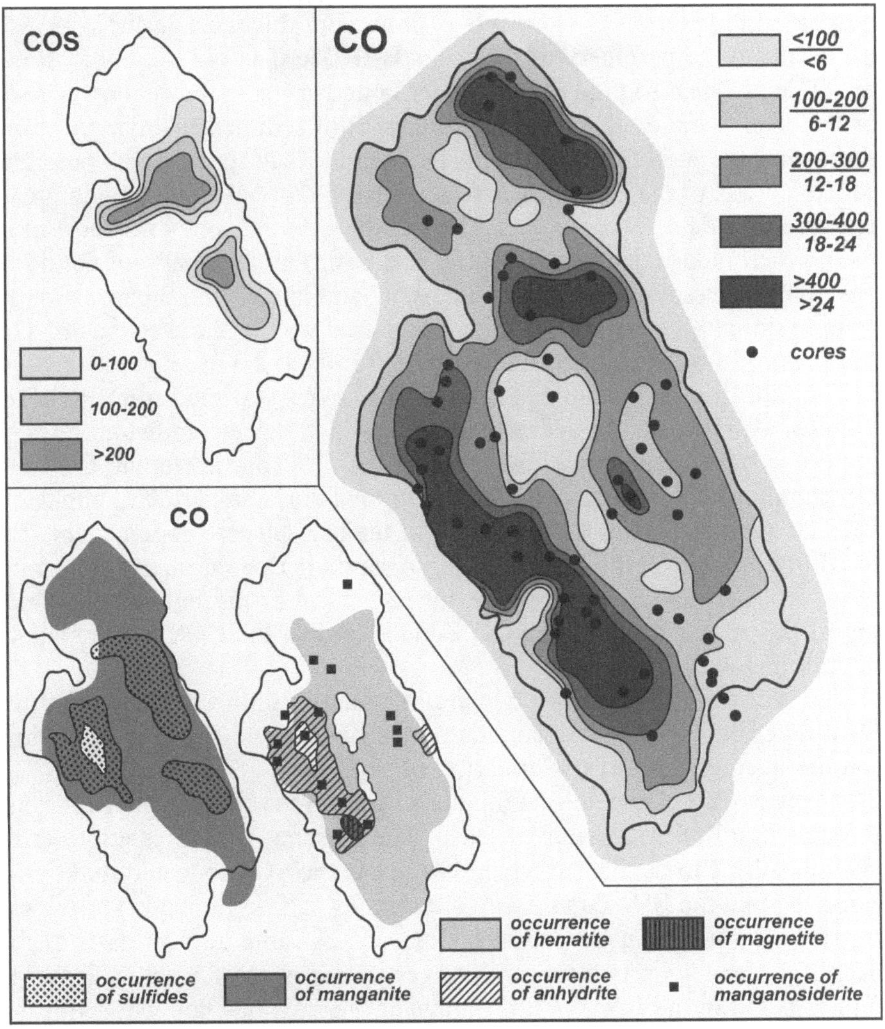

Fig. 2.9. Thickness of sediments of the CO Zone shown as nominators [cm], their accumulation rates shown as denominators [g cm^{-2} ka^{-1}], and the occurrence of some minerals. Thickness of sediments of the COS Zone [cm], which is included in the thickness of the CO Zone. After Bogdanov and Gurvich (1986b).

metal-bearing matter in these sediments is composed mainly of Fe- and Mn-oxyhydroxide minerals. Bäcker and Richter (1973) have distinguished this layer of sediments as the Central Oxidic-Silicatic Zone (COS). The sediments of the COS Zone occur in the eastern part of the deep and to the north and northwest of the central sill (Fig. 2.9).

The thickness of the sediments in the CO Zone, including the COS Zone, varies over a wide range from <1 m to >10 m and the maximum thickness is 12 m (Fig. 2.9). Sediments of minimum thickness in the zone (<1 m) occur on the periphery of the Atlantis II Deep, and in the higher areas at the depths near to the brine-seawater boundary or in the shallower parts of the deep. On the other hand, the transport of sedimentary material from these elevated parts of the floor of the deep to its deeper parts is possible. The thickness of the CO Zone within the basins of the deep is much greater and, as a rule, it exceeds 2 m. The thickness of the zone exceeds 4 m in the northern and western parts of the east basin, in most parts of the north basin, in the western part of the west basin, and in the central part of southwest basin and the maximum thickness found was in the west basin. The thickness of the COS Zone usually varies from 0 to 2.5 m or greater.

The plot that shows the distribution of thickness of the sediments (Fig. 2.9) can also provide the sedimentation rates and the accumulation rates of the dry salt-free sedimentary material of the CO Zone including the COS Zone after accepting that the CO Zone formed during a period of about 2.7 ka. The average annual accumulation of the dry salt-free sedimentary material of the zone on the bottom of the Atlantis II Deep during the time interval from 8.6 to 5.9 ka BP was about 12 thousand tons and the total amount of this material within the CO Zone including the COS Zone is about 32 million tons.

The sediments of the CO Zone are overlain by sediments of the **Upper Sulfidic (SU_2) Zone** that accumulated during a 2 to 3 ka interval of time (on the average 2.3 ka) in the period of time from 5.9 to 3.6 ka BP (Fig. 2.7). The Si-Fe gel and sulfide minerals occur in the sediments of the zone throughout all of the Atlantis II Deep. The presence of intercalations enriched in sulfide minerals is a typical feature in the SU_2 Zone and in the SU_1 Zone. Unlike the SU_1 Zone X-ray amorphous sulfide minerals are widely distributed throughout the sediments of the SU_2 Zone in most parts of the deep. Sphalerite is prevalent almost everywhere in the crystallized sulfide minerals and pyrite is much less abundant. Ferruginous laminated silicate, Fe- and Mn-carbonate, Fe-oxyhydroxide minerals, anhydrite, gypsum, and barite have been found in the sediments of the SU_2 Zone (Butuzova and Lisitzina 1986b; Fig. 2.10). Manganite has been found in the sediments on the elevated areas in the axial part of the deep and in the peripheral areas within the border steps (Fig. 2.10).

Despite their variety there is a regular lateral variation in the composition of the sediments within the zone. In the southern part of the deep the sediments contain a more varied complex of sulfide minerals including X-ray amorphous sulfides, sphalerite, pyrrhotite, pyrite, chalcopyrite and Cu-sulfosalts. Their crystallinity is higher in this part of the deep than in its other parts. Hematite is present in the sediment of the zone along with amorphous Fe-oxyhydroxides and goethite.

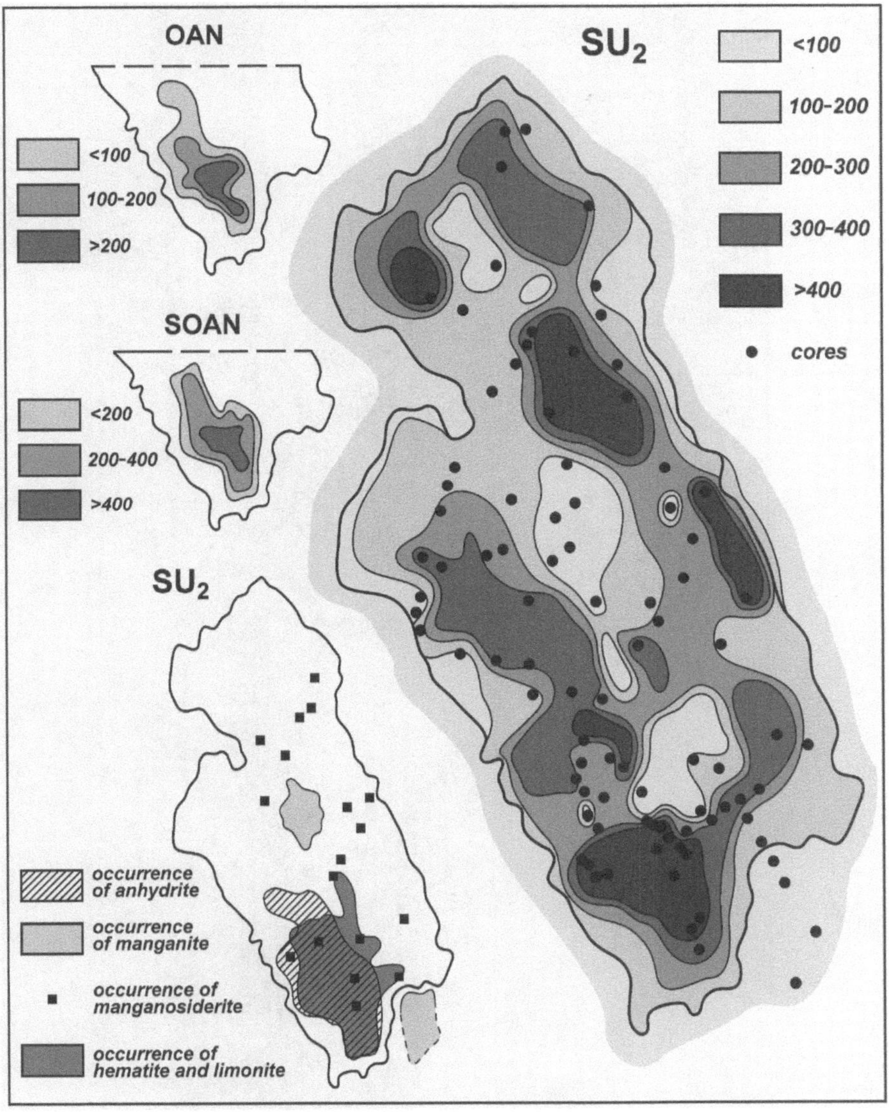

Fig. 2.10. Thickness of sediments of the SU_2, SOAN, and OAN Zones [cm] and the occurrence of some minerals. The thickness of the SU_2 Zone, including those of the SOAN and OAN zones. After Bogdanov and Gurvich (1986b).

In the southern part of the deep the sediments of the SU_2 Zone have high contents of anhydrite but gypsum is much less abundant. Anhydrite is a prevalent mineral in some of the cores that were collected in the southwestern part of the deep in sediments that correspond in their stratigraphic position to the SU_2 Zone. Bäcker and Richter (1973) have distinguished such

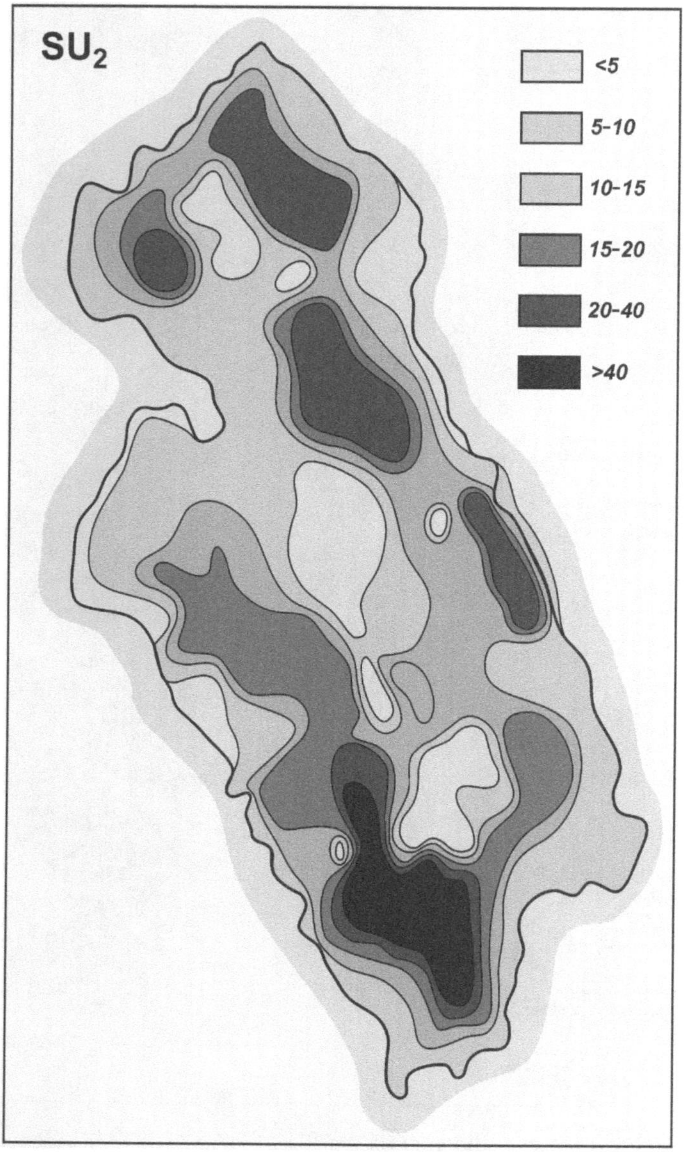

Fig. 2.11. Accumulation rate [g cm^{-2} ka^{-1}] of dry salt-free sedimentary material in bottom sediments of the SU_2 Zone (including the SOAN and OAN Zones) of the Atlantis II Deep. After Bogdanov and Gurvich (1986b).

sediments as the Sulfidic-Oxidic-Anhydritic Zone (SOAN) (Fig. 2.7). The thickness of this zone is 0 to 5 m or greater. In some areas within the stratigraphic position of the SU_2 Zone the sediments of the SOAN Zone are overlain by a layer of sediments that are enriched in oxide minerals and anhydrite and they have admixture of brecciated material. Bäcker and Richter (1973) have distinguished this layer of sediments as the Oxidic-Anhydritic Zone (OAN) (Fig. 2.7) which is up to 5 m thick (Fig. 2.10).

The thickness of the sediments in the SU_2 Zone, including those of the SOAN and OAN zones, within the Atlantis II Deep varies from <1 to >7 m (Fig. 2.10). Sediments of minimum thickness (<1 m) occur on the uplifted parts of the floor, on the border steps of the rift and on the elevated parts of the axial zone. The thickness of the zone increases in the deepest parts of the deep and reaches its maximum in the southwest basin. The plot showing the distribution of thickness of the sediments in the SU_2 Zone (Fig. 2.10) can also provide the distribution of sedimentation rate within the Atlantis II Deep during the time interval from 5.9 to 3.6 ka BP if the sediment thickness is divided by 2.3 ka.

The pattern of the distribution of accumulation rates of dry salt-free sedimentary material in the SU_2 Zone that includes the SOAN and OAN Zones shown in Fig. 2.11 is in general the same as the other patterns of the distribution, thickness and sedimentation rates of the sediments. The difference is that in the southern part of the deep, because of the greater specific weight of the material in the SOAN and OAN zones and the presence of anhydrite and lower water content, the accumulation rates are much higher than those in other parts of the deep that have similar sedimentation rates.

The average annual accumulation of dry salt-free sedimentary material in the SU_2 Zone on the bottom of the Atlantis II Deep during the time interval from 5.9 to 3.6 ka BP was about 8.8 thousand tons and the total amount of this material within the SU_2 Zone (including the SOAN and OAN zones) is about 20 million tons.

Sediments of the youngest **Amorphous-Silicatic (AM) Zone** overlying the sediments of the SU_2 Zone accumulated during the last 3 to 4 ka period (on the average 3.6 ka) (Fig. 2.7). They have the most homogeneous composition, texture, and coloring of all of the sediments within the Atlantis II Deep. They consist mainly of amorphous Fe-hydroxides and silica. Incipient products of crystallization in the mass of Si-Fe gel begin to occur at depths of 10 to 30 cm and the amount of crystallized material gradually increases downward (Butuzova et al. 1979; Butuzova and Lisitzina 1986b). Goethite, Fe-smectite, manganosiderite, X-ray amorphous sulfides and anhydrite are present in small amounts (Fig. 2.12).

In the southern part of the Atlantis II Deep the sediments of the AM Zone differ somewhat from the sediments in other parts of the deep. The contents of anhydrite and sulfide minerals in them increase notably. For this reason Bäcker and Richter (1973) have distinguished these sediments as the Sulfidic-Amorphous-Silicatic Zone (SAM) (Fig. 2.7). Hematite occurs in sediments of the SAM Zone in the southern part of the deep (Fig. 2.12).

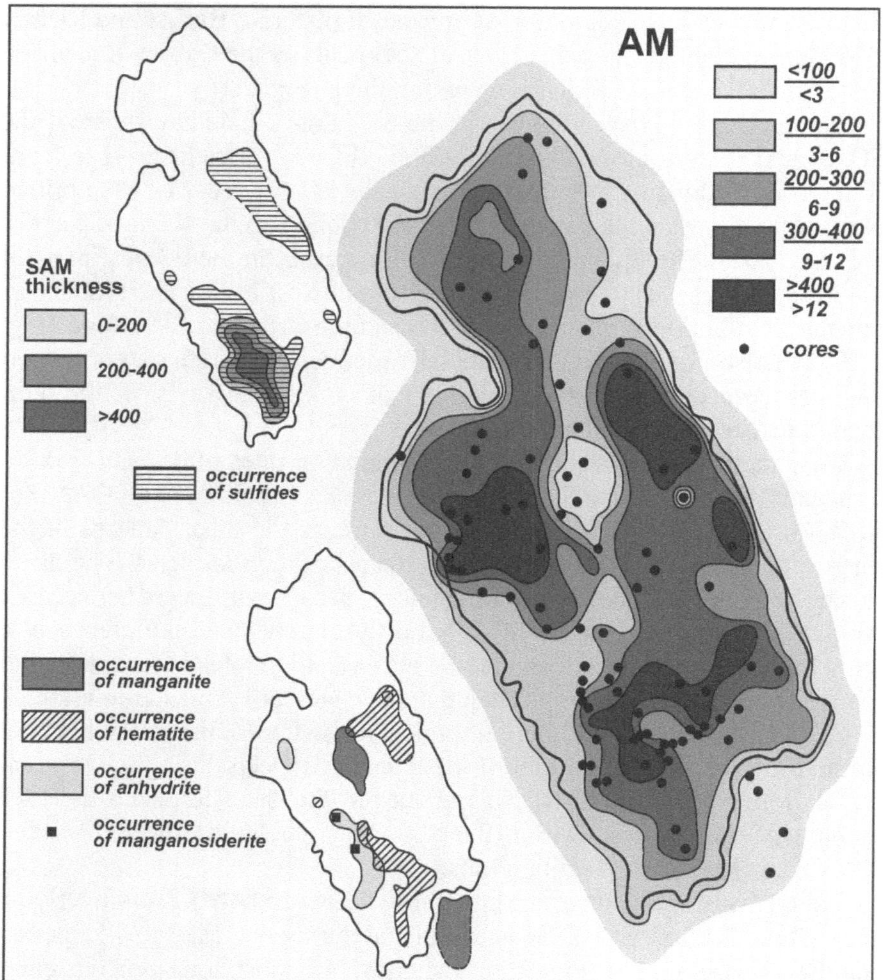

Fig. 2.12. Thickness of sediments of the AM Zone shown as nominators [cm], their accumulation rates shown as denominators [g cm^{-2} ka^{-1}], and the occurrence of some minerals. Thickness of sediments of the SAM Zone [cm], which is included in the thickness of the AM Zone. After Bogdanov and Gurvich (1986b).

Mn-oxyhydroxide minerals occur only in sediments in the AM Zone in the elevated areas underwater and in the peripheral parts of the deep, mainly above the brine mirror. The thickness of the sediments in the AM Zone varies from <1 m to >4 m and the maximum value measured is 777 cm. The maximum thickness of the SAM Zone is 475 cm. The pattern of the distribution of thickness of the sediments (Fig. 2.12) also provides the sedimentation rate and the accumulation rate of the dry salt-free sedimentary material of the AM Zone including the SAM Zone after accepting that the AM Zone formed during a period of about 3.6 ka. The average annual accumulation of the dry salt-free sedimentary material of the zone on the bottom of the Atlantis II Deep during the last 3.6 ka was about 4.7 thousand tons, and the total amount of this material within the AM Zone including the SAM Zone is about 17 million tons.

2.2.2. Mineral formation

The most comprehensive studies of the mineral-forming processes in the Atlantis II Deep have been carried out by G.Yu. Butuzova (Butuzova 1984, 1998; Butuzova and Lisitzina 1986b; Butuzova et al. 1988) and all of the available information from other specialists has been considered in her publications. Only a brief simplified description of the formation of the main mineral groups is given here. It is based on the data published by G. Yu. Butuzova with some additions. The general features of the main mineral-forming processes in the Atlantis II Deep as outlined by Butuzova (1998) are shown in Fig. 2.13.

Formation of sulfide minerals. Sulfide minerals are formed by release of metal ions from chloride complexes and their reaction with sulfide ions during intermixing of hydrothermal fluids with bottom brines. High concentrations of Fe(II) and sulfide ions as well as elevated temperatures are needed for the formation of pyrrhotite. Therefore it is prevalent in the sediments in the vicinity of the hydrothermal orifices. Pyrite is formed by the reaction of the newly formed Fe(III)-hydroxide with hydrothermal hydrogen sulfide. This results in the widespread distribution of the pyrite in the sediments of the deep. The order of relative stability of the chloride complexes ($Cu^{2+}<Zn^{2+}<Pb^{2+}$) is a controlling factor in the development and occurrence of Cu-sulfide minerals and their main distribution in the sediments in the vicinity of the hydrothermal orifices and the more widespread occurrence of Zn-sulfide minerals. Pb-sulfide minerals are not formed because of the relatively high stability of the chloride complexes of the Pb and the depletion of hydrogen sulfide during the formation of the Fe-, Cu-, and Zn-sulfide minerals.

The lack of sulfide ions and the abundance of metals in the hydrothermal fluids are distinctive features of the mineral forming process in the Red Sea. This is indicated by the presence of dissolved hydrogen sulfide in small concentrations only in the lower brine layer in the southwest basin, and by the presence of non-sulfide minerals of Fe, Zn, and even Cu such as ferrofranklinite, atacamite and chrysocolla.

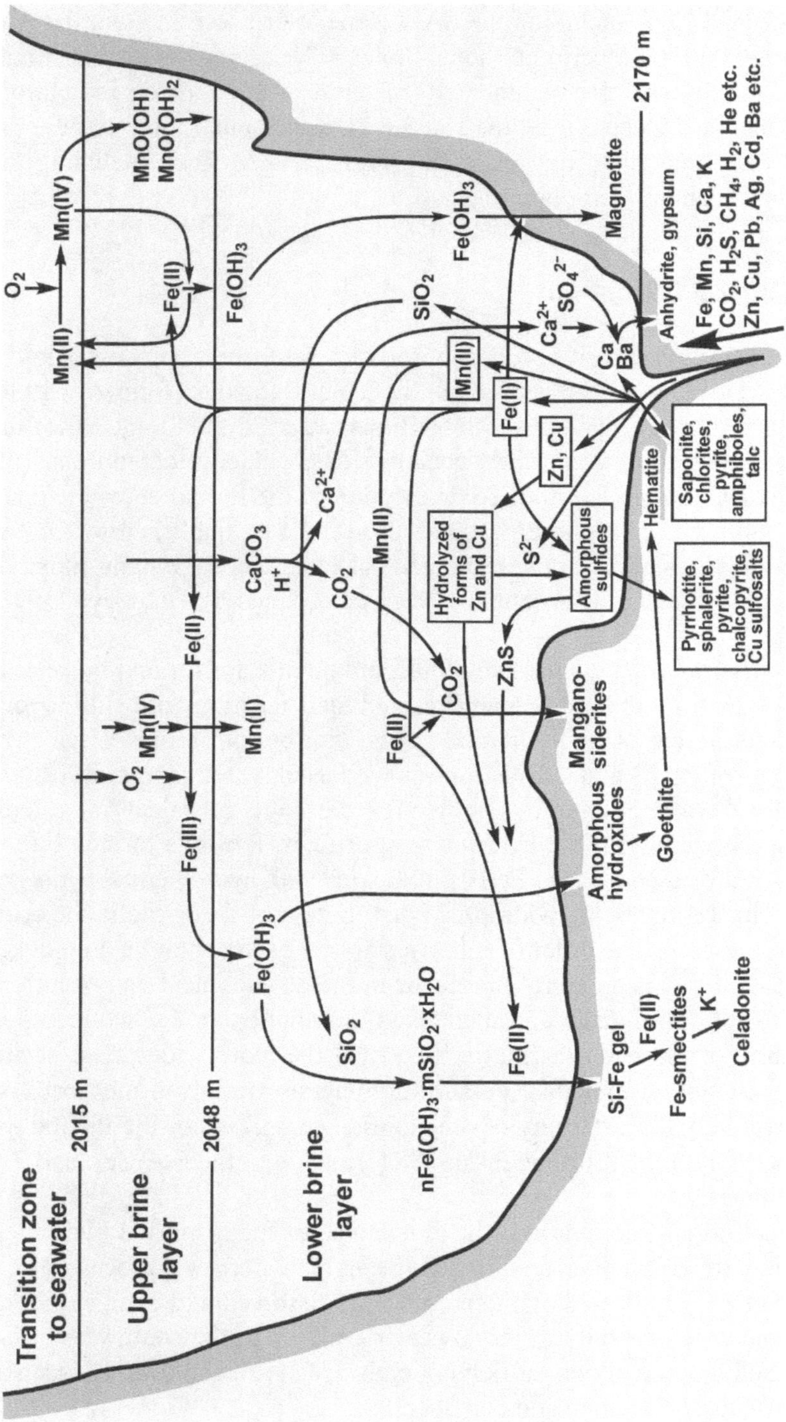

Fig. 2.13. An outline of the main mineral-forming processes in the Atlantis II Deep. After Butuzova (1998).

Formation of Fe- and Mn-oxyhydroxide minerals. Oxidation of the dissolved hydrothermal Fe(II) that enters the deep and the formation of particulate Fe(III)-hydroxide minerals takes place in the upper brine layer where dissolved oxygen is present (Table 2.2). As a result, concentration of dissolved Fe decreases and concentration of particulate Fe increases (Table 2.2, Fig. 2.6). Most of Fe-minerals in the metal-bearing matter in the sediments is derived from amorphous Fe(III)-hydroxide that is precipitated in the upper brine layer. Goethite is formed by crystallization of the amorphous Fe(III)-hydroxide and by transformation of lepidocrocite at higher temperatures. Hematite also formed by the crystallization of amorphous Fe(III)-hydroxide at higher temperatures and is found in sediments in the proximity of hydrothermal orifices. It is evident from the bacteria-like shapes of its particles that ferrihydrite was formed by the action of microbiological processes. The ferrihydrite formed in the transition zone from the brines to seawater and bacteria have not been found in the lower and upper brine layers. Two kinds of lepidocrocite were found. Particles of the first kind of lepidocrocite resemble bacteria-like forms and are similar to the ferrihydrite mineral particles. The second kind of lepidocrocite appears to be similar to goethite and hematite that were derived from amorphous Fe(III)-hydroxide.

The Atlantis Deep is a very effective trap for hydrothermal Fe. At the present time almost all of the hydrothermal Fe that enters the deep is bound in hydrothermal minerals. About 84% of the hydrothermal Fe accumulated in the sediments within the brine pool, and about 16% outside of it in a band that is about 1 km wide (Gurvich and Bogdanov 1986d).

Oxidation of dissolved hydrothermal Mn(II) that enters the deep and its precipitation as hydrated Mn(IV)-dioxide takes place in the transition zone that is enriched in dissolved oxygen, and its oxidation does not take place in the upper brine layer. The redox-potential is higher for the oxidation of Mn(II) to Mn(IV) than for the oxidation of Fe(II) to Fe(III). This is evident from lower concentrations of dissolved Mn (Mn_{diss}) and higher concentrations of Mn_{part} in the transition zone in comparison with that in both of the upper and lower brine layers (Table 2.2, Fig. 2.6). The Mn of the hydrated Mn(IV)-dioxide is reduced by the dissolved Fe(II) during sedimentation in the low-oxygen and oxygen-free brines and the Mn again enters into solution (Hartmann 1973, 1985). As a result Mn(IV)-oxyhydroxide minerals do not accumulate in the parts of the deep that are covered with brines at the present time. They are present only in sediments above the brine pool on elevated areas underwater and on the periphery of the deep (Fig. 2.12). Structures of poorly crystallized todorokite form in the transition zone where relatively high concentrations of dissolved O_2 and ambient low-alkaline conditions are developed. These structures are developed by dissociation of the hydrated Mn(IV)-dioxide and subsequent hydration, sorption and polymerization. Structures of manganite that have a high amount of crystallinity form in the upper brine layer where low concentration of dissolved O_2 and high concentration of dissolved Mn(II) exist. They are developed by processes of sorption of Mn^{2+} ions by the hydrated Mn(IV)-dioxide and subsequent polymerization and dehydration. Joint sedimentation of gel-like particles of "pre-todorokite" and "pre-manganite" compositions results finally in coagulation of particles of mixed composition that develop during diagenesis in sediment mass.

The formation of the hydrated Mn(IV)-dioxide is conducive to "pumping" of trace elements from seawater to the brines. These elements are co-precipitated during the formation of the hydrated Mn(IV)-dioxide and sorbed by it in the transition zone and then release into the upper brine layer during dissolution of the hydrated Mn(IV)-dioxide. The further scavenging and transport of the released trace elements to the sediments is carried out by the Fe(III)-hydroxide minerals in the upper brine layer where the trace elements are co-precipitated and sorbed (Gurvich and Bogdanov 1986a).

In the Atlantis II Deep the processes that bind Mn in the components of the hydrothermal minerals are less effective than the processes that bind Fe. As a result about 98% of the hydrothermal Mn, the average for the last 3.6 ka, is removed from the brine pool (Sect. 2.2.4).

Formation of silicate mineral phases. Particles of the Si-Fe gel are the most wide-spread and abundant component of the metal-bearing matter, not only in the metalliferous sediments in the Atlantis II Deep, but also in metalliferous sediments in almost all of the deeps of the Red Sea rift zone. They are formed by the polymerization of dissolved SiO_2 and its sorption on the amorphous Fe-hydroxides. The Si-Fe phase not only contains Fe(III), but also Fe(II) which entered this phase both by the process of sorption scavenging of Fe(II) by particles of Fe(III)-oxyhydroxides and silica, and by the interaction of amorphous Fe(III)-hydroxide with sulfide sulfur. Following sedimentation and in the early stages of post-sedimentation on the seafloor, extremely poorly crystallized phases of structurally disordered Fe-rich smectite minerals are formed by the alteration of the amorphous Si-Fe gel. Further crystallization and structural reorganization took place as the smectite material ages and mixed-layered phases are formed that contain up to 30% of non-inflating micaceous interlayers of tetrasilicic nontronite. A celadonite-like mica mineral that is rich in Fe and that contains 20 to 30% of inflating interlayers is the final member in the genetic series of mineral transformations from nontronite and mixed-layered nontronite-celadonite minerals in the sediments of the Atlantis II Deep. The origin of a complex of authigenic layered silicate minerals consisting mainly of Mg-bearing saponite, talc, and amphibole minerals in the sediments of the southwestern part of the deep is still not sufficiently clear.

Formation of carbonate and sulfate minerals. Manganosiderite minerals are formed by the interaction of Fe^{2+} and Mn^{2+} ions with CO_2. The CO_2 enters the deep in the hydrothermal fluids and it is also formed by the dissolution of the biogenic $CaCO_3$ in the brines at low pH. Anhydrite probably accumulates in the vicinity of the hydrothermal orifices. Obviously it is precipitated by the mixing of the hot Ca-rich hydrothermal fluids with the brines or seawater containing exogenic SO_4^{2-} ions. And the precipitation is enabled by the lowering of the solubility of anhydrite by increase in temperature.

2.2.3. Spatial-temporal evolution of chemical composition of bottom sediments and location of hydrothermal springs

Classification and significant systematization of the data for the sediments in the Atlantis II Deep are very difficult because their chemical composition varies greatly. Estimates of the average contents of the chemical elements in dry salt-free matter of the sediments within the five main stratigraphic units or zones in the vertical section and in the background Red Sea sediments are shown in Table 2.4. It can be seen from this table that the sediments from the Atlantis II Deep are enriched in Fe, S, Zn, Cu, Ba, Sr, Pb, Sb, and Co and presumably in Cd, As, Mo, V, Ag, Tl, Ge, Hg, and Au, in comparison with the background sediments, and depleted in Ca, CO_2, Al, Ti, Cr, Sc, and apparently in some other components that are typically present in the background sedimentary material that has been diluted by the metal-bearing material.

The low contents of Ca and CO_2 also result from intensive dissolution of the biogenic $CaCO_3$ in the brines, which is indicated by the considerably lower Ca/Al and CO_2/Al ratios in the metalliferous sediments in comparison with those in background sediments.

The contents of the rare earth elements (REE) in the sediments of the deep are lower than those in the background sediments. However, positive europium anomaly Eu/Eu* may be as high as 4.7 (Courtois and Treuil 1977; Butuzova and Lyapunov 1995). Evidently some part of the content of the REE has a hydrothermal origin, recognizing that the ratio of Eu/Eu* is usually higher than 10 in the high-temperature hydrothermal fluids from the ocean rift zones.

Variation in the average contents of Fe, Si, C_{carb}, Ba, and Li in the sediments of the various zones is less than 50% of their average content in the whole sedimentary layer in the Atlantis II Deep. Variations in the average contents of other elements are considerably higher. The maximum average contents of Zn, Ba, Cr, Pb, As, Mo, V, Cd, Tl, Hg, and Au occur in the sediments of the SU_2 Zone. The maximum average contents of Co, Sb, and P

Table 2.4. Average total (Tot) and excessive (Ex) contents of chemical elements in dry salt-free matter of bottom sediments from the lithostratigraphic zones of the Atlantis II Deep and in background sediments (BS) of the Red Sea rift zone[16].

Element		AM		SU_2		CO		SU_1		DOP		BS
		Tot	Ex	Tot	Ex	Tot	Ex	Tot	Ex	Tot	Ex	Tot
Si	%	11.9		11.6		9.4		13.2		11.3		
Al		0.95	0	1.09	0	1.47	0	1.37	0	2.5		3.5
Fe		33.6	32.7	26.3	25.2	34.3	32.8	33.1	31.7	23.5	21.0	3.5
Mn		0.70	0.59	0.88	0.75	4.67	4.50	1.77	1.61	1.7	1.4	0.41
Ti		0.060		0.057		0.099		0.054		0.19		0.37
Mg		0.82		3.04		0.77		0.24		1.5		1.9
Ca		3.4		4.5		2.65		1.0		6.3		17
K		0.93		0.31		0.28		0.30		0.88		
Na		2.0		3.7		1.8		1.7		1.0		
S		3.0	2.9	7.8	7.7	1.91	1.79	7.9	7.8	4.5	4.3	0.28
C_{org}		0.3		0.6		1.0		2.0		3.2		
C_{carb}		0.76		0.93		1.15		0.93		1.7		4.6
C_{tot}		1.06		1.53		2.15		2.93		4.9		
Zn		2.3	2.29	5.07	5.06	0.47	0.46	3.49	3.48	0.71	0.7	0.0318
Cu		0.49	0.48	0.84	0.83	0.21	0.20	1.09	1.08	0.21	0.2	0.0071
Pb	ppm	480	470	800	790	230	210	700	680	220	190	39
P		1100		1100		610		600		340		
Ba		960	910	1500	1440	1170	1090	1210	1130	<270	<130	195
Cd		76		144		62		13.9				
Sb		295	294.8	223	222.7	59	58.6	22	21.6			0.9
As		255	1	410		140						
Ag		45		80		16		165		<10		
Au		0.85		1.1		0.34						

in the sediments of the AM Zone are similar to those in the SU_2 Zone. The sediments in the CO Zone have maximum average contents of Mn, Fe and of C_{carb} where relatively high contents of biogenic $CaCO_3$ have been confirmed by lithological studies.

[16] Table is based on data from Bischoff (1969), Hendricks et al. (1969), Bäcker (1976), Courtois and Treuil (1977), Shanks and Bischoff (1980), Butuzova et al. (1980, 1981), Gurvich and Bogdanov (1986e), Gurvich et al. (1986b), Oudin and Cocherie (1988), Baturin (1993), Anschutz and Blanc (1995), Butuzova and Lyapunov (1995).

Table 2.4. (continued)

Element	AM		SU_2		CO		SU_1		DOP		BS
	Tot	Ex	Tot	Ex	Tot	Ex	Tot	Ex	Tot	Ex	Tot
Tl ppm	10		17.3		6						
Ge	8.2		7.5		2.7		7.8				
Hg	1.37		3.1				1.14				
Mo	140		270		182		116				
V	77		110		75		63				
Cr	18	4	23	6	20	0	13.4	0			53
Co	116	112	113	109	31	25	45	40			13.3
Ni	33	15	49	28	33	4	20	0			68
U	11.2		6		10.6		16.5		2.7		
Th	0.3		1.1		0.6		0.48		3.8		2.0
La	5.0		8.9		8.8		4.7		17		11
Ce	4.25		4.7		10.0		2.2		30		26.9
Nd	2.02		1.8		3.6		1.08				
Sm	0.47		0.45		0.93		0.3		3.5		3.1
Eu	0.20		0.26		0.31		0.28		1.3		0.61
Tb	0.07		0.06		0.10		0.03		0.5		
Yb	0.17		0.23		0.27		0.08		1.6		
Lu	0.08		0.09		0.05		0.01		0.27		
Sc	1.1		1.6		2.8		1.25		8.8		8.3
Zr							15		135		
Hf	0.8		1.2		1.3		<0.75		2.9		2.4
Sn			1.1		1						
Li	11.9		9.9		12.4		9				23.4

Sediments of the SU_1 Zone have maximum average contents of Si and of Cu, Ag, and S; the content of S is similar to that in the SU_2 Zone. The average content of Mn in the SU_1 Zone sediments is 2.5 times lower than that in the CO Zone sediments and 2.0 times greater than that in the SU_2 Zone sediments. Apparently this is because of the wide distribution of manganosiderite in these sediments (Fig. 2.8).

In comparison with the sediments from the other zones, the sediments of the DOP Zone are enriched in the Al, Ti, Sc, Zr, Hf, Th, and REE, in C_{org}, and C_{carb} (C in carbonate minerals), components that are typically present in the background sediments (Gurvich et al. 1986b). It is of special interest to note that the sediments of the DOP Zone are much richer in Fe, S, Cu,

Zn, and Pb than the background sediments. These elements are typically present in sediment material derived from hydrothermal sources and their excessive contents in the DOP Zone sediments (Table 2.4) apparently are bound with the metal-bearing hydrothermal matter. Enrichment by these elements in the sediments is evidence of the existence of a hydrothermal source for them during the formation of the DOP Zone sediments.

The data of Table 2.4 and Fig. 2.8–2.10 and 2.12 show that the chemical and mineral compositions of the sediments from the Atlantis II Deep have essentially varied through time. The sediments of the sulfide zones are enriched considerably in chalcophile elements and sulfur in comparison with the sediments of other zones. With equal contents of C_{carb} the SU_2 Zone sediments contain much more Ca in comparison with the SU_1 Zone sediments. This results from the enrichment of the SU_2 Zone sediments, or, rather, the OAN- and SOAN zone sediments, in anhydrite in comparison with the SU_1 Zone sediments. The CO Zone sediments are enriched in Mn. This reflects the presence of the Mn-oxyhydroxide minerals, mainly manganite, that are absent in the sediments of other zones within the brine pool area. Besides, the CO Zone sediments have slightly higher contents of C_{carb}, Al, and Ti in comparison with the sediments of the AM, SU_2, and SU_1 Zones because of the increased content of an admixture of background sedimentary material.

As mentioned previously the mineral composition varies considerably within the strata of coeval sediments in the Atlantis II Deep. By using the indicator minerals for some horizons it is possible to reconstruct or identify the areas of the hydrothermal orifices and to describe the variations in the composition of minerals in relation to the distance from them. The chemical composition of the metalliferous sediments is a reliable indicator of the conditions in which they formed. Unfortunately, there is not enough data for drawing maps of the distribution of the chemical elements in the sediments of the lithofacies zones[17] that have been identified. For this reason the characterization of lateral variability of the chemical composition of the coeval sediments and the average contents of the major chemical elements in the sediments of the individual physiographic parts of the Atlantis II Deep (Fig. 2.5) have been calculated (Table 2.5).

---

[17] There are maps available for the distribution of a number of economically important chemical elements in 1 m^3 of natural sediments for the whole series of metalliferous sediment of interest as mineral resources in the Atlantis II Deep. They have been used for the estimation of the metal resources in the deep (Guney et al. 1984; Mustafa et al. 1984)

Table 2.5. Average contents of chemical elements [%] in the sediments of the lithostratigraphic zones identified in the different parts of the Atlantis II Deep.

Zone	Part of the deep	Al	Si	Fe	Mn	Zn	Cu	S	C_{carb}	Ca
AM	north basin	0.80	13.1	34.3	1.1	0.71	0.40	0.80	2.2	5.6
	north passage	1.5	10.3	42.0	1.0			1.0	0.82	
	east basin	0.99	12.2	34.5	0.95	1.8	0.36	1.1	1.0	<1
	central sill	0.95	14.4	31.0	0.50	0.21		2.4		
	west basin	0.77	9.9	30.7	0.57	2.4	0.49	3.6	0.53	4.5
	southwest basin	0.60	14.0	32.2	0.48	3.9	0.79	5.5	0.08	2.0
	southeast passage	1.2	12.6	37.0	0.21	4.1	0.61	2.5	0.55	
SU_2	north basin	0.67	10.8	25.6	0.82	1.2	0.28	1.8	1.8	5.6
	north passage	1.0	17.3	16.1	0.20			3.0	0.14	
	east basin	1.7	16.5	33.8	0.86	2.8	0.48	3.4	0.44	
	west basin	0.96	7.2	23.5	0.93	5.0	0.98	11.5	0.49	1.0
	southwest basin	0.83	10.4	21.8	0.70	2.3	0.91	14.0	<0.14	9.2
	southeast passage	1.6	12.0	30.8	1.2	11.5	1.0	5.0	1.4	
CO	north basin	0.56	4.7	31.1	11.3	0.43	0.07	2.0	1.3	3.7
	north passage	2.0	9.3	31.5	1.2			0.50	2.2	
	east basin	2.25	7.2	33.3	3.7	0.28	0.65	0.17	1.5	1.4
	west basin	1.0	7.0	28.9	1.4	0.47	0.38	2.7	2.0	2.3
	southwest basin	0.60	7.6	37.5	0.68	0.47	0.20	4.2	1.1	7.7
	southeast passage	1.7	6.2	22.6	15.6	0.23	0.06	1.0	1.4	
SU_1	north basin	1.2	7.9	33.6	0.48	3.2	0.87	7.0	3.3	1.1
	north passage	1.2	4.7	38.5	0.70	5.6	1.2	9.2	0.55	
	east basin	1.25	13.4	35.5	2.5	1.3	0.57	5.6	0.90	
	west basin	0.88	15.4	31.0	1.3	4.0	1.1	9.4	0.57	1.0
	southeast passage	1.8	11.3	29.2	2.6	5.9	2.0	2.0	2.7	

To enable easier comparisons the Element/Al ratios have been calculated (Table 2.6). In this case the values reflecting enrichment or depletion of the sediments in the elements relative to the background sediments have been compared. Aluminum has been used as a proxy for the background sedimentary material. If it is assumed that during the formation of a lithofacies zone the accumulation rate of the Al within the whole deep was uniform, the ratios of Elements/Al can also reflect the average intensity of accumulation of the elements on the floor in different parts of the deep in particular time intervals.

Detrital-Oxidic-Pyritic Zone. The average contents of most elements in the DOP sediments have been calculated from the analytical data for only a few samples and they cannot be considered as representative of the whole

Table 2.6. Element/Al ratios in the sediments of the lithostratigraphic zones distinguished in the different parts of the Atlantis II Deep.

Zone	Part of the deep	Si/Al	Fe/Al	Mn/Al	Zn/Al	Cu/Al	S/Al	C_{carb}/Al	Ca/Al
	north basin	16	43	1.4	0.89	0.50	1.0	2.8	7.0
	north passage	6.9	28	0.67			0.67	0.55	
	east basin	12	35	0.96	1.8	0.36	1.1	1.0	<1
AM	central sill	15	33	0.53	0.22		2.5		
	west basin	13	40	0.74	3.1	0.64	4.7	0.69	5.8
	southwest basin	23	54	0.80	6.4	1.3	9.2	0.13	3.3
	southeast passage	11	31	0.18	3.4	0.51	2.1	0.46	
	north basin	16	38	1.2	1.7	0.42	2.7	2.7	8.4
	north passage	17	16	0.20			3.0	0.14	
SU_2	east basin	10	20	0.51	1.7	0.28	2.0	0.26	
	west basin	7.5	24	1.0	5.3	1.0	12	0.51	1.0
	southwest basin	13	26	0.84	2.8	1.1	17	<0.16	11
	southeast passage	7.5	19	0.76	7.2	0.63	3.1	0.85	
	north basin	8.4	56	20	0.77	0.13	1.6	2.3	6.6
	north passage	4.7	16	0.60			0.25	1.1	
CO	east basin	3.2	15	1.6	0.12	0.29	0.076	0.67	0.62
	west basin	6.7	28	1.4	0.45	0.37	2.6	1.9	2.2
	southwest basin	13	63	1.1	0.78	0.33	7.0	1.8	13
	southeast passage	3.7	14	9.3	0.13	0.03	0.60	0.81	
	north basin	6.6	28	0.40	2.7	0.73	5.8	2.7	0.92
	north passage	3.9	32	0.58	4.7	1.0	7.7	0.46	
SU_1	east basin	11	28	2.0	1.0	0.46	4.5	0.72	
	west basin	18	35	1.5	4.6	1.3	11	0.65	1.1
	southeast passage	6.3	16	1.4	3.3	1.1	1.1	1.5	

zone. Only on the sulfide sulfur, Zn, Cu, and C_{carb} contents there are data from some tens of samples. There are data for the north, east, west, and southwest basins and for the north passage. The sediments of the north passage have the highest contents of sulfide sulfur and Cu.

In the sediments of the DOP Zone the values of $\delta^{34}S$ in the sulfide sulfur vary from –33.75‰ to +9.6‰ (Shanks and Bischoff 1980), which is evidence of the presence of both hydrothermal sulfides, and sulfides formed by bacterial sulfate-reduction. The variation of $\delta^{34}S$ through time of the sulfide sulfur in the DOP Zone sediments in different parts of the Atlantis II Deep is shown in Fig. 2.14. The δS^{34} of the sulfide sulfur increases in sediments with increased contents of the sulfide sulfur and Cu (Fig. 2.15). Assuming that this increase in the δS^{34} is related to the proximity of hydrothermal orifices, one can conclude that 13 to 14 ka BP such orifices were present within the north passage and in the east basin. The occurrence of anhydrite in the sediments within the north passage that were dated from 14 to 16 ka BP (Bäcker and Richter 1973; Shanks and Bischoff 1980) is further evidence of hydrothermal activity in the area at this time. The anhydrite could have been precipitated from brines during their mixing with high-temperature hydrothermal fluids.

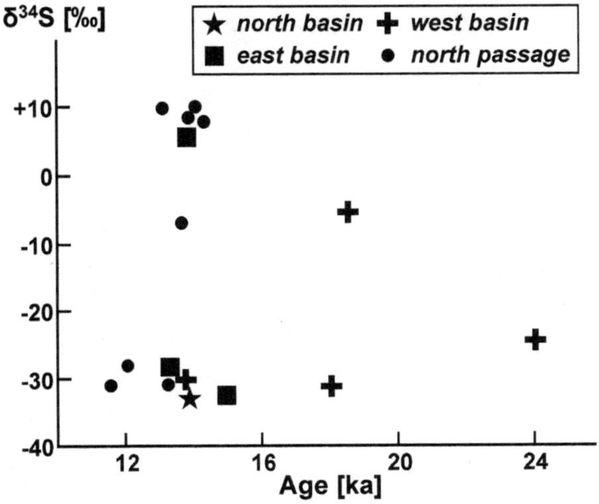

Fig. 2.14. Variation of $\delta^{34}S$ of the sulfide sulfur in time in the DOP Zone sediments from different parts of the Atlantis II Deep. Based on data from Shanks and Bischoff (1980).

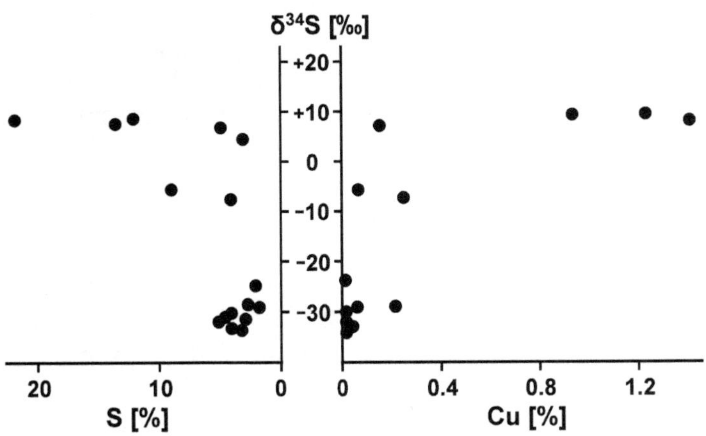

Fig. 2.15. $\delta^{34}S$ of the sulfide sulfur in relation to contents of Cu and the sulfide sulfur in the DOP Zone sediments of the Atlantis II Deep. Based on the data from Shanks and Bischoff (1980).

The isotope composition of the sulfide sulfur is evidence that the sulfide minerals in the DOP Zone sediments of the north and west basins were formed by the sulfate-reduction. The sulfide minerals in the sediments from the west basin that have an age of about 19 ka have a mixed composition. These data are an elaboration of the results of Bäcker and Richter (1973). On the basis of the distribution of limonite in the DOP Zone sediments they have suggested that the main hydrothermal activity took place in the northern part of the Atlantis II Deep at time of the accumulation of these sediments.

The accumulation of the hydrothermal sulfide minerals in the sediments of the DOP Zone during periods of intense hydrothermal activity was localized and not distributed throughout the whole deep during these periods.

Apparently sulfide minerals were formed by sulfate-reduction during these periods in other parts of the deep. It is evident that even during these periods the hydrothermal activity was not as intense as it was during the final 11.7 ka of the geological history of the Atlantis II Deep.

Lower Sulfidic Zone. Data are available for most of the elements that are considered in this section, Si, Fe, Mn, Zn, Cu, S, Ca, C_{carb}, in the sediments from all parts of the deep except for the central sill and southwest basin.

The average Element/Al ratios in the sediments of the SU_l Zone in different parts of the deep are given in Table 2.6. The highest Fe/Al, Si/Al, Cu/Al, and S/Al ratios occur in the sediments from the west basin. In the sediments from the north passage the Fe/Al, Cu/Al and S/Al ratios are a little lower, and the Zn/Al ratio is somewhat higher. According to the data of Shanks and Bischoff (1980), the sediments within the SU_l Zone in the west basin and in the north passage have the greatest enrichment in sulfides where their average contents are 17.4 and 18.8% respectively. In the sediments from the east and north basins the average contents are lower and are respectively, 13.5 and 15.3%. The $\delta^{34}S$ of the sulfide sulfur in the SU_l Zone sediments varies from +0.75‰ to +7.93‰, which is considered to be evidence of the hydrothermal origin of this sulfur. This information apparently is evidence that during the accumulation of the SU_l Zone sediments from 11.7 to 8.9 ka BP hydrothermal orifices were located at least in the west basin and in the north passage areas. Apparently this evidence is confirmed by the presence of anhydrite in the sediments in these parts of the Atlantis II Deep (Fig. 2.8).

Accepting this evidence as valid one is assured that the hydrothermal activity within the north passage and its vicinity succeeded the hydrothermal activity that existed during the formation of the sediments of the DOP Zone.

Central Oxidic Zone. Data for the sediments from all parts of the Atlantis II Deep except for those from the central sill area are included in Table

2.6. The highest ratios for Fe/Al, Si/Al, Zn/Al, S/Al, and Ca/Al and the second highest ratio for Cu/Al are in the sediments from the southwest basin. The Fe/Al and Zn/Al ratios for the sediments from the north basin are somewhat lower. The maximum ratio for Mn/Al is in the sediments from the north basin and the minimum ratio is in the sediments in the southwest basin. The Mn/Al ratio is also relatively high in the sediments from the southeast passage.

Enrichment in anhydrite results to the higher S/Al ratios in the sediments from the southwest and west basins in comparison with those in the sediments from other parts of the deep. The content of the sulfide sulfur in these sediments is low (Shanks and Bischoff 1980).

Numerous data for the Fe-oxyhydroxide minerals and for Mn-oxyhydroxide minerals indicate that in some places these are the main components of the metal-bearing matter in the sediments of the CO Zone. The Fe/Al ratios (Table 2.6) provide evidences that during the formation of the CO Zone, from 8.6 to 5.9 ka BP, the greatest accumulation of Fe minerals was in the sediments in the southwest, west and north basins and it was less intensive in the sediments in the east basin and in the southeast passage. The greatest accumulation of Mn minerals was in the sediments in the north basin and in the southeast passage. It is noted that hematite is widely spread in the sediments in the southwest basin where the Fe/Al ratio is maximum and well-crystallized hematite and magnetite occur in the northern part of the southwest basin (Fig. 2.9). The Mn/Al ratio is minimum here and manganite has not been found (Fig. 2.9). This is evidence of the differentiation of the Fe and Mn. Apparently, the main content of the Fe and the components bound in the Fe-minerals as well as anhydrite were deposited in the vicinity of the hydrothermal orifices that presumably were located in the northern part of the southwest basin, and the main part of the Mn and the components bound in its minerals were deposited at a distance from the orifices.

The presence of the Mn-oxyhydroxide minerals in the sediments of the CO Zone, not only on the uplifted parts of the floor but also within the basins, is considered to be evidence that the conditions in the Atlantis II Deep during the accumulation of these sediments were different essentially from those that existed during the accumulation of the sediments of the SU_1 Zone. Dissolved oxygen was contributed in amounts sufficient for the oxidation of a considerable amount of hydrothermal Mn(II) and the sedimentation of Mn(IV)-oxyhydroxide minerals on the floor.

During the accumulation of the sediments of the CO Zone the conditions for their sedimentation varied greatly. This is reflected in the extremely heterogeneity in the composition of the sediments in vertical sections. Manganite and sulfide minerals that indicate essentially different conditions for their formation have been found in different horizons in the CO Zone sediments in the cores that were collected in the west and east basins. They provide evidences that conditions in the bottom water fluctuated from oxidizing to reducing during the period of accumulation of the sediments in the CO Zone in the east and west basins.

Presumably, during the formation of the CO Zone sediments because of the intensive effluence of hydrothermal solutions the continuity of the brine layer was disturbed, dissolved oxygen could enter the brines from the overlying seawater. During the more active phases of hydrothermal discharge into the Atlantis II Deep Mn-oxyhydroxide minerals accumulated outside the southwest basin at some distance from the springs within most parts of the deep including the east and west basins. During some of the periods when there was a small decrease in hydrothermal activity Mn-oxyhydroxide minerals accumulated on the elevated areas and within the north basin. Possibly because of the stagnant condition near the bottom these Mn-minerals did not accumulate in sediments of the east and west basins but small amounts of sulfide minerals formed. In one sample from the CO Zone in the west basin the value of $\delta^{34}S$ measured in the sulfide sulfur was –24.65‰. This value is evidence, which indicates that the sulfides were formed by the bacterial sulfate-reduction (Shanks and Bischoff 1980).

Better intermixing of the brines with the overlying seawater during the accumulation of the CO Zone probably resulted in some increase in the pH. Under these conditions the solubility of the biogenic $CaCO_3$ in the brines decreased and small amounts of it accumulated in the sediments. The greatest accumulation of $CaCO_3$ was in the north basin.

It is noteworthy that the Fe/Si ratio in the sediments of the CO Zone that averages 3.6 is considerably higher than in the sediments of the other zones where it varies from 2.1 to 2.8. If the Fe/Si ratio of the hydrothermal solutions did not vary, the specific increase in the Fe/Si ratio in the CO Zone sediments in comparison with the sediments of the other zones can be explained by the better solubility of silica when the pH was higher. Under conditions where there was a greater amount of mixing of the brines with the overlying seawater there could have been extensive transportation of silica from the brines into seawater outside the Atlantis II Deep.

Upper Sulfidic Zone. Data are available for the sediments from all parts of the Atlantis II Deep except for those from the central sill area (Table 2.6). The highest Fe/Al and Si/Al ratios are in the sediments from the north and southwest basins. The S/Al and Ca/Al ratios are highest and the C_{carb}/Al ratio is minimum in the sediments from the southwest basin. In the sediments from the southwest basin the anhydrite carries the main part of the Ca and a significant part (about a half of the total content) of sulfur. In this particular basin of the Atlantis II Deep the SOAN and OAN Zones have been distinguished in the sediments that correspond in their stratigraphic position to the SU_2 Zone in other basins (Fig. 2.7, 2.10). The highest content of sulfide sulfur is in the sediments from the west basin where its content is a little higher than in the sediments from the southwest basin. However, in the southwest basin sediments the Cu/Al ratio is at a maximum and this ratio along with the other data apparently is evidence that there was an extensive accumulation of Cu that became a component of chalcopyrite. The Cu/Al ratio is a little lower in the sediments from the west basin. The highest Zn/Al ratios occur in the sediments from the southeast passage and from the west basin. According to Zhabina and Sokolov (1982), in this areas as well as in the east basin Zn and Cu occur in the sediments of the SU_2 Zone where they are being incorporated into only sulfide phases.

In all of the sediment samples studied from the SU_2 Zone the values of $\delta^{34}S$ in the sulfide sulfur exceed 0‰, the average is +7‰ (Kaplan et al. 1969; Shanks and Bischoff 1980). This is considered to be evidence for a hydrothermal origin of the sulfide sulfur.

In the SU_2 Zone sediments from the southwest basin the rates of accumulation of the Fe and Si are a little lower than those in the north basin sediments. However, in the sediments from the southwest basin, as well as in the sediments from the adjacent parts of the deep in the west basin and southeast passage, the rates of accumulation are the highest for Zn, Cu and for S that is mainly a component of anhydrite. These data are evidence of the existence of hydrothermal orifices in the southwest basin during formation of the SU_2 Zone sediments from 5.9 to 3.6 ka BP.

The rate of accumulation of Mn in the SU_2 Zone sediments is the highest in the north basin where the rate of accumulation of Si is also much higher than in most of the other areas of the Atlantis II Deep. However, there is no accompanying increase in the rate of accumulation of S, Zn and Cu in the SU_2 Zone sediments in the north basin. A plausible reason for this will be considered later.

Amorphous-Silicatic Zone. The formation and development of the AM Zone began 3.6 ka BP and it is continuing at the present time. There are data for the sediments of this zone from all parts of the Atlantis II Deep (Table 2.6). The most extensive accumulation of the metal-bearing matter and of the associated chemical elements was in the sediments from the southwest basin and the maximum ratios of Fe/Al, Si/Al, Zn/Al, Cu/Al, and S/Al are found in this basin. The most abundant accumulation of sulfide minerals along with Si-Fe gel (that is the principal component) in the sediments from the southwest basin was the basis for distinguishing the SAM Zone here (Fig. 2.7, 2.12).

The values of $\delta^{34}S$ in the sulfide sulfur from the AM Zone sediments vary from +3.4‰ to +9.8‰ (Kaplan et al. 1969) and the average value is +5.5‰, which is evidence for the hydrothermal origin of the sulfide sulfur. In the sediments from the southwest and west basins the sulfur accumulated not only in the sulfide minerals but also in the anhydrite. On the average the sulfide minerals contain about half and about three quarters of the total contents of the Zn and Cu respectively in the AM Zone sediments in the southwest and west basins (Zhabina and Sokolov 1982). This is evident not only from the mineralogical but also from the geochemical data.

The highest Fe/Al, Si/Al, Zn/Al, Cu/Al, and S/Al ratios, as well as the highest contents of anhydrite and hydrothermal sulfide minerals occur in the sediments in the southwest basin which is evidence of the existence of hydrothermal orifices in this part of the deep during the last 3.6 ka.

The highest ratios of Mn/Al, C_{carb}/Al, and Ca/Al, as well as high Fe/Al ratio occur in the AM Zone, SU_2 Zone and CO Zone sediments in the north basin. And the S/Al, Zn/Al, and Cu/Al ratios are low in the AM Zone sediments in this basin (Table 2.6). Apparently this is because of the different behavior of these elements during the formation and sedimentation of their mineral carriers. The sulfide minerals and anhydrite form near the floor in the vicinity of the hydrothermal orifices where there is an increase in their accumulation rates. Fe- and Mn-hydroxide minerals form at higher levels above the floor, the Fe-hydroxide in the upper brine layer and the Mn-hydroxide in the transition zone. There is an increase in the concentration of particulate Fe in the upper brine layer and Mn in the transition zone in the southwest basin above the sites interpreted as the location of the hydrothermal orifices (Hartmann 1973). Evidence shows that there is more extensive formation of particulate Fe- and Mn-hydroxide minerals in this area than in the adjacent parts of the deep. The Fe-hydroxides have a greater resistance than the Mn-hydroxides to dissolution in the brines and they settle to the floor. The places where their development was most extensive is indicated now, though nebulous because of the circulation of the brines, by their greater accumulation in the sediments. Similar indications of the development of Mn-hydroxides are not found because they are dissolved as they settle through the brines. The dissolved metals that enter the deep with the hydrothermal fluids are spread slowly throughout the area. The present locations of the hydrothermal orifices as interpreted from the data, the main current directions in the lower brine layer, as well as the average concentrations of the dissolved Fe and Mn in the lower brine layer in different parts of the deep, are shown in Fig. 2.5. It is clearly indicated that the concentrations of the dissolved Fe and Mn decrease with distance from the orifices and the greatest decrease in the concentrations of both Fe and Mn occurs in the north basin. This could have been the result of very slow inflow of the dissolved metals to the north basin. At depth in the north basin, which can be considered as a separate deep, the lower brine layer is only connected with that in other parts of the Atlantis II Deep through a narrow inlet in its western part. This restricted connection is evident from the hydrographic data (Hartmann et al. 1998; Scholten et al. 2000). This is of particular concern as the dissolved Fe, because of its high concentration, occurs only in the lower brine layer (Table 2.2). As to the dissolved Mn(II) it can enter the north basin through the upper brine layer too.

It is necessary to take notice that: (a) the bottom environment in the north basin is more oxidizing than in other parts of the Atlantis II Deep that are covered by the lower brine layer and the Eh of the interstitial water in the upper part of the sediment layer in the AM Zone in the north basin varies from +430 to +481 mV, and in other parts of the deep it does not exceed +310 mV (Brooks et al. 1969; Gurvich and Bogdanov 1986e); (b) hydrothermal sulfide minerals are absent in the AM Zone sediments in the north basin; (c) the minimum concentrations of the particulate Zn and Cu in the lower brine layer in the Atlantis II Deep occur in the north basin (Hartmann 1973). All of this is evidence of a relatively more favorable environment in the north basin for the oxidation of Fe and its accumulation as a component of the Fe-oxyhydroxide minerals than the environment in other parts of the deep.

In the AM Zone sediments that are covered by the lower brine layer the Mn-oxyhydroxide minerals are absent, and the Mn-carbonate minerals have been found only in sediments in the vicinity of sites believed to be the location of hydrothermal orifices (Fig. 2.12). The highest Mn/Al ratio in the sediments from the north basin apparently was developed by the coprecipitation and sorption of dissolved Mn by the Fe-hydroxides.

The highest Fe/Al, Si/Al, S/Al, Cu/Al, and Zn/Al ratios are in the sediments of the CO, SU_2, and AM Zones in the southwest basin, apparently where the hydrothermal fluids enter the deep. The ratios of Fe/Al, Mn/Al, and $CaCO_3$/Al in the sediments of the CO, SU_2, and AM Zones in the north basin are higher than those in the sediments of these zones in others parts of the deep distant from the locations of the hydrothermal orifices. These facts are evidences that the location of the place where the hydrothermal fluids enter the basin and the pattern of their movement within the deep did not vary greatly in the last 8–9 ka. During the formation of the DOP Zone sediments hydrothermal activity within the Atlantis II Deep obviously was episodic and took place at least 13 to 15 ka BP mainly in the north passage and partly within the east and probably in the west basins. Later, during the accumulation of the SU_1 Zone sediments, 11.7 to 8.6 ka BP, hydrothermal activity existed within the north passage and in the west basin area.

2.2.4. Evolution of accumulation rates of the main hydrothermal components in bottom sediments, intensity of hydrothermal activity and mineral formation

Estimates have shown that the average annual accumulation of the dry salt-free material in the SU_1 Zone sediments of the Atlantis II Deep was about 7.9 thousand tons, and that the total amount of this material in the zone is about 24 million tons. The average annual accumulation of the dry salt-free material in the CO Zone sediments was about 12 thousand tons, and its total amount is about 32 million tons and respectively in the SU_2 Zone 8.8 thousand tons per year and a total of 20 million tons, and in the AM Zone 4.7 thousand tons per year and a total of 17 million tons. The total amount of the dry salt-free material in the sediments of the SU_1, CO, SU_2, and AM Zones is estimated to be 93 million tons. This total amount is similar to that of 94 million tons that was estimated in the Atlantis II Project by using data from more than 600 sediment cores (Mustafa 1984).

Estimates of the average annual excessive accumulation of the main hydrothermal elements: Fe, Mn, S, Zn, Cu, and Si during the formation of the sediments in the SU_1, CO, SU_2, and AM Zones have been made by using the data listed for the average annual accumulation of the dry salt-free material and data for the average chemical composition (Table 2.4). These are shown in Table 2.7 and in Fig. 2.16.

Table 2.7. Average annual excessive accumulation and the average rates of excessive accumulation of the main hydrothermal elements in the sediments from the Atlantis II Deep during formation of the SU_1, CO, SU_2, and AM Zones.

Zone	Time of formation	Fe	Si	Mn	S	Zn	Cu	Fe	Si	Mn	S	Zn	Cu
	ka BP	$t\ a^{-1}$						$g\ cm^{-2}\ ka^{-1}$					
AM	3.6–0	1500	350	27	140	110	23	24	5.6	0.43	2.2	1.8	0.37
SU_2	5.9–3.6	2200	560	66	680	440	74	35	8.9	1.1	11	7.0	1.2
CO	8.6–5.9	3900	310	540	200	55	24	62	5.1	8.6	3.2	0.88	0.38
SU_1	11.7–8.6	2500	540	130	620	280	87	40	8.4	2.1	9.9	4.5	1.4

During the last 11.7 ka the average annual accumulation of the dry salt-free material and the average annual excessive accumulation of the main hydrothermal elements varied greatly. If one considers the excessive accumulation of Fe as a proxy for the accumulation of the metal-bearing hydrothermal matter, one is lead to conclude that the accumulation rate of the latter was at a minimum during the formation of the AM Zone, a maximum during the formation of the CO Zone, and intermediate during the formation of the SU_1 and SU_2 Zones. The values for the SU_1 and SU_2 Zones are similar. Considering that the major part, from 99.4% to 84%, of the hydrothermal Fe that entered the Atlantis II Deep accumulated on the floor within the brine pool (Craig 1969; Bignell et al. 1976b; Gurvich and Bogdanov 1986d) the rate of accumulation of the excessive Fe can be considered as a measure of the rate of contribution of the hydrothermal components to the deep, and also of the intensity of the hydrothermal activity. In this case the data in Table 2.7 and Fig. 2.16 show the temporal evolution in the intensity of hydrothermal activity in the deep. During the last 11.7 ka the intensity was at a minimum from 3.6 to 0 ka BP, at a maximum from 8.6 to 5.9 ka BP, and intermediate from 11.7 to 8.6 ka BP and from 5.9 to 3.6 ka BP.

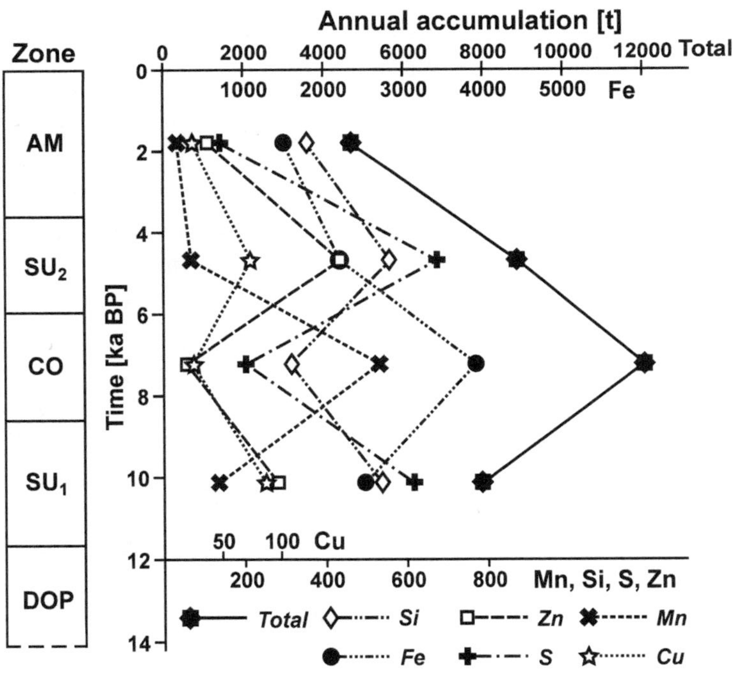

Fig. 2.16. Variations in the average annual accumulation of the dry salt-free material and the excessive accumulation of the main hydrothermal elements in the sediments of the Atlantis II Deep during formation of the lithostratigraphic zones.

When the contribution of the hydrothermal metal-bearing matter to the Atlantis II Deep sediments was relatively low in the time from 3.6 to 0 ka BP, it accumulated on the floor mainly as Si-Fe gel. The hydrothermal sulfide minerals formed only in the vicinity of the hydrothermal orifices. In the intermediate stages of hydrothermal activity from 5.9 to 3.6 ka BP and from 11.7 to 8.6 ka BP the hydrothermal sulfide minerals accumulated on the floor almost everywhere within the deep. The high-density brines prevented the sulfide minerals from becoming oxidized and no significant amount of dissolved oxygen entered the brines from the overlying seawater. When there was hydrothermal activity of maximum intensity, from 8.6 to 5.9 ka BP, the hydrothermal springs apparently were strong enough to disturb the uniformity of the brines. As a result there was extensive intermixing of the brines with the overlying seawater and the amount of dissolved oxygen that entered the brines was sufficient to enable the oxidizing not only the dissolved Fe (II) but also a significant amount of the dissolved Mn(II), which accumulated in the sediments. Under these conditions the chalcophile elements formed only small amounts of sulfide minerals and transportation of them to the sediment was not effective. Obviously these elements were transported from the deep and assimilated in the overlying seawater. A significant part of the Si was transported from the deep as well and the possible reason for this was considered in the preceding section.

The Fe-oxyhydroxide, sulfide, and Mn minerals are the basic components that determine the main mineral and chemical composition and lithological features of the metal-bearing part of the sediments from the Atlantis II Deep. From this point of view it is of interest to trace the evolution of the proportion of amounts of Fe, S, and Mn that accumulated in the metal-bearing part of the sediments at the different intensities of hydrothermal activity in the deep (Fig. 2.17).

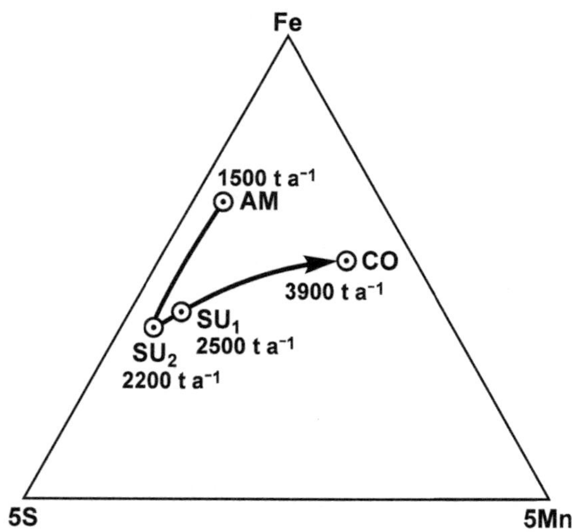

Fig. 2.17. Evolution of a proportion of the average annual accumulation of the excessive Fe, S, and Mn in the sediments of the lithostratigraphic zones in the Atlantis II Deep as the accumulation of the excessive Fe increased.

A minimum relative amount of Mn was contributed to the metal-bearing matter when the intensity of hydrothermal activity was at a minimum (AM Zone). Mn-minerals are extremely rare in the AM Zone sediments, and manganosiderite was found in only two cores (Fig. 2.12). With an increase in the intensity of the hydrothermal activity (SU_1 and SU_2 Zones) the relative contribution of Mn to the metal-bearing matter increased slightly and it accumulated in the sediments as manganosiderite (Fig. 2.8, 2.10). With a maximum intensity of hydrothermal activity (CO Zone) the relative contribution of Mn to the metal-bearing matter increased considerably, and not only manganosiderite but also manganite and todorokite were formed (Fig. 2.9; Bäcker and Richter 1973; Butuzova and Lisitzina 1986b).

As indicated in Fig. 2.17, a maximum relative contribution of S in the sum of Fe, Mn and S was during periods of hydrothermal activity of intermediate intensity (SU_1 and SU_2 Zones) under conditions that were most favorable for the accumulation of the sulfide minerals.

Fig. 2.17 shows that the proportions of the main elements in the metal-bearing matter and therefore the mineral and chemical composition of the sediments in the Atlantis II Deep depended largely on the intensity of the hydrothermal activity. The intensity of the hydrothermal activity, on the one hand, could have influenced the chemical and physicochemical processes within the brines and the retention of the chemical elements within the deep, and on the other hand influenced the amounts and proportions of the chemical elements that were contributed to the deep by the hydrothermal fluids. Therefore it is interesting to compare the evolution of the accumulation rates of the main components in the metal-bearing matter as the intensity of the hydrothermal activity increased (Fig. 2.18).

Using the accumulation rate of the excessive Fe as a proxy for the intensity of hydrothermal activity, the accumulation rates of the excessive Si, S, Zn, and Cu increase at first to their maximum and then decrease and the accumulation rate of the excessive Mn increases steadily with an increase in the intensity of the hydrothermal activity. The maximum accumulation rates of the excessive Si, S, Zn, and Cu in the sediments of the SU_1 and SU_2 Zones are related to hydrothermal activity of intermediate intensity. In the sediments of the SU_1 and SU_2 Zones the accumulation rates of these elements as well as their ratios to Fe have maximum values. With a low intensity of hydrothermal activity, which corresponds to that for the sediments of the AM Zone the ratio of Si/Fe remains equal to the ratios related to intermediate hydrothermal intensity. The ratios of S, Zn, and Cu to Fe that are related to low intensities of hydrothermal activity are lower than those related to intermediate hydrothermal intensity. This means that with a decrease in the intensity of hydrothermal activity the relative decreases in the accumulation rates of S, Zn, and Cu are greater than they are for Fe.

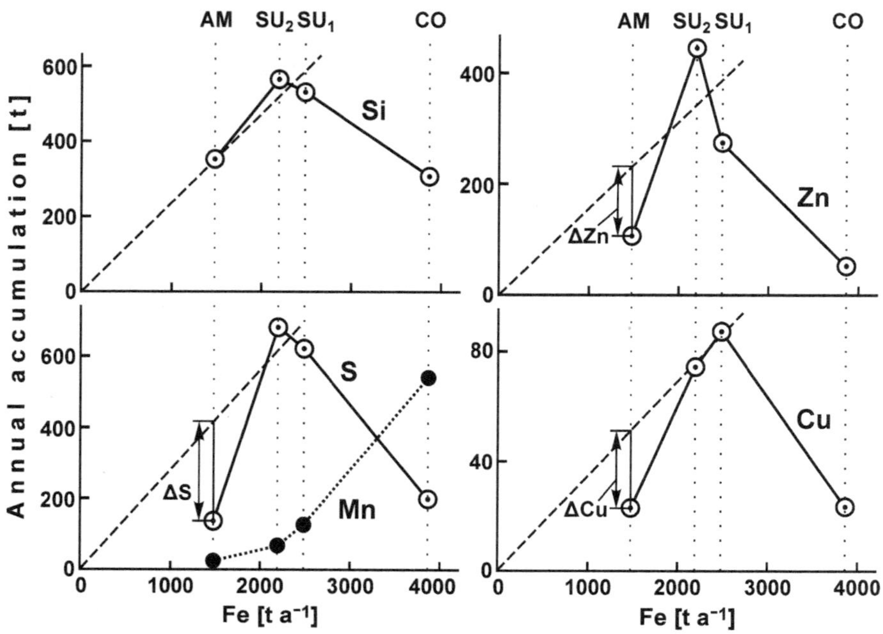

Fig. 2.18. Evolution of the average annual accumulation of the excessive Si, S, Mn, Zn, and Cu as the average annual accumulation of the excessive Fe in the sediments from the Atlantis II Deep increased.

If it is assumed that during the accumulation of the sediments of the SU_1, CO, SU_2, and AM Zones in the Atlantis II Deep concentrations of the chemical elements under consideration in the primary hydrothermal fluids did not vary or varied only a little then the variations in the rates of excessive accumulation of the elements in sediments could have resulted from: 1) losses of some amount of the elements in the crust due to their precipitation during the migration of the hydrothermal fluids from the zone of mobilization to the deep; and 2) losses of some amount of the elements after entering the deep due to their dissipation outside of the brine pool.

If the sediments of the SU_1 and SU_2 Zones formed during times of intermediate intensity in the hydrothermal activity and the proportion of the excessive Fe, Si, Zn, and Cu is close to the proportion of Fe, Si, Zn, and Cu in the primary hydrothermal fluids then there are deficiencies of S, Zn, and Cu in the AM Zone sediments that are named ΔS, ΔZn, ΔCu (Fig. 2.18). It can be assumed that if the S/Fe, Zn/Fe, and Cu/Fe ratios in the primary hydrothermal fluids do not vary then during time of low intensity in the hydrothermal activity (time of accumulation of the AM Zone sediments) some part of the S, Zn, and Cu carried by the fluids was precipitated, apparently in sulfide minerals, in the crust prior to the fluids entering the deep and a part of their Fe content was precipitated as well. The unloading of the hydrothermal fluids and deposition of their mineral content continued to take place within the deep. The sulfide minerals accumulated only in the sediments near the hydrothermal orifices. When the intensity of the hydrothermal activity was intermediate (times of accumulation of the sediments of SU_1 and SU_2 Zones) there was a maximum efficiency of unloading of mineral components within the deep. A maximum amount of sulfide minerals accumulated in the sediments at this time. Sulfide minerals accumulated in the sediments in all parts of the deep.

A possible reason for the decrease in the accumulation of the excessive Si, S, Zn, and Cu during a time of maximum intensity of hydrothermal activity (time of accumulation of the CO Zone sediments) in comparison with times of intermediate and low intensity in the hydrothermal activity (Fig. 2.18) is that on the one hand the conditions in the deep were not favorable for the accumulation of the sulfide minerals and silica, on the other hand because of greater mixing of the brines with the overlying seawater transportation of these elements from the brines into seawater outside the Atlantis II Deep was extensive.

Unlike Si, S, Zn, and Cu the accumulation of the excessive Mn in the sediments steadily increases as the intensity of the hydrothermal activity increases (Fig. 2.18). The relationship of the annual accumulation of the excessive Mn and Fe in the sediments from the Atlantis II Deep is clearly shown by an empirical equation $Mn = 4.6 \cdot 10^{0.00054Fe}$, where Mn and Fe are in tons. The correlation coefficient of the measured data of the excessive Mn accumulation and the data calculated from this equation is 0.98. It follows from this equation that with an absence of the accumulation of an excessive amount of Fe (Fe=0), 4.6 tons of the excessive amount of Mn would accumulate annually. For example if there was a decrease in the intensity of the hydrothermal activity all of the hydrothermal Fe would be precipitated within the crust but the hydrothermal Mn would enter and accumulate at the bottom of the deep.

At the present time not all hydrothermal metal-bearing matter contributed to the Atlantis II Deep accumulates in the sediments within the brine pool. Some amount of chemical elements leaves this area and accumulates in sediments around the brine pool (Bignell et al. 1976b) and also disperses in the overlying seawater. The contents of Fe and Mn in the bottom sediments at different distances from the brine pool are shown in Fig. 2.19. If the content of the Fe exceeds the background level at distances of no more than 1 km from the brine pool then the content of the Mn will exceed the background level at distances of more than 11 km. By extrapolation of the Mn contents shown in Fig. 2.19 the values above the background level will be present in the sediments that are at least 20 to 25 km from the brine pool. According to estimates, the annual accumulation of the excessive Fe outside the brine pool is about 280 tons, and of the excessive Mn is about 160 tons within a band that is 11 km wide, and about 260 tons within a band that is 22 km wide.

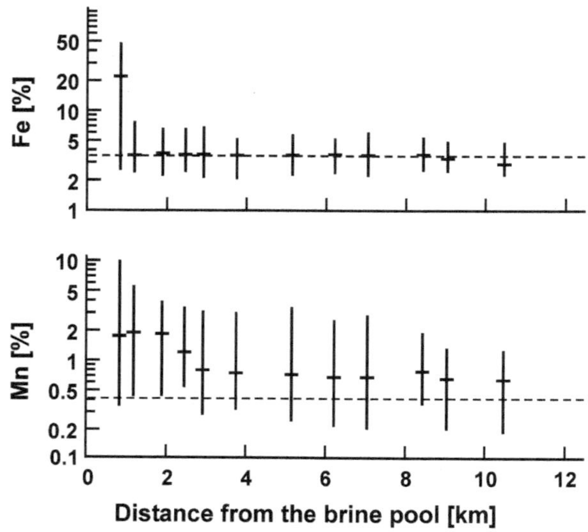

Fig. 2.19. Contents of Fe and Mn in bottom sediments at different distances from the brine pool of the Atlantis II Deep. Dash lines show background contents. Composed from Bignell et al. (1976b) with additions.

These rough estimates show that at the present time the major part, about 84% or 1500 t a^{-1}, of the hydrothermal Fe contributed to the Atlantis II Deep, accumulates within the brine pool in metalliferous sediments of the AM Zone. As for the hydrothermal Mn the predominant part (>91%) of it leaves the deep, and only 27 t a^{-1} accumulate within the brine pool in metalliferous sediments of the AM Zone. With an increase in the intensity of the hydrothermal activity and the accumulation of the hydrothermal Fe the accumulation of the hydrothermal Mn and the average Mn/Fe ratio in the sediments of the deep increases, which is clearly visible in Fig. 2.18.

It is of interest to estimate an annual budget for hydrothermal Mn in the Atlantis II Deep during the formation of the AM Zone from 3.6 ka BP to the present. This can be calculated if the following assumptions are made:
(1) The hydrothermal Fe contributed to the Atlantis II Deep accumulated completely within the brine pool (1500 t a^{-1}) and within the band 1 km wide (280 t a^{-1}) outside of it.
(2) Concentrations of the particulate and dissolved Fe and Mn in the lower and upper brine layers and in the transition zone, that are used for the esti-

mates, existed during the whole period of formation of the AM Zone, and geometric mean values from the observations that Hartmann (1973, 1985) made in 1971/1972 and in 1977 and values from Table 2.2 are used.
(3) Velocities of transfer of the dissolved Fe and Mn to the zones of their oxidation, to the upper brine layer, and to the transition zone respectively, are equal and constant: $D_{Fe} = D_{Mn} = const.$
(4) All hydrothermal Fe goes through the zone of oxidation and precipitates there[18], after the precipitation the particulate Fe of hydrothermal origin does not pass into solution again.
(5) Settling velocities of the particulate Fe and Mn precipitated are equal and constant: $P_{Fe} = P_{Mn} = const.$
(6) The amounts of hydrothermal Fe and Mn within the brines are constant.

The average vertical flux of the dissolved hydrothermal Fe up to the zone of its oxidation is:

$$F^{\uparrow}_{Fe} = \frac{H_{Fe}}{A},$$

where H_{Fe} is contribution of the hydrothermal Fe to the Atlantis II Deep, $H_{Fe} = 1500 + 280 = 1780$ [t a^{-1}]; A = 55 km^2, the area of the brine pool,

$$F^{\uparrow}_{Fe} = \frac{1780}{55} = 32.4 \text{ t km}^{-2} \text{ a}^{-1} = 32.4 \text{ g m}^{-2} \text{ a}^{-1}.$$

The average velocity of transfer of the dissolved Fe up to the zone of its oxidation is:

$$D_{Fe} = \frac{F^{\uparrow}_{Fe}}{C_{FeL} - C_{FeU}},$$

where C_{FeL} and C_{FeU} are the average concentrations of the dissolved Fe in the lower and upper brine layers[19] [g m^{-3}],

$$D_{Fe} = \frac{32.4}{100 - 5} = 0.34 \text{ m a}^{-1}.$$

The average vertical settling flux of the particulate Fe down to the sediments within the brine pool, which can be calculated based on the assumption (1) is:

$$F^{\downarrow}_{Fe} = \frac{B_{Fe}}{A},$$

where B_{Fe} is the annual accumulation of the hydrothermal Fe within the brine pool (1500 t a^{-1}),

---

[18] In the AM sediments less than 5% of the hydrothermal Fe is bound with sulfide minerals.

[19] See Table 2.2.

$$F^{\downarrow}_{Fe} = \frac{1500}{55} = 27.3 \text{ t km}^{-2} \text{ a}^{-1} = 27.3 \text{ g m}^{-2} \text{ a}^{-1}.$$

The average settling velocity of the precipitated particulate Fe from the zone where it is oxidized is:

$$P_{Fe} = \frac{F^{\downarrow}_{Fe}}{K_{Fe}},$$

where K_{Fe} is the average concentration of the particulate iron[20] [g m^{-3}],

$$P_{Fe} = \frac{27.3}{0.036} = 758 \text{ m a}^{-1}.$$

These data can be used for calculating the fluxes of the hydrothermal Mn. The average vertical flux of the particulate Mn settling down to the lower brine layer is:

$$F^{\downarrow\downarrow}_{Mn} = P_{Mn}\, K_{MnL},$$

where $P_{Mn} = P_{Fe}$ (assumption 5), K_{MnL} is the average concentration of the particulate Mn in the lower brine layer[21] [g m^{-3}],

$$F^{\downarrow\downarrow}_{Mn} = 758 \cdot 0.0015 = 1.14 \text{ g m}^{-2} \text{ a}^{-1} = 1.14 \text{ t km}^{-2} \text{ a}^{-1}.$$

For the whole brine pool the flux is: $1.14 \cdot 55 = 63$ t a^{-1}. The difference between this flux and the accumulation of the hydrothermal Mn in the sediments within the brine pool (27 t a^{-1}) is the aggregate regeneration or dissolution of the Mn in the lower brine layer and on the floor of the deep which is: $63 - 27 = 36$ t a^{-1}.

The average vertical flux of the particulate Mn settling down from the transition zone is:

$$F^{\downarrow}_{Mn} = P_{Mn}\, K_{MnT},$$

where K_{MnT} is the average concentration of the particulate Mn in the transition zone[22], g m^{-3},

$$F^{\downarrow}_{Mn} = 758 \cdot 0.009 = 6.8 \text{ g m}^{-2} \text{ a}^{-1}.$$

For the whole brine pool this flux is: $6.8 \cdot 55 = 374$ t a^{-1}. The difference of this value and the vertical flux of the particulate Mn settling down to the lower brine layer (63 t a^{-1}) gives the amount of regeneration or dissolution of the Mn in the upper brine layer, which is 311 t a^{-1}.

---

[20] See Fig. 2.6.
[21] See Fig. 2.6.
[22] See Fig. 2.6.

The average vertical flux of the dissolved Mn from the upper brine layer to the zone of its oxidation (the transition zone) is:

$$F^{\uparrow}_{Mn} = \Delta C_{Mn} D_{Mn},$$

Where ΔC_{Mn} is the difference between the concentrations of the dissolved Mn in the upper brine layer and in the transition zone[23], $D_{Mn} = D_{Fe}$ (assumption 3).

$$F^{\uparrow}_{Mn} = 88.3 \cdot 0.34 = 30 \text{ g m}^{-2} \text{ a}^{-1}.$$

For the whole brine pool this flux is: $30 \cdot 55 = 1650$ t a^{-1}.

Because of the brine flow from the Atlantis II Deep to the Chain Deeps about 2 tons of the Mn from the upper brine layer accumulates annually in the sediments of the Chain Deeps (~0.4 t in the Chain A Deep, and ~1.6 t in the Chain B Deep).

The aggregated flux of the Mn from the upper brine layer to the transition zone and to the Chain Deeps is estimated to be 1652 t a^{-1}. About 374 t a^{-1} of Mn in this flux is returned from the transition zone to the brines as particulate Mn. The difference between the vertical flux of the dissolved Mn from the upper brine layer to the transition zone and the return gives the annual amount of the hydrothermal Mn that leaves the brine pool, which is: $1650 - 374 = 1276$ t a^{-1}. About 260 t a^{-1} of this amount accumulates in the sediments that are in the band around the brine pool that is 22 km wide. The rest of the Mn, about 1016 t a^{-1}, is dispersed.

The annual contribution of the hydrothermal Mn to the Atlantis II Deep is the sum of the amounts accumulating in sediments within the brine pool of the Atlantis II Deep and the Chain Deeps and of the amount that leaves the brine pool. It is: $27 + 2 + 1276 = 1305$ t a^{-1}.

A diagram of the calculated budget for hydrothermal Mn in the Atlantis II Deep is shown in Fig. 2.20, which is generalized and based on some assumptions. Nevertheless the orders of magnitude in the budget are similar to the real ones. They are confirmed because the amount of the hydrothermal Mn leaving the brine pool and of the amount of the hydrothermal Mn accumulated in the sediments within the band 22 km wide around the brine pool are of the same order of magnitude. These amounts have been estimated by using different methods.

By using the annual hydrothermal contributions of Fe and Mn to the Atlantis II Deep, which are respectively 1780 t or $1780 \cdot 10^6$ g and 1305 t or $1305 \cdot 10^6$ g, it is possible to estimate their concentrations in the hydrother-

[23] According to Brooks at al. (1969), the concentrations of the dissolved Mn in the transition zone varies from 0.0042 to 1 mg l^{-1}. The geometric mean for these data is 0.02 mg l^{-1} = 0.02 g m^{-3}.

mal fluids. According to the estimates of Anschutz and Blanc (1996) for the period from 1965 to 1992, the average contribution of hydrothermal solutions to the deep was 2000 to 3000 $m^3\ h^{-1}$ or $17.5\cdot10^6$ to $26.3\cdot10^6\ m^3\ a^{-1}$. At this contribution the average concentration of Fe and Mn would have been 59 to 89 $g\ m^{-3}$, and 44 to 65 $g\ m^{-3}$ respectively.

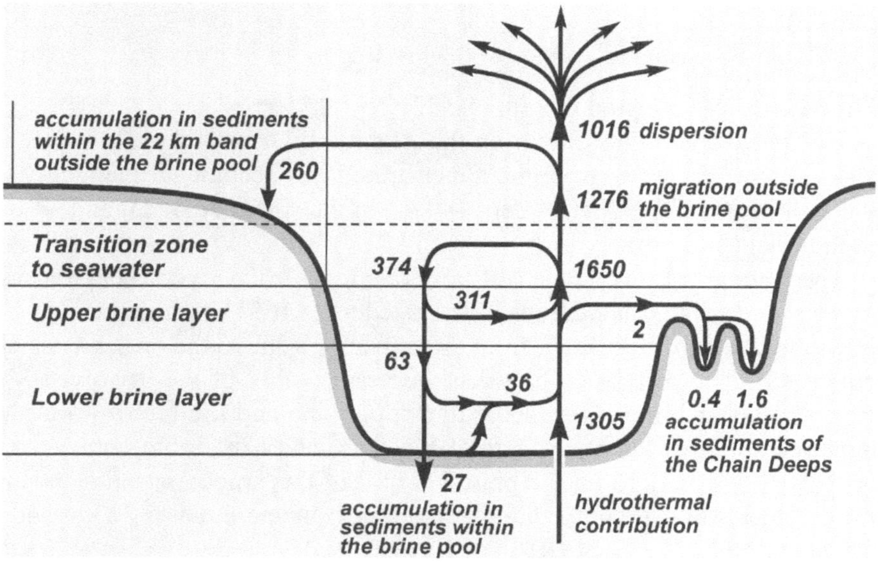

Fig. 2.20. Annual budget [tons] of the hydrothermal Mn in the Atlantis II Deep.

Comparison of these concentrations with the actual concentrations of dissolved Fe and Mn measured in the lower brine layer (Fig. 2.5) shows that they are of the same order, but apparently somewhat lower than the actual ones. The maximum concentrations of dissolved Fe_{diss} and Mn_{diss} in the lower brine layer near the hydrothermal orifices are respectively 92.5 $g\ m^{-3}$ and 92 $g\ m^{-3}$ (Fig. 2.5). It is evident that the concentration of Fe and Mn in the hydrothermal fluids would have been at least 92.5 $g\ m^{-3}$ and 92 $g\ m^{-3}$ or higher. If their concentrations in the fluids had been lower than in the lower brine layer the maxima near the orifices (Fig. 2.5) would be absent.

The estimated concentrations of the Fe and Mn in the hydrothermal fluids are of the same order as concentrations of these elements in high-temperature hydrothermal fluids in the East Pacific Rise, Mid-Atlantic Ridge, and in back-arc basins (Von Damm 1995; German and Von Damm 2004). This is further evidence that hydrothermal processes within the Red Sea rift are similar to those in other ocean rifts.

2.3. Northeast Thetis Deep

The Thetis Deep is typical of the deeps in the axial part of the Red Sea rift zone. It is located in the northern part of the Red Sea about 160 km northwest of the Atlantis II Deep. Bottom brines are absent in the Thetis Deep at the present time (Table 2.1). Metalliferous sediments occur only in the northeast basin of the deep. The NE Thetis Deep is elongate from WNW to ESE; it is about 10 km long and 3 km wide and up to 1780 m deep (Scholten et al. 1991); the coordinates for its central part are 22°47.5'N and 37°35.0'E.

2.3.1. Lithology and mineral composition of sediments from the NE Thetis Deep

Fifteen cores with a recovery of 200 to 920 cm of sediments have been investigated that were collected at depths from 1736 to 1775 m in the NE Thetis Deep (Fig. 2.21).

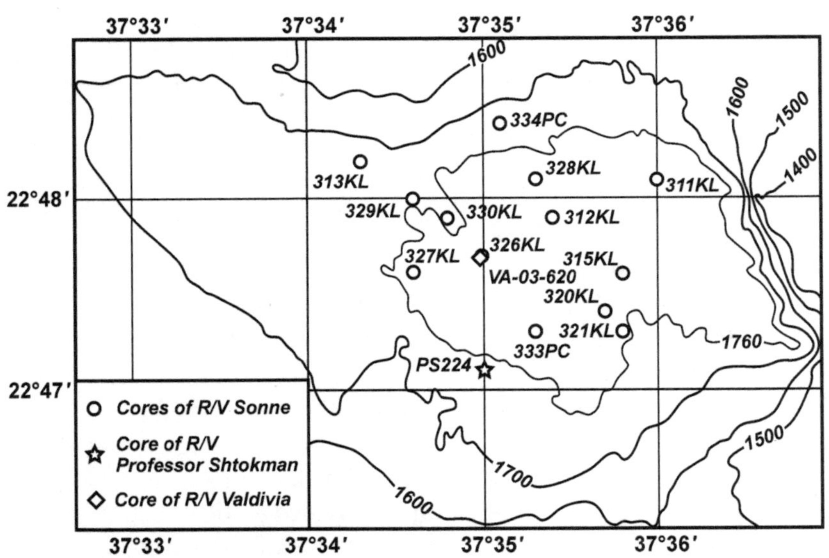

Fig. 2.21. The NE Thetis Deep and location of the sediment cores investigated. After Scholten et al. (1991) with simplification and additions.

The studies of the sediments from the NE Thetis Deep have shown alternate layers within the sediment cover that are enriched in Fe- and Mn-minerals (Bignell et al. 1976a; Bogdanov et al. 1986; Scholten et al. 1991) for the whole sediment cover at depths greater than 1736 m (the depth of the shallowest station 313KL). A generalized section of its lithology is presented in Table. 2.8.

Table 2.8. Units of the generalized lithological section of the sediment cover from the NE Thetis Deep and their approximate thickness. Compiled from Bignell et al. 1976a, Butuzova et al. (1986), and Scholten et al. (1991).

Unit	Facies	Thickness
Fe-Mn	brown manganite	80–250 cm
Fe-rich	orange lepidocrocite yellow goethite red hematite black magnetite	100–200 cm
Fe-Mn	brown manganite	300–400 cm
detrital-biogenic	grey-brown detrital-biogenic facies	
Fe-Mn	brown manganite	~150 cm
Fe-rich	orange lepidocrocite yellow goethite red hematite black magnetite	~250 cm

The basalt basement rocks are overlain by a succession of sedimentary units that consist of three main types: Fe-rich, Fe-Mn, and detrital-biogenic. These units vary considerably in their horizontal and vertical sections but their main features are fairly uniform.

Core PS224, located at 22°47.1'N, 37°35.0'E, is 510 cm long, and has been investigated most thoroughly. The main features of the sediments that were penetrated are typical of the sediment cover in the NE Thetis Deep and are described as follows.

The core has penetrated sediments with water content of 60 to 80%. The sediments are mainly dark-brown with intercalations, patches and lenses of brown, mustard yellow, red, green, grey, and black color.

The sediments in the core have been subdivided into five stratigraphic units on the basis of color, consistency, character of the laminations, and features of mineral and chemical compositions (Fig. 2.22). The descriptions reported here are from Butuzova et al. (1986, 1988a).

Clayey-carbonaceous material typical for the background Red Sea sediments and ferruginous gel with microglobular texture, lepidocrocite, and asbolane are the main components of the upper layer, 0 to 120 cm, sampled in the core. The proportions of the metal-bearing matter and background material vary within the layer. In the upper 100 cm of the layer the content

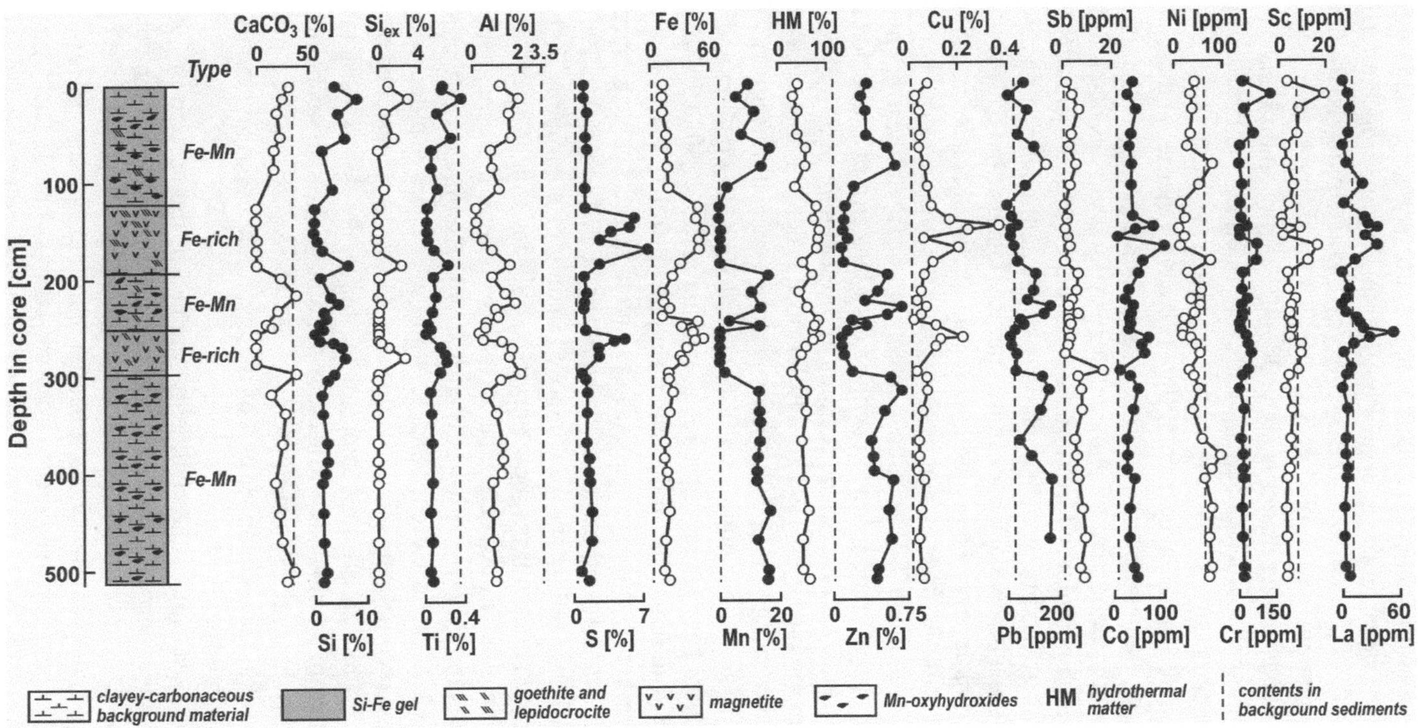

Fig. 2.22. Lithologic and chemical composition of sediments from Core PS224. After Butuzova et al. (1986) and Gurvich and Bogdanov (1986g).

of the metal-bearing matter is as high as approximately 60%, and in the lower part of the layer, in the interval from 100 to 120 cm, biogenic-terrigenous material is predominant; Fe-Mn oxyhydroxide minerals form rare microimpregnations within the interval from 100 to 120 cm, and the content of the metal-bearing matter is about 30–40%. X-ray amorphous phases are prevalent, and lepidocrocite along with very poorly crystallized manganite are rare components.

The sediments in the layer from 120 to 190 cm are denser and are composed mainly of an admixture of ferruginous gel and magnetite that is present consistently in the mass. The magnetite usually occurs as black impregnations, which commonly form thin microlayers or oriented clusters, and as a result microbanding is a characteristic of the sediments. In places the magnetite forms black clusters or individual crystals that have clear facets tens of microns in size. The content of the magnetite increases notably in the lower part in the layer. Goethite, lepidocrocite, and in places hematite occur in the sediments. The high content of the metal-bearing matter (>90%) and the extremely low content of an admixture of biogenic-terrigenous material are characteristic features of the sediment in this layer.

The layer sampled from 190 to 249 cm in the core is composed of a mixture of the background clayey-carbonaceous material, ferruginous gel, lepidocrocite, goethite-groutite, and asbolane along with goethite and traces of hematite. The content of the metal-bearing matter is 40 to 75%.

In general, the sediments in the underlying layer, from 249 to 298 cm in the core, are very similar in consistency, color, in the characteristics of their lamination and composition to those of the layer from 120 to 190 cm in the core. These sediments are composed mainly of amorphous Fe-hydroxides impregnated with magnetite crystals, only microns in size that are distributed irregularly throughout the mass. The content of the magnetite in the sediments in this layer is lower than that in the sediments in the layer from 120 to 190 cm in the core and, in some interlayers, lepidocrocite is one of the main components in the sediments. At depths represented by the core from 270 to 275 cm the content of the magnetite decreases notably, and the admixture of biogenic-terrigenous material in the sediments increases, especially beyond 287 cm in the core. Metal-bearing matter is predominant, 70 to 85 %, in the upper part of the layer, and in the lower part the content is 30 to 60%.

The lowermost part of the sediments sampled in the core from 298 to 510 cm is similar in color, structure and composition to the sediments in the layers sampled from 0 to 120 cm and from 190 to 249 cm in the core. Background clayey-carbonaceous material and ferruginous gel are the main components of the sediments from the layer sampled. The proportions of these components vary considerably along the layer. The ferruginous material consists of lepidocrocite, goethite, and hematite, and significant amounts of asbolane and goethite-groutite are present. The content of the metal-bearing matter is about 50%. In the core interval from 390 to 394 cm there is a distinct interlayer composed of background clayey-carbonaceous material that has distinct gradational laminations in which there is a minimum content of the metal-bearing matter.

Core VA-03-620, collected at the site 22°47.69'N, 37°34.96'E, investigated by Bignell et al. (1976a), is 440 cm long and has the same features as Core PS224. Sediments in both of the cores have a similar succession of layers. The layer in Core VA-03-620, sampled from 0 to 230 cm, correspond to the layer from 0 to 120 cm sampled in Core PS224 and the layer from 230 to 346 cm to that from 120 to 190 cm in Core PS224, and the layer from 346 to 440 cm to that from 190 to 249 cm in Core PS224. There are differences in the similar layers in these two cores: a) the layers in Core VA-03-620 are thicker, b) the sediments in the Fe-rich layer 230 to 346 cm contain not only magnetite and Fe-oxyhydroxide minerals but also sulfide minerals (chalcopyrite and sphalerite). Sulfide minerals have also been found in Cores 321KL, 328KL, 329KL, and 334PC (Missack 1988). The presence of sulfide minerals containing Cu in the Fe-rich layers of Core PS224 appears to be a reason of an increase in the contents of S and Cu (Fig. 2.22, Table 2.9).

2.3.2. Chemical composition of sediments from the NE Thetis Deep

The average chemical composition of the main types and facies of the sediments from the NE Thetis Deep is shown in Table 2.9, and the chemical composition of the sediments sampled in Core PS224 is shown in Fig. 2.22.

The Fe-Mn sediments differ essentially in their chemical composition from the Fe-rich sediments. Background material is present in the Fe-Mn sediments and is predominant in some places but in the Fe-rich sediments the metal-bearing matter is prevalent and the content of background material is insignificant. This is clearly indicated by the distribution of the main components of the background material.

The $CaCO_3$ content in the Fe-Mn sediments is lower than in the background sediments, but it is considerably higher than in the Fe-rich sediments where this component is almost absent (Fig. 2.22, Table 2.9). The contents of Al, Ti, Si, and Mg have a somewhat different distribution. The Fe-rich sediments from Core PS224 in the layer from 120 to 190 cm, except for its lowermost part, and in the upper part of the layer from 249 to 298 cm are depleted considerably in these elements in comparison with the

Table 2.9. Average chemical composition of the main types and facies of sediments from the NE Thetis Deep.

Sediments, facies	$CaCO_3$	Si	Al	Fe	S	Cu	Co	P	V	La	Ce	Nd	Sm	Eu	Tb	Yb	Lu	Ref.
			%								ppm							
sediments																		
Fe-rich			0.48	67.8		4670	120		350									1
Fe-rich	<2	2.04	0.64	46.76	4.0	1618	61			22.2	44.9	20.4	3.8	3.7	0.5	1.2	0.20	2, 4
Fe-rich				50		1085												5
Fe-Mn			0.80	20.5		860	20		280									1
Fe-Mn	25.2	2.67	1.03	18.38	1.5	496	40			6.0	10.0	5.8	1.1	0.32	0.17	0.64	0.11	2, 4
Fe-Mn				16.50		322												5
Intermediate	29.1	4.23	1.32	29.92	1.9	527	35			15.1	30.8	14.0	2.5	1.4	0.4	1.0	0.18	2, 4
facies																		
Magnetite	2	1.70	0.64	54.50		4101	216	3300	4226									3
Goethite	0.5	0.12	0.05	50.16		185	60	700	249									3
Lepidocrocite	1	0.34	0.14	48.97		1545	101	1400	1327									3
Hematite	27	3.70	1.41	19.86		815	60	1100	918									3
Manganite	19	2.25	0.91	13.00		759	57	900	922									3
Detrital-biogenic	62.00			1.45		102												3

Table 2.9. (continued)

Sediments, facies	Ti	K	Mg	Mn	Zn	Pb	Ni	Sr	Ba	Sb	Cr	Mo	Ga	Sc	Hf	Sn	Hg	Ref.
			%								ppm							
sediments																		
Fe-rich	0.30	0.30	2.2	0.21	0.132	80	13		190			28	6	18		15	0.11	1
Fe-rich	0.11	0.17	1.8	0.46	0.137	26	31			1.5	37			7.0	2.7			2
Fe-rich							60				54.3							5
Fe-Mn	0.054	0.30	4.6	15.5	0.750	220	55		140			15	8	6.0		64	0.42	1
Fe-Mn	0.11	0.30	2.7	16.8	0.520	122	60			5.8	26			4.6	1.2			2
Fe-Mn							88				41.6							5
Intermediate	0.19	0.31	2.6	4.8	0.225	41	28			6.3	57			9.1	1.8			2
facies																		
Magnetite	0.08	0.09	1.2	0.23	0.142	32	45	94	824		390							3
Goethite	0.01	0.13	0.8	0.51	0.173	17	35	46	43		119							3
Lepidocrocite	0.02	0.13	0.7	0.45	0.148	23	40	65	207		181							3
Hematite	0.11	0.23	2.8	6.89	0.532	94	93	531	90		131							3
Manganite	0.07	0.29	2.2	14.08	0.582	125	107	643	53		92							3
Detrital-biogenic			1.0	0.70	0.035													3

References: 1– Bignell et al. (1976a): data on Core VA-03-620; 2 – Gurvich and Bogdanov (1986g): data on Core PS224; 3 – Scholten et al. (1991): data on the cores of R/V *Sonne* (Fig. 2.21); 4 – Butuzova and Lyapunov (1995): data on Core PS224; 5 – Pierret et al. (2000): data on Core 1037 of R/V *Marion Dufresne*.

Fe-Mn sediments. There is an increase in the contents of these elements and in the contents of Ni, Cr, and Sc as well in the lower parts of the Fe-rich sediment layers (Fig. 2.22). These data apparently are evidence for an increase in the amount of detrital basalt material. There is a similar pattern in the distribution of these elements in the lower part of the Fe-rich layer from 230 to 346 cm in Core VA-03-620, where chamosite is present as well (Bignell et al. 1976a).

Usually the content of Mn in the Fe-Mn sediments is 15 to 25% and rarely is it from 4 to 10%, the content of Fe is 10 to 20%, and labile forms of Fe(II) are absent (Gurvich et al. 1986a). These sediments are enriched in Zn, Pb, Sb, and Ni.

The Fe-rich sediments contain more than 50% Fe and are depleted in Mn, which is less than 1%, and in components that are typically present in the background material. The Fe-rich sediments are enriched in S and contain up to 7.65%, whereas the content of S in the Fe-Mn sediments does not exceed 1.9% (Table 2.9, Fig. 2.22). The Fe-rich sediments are enriched in Cu, Co, Hf, and REE. The average content of Cu in the Fe-rich sediments is higher than that of Zn whereas in the Fe-Mn sediments Zn is predominant (Table 2.9).

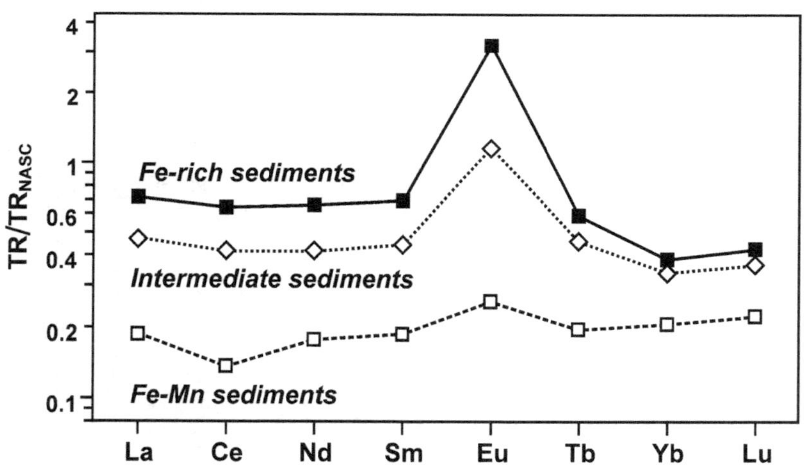

Fig. 2.23. Average NASC normalized REE patterns in the main types of sediments from the NE Thetis Deep.

A distinct positive Eu anomaly that is characteristic for high-temperature hydrothermal fluids and high-temperature hydrothermal sulfide minerals is a distinctive feature of the REE pattern in the Fe-rich sediments (Table 2.9, Fig. 2.23). In these sediments the average value of the Eu/Eu* ratio is 5.1, and the maximum value is 7.9. In the Fe-Mn sediments the average value of the Eu/Eu* ratio is 1.3. According to Pierret et al. (2000), the average Eu/Eu* ratio in the Fe-rich sediments from the NE Thetis Deep is 4.7, in the Fe-Mn sediments the Eu/Eu* ratio is 1.3. In Core PS224 the Eu/Eu* ratio varies in phase with the contents of the Fe and Cu but not with the content of Mn (Fig. 2.24). The Fe-Mn sediments have a minor negative Ce anomaly, that is typical for seawater, but in the Fe-rich sediments it is nearly absent (Fig. 2.23).

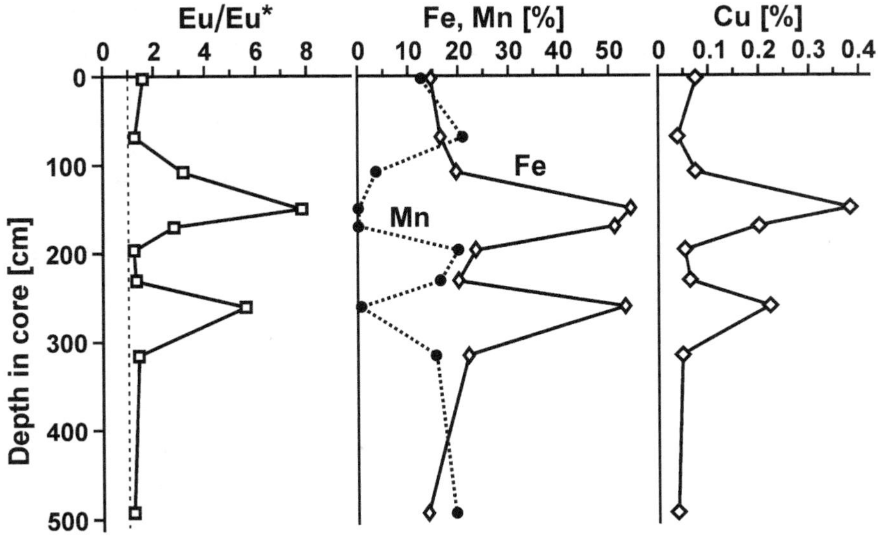

Fig. 2.24. Variations in the Eu/Eu* ratio and the contents of Fe, Mn, and Cu in sediments from Core PS224.

Unlike the sediment from the Atlantis II Deep (Table 2.4) and massive hydrothermal sulfides that contain small amounts of REE (Missack 1988; Rimskaya-Korsakova and Dubunin 2003; Fig. 1.62), the Fe-rich sediments are considerably richer in these elements. Obviously this results from effective scavenging of the hydrothermal REE during the deposition of the Fe-rich sediments.

Some features of the chemical composition of the sediments from Core VA-03-620 are similar to those in the sediments from Core PS224. These features are more pronounced in the sediments from Core VA-03-620 in which there is a higher content of metal-bearing matter and a lower content of lithogenic material (Al). The Fe-rich sediments from Core VA-03-620 are richer in Fe, Cu, and Co, and the Fe-Mn sediments are richer in Zn and Pb (Table 2.9). These differences together with the greater thickness of the Fe-rich and Fe-Mn layers in Core VA-03-620 apparently are evidence that the sediments in this core accumulated closer to a hydrothermal source.

2.3.3. Sedimentation history. Sedimentation rates. Accumulation rates of the main components of hydrothermal matter

The C-14 age of the sediments from Core PS224 has been measured (Kuptsov and Palkina 1986). Based on these data the sedimentation rates in the past ~13 ka for the sediments sampled in this core have been estimated and they are plotted in Fig. 2.25. Estimates of the sedimentation rates for the sediment sampled in Core 334PC for the last 18 ka have been made by Scholten et al. (1991) that are also based on the C-14 age dating and they are plotted in Fig 2.25. In general the parts of these cores that accumulated at the same time have similar lithology and their accumulation rates are also similar. A big difference in the sediments that are younger than ~9 ka is an artefact; if to calculate the average sedimentation rate for sediments younger than 9 ka in Core PS224 it is ≥53 cm ka^{-1} and in Core 334PC it is ≥55 cm ka^{-1}.

Sedimentation rates for the sediments sampled in these two cores have a similar order of magnitude because this parameter when other conditions are similar depends on their distance from a hydrothermal source and it decreases with the distance. For example, the sedimentation rate for the sediments in Core VA-03-620, which apparently accumulated closer to the hydrothermal source, is on an average approximately twice as high as that for the sediments in Core PS224 (Gurvich and Bogdanov 1986g).

Accumulation rates of the hydrothermal matter and its main components (Fig. 2.26) have been calculated based on the sedimentation rates for the sediments from Core PS224, their physical properties and chemical composition.

During the last 13 ka accumulation rate of the hydrothermal matter exceeded 1 g cm^{-2} ka^{-1}. The lowest value, ~2 g cm^{-2} ka^{-1}, was from 13 to 10 ka BP. At this time the accumulation rates of the excessive Mn and Fe were similar (0.5 to 0.7 g cm^{-2} ka^{-1}), and the rate of excessive accumulation of Zn (16–21 mg cm^{-2} ka^{-1}) was much higher than that of Cu (1.1–1.5 mg cm^{-2} ka^{-1}). During the time from 13 to 10 ka BP the accumulation rates of the excessive Fe, Zn, and Cu were the lowest in the entire core.

During the time period from 10 to 3.7 ka BP the accumulation rates of the hydrothermal matter and its main components were 2 to 3 times greater. The rates of accumulation of the excessive Fe and Mn were similar (1.6–2.5 g cm^{-2} ka^{-1}). The rate of accumulation of the excessive Zn (45–80 mg cm^{-2} ka^{-1}) remained much higher than that for the excessive Cu (3–8 mg cm^{-2} ka^{-1}).

The period from 3.7 to 1.4 ka BP was characterized by a sharp increase in the accumulation rate of the hydrothermal matter. On the average it was more than an order of magnitude higher than during the period from 13 to

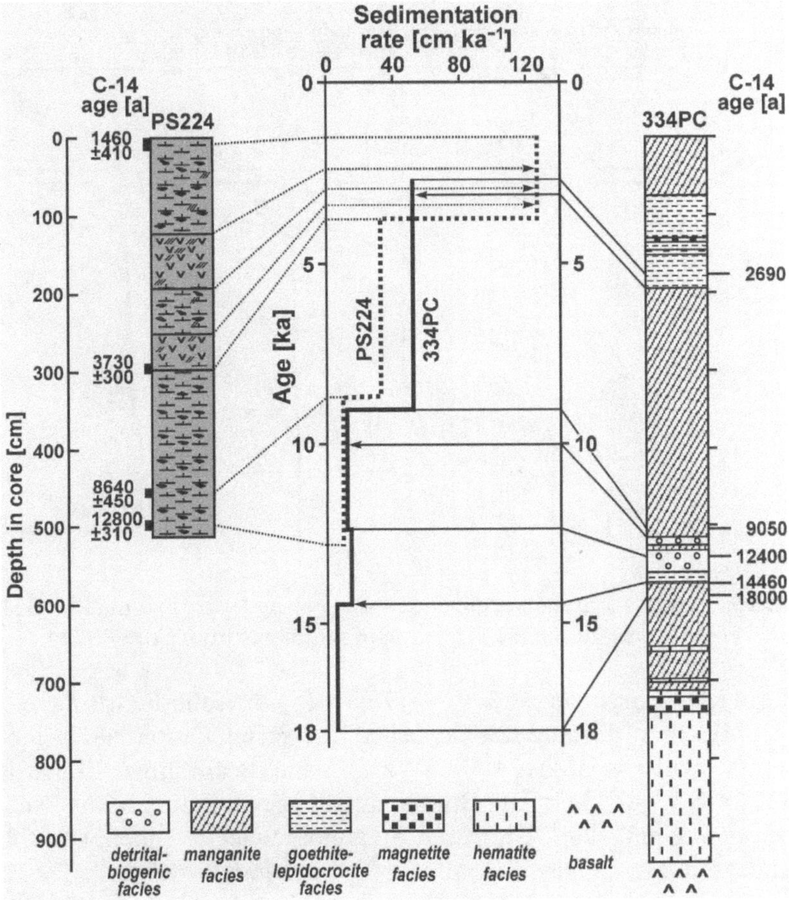

Fig. 2.25. Lithologic composition of sediments from Cores PS224 and 334PC, C-14 ages of the sediments, estimates of times when the main sediment units were accumulated, estimates of the sedimentation rates. Lithology of Core PS224 (legend is shown in Fig. 2.22) after Butuzova et al. (1986), C-14 ages for Core PS224 after Kuptsov and Palkina (1986). Lithology and C-14 ages for Core 334PC after Scholten et al. (1991).

10 ka BP. In the period of time from 3.7 to 3.4 ka BP, when the Fe-rich layer from 249 to 298 cm in the core was deposited, the accumulation rate is estimated to have been 26 g cm^{-2} ka^{-1}. The rate of accumulation of the excessive Fe, ~16 g cm^{-2} ka^{-1}, during this time interval was 6 to 10 times higher than during the period from 10 to 4 ka BP, whereas the rate of accumulation of the excessive Mn decreased to about 0.1–0.2 g cm^{-2} ka^{-1}, the minimum in the whole core. The rate of accumulation of the excessive Zn, about 50 mg cm^{-2} ka^{-1}, remained at the same level, and that for the excessive Cu (50–60 mg cm^{-2} ka^{-1}) increased by an order of magnitude.

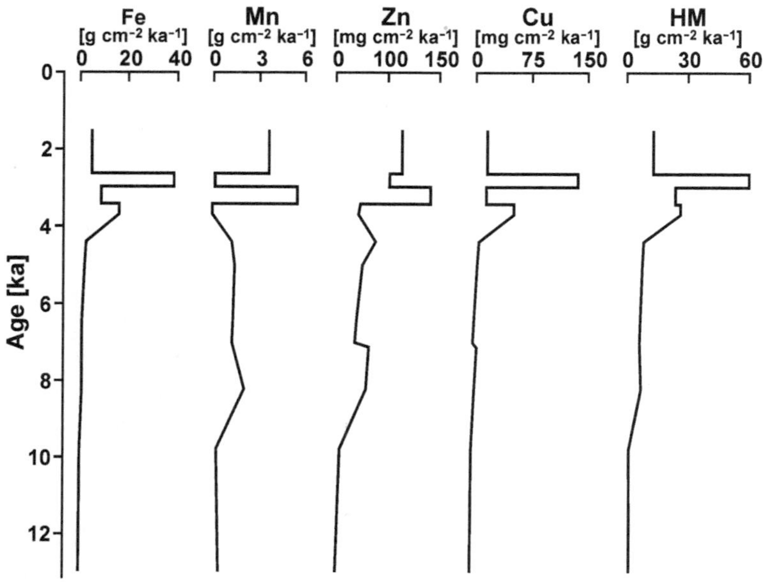

Fig. 2.26. Variations in the accumulation rates of hydrothermal matter (HM) and its main components during the formation of sediments from Core PS224.

During the time interval 3.4 to 2.9 ka BP, when the Fe-Mn layer from 190 to 249 cm in the core was deposited, the accumulation rate of the hydrothermal matter was 23–24 g cm^{-2} ka^{-1}, which was similar to that in the period from 3.7 to 3.4 ka BP. But the rate of accumulation of the excessive Fe, about 9 g cm^{-2} ka^{-1}, decreased by almost twice as much, and that for the excessive Mn increased by almost 50 times to 6 g cm^{-2} ka^{-1}. The rate of accumulation of the excessive Zn increased almost 4 times to 190 mg cm^{-2} ka^{-1}, and the rate for the excessive Cu decreased by twice as much to 20 mg cm^{-2} ka^{-1}.

The time interval from 2.9 to 2.6 ka BP, when the Fe-rich layer from 120 to 190 cm in the core accumulated, is characterized by a sharp increase in the rate of accumulation of the hydrothermal matter to 61 g cm^{-2} ka^{-1}. During this period Fe was predominant in the hydrothermal matter and its accumulation rate of 38 to 39 g cm^{-2} ka^{-1} was at a maximum. The accumulation rate of the excessive Mn, 0.36 g cm^{-2} ka^{-1}, decreased by more than 15 times. The rate of accumulation of the excessive Cu increased by 7 times to 140 mg cm^{-2} ka^{-1}, whereas the accumulation of the excessive Zn decreased almost twice as much to 110 mg cm^{-2} ka^{-1}.

During the interval 2.6–1.4 ka BP the accumulation rate of the hydrothermal matter decreased to 13 g cm^{-2} ka^{-1}. The accumulation rates of the excessive Fe and Mn were similar (4–5 g cm^{-2} ka^{-1}). The rate of accumulation of the excessive Zn was similar to that in the previous interval, 130 mg cm^{-2} ka^{-1}, and the rate of the excessive accumulation of Cu decreased by an order to 15 mg cm^{-2} ka^{-1}.

Core PS224 provides data for the accumulation of the hydrothermal matter and its components from 13 ka BP to the present. Materials on Core 334PC (Scholten 1991, Fig. 2.25) give some information for the interval of time from 18 to 13 ka BP. According to these data, the sediments of the detrital-biogenic facies accumulated in the period from ~14 ka BP to ~10 ka BP. Scholten et al. (1991) believe that hydrothermal activity became extinct during this period of time. The manganite facies sediments (Fe-Mn sediments) accumulated in the period from 18 ka BP to ~14 ka BP. Their sedimentation rate was ~7 cm ka^{-1} (Scholten et al. 1991), a minimum in both of the cores (Fig. 2.25). This indicates that the accumulation rate of the hydrothermal matter and its components was low during this time.

There are no data for the age of the sediments in the lower part of Core 334PC (Fig. 2.25). One can assume that the Fe-Mn sediments sampled in this part of the core accumulated at a relatively lower rate similar to that for the younger Fe-Mn sediments and that the lowermost sediments in the hematite facies apparently accumulated at a higher rate. Scholten et al. (1991) assumed that the accumulation of these sediments on the basalt basement began about 23 ka BP.

Estimates based on the data for Core VA-03-620 gave the following results: during the time interval 3.4 to 2.9 ka BP the average accumulation rates of the excessive Fe, Mn, Cu, and Zn were respectively: $\geq$15 g cm^{-2} ka^{-1}, $\geq$8 g cm^{-2} ka^{-1}, $\geq$30 mg cm^{-2} ka^{-1}, and $\geq$550 mg cm^{-2} ka^{-1}; and during the time interval from 2.9 to 2.6 ka BP they were: 100 g cm^{-2} ka^{-1}, 0.5 g cm^{-2} ka^{-1}, 640 mg cm^{-2} ka^{-1}, and 220 mg cm^{-2} ka^{-1} respectively, and during the time interval from 2.6 to 1.4 ka BP the respective accumulation rates were: 11 g cm^{-2} ka^{-1}, 9.2 g cm^{-2} ka^{-1}, 52 mg cm^{-2} ka^{-1}, and 460 mg cm^{-2} ka^{-1}. These rates are higher than the rates determined in Core PS224 that is evidence of the closer position of Core VA-03-620 to the hydrothermal source.

The rates of the excessive accumulation of Fe in the Fe-rich sediments and of Mn in the Fe-Mn sediments in the NE Thetis Deep are comparable to those in the sediments in the Atlantis II Deep. The rates of the excessive accumulation of Zn and Cu in the sediments from the NE Thetis Deep are lower than in the Atlantis II Deep (Table 2.7).

Similar sequences of alternate Fe-rich and Fe-Mn layers composed of hydrothermal precipitates in the cores collected in different parts of the NE Thetis Deep are evidence not only of a common source of the metal-bearing matter in these sediments, but also of uniform or nearly uniform conditions near the floor of the deep during the formation of these layers.

The accumulation rates of the main components in the metal-bearing matter clearly demonstrate that the most intensive accumulation of hydrothermal matter obviously took place during the periods when hydrothermal activity was highly intensive and conditions were most favorable for the accumulation of Fe and Cu. The periods when there was a relatively low rate of accumulation of hydrothermal matter, which obviously coincided with periods of relatively low intensity in hydrothermal activity, were the most favorable for the deposition of Mn and Zn (Fig. 2.26).

Variations in the Mn/Al and Fe/Al ratios demonstrate that the intensity of accumulation of Mn and Fe relative to the background material changed dramatically during the last 13 ka, the period when the sediments sampled in Core PS224 accumulated (Fig. 2.27).

During the periods when there was intensive hydrothermal activity there was extensive accumulation of Fe at a rate that was nearly 1.5 to 2 orders higher, than it was during the periods of hydrothermal activity of low intensity, and the accumulation of Mn was about 2 orders lower than it was during the periods of hydrothermal activity of low intensity.

In the sediments accumulated in periods of hydrothermal activity of high intensity the Mn/Al ratio (<1) is close to the background level (0.12), and the excessive Mn is practically absent. The hydrothermal Mn contributed to the NE Thetis Deep during the periods of hydrothermal activity of high intensity was dissipated almost entirely. During these periods the Fe-rich sediments accumulated that have a Fe/Mn ratio of about 100 (Fig. 2.27) and these sediments are enriched in Cu, and in some places they contain hydrothermal sulfide minerals. In the sediments accumulated during the periods when the intensity of hydrothermal activity was low the Mn/Al ratio is about 100 times higher than the background level, and the Fe/Mn ratio is close to 1, which indicates that there was a significant accumulation of hydrothermal Mn.

As an explanation for the accumulation of the Fe-rich sediments, which have a low content of Mn, Scholten et al. (1991) have assumed that a brine pool existed during the periods of highly intensive hydrothermal activity that provided an extensive contribution of dense hydrothermal fluids to the NE Thetis Deep. Their evidence is as follows: "(a) the formation of Fe-facies types like goethite, magnetite and lepidocrocite, whose precipitation from hydrothermal fluids is favored in environments more reducing than normal seawater conditions (Schwertmann and Thalmann 1976; Murray 1979; Schwertmann and Murad 1983); (b) the rapid formation of the Fe-facies types ...". In an environment such as this, hydrothermal Mn(II) was not oxidized and did not precipitate. Where there was hydrothermal activity of low intensity the dense hydrothermal fluids were diluted by seawater and the brine pool did not develop.

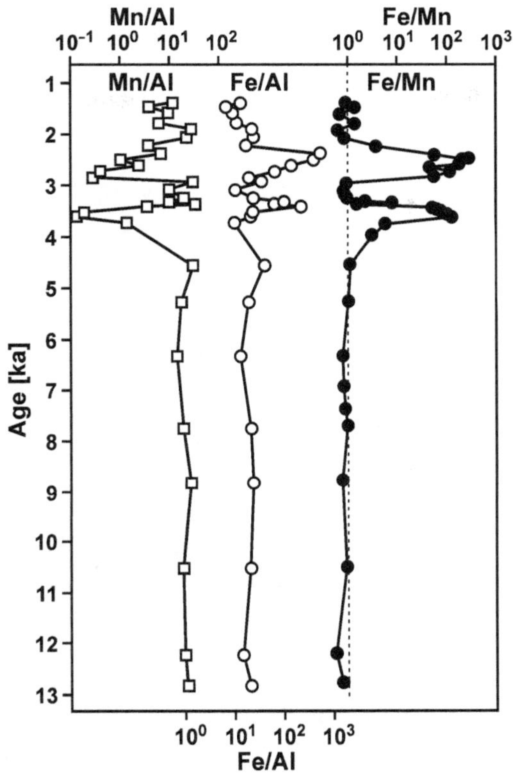

Fig. 2.27. Temporal variations of the Mn/Al, Fe/Al, and Fe/Mn ratios in sediments from Core PS224.

This assumption could be confirmed by measuring the salinity of the interstitial water in the sediments in the Fe-rich and Fe-Mn layers, or at least by measuring the content of salt in the dry sediments. Unfortunately, such measurements have not been carried out. However, a preliminary estimate can be made by using the sediments sampled in Core PS224 as they were not washed in order to remove sea-salt before chemical analyses were carried out and the content of Na was determined. The higher Na/Al ratios in the sediments of the Fe-rich layers (Fig. 2.28), in spite of their higher density and lower water contents, demonstrate indirectly that there was higher salinity in their interstitial water in comparison to that in the sediments in the Fe-Mn layers.

If a brine pool existed, it was at least 14 m thick, the maximum depth of the deep is 1780 m, and the shallowest Core 334PC containing the Fe-rich layers was collected at a depth of 1766 m. A brine pool could act as a trap for the hydrothermal Fe contributed to the NE Thetis Deep during the periods of intensive hydrothermal activity, but not for the hydrothermal Mn.

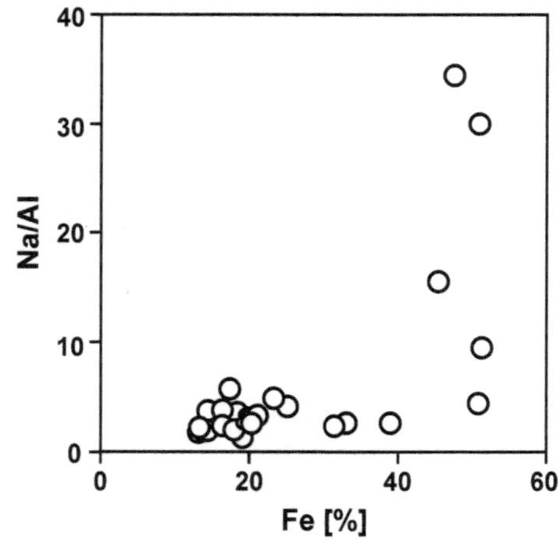

Fig. 2.28. Na/Al ratio versus Fe content in dried sediments from Core PS224.

During the periods of low intensity in hydrothermal activity the environmental conditions within the NE Thetis Deep apparently were favorable for the retention of not only the hydrothermal Fe, but also for the hydrothermal Mn. Oxidizing conditions existed in the deep and the hydrothermal Mn(II) and Fe(II) could have been oxidized and precipitated. Metalliferous sediments rich in both Fe and Mn do not accumulate in the vicinity of hydrothermal vents in the open ocean at the present time (Chap. 1). This is because of the slow rate of oxidation of Mn(II) in the cold (2°–5°C and lower) ocean water. Apparently in the NE Thetis Deep the oxidation of Mn(II) took place at a much faster rate and it could have precipitated in the vicinity of the hydrothermal vents where the temperature of the bottom water was much higher[24], and likely its deposition was assisted by microbiological activity (see Sect. 4.3).

2.4. General characteristics of mineral components in the hydrothermal matter of metalliferous sediments from the Red Sea

The hydrothermal matter in the sediments in the deeps of the Red Sea rift zone consists of a highly varied and extremely complex association of minerals. It differs from one deep to another and within the deeps it varies greatly both in vertical section and within contemporaneous strata. A comprehensive investigation of the mineral composition of the metal-bearing

[24] At present it is 22.6°C (Table 2.1).

matter in most of the deeps has been carried out by G.Yu. Butuzova or with her participation. The following is a summary of the main results of these studies as reported in the publications of Butuzova and Lisitzina (1986a) and Butuzova (1998). The occurrence of the different minerals in the sediments from the deeps is shown in Tables 2.10 and 2.11.

Fe(II) occurs in the Fe-Si gel in places where the gel is green in color and sulfide minerals are associated with it in the gel in metalliferous sediments in the Atlantis II, Albatross, Shagara, Erba, Suakin, NE Thetis, and Gypsum Deeps. A spotted distribution of the green color in the gel mass is characteristic for these occurrences.

The following Fe-minerals, goethite, hematite, magnetite, lepidocrocite, and maghemite have been found and of these minerals goethite is the most widely distributed. Hematite is abundant in the sediments in the OAN Zone in the southwest basin of the Atlantis II Deep in the areas where hydrothermal orifices may have been located. Hematite is intercalated also in some of the sediments in the NE Thetis, Erba, Albatross, and Discovery Deeps.

Magnetite occurs in the metalliferous sediments of the Atlantis II, NE Thetis, Erba, Albatross, Discovery, and Gypsum Deeps. It has been found also in hydrothermal deposits in the Red Sea rift zone at 18°N (Bogdanov and Gurvich 1986a). Magnetite is one of the main components of the metal-bearing matter in the sediments in the NE Thetis Deep (Sect. 2.3.1) and it occurs locally in the Atlantis II Deep. It has been found in the sediments of the CO Zone in the vicinity of sites where hydrothermal orifices may have existed in the southwest basin (Fig. 2.9). Coarse-grained crystals of magnetite are associated with hematite in the sediments in local places in this zone, and in the sediments in other deeps magnetite occurs as fine (1–5 μm) crystals and in particles that have poorly defined crystallographic forms, and in colloform clots.

Lepidocrocite is associated with goethite and siderite in some of the interlayers in the sediments of the NE Thetis and Discovery Deeps. It has been found also in a few interlayers in some of the sediment cores from the Atlantis II Deep.

A crystalline ferromagnetic mineral phase that appears to be a type of maghemite (γ-Fe_2O_3) occurs in small amounts (0.n%) in some places in the sediments of the NE Thetis Deep. This mineral has been found also in some of the cores from the Atlantis II Deep.

Mn-oxyhydroxide minerals occur in various amounts in the sediments from most of the deeps. In some parts of the NE Thetis, Discovery, Shagara, and Atlantis II Deeps they form interlayers from 15–20 cm up to 3–4 m thick (in the NE Thetis Deep). These interlayers are dark cherry-brown and

Table 2.10. Occurrence of oxide-, hydroxide-, and silicate minerals in metalliferous sediments from the deeps of the Red Sea rift zone. After Butuzova and Lisitzina (1986a).

Deep, zone	Oxides and hydroxides											Silicates			
	ASi	AFe	Goe	Lep	Hem	Mgn	Mgh	AMn	Mng	Tod	COx	FSm	Chr	Chl	Am
Suakin	++	+++	+		+?	+		+							
Port Sudan	++	+++	+					+							
Erba	++	+++	++		+	++	+	+							
Shagara	++	+++	++					++	+	+					
Albatross	++	+++	++		+	++						+			
Discovery	++	+++	++	+	+	++		++	+	+		+?			
Valdivia	++	+++	+												
Atlantis II, southern part															
SAM	++	+++	+		+							+			
SOAN	++	+++	++		+							+		++	+
OAN	++	+++	++	+	++							+		++	++
CO	+++	+++	++	+	++	+	+	+				++			
SU_1	+++	+++	++		+							+	+		
Atlantis II, northern and central parts															
AM	+++	+++	+	+	+						+	+			
SU_2	+++	+++	+									++			
CO	+++	+++	++	+	++		+	+	++	+		++			
SU_1	++	+++	+									++			
Hadarba	++	+++	+					+++	+	+?					
NE Thetis	+	+++	++	++	++	+++		++	+						
Nereus	++	+++	+	+											
Vema	++	+++	+												
Gypsum	+	+++	++	++		++						++			
Kebrit	+	+	+												

Abbreviations: ASi – amorphous silica; AFe – amorphous Fe-hydroxides; Goe – goethite; Lep – lepidocrocite; Hem – hematite; Mgn – magnetite; Mgh – maghemite; AMn – amorphous Mn-hydroxides; Mng – manganite; Tod – todorokite; COx – complex Fe- and Mn-oxide minerals; FSm – Fe-smectites, celadonite; Chr – chrysocolla; Chl – chlorite, saponite; Am – amphiboles, talc.
+++ – abundant, ++ – common, + – rare.

in many places dark grey to black in color and characteristically have an unctuous consistency. The maximum values of Eh (+500 ÷ +600 mV) for metalliferous sediments in the Red Sea have been measured in these interlayers. Mn-oxyhydroxide minerals occur as colloform clots and microglobular masses and in some places they have a characteristic microdendritic form. The mineral composition of the Mn-rich interlayers is complex and the proportions of individual mineral phases vary considerably in the different deeps. For example, in the Shagara and Discovery Deeps X-ray amorphous

Table 2.11. Occurrence of sulfide-, sulfate-, carbonate-, and chloride minerals as well as native metals in metalliferous sediments from the deeps of the Red Sea rift zone. After Butuzova and Lisitzina (1986a) and Butuzova (1998).

Deep, zone	Sulfides							Sulfates			Carbonates			Chlorides	Native metals		
	ASu	Sph	Py	Ch	Pyr	CuS	Gr	Anh	Gy	Bar	Sid	Rho	MnS	At	Al	Fe	Pb
Suakin	++		+				+	+	+		+	+	+				
Port Sudan	+							+	+								
Erba	+		+				+		+				+				
Shagara	+		+				+										
Albatross	+		+				+										
Discovery	+						++		+				+				
Valdivia																	
Atlantis II, southern part																	
SAM	+++	+						++	+				+				
SOAN	++	++	++	++	++	++		+++	++	+			++		+	+	+
OAN								+++	++	+			++				
CO								+				+	+				
SU_1	++	++	+	+				+					++		+	+	+
Atlantis II, northern and central parts																	
AM	++	+															
SU_2	++	++	+					+	+	+	+		++	+			
CO	+							+					++	+?			
SU_1	++	+	++	+				++	+	+			++	+			
Hadarba													+				
NE Thetis	+	+?						+									
Nereus	+																
Vema																	
Gypsum	+		+				+?	+	++	+	++						
Kebrit	+																

Abbreviations: ASu – amorphous sulfides of Fe, Zn, and Cu; Sph – sphalerite, marmatite; Py – pyrite; Ch – chalcopyrite; Pyr – pyrrhotite; CuS – Cu-sulfosalts; Gr – greigite; Anh – anhydrite; Gy – gypsum; Bar – barite; Sid – siderite; Rho – rhodochrosite; MnS – manganosiderites; At – atacamite.
+++ – abundant, ++ – common, + – rare.

Mn-hydroxide minerals and in places poorly crystallized todorokite, and in the NE Thetis Deep asbolane minerals appear to be the main mineral phases. Essentially different morphological and structural features have been distinguished in some mineral phases in the Mn-rich interlayers in the sediments from the Atlantis II Deep. Most of them have a microglobular texture, the globules are about 2 to 6 μm in size, and this crystalline phase appears to consist of todorokite. Numerous micronodules, 0.2 to 0.5 mm and rarely up to 1 mm in size, occur in the microglobular mass that consists mainly of well-crystallized manganite. Fe and trace elements are almost absent in these micronodules. Ferrofranklinite, a mineral similar to magnetite that belongs to the spinel complex oxide group, with the formula (Fe, Zn, Mn)(Fe, $Mn)_2O_4$, has been found in small amounts in the sediments of the Atlantis II Deep.

One of the main mineral components in the Red Sea metalliferous sediments are layered silicate minerals of the Fe-smectite group. They are most widely distributed in the Atlantis II Deep, and occur in significant amounts in the metalliferous sediments in the Gypsum and Albatross Deeps. Eh values vary from +50 mV to +150 mV in the interlayers that are rich in silicate minerals. Mixed-layered micaceous-montmorillonite minerals are formed during ageing of the gel and scavenging of K^+ from the interstitial water and a mineral very similar to celadonite is the end-member in this series of newly formed minerals. Various silicate minerals occur in the sediments in the southwest basin of the Atlantis II Deep, apparently in the vicinity of hydrothermal orifices that once existed in the area. Along with the layered silicate minerals of the nontronite-celadonite series that are commonly present in the metalliferous sediments, talc, chlorite, saponite, and a mineral belonging to the amphibole group have been found. Chrysocolla, which belongs to rare authigenic silicate minerals, occurs sporadically in the sediments of the Atlantis II Deep in bright green grains that are tens of microns in size. Crystals of pyroxenes, plagioclases, garnet, quartz and ilvaite have been found in the sediments from the southwest basin of the Atlantis II Deep.

The various sulfide minerals are most abundant in the sediments from the Atlantis II Deep. The following sulfide minerals and material have been identified: X-ray amorphous sulfides of Zn, Cu, and Fe, sphalerite, marmatite, pyrite, pyrrhotite, chalcopyrite, and also sulfide minerals belonging to the Cu-sulfosalt group that have structures similar to those of colusite-germanite. Interlayers and parts of the sediments that are highly enriched in sulfide minerals characteristically have the lowest Eh, from +50 to –90 mV, measured.

Sulfide minerals are not as common in the metalliferous sediments from the other deeps. They occur in small amounts within some of the interlayers in the sediments of the Albatross, Discovery, Erba, Suakin, and NE Thetis Deeps. While sulfide minerals of various composition and complex sulfide minerals of nonferrous metals are prevalent in the Atlantis II Deep, various Fe-sulfide minerals have a prominent place in the other deeps where X-ray amorphous sulfide matter, pyrite, greigite, smythite and kansite, have been diagnosed. Sphalerite is present in some interlayers in the Albatross and NE Thetis Deeps, and Cu-sulfide minerals are present in the

Erba and NE Thetis Deeps. Pyrite is the most common sulfide mineral. Well-crystallized greigite (melnikovite) has been found in significant amounts in some of the interlayers in the sediments of the Discovery Deep, and also in small amounts in sediments of the Suakin, Albatross, and Valdivia Deeps. Specific Pb-sulfide minerals have not been found. However, microroentgenospectral analyses indicate the presence of this element within the composition of the sulfide mineral phases in the sulfide-bearing interlayers where there is an increase in the content of Pb.

Fe- and Mn-carbonate minerals, in a series ranging from siderite to rhodochrosite, are characteristic minerals in the sediments in a number of the deeps. As a rule, they consist of complex Fe and Mn carbonate minerals in which the contents of these elements vary considerably, i.e. the manganosiderite minerals. These minerals are widely distributed in the sediments of the Atlantis II Deep and are dispersed in the sediment mass in all of the lithostratigraphic zones and are the main mineral carriers of Mn in the sediments in the deep. The complex Fe and Mn carbonate minerals occur in some of the interlayers in the sedimentary strata in the Suakin, Erba, Discovery, and Hadarba Deeps. Carbonate minerals consisting mainly of siderite are widely distributed in the sediments of the Gypsum Deep. In some interlayers they are the main component in the metal-bearing matter.

The main sulfate minerals, anhydrite, gypsum, and barite are widely distributed in the sediments in the Atlantis II Deep. Anhydrite, the most widely distributed of the sulfate minerals, is present in well-formed crystals that vary from 0.1 to 0.5 mm in size and form one of the most distinct crystalline mineral phases in the sediments. The anhydrite crystals are irregularly dispersed in the sediments where they accumulate in aggregate masses in many places, and the sediments in the southern part of the Atlantis II Deep have the highest enrichment in anhydrite. In the southwest basin of the deep the anhydrite forms layers that are up to 2.5 m thick that contain admixtures of sulfide minerals and silica gel. Gypsum usually occurs in large idiomorphic crystals and in intergrowths, twins, druses, and spherulites that are 0.1 to 1 mm and rarely >1 mm in size, that are most widely distributed in the sediments in the Gypsum Deep. There are gypsum "roses" up to 10 cm in size in the Kebrit Deep (Puchelt and Laschek 1984). The occurrence of barite has been confirmed in some sediment samples in the Atlantis II and Gypsum Deeps. It occurs as microcrystalline aggregates and individual spherulites that are about 0.01 mm in size in the Atlantis II and Gypsum Deeps, and also as large idiomorphic crystals and cross-shaped twin crystals, 0.1 to 0.01 mm in size, in the Atlantis II Deep. At some of the stations in the Atlantis II Deep barite crusts up to 10 cm thick overlying the basalt rocks were found (Sval'nov et al. 1984).

The occurrence of Cu-oxychloride (atacamite) has been confirmed in the saltless part in some of the cores of metalliferous sediments collected in the Atlantis II Deep. The individual grains, 0.01 to 0.1 mm in size, are bright green. Fe-chloride minerals also occur in some places.

Particles of native metals, Al°, Fe°, and Pb°, have been found in the sediments in the southwest basin of the Atlantis II Deep (Butuzova 1998).

The contents of the metal-bearing matter in the metalliferous sediments from the Red Sea rift zone and the main groups of minerals that compose it are shown in Fig. 2.29 with a time-scale for their deposition. It is evident that the enrichment of the sediments in the hydrothermal metal-bearing matter in the deeps was irregular in distribution both in time and along the strike of the rift. The content of the metal-bearing matter differs considerably not only from deep to deep but also within the deeps. This variation in the content of the metal-bearing matter is direct evidence of the irregularity in the contribution of the hydrothermal matter. There is no distinct synchronism in the periods of accumulation of the hydrothermal metal-bearing matter in the sediments from deep to deep, at least for the last 40 ka. In some deeps cyclicity in the formation of the metalliferous sediments can be detected.

The mineral composition of the hydrothermal metal-bearing matter varies from deep to deep, and also both vertically and laterally in the successions of sedimentary strata within the deeps. Amorphous Fe-hydroxide minerals and SiO_2 are present almost everywhere in the metalliferous sediments in the Red Sea Deeps and other minerals that were formed by hydrothermal processes are distributed locally and do not occur in all of the deeps (Tables 2.10, 2.11). Of these mineral groups the sulfide minerals of Fe, Cu and Zn are of greatest significance, followed by the oxyhydroxide minerals of Mn, the sulfate minerals of Ca and Ba, the carbonate minerals of Fe and Mn, and also some other minerals.

It is important to emphasize that often in the environmental conditions of the homogeneous stable brines the metalliferous sediments differ essentially in the proportions of hydrothermal minerals developed in them. This heterogeneity in the composition of the metalliferous sediments apparently is related to the distance of their sedimentation sites from the hydrothermal source or sources. This is clearly illustrated in the southwest basin of the Atlantis II Deep where the locations of the hydrothermal sources are indicated with some certainty and the sediments from the AM Zone are enriched in sulfide minerals and anhydrite and are unlike the AM Zone sediments in the other parts of the deep. Also in the southwest basin well crystallized Fe-oxyhydroxide minerals are more abundant than in the sediments in other parts of the deep.

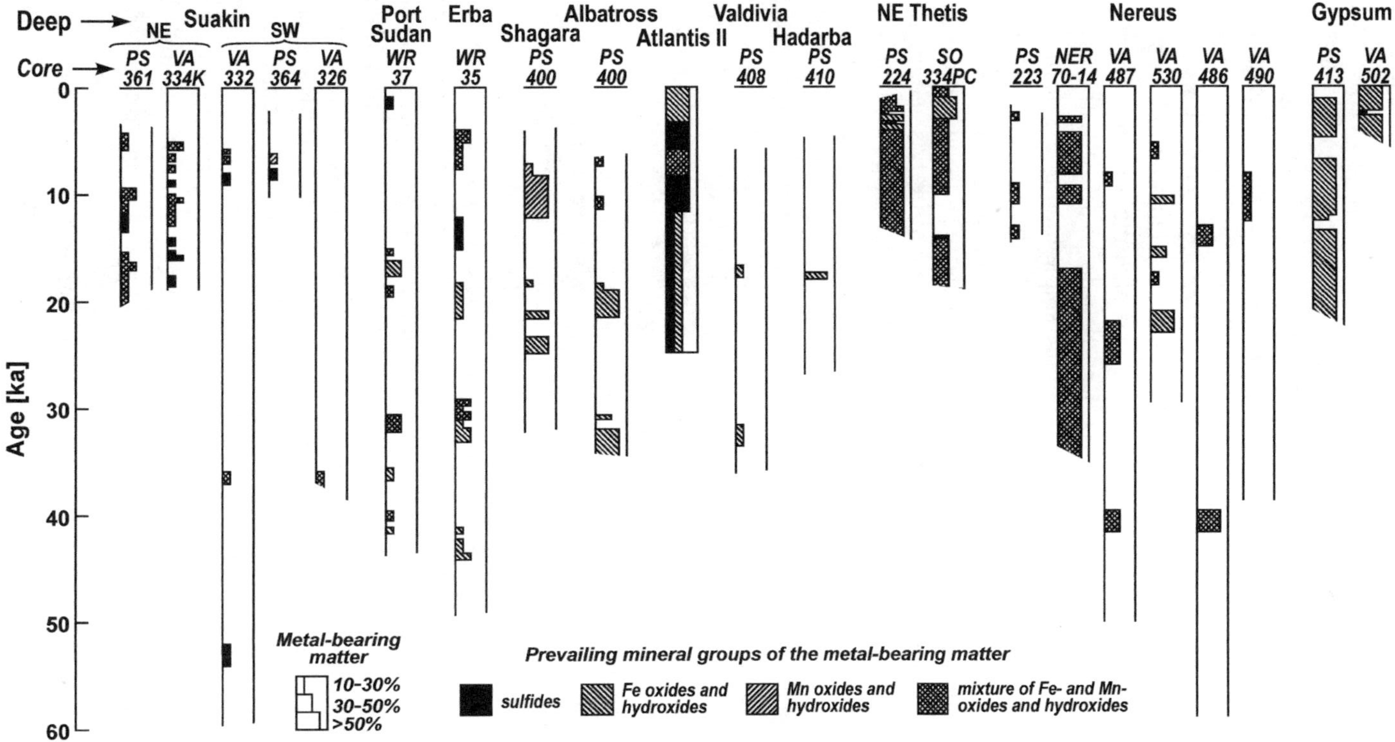

Fig. 2.29. Contents of the metal-bearing matter and composition of the prevailing mineral groups (except for Si-Fe gel) in metalliferous sediments from the deeps of the Red Sea rift zone. Composed on the data from Baumann et al. (1973), Bignell and Ali (1976), Bignell et al. (1976a), Gurvich and Bogdanov (1986b), Scholten et al. (1991). *PS, VA, WR, SO, NER* – stations of R/V *Professor Shtokman*, *Vema*, *Wando River*, *Sonne*, and *Nereus*, respectively. Positions of the deeps are shown in Fig. 2.2.

In general the mineral composition of the metalliferous sediments in the southwest basin of the Atlantis II Deep is similar to that of the massive hydrothermal bodies formed directly at orifices of high-temperature (>300°C) hydrothermal springs within the hydrothermal fields in the mid-ocean ridges (Gurvich and Bogdanov 1986b). Regardless of the place of discharge of the hydrothermal fluids, whether in the environmental conditions of the open ocean and into ocean water, or in the environmental conditions of the Red Sea and into the bottom brines, metal-bearing accumulations of similar mineral composition form directly at the hydrothermal orifices. The differences are simply that, firstly, in the Red Sea deeps filled with bottom brines hydrothermal sulfide minerals are more stable than in the open ocean; secondly, in the environmental conditions of the bottom brines around the hydrothermal orifices relatively large fields of sulfide metalliferous sediment accumulate, while in the open ocean the massive sulfide bodies that accumulated are succeeded laterally by oxidized metalliferous sediments, and it is only within the hydrothermal fields that they merge with sulfide sediments that usually are composed of products of the disintegration of the massive hydrothermal bodies. In the Atlantis II Deep that is filled with hot dense brines particulate sulfide minerals form in the vicinity of the hydrothermal orifices. They are disseminated throughout the whole area of the deep. Particulate sulfide minerals also form in the vicinity of the hydrothermal orifices in the open ocean, however in the oxidizing conditions of the bottom ocean water most of them are oxidized and only a small amount accumulates in the sediments.

The distribution of the Mn-oxyhydroxide minerals in sediments is illustrative and of further interest. In the thick stratified hot dense bottom brines where dissolved oxygen is absent these minerals do not form or accumulate in the bottom sediments. Hydrothermal Mn(II) contributed by hydrothermal fluids enriches the bottom brines. The main part of the hydrothermal Mn contributed to the brines is dispersed and it accumulates in the periphery area of the deep outside of the brine table. In contrast the hydrothermal Fe is trapped within the deep and accumulates almost completely in the sediments. This spatial separation in the areas of accumulation of hydrothermal Fe and Mn is one of the most characteristic features of the sedimentation environment in the stratified hot dense brines of the Red Sea. In the case of the entry of dissolved oxygen from the overlying Red Sea water into the bottom brines, either by diffusion processes, or by the mixing of the brines with seawater, there is less differentiation of Fe and Mn. Probably the least amount of differentiation of these elements takes place in the oxidizing environment and conditions in the bottom water in the brine-free deeps. This is evident from the Fe-Mn sediments that accumulated in the NE Thetis Deep.

2.5. General characteristics of chemical composition of the metal-bearing matter in metalliferous sediments from the Red Sea

The average contents of chemical elements in metalliferous sediments from the deeps of the Red Sea rift zone are shown in Table 2.12. Even the average data clearly reflect high variability in sediment composition from deep to deep. The average content of the metal-bearing matter in the sediments is also highly variable. The proportions of the average contents of Fe, Mn and Al can be seen in the triangular diagram in Fig. 2.30.

The sediments in the Thetis, Atlantis II, Gypsum, and Discovery Deeps have the highest contents of the metal-bearing matter, and the sediments in the Valdivia, Port Sudan, and SW Suakin Deeps have the lowest contents. Almost all of the sequences of sediment strata recovered in the cores from the Thetis, Atlantis II, Gypsum, and Discovery Deeps are metalliferous whereas the sequences of sediment strata recovered from the other deeps have alternate layers of metalliferous and non-metalliferous sediments that is clearly shown in Fig. 2.29.

The chemical composition of the metal-bearing matter is very different in the metalliferous sediments that accumulated under conditions of different intensity of hydrothermal activity and in the varied conditions in the bottom water or brines. As Fe is the main element in the metal-bearing matter in the metalliferous sediments the ratios of the excessive contents of chemical elements to the excessive content of Fe can be used to indicate the chemical composition of this matter and show the relative amount of enrichment of the various elements. The ratios that were calculated for the average contents of the elements are shown in Table 2.13. For comparison, the ratios for the distal metalliferous sediments from the Southeast Pacific and for the proximal metalliferous sediments from the TAG hydrothermal field are also shown in this table.

In the metalliferous sediments from the NE Suakin, SW Suakin, Shagara, Atlantis II (CO Zone), Wando, Hadarba, Nereus, and NE Thetis Deeps the Mn/Fe ratios exceed 0.1 (Fig. 2.30) and are comparable to the Mn/Fe ratio for the metalliferous sediments from the Southeast Pacific. It is evident that the conditions that existed in the deeps named were favorable for the oxidation of the hydrothermal Mn(II) to Mn(IV) and for its accumulation in the sediments. Apparently, the low Mn/Fe ratios in the sediments from the other deeps and from the TAG hydrothermal field are evidence of the removal of the Mn.

The Ni/Fe ratios vary over a wide range and in many cases they are greater than the ratio in the metalliferous sediments from the TAG hydrothermal

Table 2.12. Average chemical compositions of bottom sediments in the deeps of the Red Sea rift zone.

Deep	$CaCO_3$	Si	Al	Ti	Fe	Mn	Zn	Cu	Pb	Sb	Ba	Ni	Co	Cr	Sc	La	Hf	Th
	%								ppm									
NE Suakin	33		3.75		9.7	1.92	0.14	0.011	40	3.6		75	31	48	7.0	10.4		
SW Suakin	41		4.4		6.4	1.4	0.067	0.0078	24	2.9		77	23	52	7.6	12		
Port Sudan	33		3.8	0.29	5.9	0.40	0.055	0.0086	36	5.3		53	32	71	10	17		
Erba	36		2.4	0.20	14.5	0.64	0.078	0.033	70	1.5	250	115	43	45	6.8	23	2.1	1.2
Shagara	46.7		2.4		7.4	3.1	0.053	0.0090	55	1.1	280	120	18	56	7.5	11.5	2.4	1.8
Albatross	38.3	11.2	2.3	0.20	19.6	0.35	0.054	0.019	16			59	20.5	42	4.2		1.8	2.1
Atlantis II, AM		11.9	0.95	0.060	33.6	0.70	2.3	0.488	480	295	960	33	116	18	1.1	5.0	0.8	0.3
Atlantis II, SU_2		11.6	1.09	0.057	26.3	0.88	5.07	0.84	800	223	1500	49	113	23	1.6	8.9	1.2	1.1
Atlantis II, CO		9.4	1.47	0.099	34.3	4.67	0.475	0.205	230	59	1170	33	31	20	2.8	8.8	1.3	0.6
Atlantis II, SU_1		13.2	1.37	0.054	33.1	1.77	3.49	1.09	700	22	1210	20	45	13.4	1.25	4.7	<0.8	0.48
Atlantis II, DOP	13	16.4	3.00	0.14	22.7	0.14	0.92	0.37	330		<270				8.8	17	2.9	3.8
Discovery		11.2	1.75	0.125	23.7	2.11	1.34	0.33	316	29.5	3200	75	120	41	7.4	6.9	1.4	
Atlantis					31	0.4	0.09	0.015										
Valdivia	43		4.1		4.5	0.3	0.024	0.0076	21	1.6		50	22	38	6.8	13.3	2.0	2.8
Wando					14.5	1.7	0.17	0.03										
Hatiba					22	0.5	0.08	0.25										
Hadarba	58		3.0	0.23	6.3	1.3	0.057	0.012	32	5.4		77	27	32	6.0	14	2.8	2.8
Thetis	19	2.7	0.95	0.12	28.3	10.6	0.37	0.082	77	4.6		49	45	33	5.6	13.3	1.7	0.8
Nereus	48		2.6	0.20	7.6	1.7	0.066	0.019	35	0.9	310	82	25	58	11.1	15	2.2	0.7
Vema	58		2.5	0.20	10.5	0.29	0.025	0.0025	20	2.0		46	10.6	32	3.6	10	1.9	2.1
Gypsum	14.6		1.6	0.09	39	0.36	0.16	0.089	25	4.1		35	7.8	34	3.2	13	1.3	1.4
Background sediments	38		3.5	0.37	3.5	0.41	0.032	0.0071	39	0.9	195	68	13.3	53	8.3	11	2.4	2.0

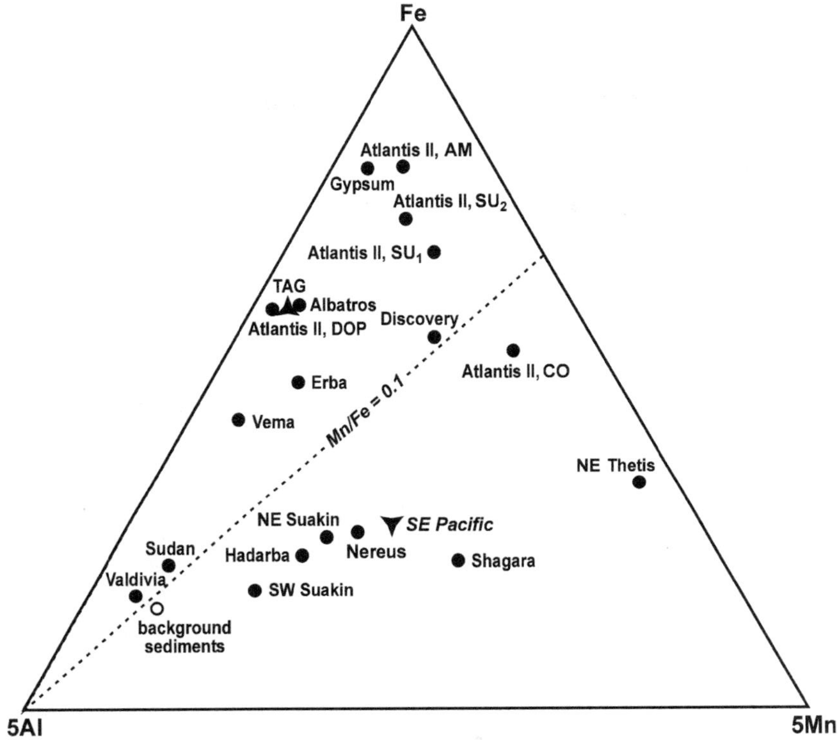

Fig. 2.30. Triangular diagram showing proportions of the average contents of Fe, Mn, and Al in metalliferous sediments from the deeps of the Red Sea rift zone, Southeast Pacific and TAG hydrothermal field.

field. However, even the maximum Ni/Fe ratios in the Red Sea metalliferous sediments are much lower than those in the metalliferous sediments from the Southeast Pacific. The concentration of Ni in the hydrothermal fluids is relatively low (Sect. 4.2), and seawater is the main source of it in the metal-bearing matter of the metalliferous sediments in the Red Sea deeps as well as in the metalliferous sediments of oceans. Apparently the inverse relationship of the Ni/Fe ratio and the accumulation rate of the hydrothermal Fe (Fig. 2.31) is evidence of this. The Ni/Fe ratios evidently indicate that in many deeps in the Red Sea rift zone the contribution of Ni to the sediments from seawater can be significantly important although it does not result in as much enrichment of this element in the metalliferous sediments of the Red Sea deeps as in the metalliferous sediments of the Southeast Pacific.

The Cr/Fe ratio has an inverse relationship to the accumulation rate of the hydrothermal Fe in the metalliferous sediments from the deeps filled with bottom brines (Fig. 2.31). Apparently this evidence indicates that seawater

Table 2.13. Average accumulation rates of hydrothermal Fe [g cm^{-2} ka^{-1}] and average Element/Fe$^{+)}$ and Element/Ni ratios in the metal-bearing matter of metalliferous sediments from the deeps of the Red Sea rift zone, Southeast Pacific and TAG hydrothermal field.

Deep	Fe_h*	Mn/Fe	Zn/Fe	Cu/Fe	Pb/Fe	Sb/Fe	Ba/Fe	Ni/Fe	Co/Fe	Cr/Fe	La/Fe	Zn/Ni	Cu/Ni	Pb/Ni	Sb/Ni	Ba/Ni	Co/Ni	Cr/Ni	La/Ni
NE Suakin	1.1	2300	170	7.9	3.2	0.42		3.2	3.0	0.95	0.26	53	2.5	1.0	0.13		0.94	0.3	0.08
SW Suakin	0.38	3200	120	4.0	0.8	0.65		4.6	3.4	1.1	0.60	26	0.9	0.16	0.14		0.74	0.23	0.13
Port Sudan	0.31	240	95	9.5	5.7	1.5			6.9	9.3	2.6								
Erba	0.78	340	49	23	4.6	0.08	12	6.4	2.9	1.5	1.4	8	3.6	0.72	0.012	1.9	0.45	0.23	0.22
Shagara	0.4	5100	66	9.7	7.7	0.12	32	15	2.0	5.3	1.1	4	0.6	0.50	0.008	2.1	0.13	0.35	0.07
Albatross	~1	86	22	8.7	0.3			1.5	0.79	0.93		15	5.9	0.19			0.53	0.63	
Atlantis II, AM	24	190	700	150	14	9.0	28	0.59	3.4	0.17	0.086	1200	250	24	15	47	5.8	0.28	0.14
Atlantis II, SU_2	35	310	2000	330	31	8.7	57	1.3	4.3	0.43	0.25	1500	250	24	6.6	43	3.3	0.33	0.19
Atlantis II, CO	62	1400	140	61	6.7	1.8	33	0.37	0.81	0.11	0.16	380	170	18	4.8	91	2.2	0.31	0.45
Atlantis II, SU_1	40	520	1100	340	22	0.68	36	0.11	1.3		0.049	9600	3100	200	6.2	330	12		0.44
Atlantis II, DOP		0	440	180	15		7.2				0.49								
Discovery	1.5	870	590	150	14	1.3	140	2.2	5.1	0.97	0.13	260	65	6.1	0.58	62	2.3	0.43	0.06
Atlantis		130	29	4.8															
Valdivia		0	0	10	0.9	0.55		0	7.0	0	2.6								
Wando		1200	120	21															
Hatiba		230	36	110															
Hadarba	0.36	2500	90	18	4.3	1.2		8.4	4.5	0	1.7	11	2.2	0.51	0.14		0.54		0.21
Thetis	3.3	3800	130	29	2.6	0.16		1.3	1.5	0.82	0.40	100	23	2.0	0.12		1.2	0.63	0.31
Nereus		2600	85	27	4.0	0.07	36	7.9	3.1	5.1	1.6	11	3.4	0.50	0.01	4.5	0.39	0.65	0.20
Vema		90	10	0	0.9	0.18		1.2	0.4	0.51	0.49	7.9	0	0.76	0.14		0.35	0.42	0.40
Gypsum	2.5	54	37	23	0.4	0.10		0.23	0.07	0.35	0.23	160	100	1.8	0.44		0.31	1.5	1.0
SE Pacific		3400	24	71	9.7	0.70	363	55	15	0.08	8.9	0.4	1.3	0.18	0.01	6.6	0.27	0.002	0.16
TAG		140	220	510	7.0		25	0.7	4.5	1.4		310	675	9.3		33	6.0	2.0	

$^{+)}$ Element/Fe ratios have been multiplied by 10000.

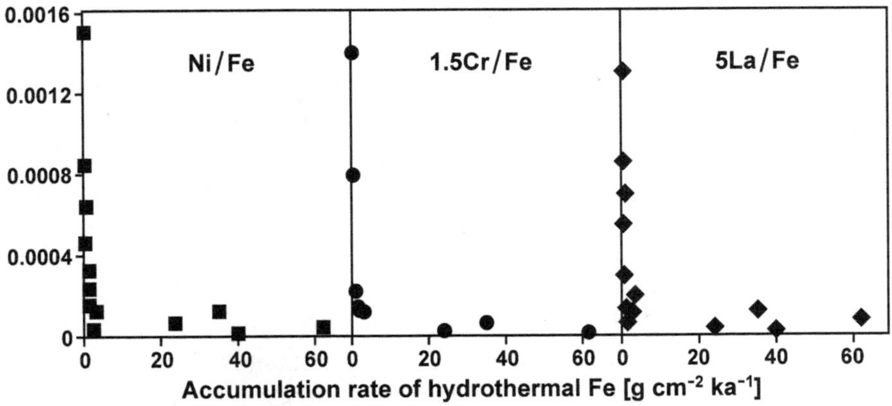

Fig. 2.31. Plots of the average Ni/Fe, Cr/Fe, and La/Fe ratios in the metal-bearing matter of metalliferous sediments from the deeps of the Red Sea rift zone in relation to the average accumulation rates of hydrothermal Fe in these sediments.

is the main source of the excessive Cr in the metal-bearing matter of these sediments. The accumulation of Cr is practically absent in the metalliferous sediments in the Southeast Pacific, which may be attributed to different redox conditions. The environment is oxidizing in the open ocean and it is reducing in the Red Sea deeps that are filled with brines. The reduction of Cr(VI) to Cr(III) can occur in this kind of environment (Krauskopf 1956). In seawater the Cr(VI) in oxyanions can be co-precipitated with the newly formed Fe-hydroxide minerals (Rudnicki and Elderfield 1993; German and Von Damm 2004). However, this process is transient and results in a noticeable accumulation of Cr only in sediments in the vicinity of hydrothermal vents but it is of little significance in distal metalliferous sediments. The Cr(III) is a hydrolysate that can be sorbed from solution by the Fe-hydroxide minerals. This probably accounts for the noticeable accumulation of Cr in the metal-bearing matter in the sediments in some of the Red Sea deeps that are filled with the bottom brines.

The La/Fe ratio has an inverse relationship to the accumulation rate of the hydrothermal Fe in the metalliferous sediments from the deeps of the Red Sea rift zone (Fig. 2.31). Apparently this evidence indicates that seawater is the main source of the excessive REE in the metal-bearing matter of these sediments. It was shown above in Sect. 2.2.3 that hydrothermal REE occur in the sediments in the Atlantis II Deep. However they are of secondary importance in the bulk contents of these elements. Only in some of the metal-bearing layers that are enriched in magnetite in the metalliferous sediments from the Thetis Deep could the REE have been derived mainly from hydrothermal sources (Fig. 2.23). However, the thickness of these layers is less than a quarter of the sediment thickness sampled in the cores and in most of the sediments sampled the REE derived from hydrothermal sources are present only in minor amounts.

The Zn/Fe ratios in the metal-bearing matter of the metalliferous sediments in most of the deeps of the Red Sea rift zone exceed the ratios found in distal metalliferous sediments from the Southeast Pacific (Table 2.13), in which ocean water is the main source of the Zn (Sect. 1.1.7). But they are lower than the Zn/Fe ratios in proximal metalliferous sediments from the TAG hydrothermal field where the Zn is mainly of hydrothermal origin. Only in the sediments from the AM, SU_2, SU_1, and DOP Zones of the Atlantis II Deep and in the sediments from the Discovery Deep do the Zn/Fe ratios exceed those in the TAG proximal metalliferous sediments. This evidence indicates that the Zn in the metal-bearing matter of the metalliferous sediments from these deeps is mainly from hydrothermal sources. If it is assumed that Ni is a representative proxy for indicating the contribution of elements from seawater, the Zn/Ni ratios will (Table 2.13) reflect the relative contribution of Zn from hydrothermal solutions to the metal-bearing matter of the sediments. The maximum Zn/Ni ratios occur in the sediments from the Atlantis II Deep. The Zn/Ni ratios are lower in the sediments from the Discovery, Gypsum, and Thetis Deeps and are comparable to those in the sediments from the TAG field, but essentially they exceed the ratios in the metalliferous sediments from other deeps and especially in the metalliferous sediments from the Southeast Pacific where there is a minimum role of hydrothermal Zn in the excessive accumulation of the element.

The highest Cu/Fe ratios also occur in the sediments of the AM, SU_2, SU_1 and DOP Zones from the Atlantis II Deep and in the sediments from the Discovery Deep. However, unlike the Zn/Fe ratios, they are lower than the Zn/Fe ratio in the metalliferous sediments from the TAG hydrothermal field (Table 2.13). Obviously this results from specific metallogenic differences in the hydrothermal sources. In the TAG hydrothermal field they are rich in copper where the concentration of Cu is three times higher than the concentration of Zn (Charlou et al. 2002) and in the deeps of the Red Sea rift zone they are most likely to be rich in zinc (Table 2.2). In the metal-bearing matter of metalliferous sediments from all of the deeps, except for the Hatiba, the contents of the Zn exceed the contents of the Cu. In the metalliferous sediments in almost all of the deeps, and especially those where there is a high content of metals in the sediments as in the Atlantis II, Discovery, Thetis, and Gypsum Deeps, the Cu/Ni ratio is higher than in the distal metalliferous sediments from the Southeast Pacific where the contribution of Cu from ocean water is the most pronounced. And only in the sediments from the Atlantis II and Gypsum Deeps are the Cu/Ni ratios similar to those in the metalliferous sediments in the TAG hydrothermal field where the Cu is mainly of hydrothermal origin. In general, based on the Cu/Ni and Zn/Ni ratios in the metal-bearing matter of the metalliferous sediments from the Red Sea deeps, a hydrothermal source for the excessive accumulation of the Cu and Zn is prevalent and seawater is of minor importance.

The Pb/Fe ratios in the metal-bearing matter of the metalliferous sediments from the Southeast Pacific and TAG hydrothermal field are similar. However, in the metal-bearing matter of the metalliferous sediments from the Southeast Pacific, seawater is the main source of Pb, and in the metal-bearing matter of the metalliferous sediments from the TAG hydrothermal field hydrothermal sources are prevalent (Godfrey et al. 1994). This is also evident from the much higher Pb/Ni ratios in the metalliferous sediments from the TAG hydrothermal field. Only in the metalliferous sediments from the Atlantis II and Discovery Deeps the ratios of Pb/Fe are close to or exceed those in the ocean metalliferous sediments. The ratios Pb/Ni in the metalliferous sediments from most of the Red Sea deeps are higher, and usually much higher, than the Pb/Ni ratio in the metalliferous sediments from the Southeast Pacific. The Pb/Ni ratios are higher in the sediments from the Atlantis II Deep than in the metalliferous sediments from the TAG hydrothermal field, and in the metalliferous sediments from the Discovery Deep the ratio is comparable to that in the metalliferous sediments from the TAG field. Obviously this evidence indicates mainly a hydrothermal source for the Pb in the metal-bearing matter of the metalliferous sediments in the Red Sea deeps.

The Sb/Fe ratios in the metal-bearing matter of the metalliferous sediments from the Red Sea deeps are both higher and lower than the ratio in the metalliferous sediments from the Southeast Pacific, in which ocean water is the main source of the excessive Sb in the metal-bearing matter (Sect. 1.1.7). The process of scavenging of chemical elements from seawater is most pronounced in the Southeast Pacific. Therefore, it can be concluded that in the metal-bearing matter of the metalliferous sediments from the Red Sea deeps where the Sb/Fe ratios exceed or are close to those in the metalliferous sediments from the Southeast Pacific, hydrothermal solutions are the main source of the excessive Sb. Low Sb/Fe ratios in sediments in some of the deeps apparently is evidence that indicates the removal of hydrothermal Sb. Presumably, some part of the excessive Sb is contributed to these sediments from seawater. However, this process is of minor importance, because in some deeps low Sb/Fe ratios occur in the metalliferous sediments with relatively low accumulation rates of the hydrothermal Fe (Table 2.13).

The Co/Fe ratios in the metal-bearing matter of the metalliferous sediments from the Red Sea deeps vary over a wide range (Table 2.13). All these ratios are lower than those in the metalliferous sediments from the Southeast Pacific. The Co/Fe ratios in the metalliferous sediments from some of the deeps are similar to those in the metalliferous sediments from the TAG hydrothermal field and in the others they are much lower. Similar values occur in the sediments from the deeps where there are both high and low temperature brines. The highest Co/Ni ratios occur in the sediments from the Atlantis II, Discovery, and Thetis Deeps. They are similar to the ratios in the metalliferous sediments from the TAG hydrothermal field. In sediments from the other deeps the Co/Ni ratios are a little lower, comparable, or a little higher than the Co/Ni ratio in the metalliferous sediments from the Southeast Pacific (Table 2.13), in which ocean water is the main source of the excessive Co. Apparently this is evidence of the dual nature of the Co in the metal-bearing matter of the metalliferous sediments from the Red Sea deeps. In the sediments that have high Co/Fe and Co/Ni ratios the excessive Co is mainly of hydrothermal origin. In the sediments that have high Co/Fe ratios and low Co/Ni ratios the excessive Co is derived mainly from seawater.

The Ba/Fe ratios in the metal-bearing matter of the metalliferous sediments from the Red Sea deeps are much lower than in the metalliferous sediments from the Southeast Pacific and comparable to the ratios in the metalliferous sediments from the TAG hydrothermal field. In the metalliferous sediments from the southeast Pacific the major part of the Ba accumulates as a component in the mineral barite (Fig. 1.23) during sedimentation that is related to biological processes (Gurvich et al. 1979). In the metalliferous sediments from the TAG hydrothermal field Ba is associated mainly with material that sinks from the nonbuoyant plume (Sect. 1.6.1). If any of the Ba in the metalliferous sediments from the Red Sea deeps accumulates as a component of barite by biological processes it is in insignificant minor amount. However hydrothermal barite occurs in these sediments in small amounts (Sect. 2.4). The concentration of dissolved Ba in the brines of the Atlantis II Deep is 50 times greater than in seawater (Table 2.2). It is evident that Ba is contributed to the brines both from hydrothermal sources and from seawater by the scavenging of the Mn-oxyhydroxides that form in the transition zone between the brines and seawater. During sedimentation the Mn-oxyhydroxides are dissolved as they settle through the brines and the Ba, like many other trace elements that are scavenged in the transition zone, enters the brines. Probably, in the Atlantis II and Discovery Deeps, both mechanisms lead to the enrichment of the brines in Ba, which enters the sediments from the brines. In the metalliferous sediments in the deeps where there is less intensive hydrothermal activity as in the Erba, Shagara, and Nereus Deeps, the Ba/Fe ratios are comparable to those in the metalliferous sediments from the Atlantis II and Discovery Deeps, but the

Ba/Ni ratios in these sediments are much lower (Fig. 2.13). This indicates that hydrothermal fluids are much less significant as a source of Ba in the metal-bearing matter of the metalliferous sediments from these deeps.

The assumptions made about the sources of the chemical elements in the metal-bearing matter of the metalliferous sediments from the Red Sea deeps are also evident in the factor analysis of the Element/Fe ratios, except for the Ba/Fe ratios because of the few data available. Three reasonable factors that account for 98.1% of the variance have been revealed (Fig. 2.32).

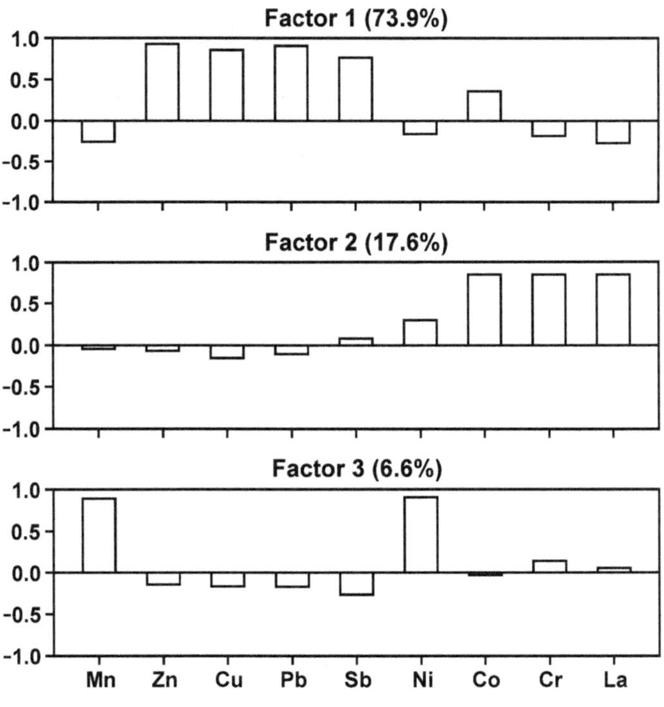

Fig. 2.32 Factor loadings of three varimax rotated factors for the Element/Fe ratios in the metal-bearing matter of metalliferous sediments from the deeps of the Red Sea rift zone.

The first factor accounts for 73.9% of the variance. The association of Zn, Cu, Pb, and Sb that have high positive loadings is unequivocally evidence of their hydrothermal nature. Co has a significant positive loading. However, it also has a high positive loading in the second factor.

The second factor accounts for 17.6% of the variance. Co, Cr, and La have high positive loadings and the loading for Ni is significant. This association of the elements in the metal-bearing matter of the metalliferous sediments is evidence of their source in seawater. The loadings of Co in both the first and the second factors demonstrate its dual nature. In the sediments that accumulated during intensive hydrothermal activity hydrothermal Co prevails in the metal-bearing matter, and in the sediments that accumulated during less intensive hydrothermal activity Co derived from seawater is prevalent.

The third factor accounts for 6.6% of the variance. Mn and Ni have high positive loadings. Apparently this is evidence that indicates that the Ni that enters the metal-bearing matter of the metalliferous sediments from seawater is associated mainly with Mn-minerals.

Comparison of the formation of metalliferous sediments that accumulated in rift zones of the open ocean and of the Red Sea shows that there are certain features that are similar and some that are different. The similarities probably resulted because of similarities in the composition of the metal-bearing hydrothermal fluids from which they were derived. This is demonstrated by the similarity in the compositions of the hydrothermal accumulations in the vicinity of the orifices of high-temperature springs in zones where hydrothermal fluids are intermixed with bottom water or brines. As for the metalliferous sediments that formed at some distance from the orifices, their compositions in the Red Sea and in the open ocean are considerably different. These differences result from different conditions for the sedimentation of elements contributed from hydrothermal sources in the highly mineralized brines confined in the Red Sea deeps and in ocean water, and the various factors affecting their differentiation in environmental conditions that are very dissimilar.

CHAPTER 3 ANCIENT METALLIFEROUS SEDIMENTS IN THE OCEANS

Exploration of the recent metalliferous sediments has shown that accumulation of metalliferous strata begins near the axes of the spreading ridges on the young basalt ocean crust. As a rule the metalliferous strata in the places where they accumulated occupy the lowermost part of the sedimentary cover of the oceans and are a part of the basal sedimentary layer and should be considered as a separate basal metalliferous formation. Recognizing that the ancient metalliferous sediments occur in the lower part of the sequence of sedimentary strata and are overlain by strata of younger non-metalliferous sediments has required a special strategy in the search for ancient metalliferous sediments in the oceans. In the first legs of the drill ship *Glomar Challenger* metalliferous sediments were found in the lowermost part of the sedimentary sections in some of the holes that reached basalts (Von der Borch and Rex 1970; Von der Borch et al. 1971; Boström 1973; Dymond et al. 1973; Lange 1974; Horowitz and Cronan 1976 et al.). In the ensuing years the basal metalliferous sediments were sampled repeatedly in the Pacific, Indian and Atlantic Oceans. However, methodical investigations were not carried out[25]. In many cases the studies included only descriptions of the lithology, and the metalliferous sediments were not recognized. In other cases sediments were considered to be metalliferous that contained a very small admixture of hydrothermal matter and less than 10% of Fe in their abiogenic part; they were transitional to metalliferous sediments.

Many publications provide information on the composition of bottom sediments, including the metalliferous sediments that were sampled in the deep-sea drilling programs in the Pacific, Indian and Atlantic Oceans. The information was published mainly in Initial Reports of the Deep-Sea Drilling Project (1970–1987) and in Proceedings of the Ocean Drilling Program (1988–2004), and only a small part of it was published in journals and books. Many samples of metalliferous and transitional sediments pro-

[25] Levin et al. (1985, 1987) have investigated the buried metal-bearing material in the ancient sediments of the Pacific Ocean. However separate consideration of the metalliferous sediments is not given in their work.

vided by the DSDP Directory on our request have been studied in the P.P. Shirshov Institute of Oceanology.

3.1. Occurrence of ancient metalliferous sediments, thickness of metalliferous strata

There is considerable variation in the occurrence of metalliferous sediments in the sedimentary cover of the oceans. Metalliferous sediments were found in the basal parts of the sedimentary layer in many of the holes of the deep-sea drilling program that reached the basaltic crust. Metalliferous sediments were found in 18% of the holes drilled that reached the basaltic crust in the Atlantic Ocean, in 33% in the Indian Ocean, and in 48% in the Pacific Ocean. Even these approximate data indicate that occurrence of basal metalliferous sediments increase in number in relative order from the Atlantic to the Indian to the Pacific Ocean and the occurrence of recent metalliferous sediments increases likewise. Considering that in the South Pacific where the largest field of recent metalliferous sediments is located but there are very few drilling stations, one can say with good reason, that the occurrence of basal metalliferous sediments in the Pacific Ocean is underestimated.

The thickness of the basal metalliferous sediments recovered during the deep-sea drilling program varies from zero to some tens of meters, and in some cases up to a little more than one hundred meters. The distribution of basal metalliferous sediments and thickness of the basal metalliferous strata in the World Ocean are shown in Fig. 3.1. The largest areas of metalliferous strata occur in the basal layer of the sedimentary cover of the Pacific Ocean. They occur to a lesser extent in the basal layer of the sedimentary cover of the Indian Ocean, and to a considerably lesser extent in the basal layer of sedimentary cover in the Atlantic Ocean.

The greatest variation in the thickness of the metalliferous strata, from minimum to maximum values, occurs in the Pacific Ocean. The thickest (>50 m) strata are located in the Pacific Ocean south of the equator within a sublatitudinal belt from ~10°S to ~40°S. A similar sublatitudinal belt also exists north of the equator from ~5°N to 30°–40°N. However, within this belt the thickness of the metalliferous strata is not as great as that within the southern belt and in rare cases it exceeds 30 m. A belt of metalliferous strata of reduced thickness occurs between these two belts. In the east it begins in the equatorial part of the ocean and extends in a latitudinal direction approximately to where it crosses the East Pacific Rise. Further on it alters in direction to the northwest along the motion of the Pacific lithosphere plate (Winterer

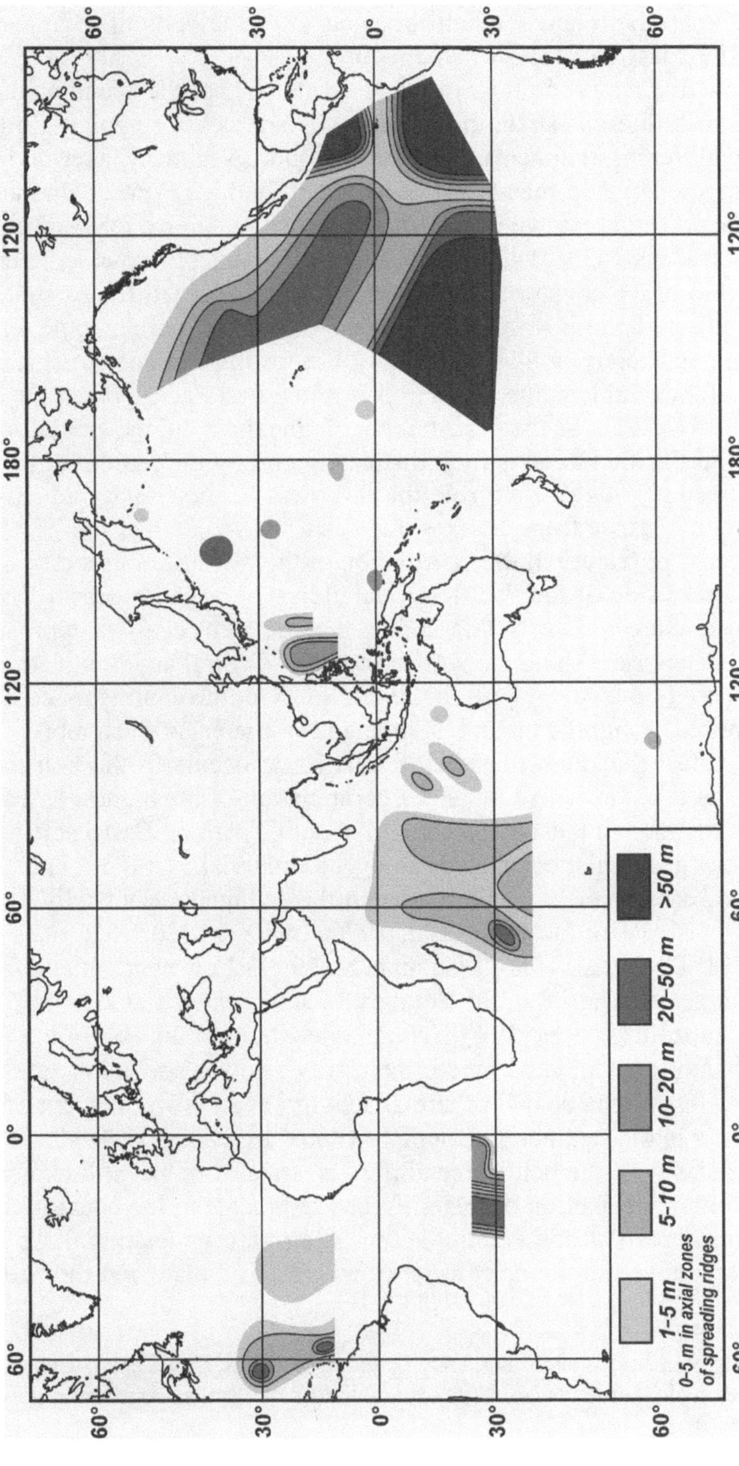

Fig. 3.1. Thickness of basal metalliferous strata in the World Ocean.

1973). The northern and southern boundaries of the belts of sediments of greatest thickness also follow this direction.

A distinctive feature in the Southeast Pacific is that the accumulation of the basal metalliferous strata in large areas is not as yet complete. In these areas metalliferous sediments compose the entire sediment layer and they are not covered by non-metalliferous sediments and are exposed on the bottom surface. The areas where the metalliferous strata are exposed on the bottom surface coincide with the areas of modern metalliferous sediments[26]. To the north of the equator in the Pacific Ocean the metalliferous strata are exposed on the bottom surface only in the proximity, within some tens of kilometers and rarely within 100 to 200 km, of the axes of the spreading ridges, as for example in the northern part of the East Pacific Rise.

A typical feature in the distribution of thickness of the metalliferous strata in the Pacific Ocean is their distinct symmetry in relation to the axes of the spreading ridges. As a rule the thickness of the strata gradually increases with distance from the axes.

An important feature in the distribution of the metalliferous strata is that on the eastern side of the Pacific Ocean along the western coast of North America they are relatively thin and as the continent coast is approached the strata disappear. Along the western coast of Central and South America the thickness of the strata, even in the proximity of the continent and of the deep trenches, is significant and is some tens of meters or even more.

The greatest thickness of metalliferous strata occurs in the belt that is south of the equator, where there is a combination of the highest spreading rates and the most intensive hydrothermal activity in the East Pacific Rise and the low accumulation rates of abiogenic material, <100 mg cm^{-2} ka^{-1}, and a high content of biogenic matter in the sediments (Donnelly 1980a; Dymond et al. 1976; Boström et al. 1976; Lisitzin 1980; Heath and Dymond 1981; Lyle et al. 1986; Lisitzin et al. 1990). Low accumulation rates of the abiogenic material in the Southeast Pacific existed at least throughout the Cenozoic (Lisitzin 1980). The thinner strata in the belt to the north of the equator accumulated where there was a lower spreading rate (and apparently less intensive hydrothermal activity) and a lower content of biogenic matter in the sediments (Donnelly 1980a; Lisitzin et al. 1990).

The existence of the belt of metalliferous strata of reduced thickness resulted from the deposition of metalliferous sediments in the equatorial and subequatorial parts of the Pacific Ocean. After their deposition these sediments were moved to the northwest with the Pacific plate and the belt tra-

---

[26] Ancient metalliferous sediments may be exposed where hiatuses are present. For example, such sediments occur in some parts of the Bauer Depression (Lisitzin et al. 1976).

ces the direction of this movement. The thinner metalliferous strata in the equatorial and subequatorial parts of the ocean were deposited due to relatively high accumulation rates of diluting abiogenic material as happens now and in the past (Lisitzin 1978, 1980; Bogdanov et al. 1979c).

Metalliferous strata are essentially absent in the axial parts of the spreading ridges where the most recent basement rocks occur. With an increase in the age of the basement rocks on either side of the axes the time increased during which the metalliferous sediments could accumulate on the bottom and there was an increase in the thickness of the metalliferous strata up to the end of the period of their accumulation.

In general the metalliferous strata in the Indian Ocean are thinner than in the Pacific Ocean and only in one hole does the thickness exceed 20 m, whereas in the Pacific Ocean strata of this thickness are common. In the Indian Ocean the areas where the metalliferous strata are the thickest occur in the belts located from 5°–10°S to 35°–40°S latitude. Most of the areas where the basal metalliferous strata have been found are around the Rodriguez Triple Junction (Fig. 3.1).

In the area of the Rodriguez Triple Junction basal metalliferous strata are exposed on the ocean floor. As in the Pacific Ocean the thickest basal metalliferous strata in the Indian Ocean are distributed symmetrically to the axes of the spreading ridges. There is a comparatively small area in the West Australian Basin that has basal metalliferous strata that are up to 8 m thick. Obviously they formed along a spreading ridge that existed between the Indian and Australian continents during Late Cretaceous to Early Paleogene time (Kennett 1982).

As mentioned previously basal metalliferous strata have been found in only a relatively small number of holes in the Atlantic Ocean in the deep-sea drilling program. Most of them are located in the southern part of the ocean from ~25° to 30°S latitude near the axis of the Mid-Atlantic Ridge. Basal metalliferous strata have also been found north of the equator from ~10°N to 35°N latitude. The metalliferous strata in the Atlantic Ocean are up to tens of meters thick. Like those in the Pacific and Indian Oceans, the thickness of the basal metalliferous strata in the Atlantic Ocean is distributed symmetrically to the axis of the spreading ridge (Fig 3.1).

3.2. Age and duration of accumulation of metalliferous strata

The age of the metalliferous strata and duration of period of time when they formed (duration of accumulation) are their major parameters. The ages of the metalliferous strata studied have been obtained from the published results of studies of the sediment cores that were collected during the deep-sea drilling program (Initial Reports … 1970–1987; Proceedings … 1988–2004) and also from some other publications. An overwhelming majority of the data for the sediment ages has been obtained by biostratigraphy methods. In many cases the age data have enabled the estimation of sedimentation rates. Use of the data on the thicknesses of the metalliferous strata and on the sedimentation rates has enabled calculation of the duration of accumulation of these strata.

The age of the basal part of a metalliferous stratum has been used to represent its age, which is the time when the accumulation of the stratum began. This time, within the limits of accuracy of the age determination methods used, usually coincides or is somewhat less than the age of the basalt basement that underlies the metalliferous stratum.

The most ancient metalliferous sediments recovered in the modern ocean accumulated in Upper Jurassic time, about 150 Ma BP, and the youngest are accumulating at the present time. In all of the oceans the age of the metalliferous strata increases with distances from the axes of the spreading ridges due to the increase in the age of the basalt basement.

Occurrence of basal metalliferous sediments of different age is shown in Fig. 3.2. In the open part of the Pacific Ocean, mainly west of the East Pacific Rise, the occurrence of the metalliferous sediments within the basal layer of the sedimentary cover demonstrates determinate temporal variations. Even without consideration of the Upper Jurassic basal metalliferous sediments[27], one can say that the occurrence of the metalliferous sediments decreased at least from Early Cretaceous, 140–120 Ma BP, until Middle Cretaceous time, 100–80 Ma BP. There was an increase in the occurrence of basal metalliferous sediments in Middle-Late Paleogene time, 50–30 Ma BP, when it reached a maximum. Later, until 20–10 Ma BP, the occurrence decreased again. The time period from 10 to 0 Ma BP is characterized by some increase in the occurrence of metalliferous sediments in comparison with the period from 20 to 10 Ma BP. These variations in the occurrence of the basal metalliferous sediments in time cannot be considered as a quantitative criterion of the influence of hydrothermal process in the sedimentation that took place in the open part of the Pacific Ocean. However, the monotonous variations in the occurrence as determined during the time intervals referred to obviously may reflect temporal variations in the influence of the hydrothermal process.

The most ancient metalliferous sediments recovered in the Indian Ocean during the deep-sea drilling program in DSDP Hole 260 are Albian in age, according to biostratigraphic dating. However, the metal-bearing hydro-

[27] I have information only on one hole.

thermal matter entered these sediments owing to a basalt sill that was intruded into the sediment layer at a later time (Gurvich et al. 1988). Therefore the age of the metal-bearing matter in this hole is uncertain and cannot be confirmed as Albian. The most ancient basal metalliferous sediments recovered during the deep-sea drilling program in the Indian Ocean certainly are of Maastrichtian age. According to the available data (Fig. 3.2), the basal metalliferous sediments that occur in the Indian Ocean are mainly of Maastrichtian to Early Eocene in age and there is a minor maximum indicated in their occurrence in the interval from 60 to 70 Ma.

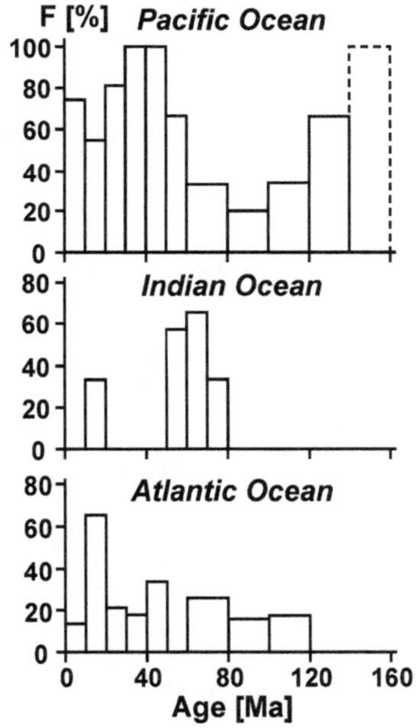

Fig. 3.2. Occurrence of metalliferous sediments of different age in the basal layer of the sedimentary cover of the oceans.

Despite the absence of information from samples obtained in the Indian Ocean during the deep-sea drilling program for basal metalliferous sediments younger than 10 Ma, some estimation can be based on samples from the Cruise 25 of R/V *Dmitry Mendeleev* (Metalliferous sediments… 1987) and from Cruise 92 of R/V *Sonne* and Cruise 33/2 of R/V *Meteor* (Kuhn et al. 2000). Metalliferous sediments collected in these expeditions in the area of the Rodriguez Triple Junction have ages ranging from recent to >2.5 Ma. Recognizing that the sediment cores did not extend to the basalt basement it is most likely that, at least in the area of the Rodriguez Triple Junction and in the adjoining areas, the Early Pliocene and probably Miocene basal metalliferous sediments occur on the basement. It is seen from

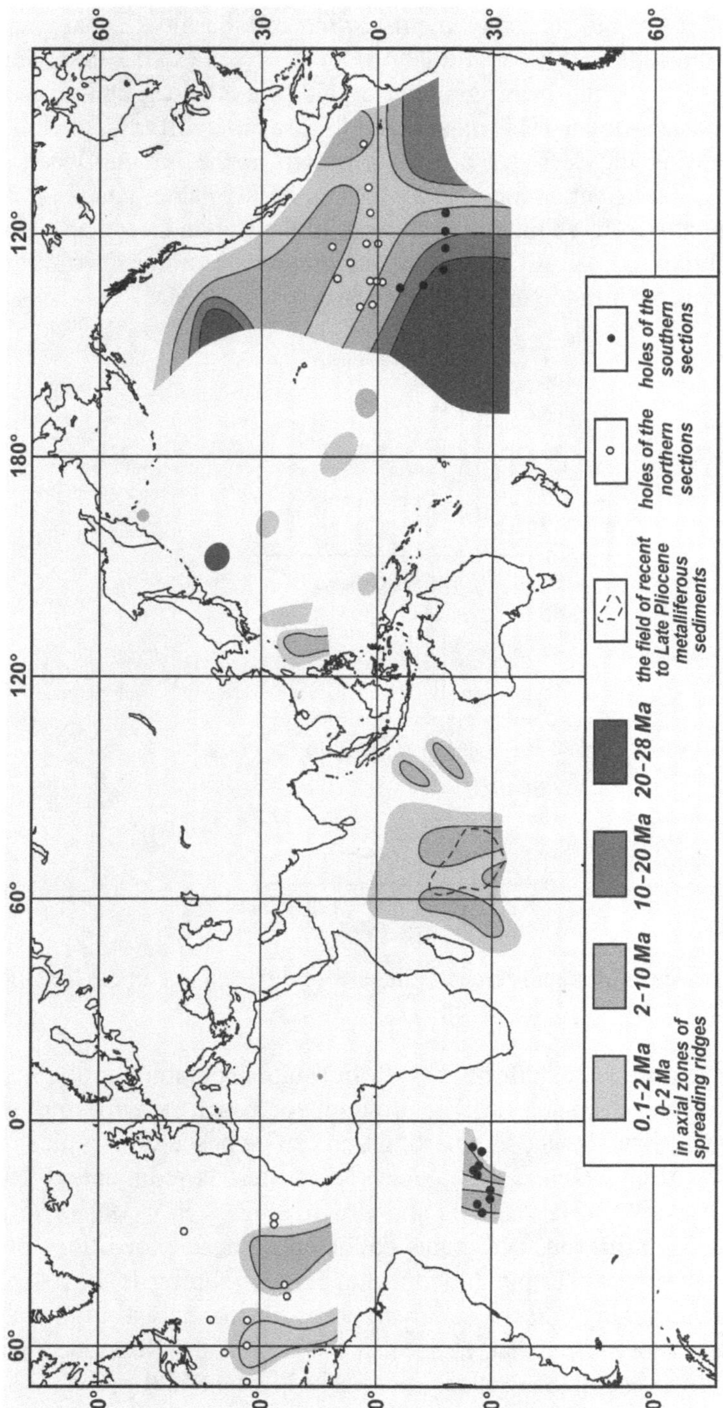

Fig. 3.3. Duration of accumulation of basal metalliferous strata in the World Ocean.

Fig 3.2 that there was no significant accumulation of metalliferous sediments in the past 160 Ma in the Indian Ocean. There was more intensive accumulation of these sediments in the time period from 80 to 50 Ma BP and the maximum accumulation corresponds to the time period from 70 to 60 Ma BP. Basal metalliferous sediments have accumulated during the last 20 Ma, but apparently during this time there were fewer occurrences than in the time period from 80 to 50 Ma BP.

Occurrence of basal metalliferous sediments in the Atlantic Ocean during most of its geologic history did not exceed 10–20%, if any. Only in the period of time from 20 to 10 Ma BP has there been an increase in their accumulation (Fig. 3.2), but these data require further confirmation.

The distribution of duration of accumulation of the basal metalliferous strata in the World Ocean, calculated on the data from the analyses of core samples from the deep-sea drilling program, are shown in Fig. 3.3. The specific duration of time that was calculated varied from several tens of thousands of years to tens of millions of years and the maximum was 28 Ma. Theoretically the minimum duration of time of accumulation may be shorter than the time calculated but a zero means that basal metalliferous sediments have not accumulated or that the sediment strata were non-metalliferous or there was no sedimentary cover on the basement rock.

In the Pacific Ocean the duration of accumulation of basal metalliferous strata varied greatly and was from zero to more than 20 Ma. The pattern of the distribution of duration of accumulation of basal metalliferous strata in the Pacific Ocean is very similar to that for the distribution of their thickness (compare Fig. 3.3 and Fig. 3.1). Also there are two belts to the south and to the north of the equator where there was a greater length of time indicated of the accumulation of the basal metalliferous strata. Between these two belts there is a belt where there was a shorter duration of accumulation. This belt extends from the equatorial part of the East Pacific in a latitudinal direction to the intersection with the East Pacific Rise, and extends further in a northwestern direction. The distribution of the duration of accumulation of the basal metalliferous strata is also symmetrical to the axes of the spreading ridges and in the axial zones they are zero or near zero and increase away from the axes. A short duration of accumulation is characteristic for the thin metalliferous strata in the northeast part of the ocean, along the western coast of North America north of 22°N, where there is no subduction zone, and a long duration of accumulation is characteristic for the thick basal metalliferous strata that formed in the pelagic zone and now are located in the eastern part of the ocean along the western coast of Central and South America, where subduction zones exist. In the back-arc basins of the Pacific Ocean the duration of accumulation of the basal metalliferous strata that have thickness from 10 to 15 m was up to several millions of years.

The duration of accumulation of the basal metalliferous strata sampled during the deep-sea drilling program in the Indian Ocean was much shorter than for the metalliferous strata in the Pacific Ocean. Strata that accumulated in a period longer than 5 Ma have been sampled in only one of the holes drilled in the West Australian Basin. Probably a long period of time, several millions of years, was required for the accumulation of the strata that are associated with the recent metalliferous sediments in the area of the Rodriguez Triple Junction and some distance from it. These strata may have accumulated during a period longer than 2–3 Ma because within the field of the recent metalliferous sediments, where the accumulation of the metalliferous strata is not as yet complete, sediment cores provided samples of metalliferous sediments that are up to >2.5 Ma in age and did not extend to the basalt basement.

The area where basal metalliferous strata have accumulated for more than 1 Ma and are accumulating at the present time is located within a belt that extends from 5°–10°S to 35°–40°S latitude. The pattern of the distribution of duration of accumulation of the basal metalliferous strata is very similar to the pattern of the distribution of thickness (compare Fig. 3.3 and Fig. 3.1).

According to the data available, the duration of accumulation of the basal metalliferous strata in the Atlantic Ocean is somewhat longer than in the Indian Ocean and much shorter than in the Pacific Ocean. The maximum duration of accumulation of the basal metalliferous strata is 10 Ma and the duration most frequently occurred is several millions of years. The locations of the metalliferous strata that accumulated over a long period of time correspond to the latitudinal belts where the thickest strata occur. In general the pattern of the distribution of the duration of accumulation of the basal metalliferous strata in the Atlantic Ocean is very similar to the pattern of distribution of their thickness (compare Fig. 3.3 and Fig. 3.1).

The main factors influencing on the duration of the accumulation and on the thickness of the metalliferous strata will be considered in Chap. 4. The rates of accumulation of the hydrothermal metal-bearing matter in the ancient metalliferous sediments will be considered in Chap. 5.

3.3. Chemical composition of ancient metalliferous sediments

The average chemical compositions of basal metalliferous sediments from the Pacific, Atlantic and Indian Oceans are shown in Table 3.1. For the Pacific and Atlantic Oceans the average values reflect practically the whole age range, from 140–120 Ma and from 150–140 Ma respectively.

The average composition of the basal metalliferous sediments from the Indian Ocean represents mainly the age interval 75 to 50 Ma; the basal sediments accumulated outside of this period of time have been studied from only one hole. The average composition of the basal metalliferous sediments in the back-arc basins of the Pacific Ocean is only approximate because of the paucity of data. The age range for these sediments is from 55 to 0.1 Ma. However, the composition of the basal metalliferous sediments has not been studied in sediments from all of the intervals in this age range.

Comparisons of the average compositions of ancient metalliferous sediments from different parts of the World Ocean with the average composition of the most thoroughly studied recent metalliferous sediments from the Southeast Pacific are shown in Fig. 3.4. The average composition of the ancient metalliferous sediments from the Pacific Ocean has features that are both similar and dissimilar to the average composition of recent metalliferous sediments. The average contents of the main hydrothermal element Fe, and of others including Al, Ti, Mo, Th, and La in the ancient and recent metalliferous sediments are similar. The contents of Zn and B are a little higher, and the contents of Cr, Li and Rb are approximately twice as high in the ancient metalliferous sediments as in the recent metalliferous sediments. The average contents of other elements in the ancient metalliferous sediments from the Pacific Ocean shown in Table 3.1 are lower than their contents in the recent metalliferous sediments from the Southeast Pacific. The contents of Mn, Cu, Ni, V, Y, Ce and Sc in the ancient metalliferous sediments are nearly twice as low, and the contents of Pb, Sb, Zr, Hf, Co and Ba are 2 to 3 times lower than in the recent metalliferous sediments from the Southeast Pacific.

The elements with Type 1 distribution in the recent metalliferous sediments of the Southeast Pacific (Sect. 1.1.4), Fe, Mn, Zn, V, and Mo, have average contents in the ancient metalliferous sediments of the Pacific Ocean that are close to or exceed their average contents in the recent metalliferous sediments of the Southeast Pacific. And the average contents of Fe, Mn, Zn, and Mo in the ancient metalliferous sediments from the southern part of the Pacific Ocean are higher than in the recent metalliferous sediments of the Southeast Pacific.

The average contents of the elements with Type 2 and Type 3 distribution in the recent metalliferous sediments of the Southeast Pacific, Cu, Ni, Zr, Y, Co, Ce, Sc, Sb, Hf, and Ba, have average contents in the ancient metalliferous sediments from the southern and northern parts of the Pacific Ocean lower than in the recent metalliferous sediments from the Southeast Pacific. The differences in composition as determined can be attributed to natural factors and to differences in sampling methods used in the investigation of the recent and ancient metalliferous sediments. Apparently the

Table 3.1. Average contents of chemical elements in abiogenic matter of basal metalliferous sediments from holes drilled in the deep-sea drilling program. Compiled from original data and data from Initial Reports ... (1970–1987), Dymond et al. (1973, 1976), Boström et al. (1974), Lange (1974), Horowitz and Cronan (1976), Bloch (1978), Stoffers et al. (1983), Adachi et al. (1986), Ruhlin and Owen (1986), Barrett et al. (1987), Barrett and Jarvis (1988), Gurvich et al. (1988, 1995b), Proceedings … (1988–2002), Cronan and Hodkinson (1997).

Element	Pacific Ocean *(L. Jurassic – Pleistocene)*	Atlantic Ocean *(L. Jurassic – Pleistocene)*	Indian Ocean** *(Maastrichtian – E. Miocene)*	Back-arc basins of the Pacific Ocean *(E. Paleogene – Pleistocene)*
Fe %	19.2 (16.7; 23.0)*	12.8	11.8	12.3
Mn	4.0 (3.1; 5.9)	2.3	1.6	3.6
Al	4.0 (4.6; 3.0)	6.7	6.2	5.5
Ti	0.28 (0.35; 0.21)	0.48	0.40	0.30
Mg	2.5	2.9	2.1	2.9
K	1.6	2.0	1.7	1.3
P	0.71	0.4	0.43	0.27
Ba	0.53 (0.52; 0.62)	0.14		0.37
Cu ppm	820 (690; 1120)	270	230	420
Zn	560 (470; 730)	330	330	450
Pb	85	140	120	120
Ni	520 (350; 790)	320	270	360
Co	80 (59; 120)	86	55	76
Cr	67	88	100	38
V	410	310	240	280
Mo	82 (70; 135)			
Li	69	59	41	62
Rb	65	86		110
Sr	510	1100	850	820
Hg	0.5	0.15		1.2
Sb	5.2	4.4		>0.5
La	140	150		83
Ce	62	120		35
Y	230			
Sc	17	15		7.5
Th	4.4	9.7		4.5
Zr	190			380
Hf	2.2	3.0		5.3
Ta	0.8	1.4		0.3
B	320	190		

* – average content for the whole ocean (average content for the northern part; average content for the southern part of the ocean); ** – without sediments from the holes that penetrated sills. Contents of Sr in natural dry sediments are shown.

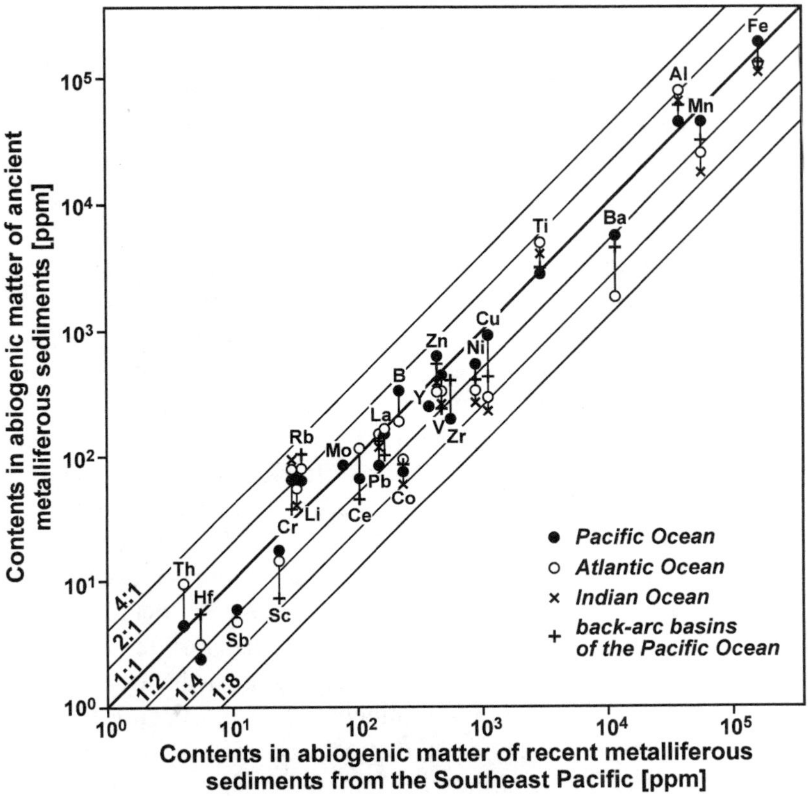

Fig. 3.4. Comparison of the average composition of ancient metalliferous sediments with the average composition of recent metalliferous sediments from the Southeast Pacific.

most important natural factor affecting the enrichment of the ancient metalliferous sediments in hydrothermal components is that in the Pacific Ocean the average spreading rate in recent time is considerably slower than the average spreading rate for the last 150 Ma. According to Zonenshain and Khain (1989), the recent crustal accretion rate in the Pacific Ocean is 1.5 times lower than the average rate for the last 150 Ma and almost twice as low as in Cretaceous time.

The higher contents of elements with Type 1 distribution (Sect. 1.1.4) and the lower contents of elements with Types 2 and 3 distribution (Sect. 1.1.4) in the ancient metalliferous sediments than in recent metalliferous sediments can be attributed also to the sampling. During the sampling of the ancient metalliferous sediments the parts of the samples of sediments that had accumulated close to ridge axes were usually greater than during the sampling of the recent metalliferous sediments. Very often only the samples of the ancient metalliferous sediments from the lowermost parts of the metalliferous strata accumulated in axial and near-axial parts of the spreading ridges were of greatest interest to the researchers. During the coring of the recent metalliferous sediments the researchers had their greatest difficulties in the axial and near-axial parts of the spreading ridges[28] because the sediment cover in these areas is very thin or completely absent. As a result the relative proportions of sample material from the axial and near-axial parts of the ridges were decreased while the amounts of sample material from the flanks of the ridges were considerably increased.

The average compositions of the ancient metalliferous sediments from the Atlantic and Indian Oceans and from the back-arc basins of the Pacific Ocean are considerably different (Fig. 3.4). These metalliferous sediments contain much less metal-bearing matter including Fe and Mn compared to the metalliferous sediments from the Pacific Ocean and they have lower average contents of Cu, Ni, V, Co, Sc, and Ba. The contents of Zn in ancient metalliferous sediments from the Atlantic and Indian Oceans are also lower, but in the ancient metalliferous sediments from the back-arc basins the average content of Zn is close to that in the ancient metalliferous sediments from the Pacific Ocean. The average contents of Zr and Hf in the ancient metalliferous sediments from the back-arc basins are much higher than in the ancient metalliferous sediments from the Pacific Ocean. The average contents of Al, Rb, and Th, that occur mainly in lithogenic matter, in the ancient metalliferous sediments from the Atlantic and Indian Oceans and from the back-arc basins are higher than in the ancient metalliferous sediments from the Pacific Ocean that have a higher content of hydrothermal metal-bearing matter. The same is true for the average contents of Cr and Ti in the ancient metalliferous sediments of from Atlantic and Indian Oceans. Because of the lower amount of basic material in the lithogenic matter of the metalliferous sediments from the back-arc basins located in the vicinity of continents and islands these sediments have lower average contents of Ti, Cr, and Sc.

Comparison of the chemical composition of the ancient and recent metalliferous sediments shows that the differences observed in them are explainable in terms of the factors or processes recognized that control the distribution and accumulation of the elements in the recent metalliferous sediments. This means that the recent and ancient metalliferous sediments have similar origin.

---

[28] Lisitzin at al. (1976) gave a full description of the coring difficulties in the axial part of the East Pacific Rise.

3.4. Evolution of chemical composition of ancient metalliferous sediments

Chemical composition of metalliferous sediments is related to and influenced by many factors, such as: the spreading rate, intensity of hydrothermal activity, distance from a spreading axis, the bottom relief, deep and bottom currents, redeposition of sediment matter, accumulation rate of diluting abiogenic matter and its composition, composition of ocean water, physical and chemical conditions in the environment of the deep waters, bottom waters, and interstitial waters, and others. The nature and influence of some of the factors is known or interpreted but nothing or almost nothing is known about other factors. Some of the factors are of global significance while others are of regional or local character and significance.

Investigation of the evolution of the chemical composition of the metalliferous sediments is an extremely painstaking task. Many hundreds or even thousands of surface and near-surface samples have been used in the study of recent metalliferous sediments, and an adequate study of the ancient metalliferous sediments requires at least the same number of samples to be obtained by deep-sea drilling for the investigation of each isochronal layer or stratigraphic unit. There is no opportunity for study of ancient metalliferous sediments submerged in subduction zones as nature has made this impossible. The data available from the analysis of a significant number of samples of the ancient metalliferous sediments, some thousands, are obviously insufficient for dealing with the questions that arise in the course of this research work. The data available have enabled calculations to be made of the average contents of chemical elements in metalliferous sediments that were deposited in time intervals of 5 Ma to 20 Ma (Tables 3.2–3.4). The average contents listed in the tables should be considered as preliminary data.

The average contents of elements in the ancient metalliferous sediments from the Pacific Ocean (Table 3.2) with 5 Ma intervals cover only the range of time from 45 to 0 Ma BP. The average data on metalliferous sediments that accumulated earlier have been calculated for the time intervals with the duration from 10 Ma to 20 Ma. These are mainly data for the sediments from the northern part of the Pacific Ocean. There are only a few data for the Atlantic and Indian Oceans because of the small number of occurrences of ancient metalliferous sediments. The average contents of elements in the ancient metalliferous sediments of the Atlantic Ocean (Table 3.3) have been calculated for the time intervals from 5 Ma to 20 Ma. There are no data for sediments in the periods of time from 20 to 15, 75 to 50, 100 to 85, 140 to 120 Ma BP. The average chemical composition has been calculated for the ancient metalliferous sediments from the Indian Ocean in the time intervals from 5 Ma to 10 Ma (Table 3.4).

Table 3.2. Average chemical composition of abiogenic matter of ancient metalliferous sediments from the Pacific Ocean.

Element	Age of sediments, Ma							
	5-0	10–5	15–10	20–15	25–20	30–25	35–30	40–35
Fe %	21.9	20.0	20.6	20.6	21.0	21.7	18.4	18.9
Mn	5.4	4.6	4.5	4.5	4.8	5.8	5.0	5.1
Al	3.9	3.0	3.7	3.3	3.5	3.6	4.0	3.8
Ti	0.27	0.23	0.29	0.21	0.24	0.25	0.12s	0.34n
Mg	2.9	2.7	3.0	2.65	2.4	2.1s	2.1	4.6s
K	1.1	0.76	0.98	1.6	1.4	0.88s		1.7s
P	0.72	0.58	0.72	0.62	0.58	1.0s	1.4s	1.2s
Ba	0.89s	0.64	0.52	0.36	0.74	0.48	0.30	0.76
Cu ppm	1080	970		870	935	1100	720	1150
Zn	610	850	640	570	640	710	550	600
Pb	175s	110n	110n	40n	125n	86n	65n	71n
Ni	730	820	750	730	570	520	440	650
Co	160				110	105	85	92
Cr	50n	52n	75n	56n	45	43	69	46
V		600n	280n	320n	330	340n	430n	420n
Mo	140	84n	60n					
Li	40n	73n	72n	40n	96n	84n	42n	110n
Rb	49n		130n		62n	45n	49n	35n
Hg		1.2n	0.8n		0.4s	0.4s	0.1s	0.3s
Sb		4.3s	4.0s		2.1s	5.7	4.8	8.6
La	120	120	160s	145s	120s	160	91	160
Ce	40	36		96s	66s	63	63	55
Y	270	320	280		240s		170	200
Sc	20s	17s	17s	12s	12s	19s	19	23
Th		4.1	3.4s		5.4	5.6	3.8	
Zr	125s	150	120s		210	280	250	210
Hf		1.9s	1.6s			2.7	2.6	2.0
Ta							0.65s	
B						530		120s

n – data for sediments north of the equator; s – data for sediments south of the equator; no mark – data for the whole ocean.

The evolution through time of proportions of contents of Fe, Al, and Mn in metalliferous sediments from the Pacific, Atlantic, and Indian Oceans as well as from the back-arc basins of the Pacific Ocean is shown in Fig. 3.5.

The high content of the metal-bearing matter in the metalliferous sediments from the Pacific Ocean and the minimum of variation in their composition since Eocene time are conspicuous. The composition of the Cretaceous metalliferous sediments in the Pacific Ocean, unlike that of sediments of

Table 3.2. (continued)

Element		Age of sediments, Ma						
		45–40	55–45n	75–65	85–75n	100–85n	120–100n	140–120n
Fe	%	21.1	22.3	11.0	21.1	11.8	13.3	24.5
Mn		5.1	6.4	0.39	4.5	0.62	2.2	1.2
Al		3.1	3.0	7.0	3.9	6.2	4.4	3.4
Ti			0.16	0.54	0.48		0.40	0.23
Mg				3.2	1.4		2.4	1.0
K				2.3	2.1		2.8	
P				0.25	0.46		0.51	0.45
Ba		0.40	0.22	1.08	0.22	0.53	0.46	0.35
Cu	ppm	1040	1040	230	830	110	300	1280
Zn		560	550	260	610	450	370	490
Pb		86	83		31	53	80	66
Ni		650	715	90	335	200	265	305
Co		72	100	39	81	33	58	83
Cr		27	25	63	97	75	98	190
V		420 n	650	240			460	
Mo							105	
Li		70	48	63 n	33	110	90	
Rb		42 s	60	59		38	120	100
Hg		0.5 s	0.1					
Sb		5.7	10	1.8				
La		150	160					150
Ce		60	63					72
Y		260 s	210				145	
Sc		13	11	30 s				11
Th		2.5	2.6	7.6 s				4.3
Zr		260	250	100			170	
Hf		2.2	2.1					2.4
Ta			1.2	0.4				0.9
B								

other ages, is highly variable and the contents of the metal-bearing matter in these sediments are less than in the Cenozoic metalliferous sediments.

In general the content of Mn in the metal-bearing matter of the Cretaceous metalliferous sediments in the Pacific Ocean is considerably lower compared to that in the younger sediments. The lower content of the metal-bearing matter in the Cretaceous metalliferous sediments probably was caused by different conditions for the dispersion of the hydrothermal metal-bearing matter during the Cretaceous and especially in Middle Cretaceous time. According to the calculations made by Vogt (1989), because of

Table 3.3. Average chemical composition of abiogenic matter of ancient metalliferous sediments from the Atlantic Ocean.

Element	Age of sediments, Ma							
	10–5s	15–10	30–20s	40–30s	50–40s	85–75n	120–100n	150–140n
Fe %	11.8	10.9	15.8	12.4	15.6	10.6	14.4	11.3
Mn	1.4	1.3	6.6	3.3	2.4	0.9	1.9	0.6
Al	7.7	8.0	5.0	6.0	6.0	6.7	6.7	7.4
Ti	0.55	0.47	0.39	0.64	0.32	0.50	0.36	0.65
Mg	3.0	2.6	2.4	3.2	3.4	3.4	2.6	
K	0.4	1.6	1.8	0.3	2.6	3.4	3.8	
P		0.2	0.8	0.4	0.4	0.5	0.1	
Ba ppm	480	620	1960	3840	970	330		
Cu	275	205	360	420	420	50	240	150
Zn	130	210	700	440	540	100	360	130
Pb	120		210	170	170	80	130	60
Ni	200	250	560	460	360		320	120
Co	110	90	100	90	65	72	94	55
Cr	36	82	115	80	84	90	120	100
V	300	280		250		210	520	
Li			145		105		41	45
Rb			90		58	110		
Hg			0.2	0.1				
Sb			5.8	3.6	3.8			
La			180	120	160			
Ce			96	120	150			
Sc			14	15	15			
Th			8.0	10	11			
Hf			2.8	2.9	3.4			
Ta					1.6	1.3		
B	240	280s		57				

n – data for sediments north of the equator; s – data for sediments south of the equator; no mark – data for the whole ocean.

the lower vertical gradients of temperature and salinity in the ocean water at that time, the buoyant hydrothermal plumes could rise much higher above the ocean bottom than at present and even reach the ocean surface. This could result in much wider dispersion of the hydrothermal metal-bearing matter and, accordingly, to a decrease in the average contents of chemical elements in the metalliferous sediments. The depletion of metal-bearing matter of the Cretaceous metalliferous sediments in Mn obviously was caused by weak vertical circulation and weak aeration of ocean water during Cretaceous time (Lisitzin 1980). At this time the average concentration of dissolved oxygen in the bottom water of the Pacific Ocean decreased

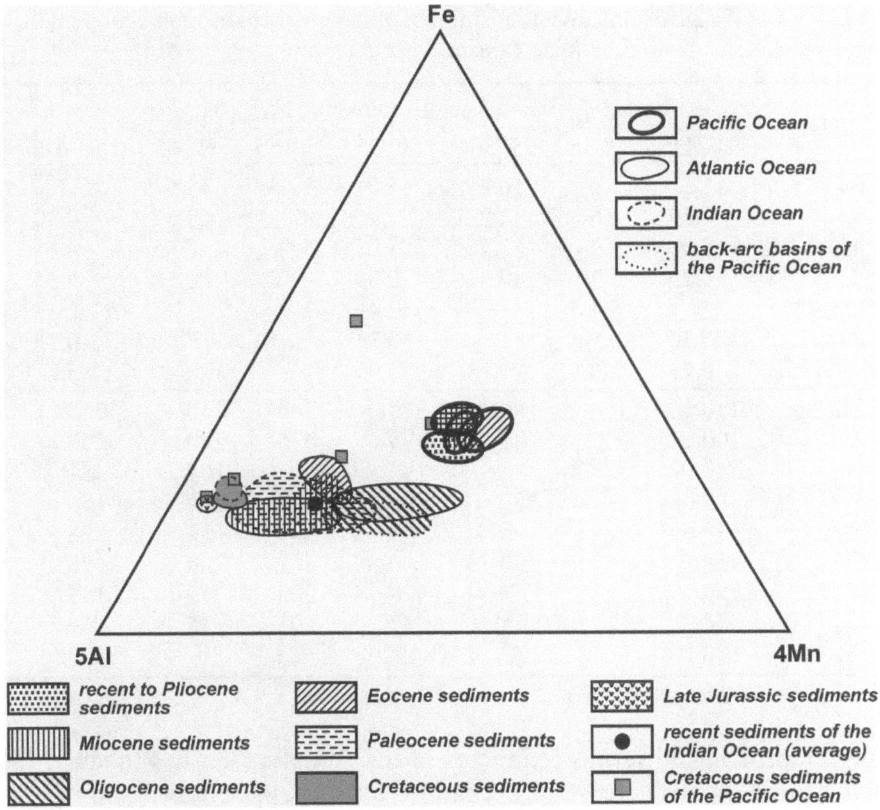

Fig. 3.5. The proportions of contents of Fe, Mn, and Al in recent and ancient metalliferous sediments.

to <0.15 ml l^{-1} (Vogt 1989), and was more than 20 times lower than its concentration at the present time.

The proportions of contents of Fe, Mn, and Al in the metalliferous sediments from the Atlantic Ocean have varied essentially through time. The Cretaceous and especially the Upper Jurassic metalliferous sediments have a minimum content of the metal-bearing matter, and the amount of Mn in this matter is minimal. The minimum content of the metal-bearing matter obviously was attributable to the high amount of diluting abiogenic matter that accumulated during that time at an increasing rate in the central part of the young narrow ocean and the low amount of Mn obviously was related to the stagnant conditions in the bottom water (Lisitzin et al. 1980a). There is no information on the Paleocene metalliferous sediments, and later during Eocene and Oligocene time the content of the metal-bearing matter and the amount of Mn in this matter in the metalliferous sediments increased gradually. The amount of diluting abiogenic material increased again in

Table 3.4. Average chemical composition of abiogenic matter of ancient metalliferous sediments from the Indian Ocean.

Element		Age of sediments, Ma						
		20–15	40–35	55–50*	60–55	65–60	75–65	110–100*
Fe	%	11.8	13.6	10.4	13.0	12.3	11.5	16.7
Mn		0.59	4.6	1.2	1.4	4.0	0.62	0.30
Al		5.5	3.9	7.15	5.95	5.6	6.65	4.8
Ti		0.25	0.30	0.33	0.34	0.50	0.56	0.35
Mg		1.9	4.4		2.4	2.0		2.8
P		0.30	0.52		0.34			0.33
K		0.7	2.1		1.9	2.5		5.2
Cu	ppm	110		140	310	360	230	380
Zn		100		480	375	430	270	250
Pb		80			190		105	37
Ni		99		280	405	460	130	165
Co		30		65	63	84	34	28
Cr		38		90	100	86	200	70
V		100			420	200	250	185
Li		56		34	32	59	25	50
Ba		190						400

* – sediments near basalt sills.

the Miocene metalliferous sediments of the Atlantic Ocean. Probably this was caused by an increase of the rate of terrigenous sedimentation (Lisitzin et al. 1980a) but on the other hand it could have been caused by a decrease in the crustal accretion rate since in Early-Middle Miocene time this was at a minimum for all Cenozoic time in the Atlantic Ocean (Zonenshain and Khain 1989). The amount of Mn was lower in the metal-bearing matter of the Miocene metalliferous sediments in comparison with that of the Oligocene sediments and was similar to that of the Eocene metalliferous sediments.

In the Oligocene and Miocene metalliferous sediments from the back-arc basins of the Pacific Ocean that were studied, the contents of major elements are similar to those in the contemporaneous metalliferous sediments from the Atlantic Ocean. But the accumulation rates of these elements in the sediments from the back-arc basins are several times higher.

The contents of major elements in the Cretaceous or Late Cretaceous metalliferous sediments from the Indian Ocean are very similar to those in the Cretaceous metalliferous sediments from the Atlantic Ocean. In the Cretaceous metalliferous sediments from the Indian Ocean the average content of the metal-bearing matter and the amount of Mn in this matter are minimal. In the Paleocene and recent metalliferous sediments the content of the metal-bearing matter and the amount of Mn in this matter is greater. The extensive occurrences of Late Cretaceous, Paleocene, and recent metalliferous sediments obviously developed during periods of time when the highest crustal accretion rates prevailed in the Indian Ocean (Zonenshain and Khain 1989).

The variations in the contents of major elements in the metalliferous sediments are reflected in the contents of the trace elements, which are subdivided into three groups: 1) the elements bound mainly in the metal-bearing matter, 2) elements bound mainly in the lithogenic matter, and 3) elements bound approximately in equal amounts in both the metal-bearing and lithogenic matter. In the study of the metalliferous sediments the trace elements that are bound mainly in the metal-bearing matter are of greatest interest.

Because of the paucity of data it is not possible to consider the evolution through time in the chemical composition of the metalliferous sediments in all of the oceans. This can be done, to some extent only, for a few of the elements in the Pacific Ocean during the last ~50 Ma period (Table 3.2). The graphs showing the variations in the average contents of chemical elements in the metalliferous sediments from the Pacific Ocean during this period of time are shown in Fig. 3.6.

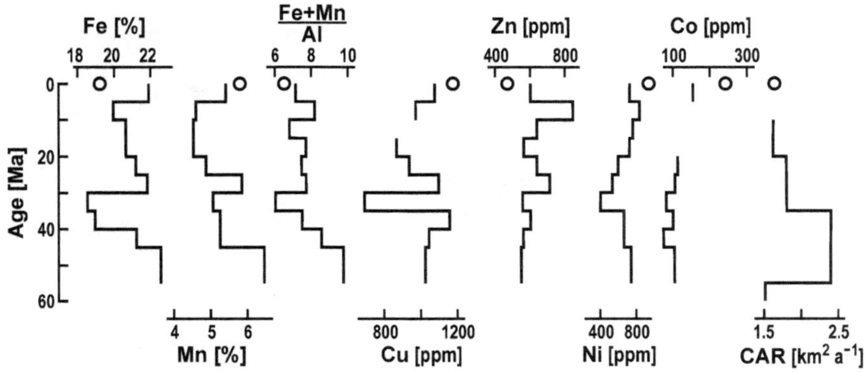

Fig. 3.6. Variation through time in the average chemical composition of abiogenic matter of metalliferous sediments from the Pacific Ocean. CAR – crustal accretion rate in the Pacific Ocean from Zonenshain and Khain (1989).

It is evident from these graphs that the average composition of the metalliferous sediments in the Pacific Ocean from the Early Eocene to the present has not changed significantly. The Pliocene-Pleistocene-, Early Oligocene-, and Early-Middle Eocene sediments have somewhat higher contents of Fe, Mn, and Cu. Cronan (1976) commented on the maximum contents in the Early Oligocene and Early-Middle Eocene in the East Pacific metalliferous

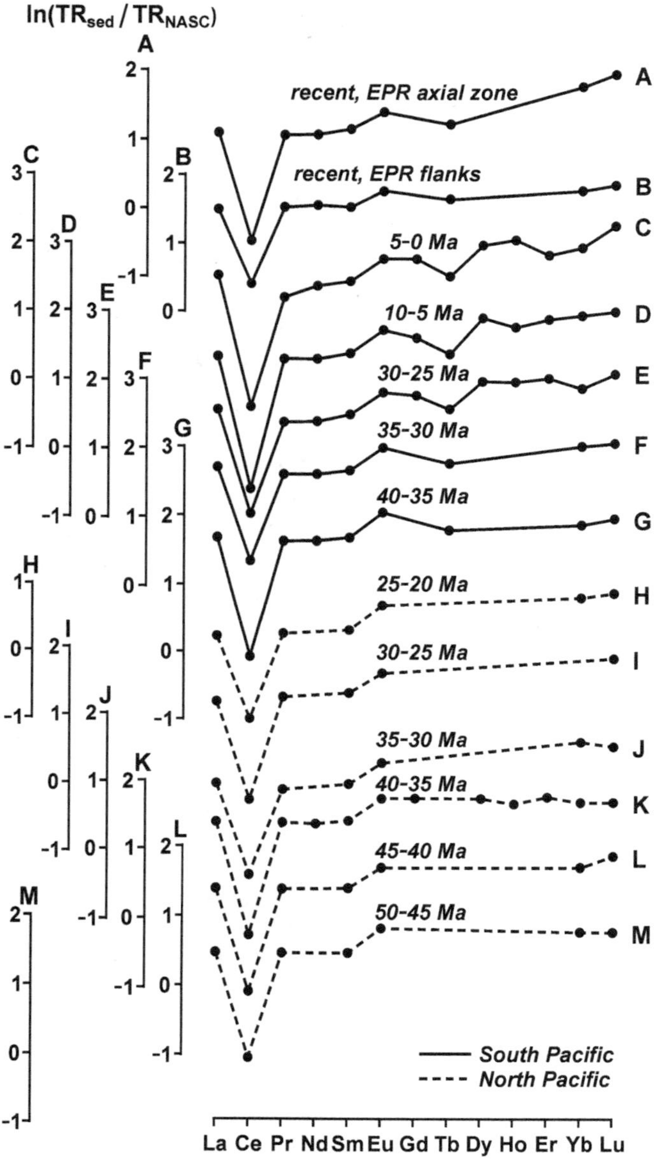

Fig. 3.7. Variations through time in the shale normalized contents and patterns of the REE in abiogenic matter of metalliferous sediments from the Pacific Ocean.

sediments and he assumed that they corresponded to stages of maximum contribution of endogenic matter. The (Fe+Mn)/Al ratios show only the maximum in the Early-Middle Eocene sediments which coincides with the period when there was a maximum rate of crustal accretion in the Pacific

Ocean in Cenozoic time. The maxima as mentioned cannot be considered to correspond specifically to stages of maximum hydrothermal contribution to the ocean since the contents of the metal-bearing matter in metalliferous sediments depend not only on the contribution from hydrothermal sources but also on the rate of accumulation of diluting abiogenic matter. Therefore satisfactory conclusions could be based only on the analysis of the accumulation rates of the hydrothermal metal-bearing matter.

The evolution through time in contents and patterns of the rare earth elements in metalliferous sediments from the Pacific Ocean can be determined from the data available (Fig. 3.7). It was shown in numerous publications that the patterns of the REE in the recent metalliferous sediments from the Pacific Ocean are similar to the patterns of the REE dissolved in ocean water. This evidence indicates that ocean water is the main source of these elements in the recent metalliferous sediments (see Sect. 1.1.7). This similarity is more evident in the metalliferous sediments sampled in the vicinity of the axes of the spreading ridges but not within hydrothermal fields. For example, the content of the REE in the recent metalliferous sediments from the East Pacific Rise at 39°S (Piper and Graef 1974) is shown on Fig. 3.7. In the metalliferous sediments in the axial zone the negative cerium anomaly and enrichment in heavy REE is more pronounced. These features, and especially the negative cerium anomaly, are characteristic of the pattern of the REE dissolved in ocean water.

The negative cerium anomaly was detected in the initial study of the REE in the Oligocene and Eocene metalliferous sediments from the north Pacific (Dymond et al. 1973). The similarities in the pattern of the REE in the Eocene to Pleistocene and in recent metalliferous sediments from the Pacific Ocean were confirmed in the subsequent publications and reports by Dymond et al. (1976, 1977), Jarvis (1985), Ruhlin and Owen (1986), et al. Variations in the pattern of the REE in the metalliferous sediments from the South- and North Pacific from Lower Eocene time until the present can be traced from the original and available published data (Fig. 3.7). The pattern has not changed essentially in the last 50 Ma. Some of the variations during this time do not exceed the variations observed in the contents and patterns of the REE in the recent metalliferous sediments. Therefore, it can be concluded with confidence that the pattern of the REE dissolved in the Pacific Ocean water has not changed in the last 50 Ma.

CHAPTER 4 FORMATION OF COMPOSITION OF METALLIFEROUS SEDIMENTS, FORMATION OF METALLIFEROUS STRATA

Metalliferous sediments are formed in the deep ocean by the submarine discharge of high-temperature hydrothermal fluids and the interaction of these fluids and of the products of the interaction with bottom ocean water. In order to understand the composition of the metalliferous sediments it is important to trace the processes involved in their formation from the influx of the hydrothermal fluids into the ocean to the accumulation of the metalliferous sediments on the ocean floor.

4.1. Sources of metal-bearing matter in metalliferous sediments

Most of known accumulations of hydrothermal and hydrothermal-sedimentary materials, including the metalliferous sediments, are confined to the mid-ocean rift system and back-arc spreading areas of the World Ocean. The mid-ocean rift system is more than 60 thousand kilometers long and extends through all of the oceans. Sources of the metal-bearing matter in the metalliferous sediments are associated with volcanic rocks in the ocean crust. The hydrothermal fluids originate as a product of the circulation of ocean water within the spreading ridges. According to the estimates of Wolery and Sleep (1976), $(0.13\div0.9)\cdot10^{12}$ tons of ocean water are involved annually in the spreading ridges in the hydrothermal circulation. Later estimates by Elderfield and Schultz (1996) of the annual hydrothermal flux in the axial zones of the spreading ridges indicated that it was $4.2\cdot10^{12}$ m^3, and in the estimates of Johnson and Pruis (2003) it was $2.3\cdot10^{12}$ m^3. The annual flow of hydrothermal high-temperature fluids is much less. Holland (1987) estimated that it was $(4\pm2)\cdot10^{10}$ tons. According to the estimates of Mortlock et al. (1993) and of Chan et al. (1992, 1993), the annual on-axis hydrothermal contribution is $(0.6\div2)\cdot10^{10}$ tons and $(1.1\div1.7)\cdot10^{10}$ tons respectively. The annual flow of high-temperature fluids within ±1 km of the spreading axes, according to the estimates of Pelayo et al. (1994) and of

Baker et al. (1996), is $0.4 \cdot 10^{10}$ tons and $1.8 \cdot 10^{10}$ tons respectively. The amount of fluids from magmatic sources is considerably less. The amount of magmatic fluid released annually within the mid-ocean ridges is estimated to be $(1.1 \pm 0.3) \cdot 10^8$ tons (Ito et al. 1983).

High-temperature hydrothermal fluids are the main source of the metal-bearing matter in the metalliferous sediments. The low-temperature diffusive flows of hydrothermal matter are insignificant. In spite of the fact that within the hydrothermal fields the low-temperature diffusive flows contribute 65 to 90% of the heat (Rona and Trivett 1992; Schultz et al. 1992; Elderfield et al. 1993b; Massoth et al. 1994), and about 97% of the water, their contribution of hydrothermal Fe and Mn is estimated to be less than 10%, and less than 40% of total contribution of these metals from hydrothermal sources. An overwhelming amount of the hydrothermal Fe (>90%) and most of the hydrothermal Mn (>60%) are contributed to the ocean from high-temperature vents (Massoth et al. 1994).

The occurrence of hydrothermal activity in the ocean is not uniform and in general it is related to the volcanism, which is reflected by the spreading rate of the crust at the ridges. A direct relationship exists between the high-temperature venting and the incidence of hydrothermal plumes and the spreading rate. This relationship between the probability of venting and the spreading rate as confirmed by Baker and Hammond (1992) is shown in the equation: $y = x/30$, where y – venting probability, x – full spreading rate in cm a^{-1}. A close relationship in the incidence of hydrothermal plumes and the spreading rate is illustrated by a least squares regression fit: $Y = 0.032 + 0.036X$, where Y represents plume incidence, and X represents the full spreading rate in cm a^{-1}; $R^2 = 0.93$ (Baker et al 2004)[29]. The faster the spreading rate, the more frequent is the occurrence of hydrothermal activity and the greater incidence of plumes. Apparently this dependent relationship can be considered both in space and in time. The intensity of hydrothermal activity in individual hydrothermal fields from rifts, which have different spreading rates, may be similar or different.

There are many publications that are devoted to the formation and genesis of high-temperature hydrothermal fluids in the spreading centers in the oceans. Comprehensive reviews of the formation of high-temperature hydrothermal fluids have been published by Von Damm (1995) and German and Von Damm (2004). Only a general description of this process is given here, as it is not a part of the main problems that are considered in this book.

Axial high-temperature basalt hosted hydrothermal circulation systems occur in the axial parts of spreading ridges in areas that lie above magma-

[29] Data on the hot spots affected Reykjanes and Southeast Indian Ocean Ridges and data on the Gakkel Ridge have not been included in the calculation.

tic chambers and the following processes are active in the systems. In the rift zone areas there is interaction between the hot volcanic rocks and the cold ocean water and with the brines in the Red Sea rift zone. In the marginal areas of the rift valleys and in their marginal scarps ocean water percolates downward through fractures in the ocean crust to depths of 1.5 to 3 km below the surface of the ocean bottom where the roofs of the magmatic chambers are located. In this part of the crust the circulating water that has descended from the ocean is heated to temperatures of ~400°C and becomes reduced because of its interactions with minerals in the extrusive basalt or dolerite rocks of the dike series. With the changes in pressure and temperature (PT-conditions) Mg^{2+}, SO_4^{2-} and OH^- ions are extracted from the water and H^+ ions are formed. A part of the sulfate sulfur of the ocean water is reduced to sulfide sulfur. This alteration of the ocean water and the high temperatures cause the intensive leaching of many chemical elements from the basalt rocks including Fe, Mn, Cu, Zn, Ca, Ba, Li, K, Rb, Si etc. The hydrothermal fluids formed at high temperatures become highly buoyant and ascend in the hydrothermal circulation systems, rising rapidly to the ocean floor, and eject into the deep ocean water.

The basalt rocks lose from 8 to 16% of their weight in this leaching process. The loss of Si accounts for about a half of the total loss in the basalt rocks, the loss of Ca accounts for about one fifth of the loss and the loss of all other elements severally does not exceed 10%. The total loss of trace elements is 1 to 2 orders of magnitude less than that of the major elements. However, this loss is significantly large in comparison to the initial amounts of trace elements in the basalt rocks, and under favorable conditions it can be as high as one third to a half of their original contents (Lisitzin 1993). Not all of the leached material is removed by the ascending fluids in the hydrothermal circulation systems and transferred to the deep ocean. Some part of it is precipitated in the ocean crust by the following processes. The metals are present as components in the chloride complexes in the hydrothermal fluids (Shanks and Bischoff 1977). Experimental studies have confirmed the possibility of joint transport of the chloride complexes of the metals and sulfur in their reduced form in the hydrothermal fluids (Ovchinnikov and Masalovich 1981).

The changes in the PT-conditions during the migration of the hydrothermal fluids from the zone of leaching or mobilization of the chemical elements to the surface of the ocean floor are insufficient to cause the destruction of the chloride complexes of the metals and their precipitation in sulfide minerals. The hydrothermal fluids have temperatures close to those at which the mobilization of the chemical elements took place when they reach the orifices of the hydrothermal vents in the ocean floor, if they do not mix with cold ocean water during their migration to the surface of the ocean floor. A decrease in the acidity (an increase in pH) of the fluids is necessary for the precipitation of the metals. Such conditions can exist in the ocean crust by the interaction of the hot fluids with the surrounding rocks in separate fracture systems and cavities that are separate from but connected to the main upward flow of fluids. Under these conditions disseminated and vein mineral occurrences form in the ocean crust.

Most of the metal content is precipitated in the zone where the hydrothermal fluids meet and mix with the cold bottom ocean water that is enriched in oxygen. Under these circumstances the two conditions necessary for the destruction of the chloride complexes of the metals and the formation of their solid phases are satisfied by the cooling and decrease in acidity of the solutions. If the mixing of the hydrothermal fluids and ocean water takes place at the orifices of the hydrothermal vents at the surface of the ocean floor, deposition of massive hydrothermal sulfides takes place. The mineral and chemical composition of the metalliferous sediments that are deposited at a distance from the hydrothermal vents is controlled by the environmental conditions in bottom waters.

If the mixing of the hydrothermal fluids and the ocean water takes place in subsurface conditions a major part of their metal and sulfide sulfur content is precipitated in the cavities and fractures in the crust as the sulfide minerals. The solutions, following their depletion in the Fe, Cu, Zn, and other metals that were precipitated in the subsurface fracture zones, enter the ocean and form Fe-Mn and Mn hydrothermal crusts at and in the vicinity of the orifices of the hydrothermal vents (Gurvich et al. 1984). In cases where the hydrothermal solutions migrate through sulfide minerals, that had accumulated in subsurface rocks, cavities, and fracture zones, they remobilize metals that were deposited in them and become enriched in elements such as Zn, Ag, Au, etc.; when they enter the ocean they deposit the sulfide minerals that are enriched in these elements in the vicinity of the orifices of the hydrothermal vents (Bogdanov 1996).

The ocean water penetrates much deeper, probably to the lower part of the ocean crust or to the upper mantle zone, in the high-temperature hydrothermal circulation systems that are hosted in ultramafic rocks in the slow spreading rift zones, than in the axial circulation systems hosted in basalt rocks, and may be involved in the serpentinization of the ultramafic rocks. The serpentinite rocks formed are less dense (~2.6 g cm^{-3}) than the surrounding basalt (~3.0 g cm^{-3}) or peridotite (~3.3 g cm^{-3}) rocks. Uplift of a serpentinized block of rock takes place if a very large mass of rock is altered and serpentinized (Zonenshain et al. 1989). Uplifted blocks occur within the abnormal marginal scarps of rift valleys such as the eastern marginal scarps in the areas of TAG and Logachev hydrothermal fields or along an axial fracture in a rift zone, like that in the area of the Rainbow hydrothermal field (Rainbow ridge).

Hydrothermal fluids in the deep circulation systems hosted in ultramafic rocks differ from the axial circulation systems hosted in basalt rock. Significant changes in the hydrostatic pressure during the migration of the fluids through the Earth's crust to the surface of the ocean floor cause repeated phase separation of the fluids. As a result, when hydrothermal fluids that are derived from the circulation systems hosted in ultramafic rocks enter the ocean their salinity and the concentrations of many chemical elements vary greatly. Leaching of chemical elements from the ultramafic rocks results in the enrichment of the fluids in some of the elements that are present in these rocks in high concentrations such as Co, Ni, etc.

4.2. Chemical composition of hydrothermal fluids

High-temperature hydrothermal fluids have been investigated from many of the hydrothermal fields in rift zones in the mid-ocean ridges as well as in the spreading areas of the back-arc basins in the southwestern and western parts of the Pacific Ocean. Most of them have temperatures higher than 200°C and the maximum temperature measured was 405°C (German and Von Damm 2004). Unlike the ocean water, all primary high-temperature hydrothermal fluids, in which there was no mixing of cold ocean water, have a low pH (2.0÷5.9) and are reducing. Mg^{2+} and SO_4^{2-} ions, and dissolved O_2, i.e. components characteristic of ocean water, as well as U and Mo are practically absent in the primary hydrothermal fluids. Frequently they have high concentrations of H_2S, which is absent in the ocean water. Sampling of the primary high-temperature hydrothermal fluid is a very delicate and difficult procedure and usually there is some admixture of ocean water present in the samples. In the calculation of the chemical composition of the primary hydrothermal fluids based on the composition of the sampled hydrothermal solutions it is assumed that Mg was absent in the primary hydrothermal fluids. The chemical compositions of primary hydrothermal fluids from many hydrothermal fields including recent data for the range in content of the chemical components in all of the known primary hydrothermal fluids have been summarized by Von Damm (1995) and German and Von Damm (2004).

The average chemical compositions of primary high-temperature hydrothermal fluids from basalt hosted hydrothermal systems of the East Pacific Rise, Mid-Atlantic Ridge, Indian Ocean, and back-arc basins of the Southwest and West Pacific as well as from ultramafic hosted hydrothermal systems of the Mid-Atlantic Ridge, normalized to the average chemical composition of the ocean water, are shown in Fig. 4.1.

The lines plotted showing the enrichment of the hydrothermal fluids relative to the ocean water are similar for the general features. But the hydrothermal fluids from the basalt hosted hydrothermal systems of the back-arc basins and from the hydrothermal systems hosted in the ultramafic rocks in the Mid-Atlantic Ridge have some significant differences.

In the hydrothermal fluids from the basalt hosted systems of the mid-ocean ridges the enrichment in Mn and Fe relative to that in the ocean water is much greater than the enrichment in other elements, and the enrichment ratios are $2{\cdot}10^6{\div}6{\cdot}10^6$. For Pb, Zn, Cu, and Co the enrichment ratios are considerably less, $8{\cdot}10^3{\div}20{\cdot}10^3$, and the ratios for Ga, In, Ag, and Be are $1{\cdot}10^3 \div 4{\cdot}10^3$.The enrichment ratios for Tl, Ba, Al, Ni, Cd, and Si are 100÷800. The average concentrations of the rare earth elements in the hydrothermal fluids exceed those in the ocean water by a wide range, from ~10 to ~3000. Ce and Eu have the highest enrichment ratios, 1000÷3000. The enrichment in Ce reflects its strong negative anomaly in the ocean water ($Ce/Ce^*_{NASC} \approx 0.1$), and the enrichment in Eu results from its strong positive anomaly in the high-temperature hydrothermal fluids ($Eu/Eu^*_{NASC} \approx 15$). The enrichment ratios for Li and Rb are 10÷30, and for As, Sb, and W they are 2÷8. Average concentrations of Ca and K in the hydrothermal fluids exceed those in the ocean water by 1.5÷3 times. The average concentrations of Sr, Br, Cl and Na in the hydrothermal fluids and in ocean water are similar and their ratios vary from 0.9 to 1.2. Mo, U, SO_4, and Mg are absent (or practically absent) in the hydrothermal fluids and their enrichment ratios are zero.

The concentrations of many chemical elements in the high-temperature hydrothermal fluids in the basalt hosted hydrothermal systems of the back-arc basins and in the ultramafic hosted hydrothermal systems of the Mid-Atlantic Ridge are similar to those in the high-temperature hydrothermal fluids in the basalt hosted hydrothermal systems of the East Pacific Rise, Mid-Atlantic Ridge, and Indian Ocean; they are within the ranges mentioned. However, the hydrothermal fluids in the back-arc basins have 10 to 20 times higher concentrations of Pb, Zn, Cd, and Sb, ~50 times higher concentration of As, and ~200 times higher concentration of W, and they have the highest concentrations of Mn, B, K, and Rb. Besides, the hydrothermal fluids in the back-arc basins have average concentrations of Fe and Cu that are 2 to 4 times lower than those in other hydrothermal fluids.

Much higher concentrations of Co, light rare earth elements, and Eu, are the most prominent features in the chemical composition of the high-temperature hydrothermal fluids from the ultramafic hosted systems, relative to those from the basalt hosted systems. The average concentration of Co is ~40 times higher, of La – 20 to 40 times higher, of Ce – 3 to 6 times higher,

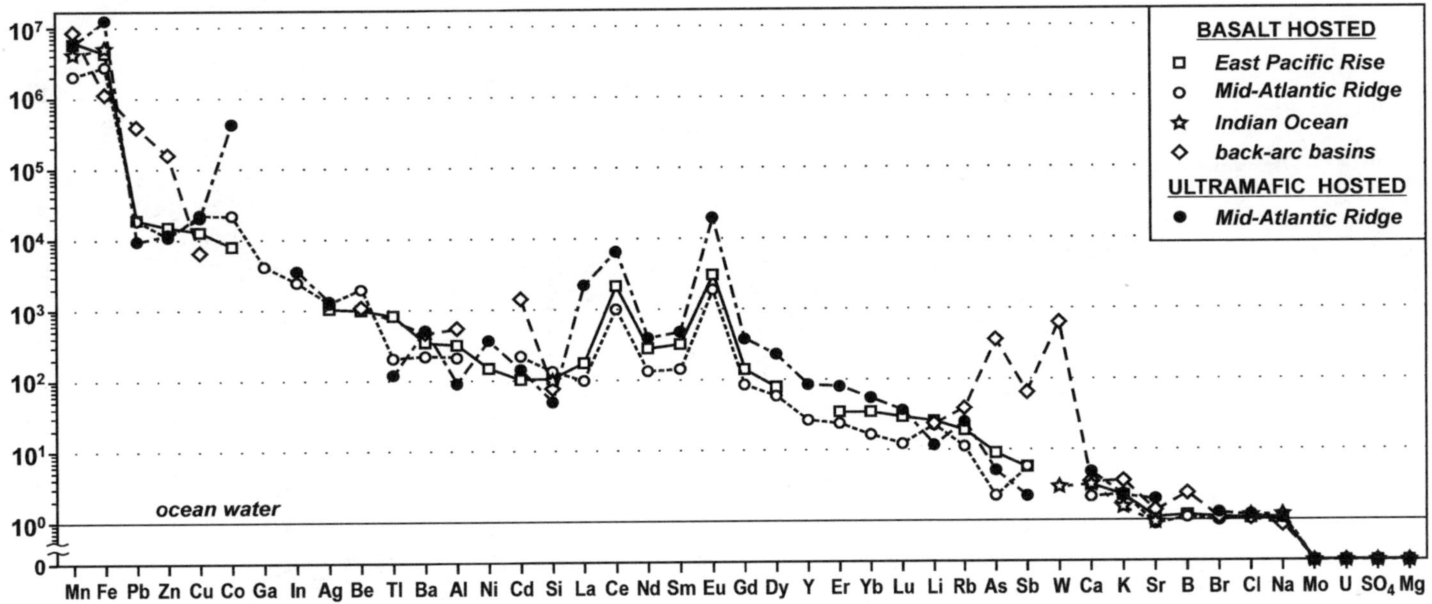

Fig. 4.1. Chemical compositions of primary hydrothermal fluids normalized to the composition of the ocean water. Compiled from data of Gordeev and Lisitzin (1979), Von Damm et al. (1985a,b, 2003), Michard and Albarede (1985, 1986), Michard et al. (1983, 1984), Tunnicliffe et al. (1986), Thompson (1983), Edmond and Von Damm (1985), Edmond (1987), Edmond et al. (1990, 1995), Spivack and Spivack et al. (1987), Von Damm and Bischoff (1987), Campbell et al. (1987, 1988b), Bowers et al. (1988), Campbell and Edmond (1989), Palmer and Edmond (1989), Von Damm (1990, 1995), Fouquet et al. (1991), Sudarikov (1992), Elderfield et al. (1993a), Chan et al. (1993), Butterfield and Massoth (1994), Klinkhammer et al. (1994, 1995) Massoth et al. (1994), Mitra et al. (1994), Trefry et al. (1994), German and Angel (1995), James et al. (1995a), Charlou et al. (1996, 2002), Lein et al. (2000), Metz and Trefry (2000), Gamo et al. (2001), Douville et al. (1999a,b, 2002), Kishida et al. (2004), Rehkämper and Nielsen (2004).

and of Eu – 6 to 10 times higher. These hydrothermal fluids also have the highest concentrations of Fe, Ni, other rare earth elements, Ca, and Sr. Besides, they have the lowest average concentrations of Al, Si, Li, Tl, Pb, and Sb.

4.3. Hydrothermal plumes

The velocity of hydrothermal solutions as measured ranges from 30 to 240 cm s^{-1} where they are ejected from the orifices of the high-temperature hydrothermal vents into the ocean (Spiess et al. 1980; Converse et al. 1984; McConachy et al. 1986; Rona et al. 1986, 1990; Little et al. 1987; Backer et al. 1990; Edmond et al. 1990; Delaney et al. 1992; Rona and Trivett 1992; Thompson et al. 1992; Bemis et al. 1993; Elderfield et al. 1993; Backer 1994; Butterfield and Massoth 1994; Massoth et al. 1994). These high-temperature solutions generally have a lower density than the cold ocean water and are buoyant in it. As they rise they mix with the ocean water and form buoyant hydrothermal plumes. As the plumes rise, the dilution of the hydrothermal solutions with the ocean water continues, they become less buoyant, their velocity in rising decreases progressively (Fig. 4.2), and the plume area increases (Fig. 4.3).

The amount of dilution of the hydrothermal solutions by ocean water increases with their height above the ocean bottom and differs from one hydrothermal field to another. Fifty meters above the active hydrothermal mounds in the TAG field the dilution factor is about 500, and 50 meters above the Monolith vent in the North Cleft segment of the Juan de Fuca Ridge it is about 10000 (Fig. 4.4). Differences such as this are caused by different heat flow from the vent sites, and at the same height above the ocean bottom higher heat flow from the vents results in smaller dilution factors (Little et al. 1987).

Eventually, at some maximum height above the ocean floor, the densities of the water mass in the buoyant plume and in the surrounding ocean water become equal and the plume ceases to rise and the mass of water in it begins to spread laterally. The height of plume rise above a typical high-temperature vent is from ~150 m to ~400 m, and the higher the heat flux at the vent the higher will the plume rise (Lupton 1995). A hydrothermal plume that spreads laterally is referred to as a nonbuoyant plume or a neutrally buoyant plume. Usually they are about 200 m in thickness.

Dilution of the hydrothermal solutions with ocean water continues to take place in the nonbuoyant plumes. The dilution factor at the beginning of the lateral outspreading of a plume is reported to be many thousands, and

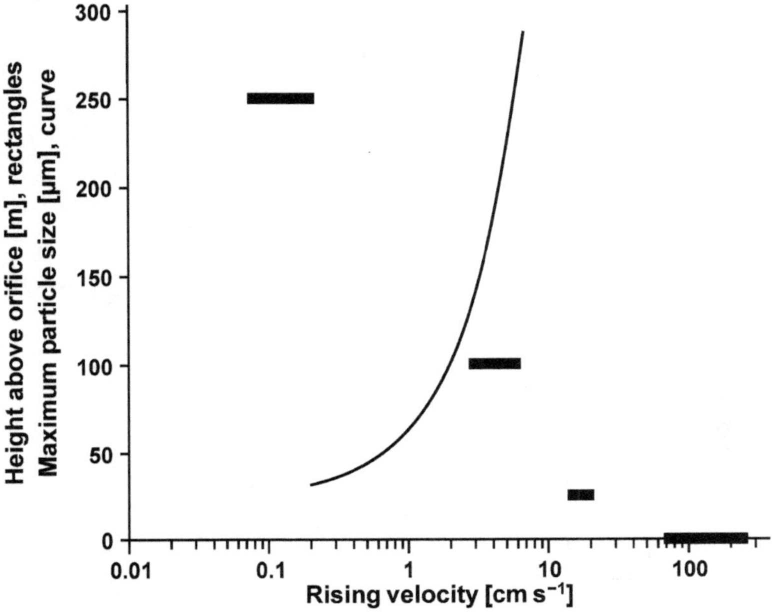

Fig. 4.2. Rising velocity of buoyant plume versus the height above the orifice and the maximum size of the floating pyrite, pyrrhotite, and sphalerite particles (specific gravity >2.8 g cm^{-3}). Based on data from Converse at al. (1984), Little et al. (1987), and Dymond and Roth (1988).

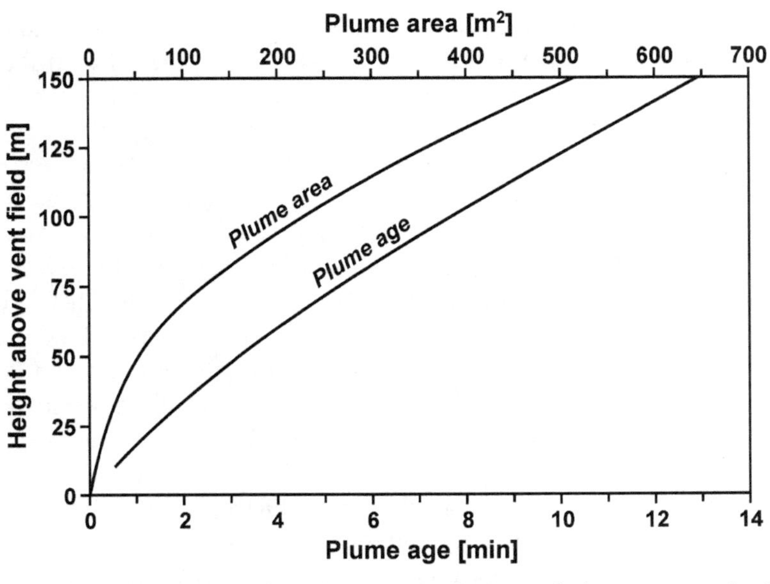

Fig. 4.3. The age and area of the buoyant hydrothermal plume versus the height above the TAG vent field. Composed from Rudnicki and Elderfield (1993).

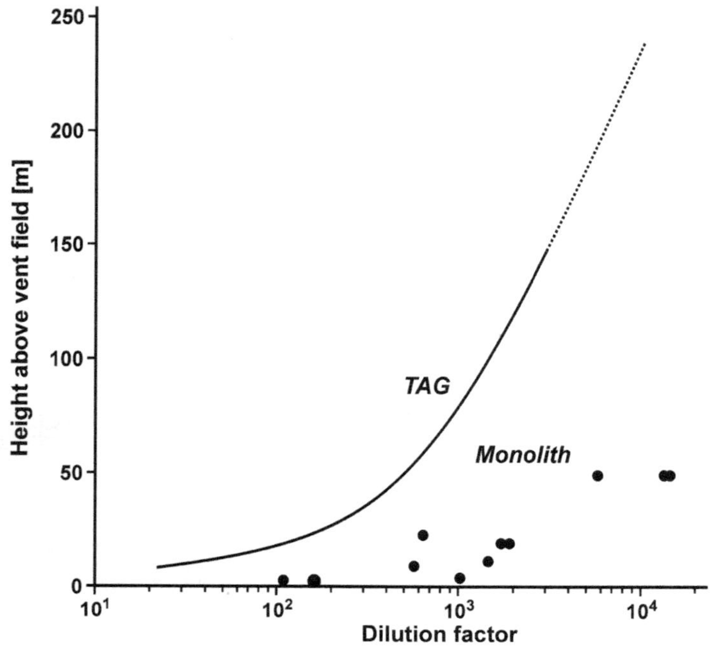

Fig. 4.4. Dilution factors in the buoyant plumes above the TAG hydrothermal field and Monolith vent versus their height. Based on data from Rudnicki and Elderfield (1993) and Feely et al. (1994) with additions.

in the peripheral parts of a plume where in-situ measurements indicate a mixing of hydrothermal fluids with the ocean water, the dilution factor is found to be millions, as it is in the plume above the TAG field.

In the TAG hydrothermal field the rise of the buoyant plume, from the hydrothermal vent to when it ceases to rise, takes about 40 minutes (Rudnicki 1990). At the beginning of the formation of the nonbuoyant plume, with its lower boundary at ~240 m above the vent and the upper boundary at ~420 m above the vent, the dilution factor is about 6000 to 10000 (Rudnicki 1990; Rudnicki and Elderfield 1992, 1993). In the North Cleft segment of the Juan de Fuca Ridge the lower boundary of the nonbuoyant plume is ~100 m and the upper boundary is ~250 m above the ocean bottom (Feely et al. 1994). In the rift zone in the northern part of the Mohns Ridge, Norwegian Sea, the lower boundary of the nonbuoyant plume is ~750 m above the ocean bottom (Bogdanov et al. 1997c). The nonbuoyant plumes found along the Gakkel Ridge in the Arctic Ocean are located extremely high above the inferred vent sites and the thickness of these plumes is up to ~500 m (Edmonds et al. 2003; Baker et al. 2004). Such unusual parameters for the plumes apparently result from the extraordinarily weak vertical density gradient in the deep Arctic Ocean (Baker et al. 2004).

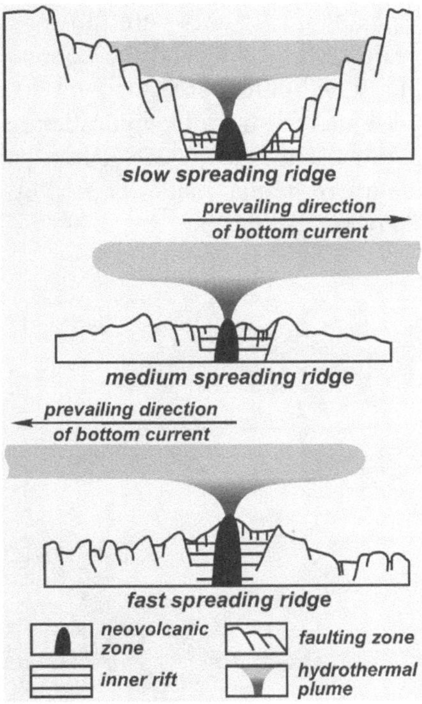

Fig. 4.5. Possible outspread of nonbuoyant plumes in different rift zones.

Nonbuoyant plumes spread out and diverge from the vent fields in the prevailing direction of the bottom currents and are affected greatly by their directions and velocities. The morphology of the ocean bottom in the rift zones has an important influence on the spreading and migration of the nonbuoyant plumes. Slow spreading rift zones where the full spreading rate is <4 cm a^{-1}, usually have axial valleys that are 500 m or more deep and several kilometers to several tens of kilometers wide and extend along the spreading axes (Fig. 4.5). Under these conditions the nonbuoyant plumes rarely extend beyond the rift valleys, and if the direction of flow of the bottom currents is favorable they may spread only along the valleys. As a result, when the spreading rates are slow almost all of the hydrothermal matter carried in the nonbuoyant plumes eventually settles on the floor of the rift valleys. The nonbuoyant plumes of the slow-spreading Mohns Ridge and Gakkel Ridge are the only known exceptions to this in the modern ocean. During the Cretaceous and especially in Middle Cretaceous time because of the lower vertical gradients of temperature and salinity in the ocean water, the buoyant hydrothermal plumes could rise much higher above the ocean bottom than at present (Vogt 1989) and the nonbuoyant plumes could extend beyond the rift valleys of slow-spreading ridges.

Medium and fast spreading rift zones with full spreading rates >4 cm a^{-1} have no deep axial valleys (Fig. 4.5) and the nonbuoyant plumes or their parts may spread latterly and unhindered for long distances, hundreds and several thousands of kilometers, from the hydrothermal fields (Lupton and Craig 1981; Klinkhammer and Hudson 1986; Lupton 1995; Lupton et al. 2004). As a result, the hydrothermal matter carried by the nonbuoyant plumes may be dispersed over vast areas.

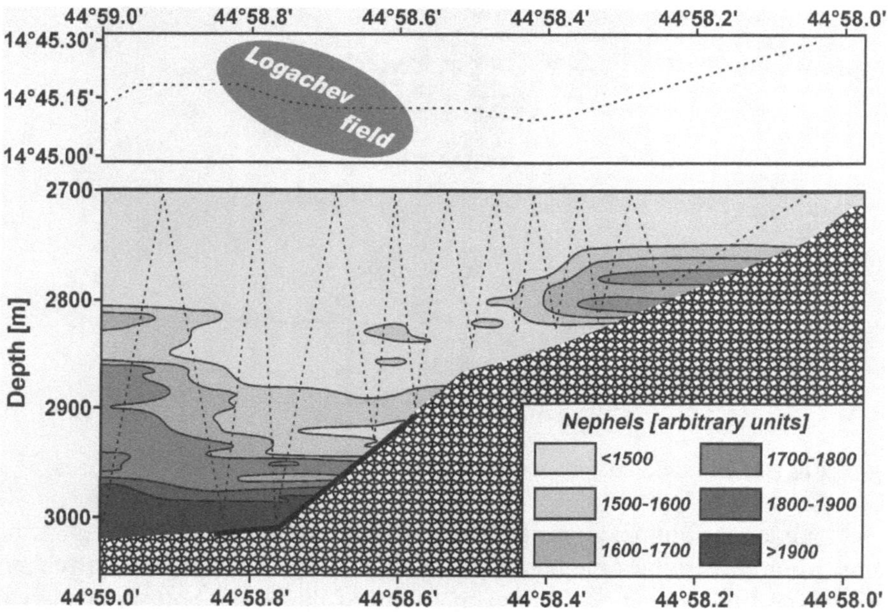

Fig. 4.6. Section in the bottom water layer in the vicinity and above the Logachev hydrothermal field showing particle concentrations measured by nephelometry.

In 1995 hydrothermal plumes were found in the Logachev hydrothermal field, in which the buoyancy of some parts was very low, several meters or less, or absent (Bogdanov et al. 1995c, 1997a,b; Sect. 1.6.3). They have been referred to as ground plumes. The nephelometry measured for a section above the Logachev field is shown in Fig. 4.6. The measurements showed a maximum concentration of particles in the deepest water and evidence of ground plumes derived from the hydrothermal vents at depths of 3005 and 2960 m (Sect. 1.6.3). These plumes result from the injection of high-temperature hydrothermal fluids that are denser than the ocean bottom water. A typical nonbuoyant plume is present in the eastern part of the section. It is derived from typical black smoker vents at a depth of 2940 m and the lower boundary of this plume is ~120 m above this depth (Sect.

1.6.3). The ground plumes only play a significant role in the formation of metalliferous sediments in the vicinity of hydrothermal vents.

Many of the hydrophysical and geochemical parameters of hydrothermal plumes are different than those of ambient ocean water. The plume acts as a barrier and hydrothermal material can only penetrate above the level of this barrier after this specific water mass is dispersed. The migration of hydrothermal material from the orifice of a hydrothermal vent, its presence in the plume, and the very existence of a plume, and its subsequent dispersion or settling to the ocean floor depend to a great extent on hydrophysical processes. However, chemical, physico-chemical and biological processes also play an extremely important role in the transformation of the material from hydrothermal sources.

Only a part of chemical element content carried in the hydrothermal metal-bearing solutions is deposited directly in the massive hydrothermal bodies at the orifices of the hydrothermal vents when these solutions enter the ocean. According to Rona (1984), about 5% of the hydrothermal metal-bearing matter that enters the ocean in the hydrothermal solutions accumulates within the massive bodies at the vents, and about 95% is deposited outside of them, and according to Lisitzin (1993), the relative amounts are 2–5% and 95–98% respectively. About 200 tons of the metal-bearing matter or about 60 tons of Fe accumulate annually within the Active hydrothermal mound of the TAG hydrothermal field, according to the estimates of Bogdanov (1997). The annual hydrothermal contribution of Fe to the ocean from the vents in this mound is estimated to be approximately 2200 tons (Rudnicki 1990). That is about 3% of the annual hydrothermal contribution of Fe accumulates in the mound. According to Feely et al. (1994), more than 99% of the hydrothermal Fe and more than 99.9% of the hydrothermal Mn entering the ocean in the hydrothermal fields of the North Cleft segment of the Juan de Fuca Ridge eventually leave these fields. More than 90% of the particulate material in the hydrothermal plume above the vent field at the Endeavor Ridge is transported more than 2 km away from the field (Dymond and Roth 1988). A predominant amount of the hydrothermal matter that enters the ocean is dispersed by the bottom currents over huge distances and its source cannot be recognized and its identity is lost as it is assimilated into the ocean (Bogdanov et al. 1983). Only in the deeps of the Red Sea rift zone that are filled with dense brines can most of the bulk of the hydrothermal metal-bearing matter be accumulated. On the average about 84% of the hydrothermal Fe contributed annually in the last 3.6 ka has accumulated in the sediments within the brine area of the Atlantis II Deep (Sect. 2.2.4).

Immediately after venting from the ocean crust, and sometimes in the crust below the ocean floor, acid reducible high-temperature hydrothermal solutions begin to mix with the low-alkaline oxidizing cold ocean water. During the mixing process the chloride complexes decompose and the metal ions released combine with the sulfide ions and form Fe-, Cu-, and Zn-sulfide minerals. Most of the Cu and Zn and a considerable part of the Fe passes into a solid phase. When the hydrothermal solutions come into contact with the ocean water, sharp gradients in the temperature and in the Eh and pH conditions develop, and a number of the chemical elements dissolved in the ocean water are precipitated. Barite and anhydrite are formed under these conditions. Apparently the Ba and Ca are derived from the hydrothermal solutions and the SO_4^{2-} is derived from the ocean water.

Table 4.1. Mineral composition of particles from black smoke, of material that settles from the black smoke, and of the metalliferous sediments that accumulated near the hydrothermal vents at the 21°N EPR hydrothermal field. According to Mottl and McConachy (1990) and Haymon and Kastner (1981).

Mineral	Ideal formula	Smoke	Settling material	Metalliferous sediments
Pyrrhotite	$Fe_{1-x}S$	+++++	++++	+
Pyrite	FeS_2	+++++	+++++	+++ to ++++
Sphalerite	ZnS	+++++	+++++	++ to +++++
Chalcopyrite	$CuFeS_2$	+ to ++++	++	+++ to +++++
Fe–S±Si phases		+++	++	+++
Fe–Si±S phases		+++	++	++
Anhydrite	$CaSO_4$	++	++	– to ++
Amorphous silica	SiO_2	++	++	+ to ++
Fe-oxyhydroxides		++	++	++
Goethite	FeOOH	++	+ to ++	– to +++
Sulfur	S	+ to ++	++	– to ++
Organic matter		+ to ++	+	+
Isocubanite	$CuFe_2S_3$	+	++	+++ to +++++
Wurtzite	ZnS	+	++	+ to +++
Barite	$BaSO_4$	+	+	+ to ++
Covellite	CuS	+	+	+
Marcasite	FeS_2	+?	+	+ to +++
Melnikovite	FeS_2	–	+ to +++	+ to +++
Bornite	Cu_5FeS_4	–	+	+ to ++
Galena	PbS	–	– to +	– to +
Mg-Ca-silicates		+	–	+
Aluminosilicates		+	–	+
Natrojarosite	$NaFe_3(SO_4)_2(OH)_6$	–	–	– to +++

– absent; + trace; ++ minor; +++ moderate; ++++ abundant; +++++ very abundant

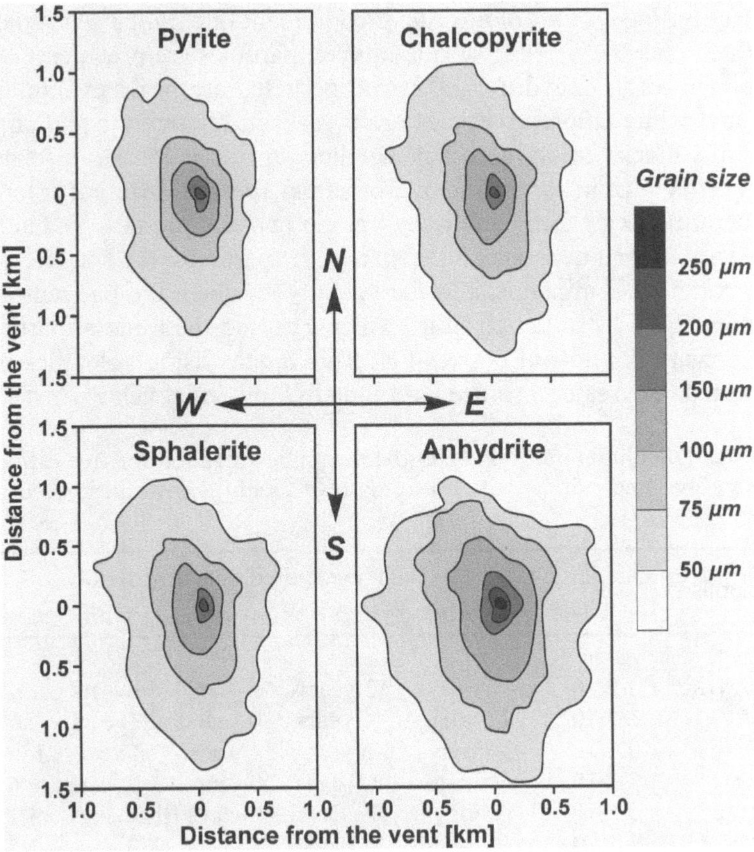

Fig. 4.7. Sedimentation of sulfide and sulfate minerals of given grain size from the nonbuoyant plume above the Monolith vent. After Feely at al. (1994).

Mineral particles form immediately after the venting of the high-temperature hydrothermal solutions into the ocean water. Their presence makes the hydrothermal effluents appear smoky. The size of the mineral particles varies from <10 μm to >500 μm (Feely et al. 1987). Pyrrhotite, Fe-rich sphalerite, pyrite, non-identified Fe-S-Si phases, chalcopyrite, amorphous SiO_2, native S, Fe-oxyhydroxide minerals including goethite, and anhydrite are the main minerals in the smoke. Barite, isocubanite, wurtzite, covellite, marcasite (?), galena, magnetite, talc, graphite as well as silicate and aluminosilicate minerals that were not identified may be present in small amounts and organic matter is commonly present (Jedwab and Boulegue 1984; Lewison et al. 1986; Feely et al. 1987, 1990; Capmbell et al. 1988; Trocine and Trefry; 1988; Walker and Baker 1988; Roth and Dymond 1989; Mottl and McConachy 1990; Sudarikov et al. 1992; Table

4.1). Depending on the color of the predominant mineral particles the smoke effluent may be black if sulfide mineral particles are prevalent, white if amorphous SiO_2, anhydrite and barite particles are prevalent, or gray if black and white colored particles are present in relatively equal amounts. The terms black, white, and gray smokers are used for the discharge of smoke effluent from submarine hydrothermal vents of different color.

Mineral particles form and ascend in the buoyant plumes. The height to which they ascend depends on the size of the particles, the specific gravity of the component minerals, and the velocity at which the buoyant plumes are rising (Fig. 4.2). Particles of smaller size and the mineral particles of lower density in buoyant plumes that have higher rising velocities will be transported the greater distances from the hydrothermal vents.

Table 4.2. Dissolution rates and approximate times required for five different sizes of sulfide- and sulfate particles to dissolve totally in seawater. According to Feely et al. (1987).

Minerals	Dissolution rate [cm s^{-1}]	Time required to totally dissolve				
		2 µm	10 µm	30 µm	50 µm	100 µm
Sulfides						
Chalcopyrite*	$1.2 \cdot 10^{-12}$	964 days	13 years	40 years	66 years	132 years
Pyrite*,†	$5.3 \cdot 10^{-12}$	218 days	3.0 years	9.0 years	15 years	30 years
Pyrite*,‡	$1.1 \cdot 10^{-11}$	106 days	1.5 years	4.4 years	7.3 years	14.6 years
Sphalerite*	$2.4 \cdot 10^{-11}$	48 days	241day	2.0 years	3.3 years	6.6 years
Marcasite*	$5.5 \cdot 10^{-11}$	23 days	113 days	340 days	1.5 years	3.1 years
Pyrrhotite*	$1.1 \cdot 10^{-10}$	11 days	55 days	164 days	273 days	1.5 years
Sulfates						
Barite*	$2.0 \cdot 10^{-11}$	58 days	289 days	2.4 years	4.0 years	7.9 years
Anhydrite*	$2.5 \cdot 10^{-9}$	11.7 hours	2.3 days	7.0 days	12 days	23 days
Anhydrite⁺	$3.2 \cdot 10^{-8}$	0.9 hours	4.3 hours	13 hours	21.7 hours	1.8 days

* Based on in situ dissolution studies on the Southern Juan de Fuca Ridge.
† Supplied by C.M. Taylor Inc.
‡ Supplied by USGS.
⁺ Based on laboratory dissolution studies at 1 atm pressure and 2°C

The sedimentation is characteristically different for particles consisting of different minerals and particles of different size that are transported in buoyant and in nonbuoyant plumes. This has been demonstrated clearly in calculations by Feely et al. (1987, 1994) for pyrite, chalcopyrite, sphalerite, and anhydrite particles in the plume from the Monolith vent (Fig. 4.7). Pyrite particles greater than 50 µm settle out of the plume within a distance of 1300 m from the vent, chalcopyrite and sphalerite particles with lower specific gravity settle within 1500 m, and the anhydrite particles of lower specific gravity than that for chalcopyrite and sphalerite within 2000 m. Coarser particles settle closer to the vent, and finer particles further from it.

The particles of different minerals resist their dissolution in the ocean water at different rates and the coarser particles are usually more resistant (Table 4.2). Of the minerals listed in the table, chalcopyrite and pyrite are the least readily dissolved and anhydrite is the most readily dissolved. Among the minerals shown in this table, despite their settling to the bottom, the more dissolvable minerals are depleted and the less dissolvable minerals are enriched in the metalliferous sediments that accumulated near the hydrothermal vents in comparison with the material of the black smoke (Table 4.1).

4.3.1. Processes in buoyant hydrothermal plumes

Before considering the behavior of chemical elements in buoyant plumes it is reasonable to consider changes in the role of hydrothermal fluids and ocean water in the concentration of chemical elements in mixtures of hydrothermal fluids and ocean water that have different dilution factors, assuming that there is conservative mixing of them. With an increase in the dilution factor[30] the role of the hydrothermal fluids decreases in the mixture. The dilution factors at points where the roles of the hydrothermal fluids and the ocean water become equal, $D_{50\%}$, for a number of chemical elements can be seen in Fig. 4.8, and for iron $D_{50\%}$ is $\sim 4 \cdot 10^6$, for Mn $\sim 5 \cdot 10^6$, for Pb $\sim 2 \cdot 10^4$, for Cu and Zn $\sim 1.5 \cdot 10^4$, for Co $\sim 1 \cdot 10^4$, for Eu $\sim 3 \cdot 10^3$, for Ce $\sim 2 \cdot 10^3$, for Be and Ag $\sim 1 \cdot 10^3$, for Tl ~600, for Ba and Al ~300, for Cd ~150, for Si and Ni ~100, for La ~60, for Lu ~25, for As ~7, for Sb ~6, and for P ~0.3 (the concentration of P in ocean water is higher than in the hydrothermal fluids). Because of the settling and removal of some amount of the chemical elements from the plumes the real dilution factors, at which the roles of hydrothermal fluids and the ocean water become equal, are some lower for the elements with higher concentrations in hydrothermal fluids, and some higher for the elements with higher concentrations in ocean water.

Fig. 4.8 shows the role of the ocean water as it becomes prevalent for P, V, Sb, and As in areas very close to hydrothermal orifices, for REE (except for Ce and Eu), Ni, Si, Cd, Al, and Ba in lower parts of the buoyant plumes, for Tl, Ag, Be, Ce, and Eu in middle parts of the buoyant plumes, for Co, Zn, Cu, and Pb in upper parts of the buoyant plumes or transition zones from buoyant to nonbuoyant plumes, and for Fe and Mn in nonbuoyant plumes.

---

[30] Ratio of amount of the ocean water to amount of the hydrothermal fluid in the mixture.

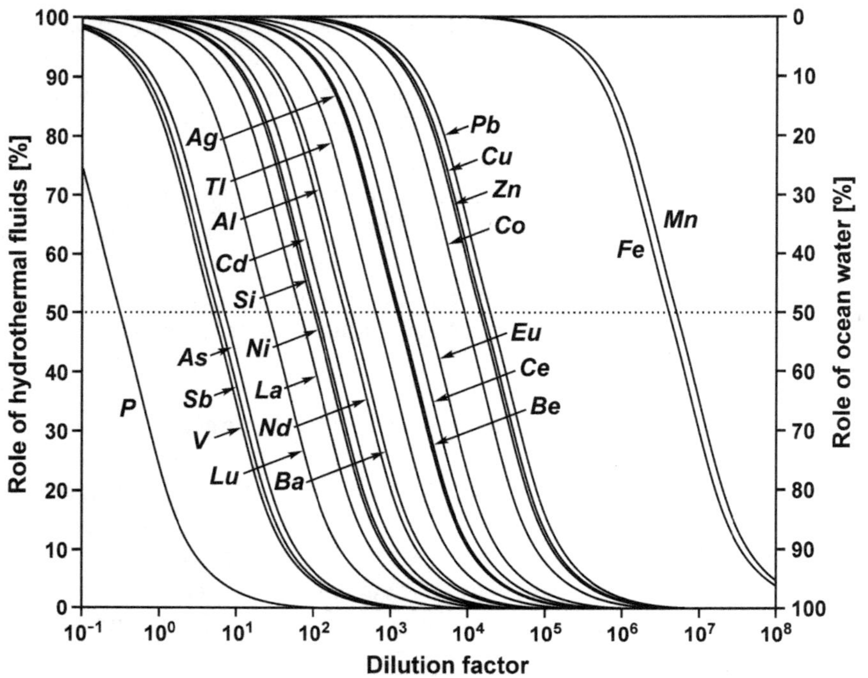

Fig. 4.8. The role of hydrothermal fluids and ocean water in the concentration of chemical elements in mixtures of hydrothermal fluids and ocean water in relation to the dilution factor at their conservative mixing. Calculations based on the average ratios of element concentration in East Pacific Rise and Mid-Atlantic Ridge basalt hosted primary hydrothermal fluids and ocean water (Fig. 4.1).

Fe-minerals are predominant in the particulate matter in the buoyant plumes of the high-temperature hydrothermal vents and this element is considered first. Rapid precipitation of Fe(II) ions and the formation of particulate sulfide minerals take place immediately after the venting of the hydrothermal fluids into the ocean. Rudnicki and Elderfield (1993) have estimated that in the TAG hydrothermal field this process takes about 15 seconds and occurs in the lower 8 meters of the buoyant plume where the dilution factors are up to 20–25. Particles of native sulfur and sulfide minerals of Cu, Zn, and Pb form simultaneously, and H_2S is almost completely depleted. About 20% of the H_2S derived from the fluids is used in the formation of sulfide minerals and native sulfur and about 80% of it is oxidized (Mottl and McConachy, 1990). The oxidation of the remaining dissolved Fe(II) ions and the formation of particles of Fe(III)-oxyhydroxides follow in the next stage. Above the Active mound in the TAG hydrothermal field about 35% of the Fe(II) ions that remain in solution after the sulfide minerals form are precipitated by oxidation within one minute of venting as the plume rises to ~20 m, and the dilution factor reaches 100 to 200 (Rudnicki and Elderfield 1993), while about 38% of the hydrothermal Fe remains in solution. This is similar to data for a buoyant plume from the 21°N EPR hydrothermal field. According to Mottl and McConachy (1990), within 22 m of the plume 35±25% of the Fe remains in solution, sulfide minerals are predominant and small amounts of Fe-oxyhydroxides and goethite are present in the particulate matter. After this, within 10 minutes of venting of the fluids, the plume rises to ~120 m and the dilution factor increases to about 2000, and about 90% of the Fe(II) ions that remain in solution after the formation of the sulfide minerals are precipitated by oxidation. At the TAG hydrothermal field the process of oxidation of Fe(II) ions and precipitation of Fe-hydroxides in the buoyant plume is complete when the plume is about 150 m above the vents and the dilution factor is about 3000.

The concentration of particulate Fe in buoyant plumes varies over a wide range. In the lower 14 m of the buoyant plume above the 21°N EPR hydrothermal field it has been measured from 290 to 31000 μg l^{-1} (Mottl and McConachy 1990), in the lower 50 m of the buoyant plume of the Monolith vent, North Cleft segment of the Juan de Fuca Ridge – from 3.1 to 675 μg l^{-1} (Feely et al. 1994), in the buoyant plume above the TAG hydrothermal field – from ~20 to ~2700 μg l^{-1} (Rudnicki Elderfield 1993; Mitra et al. 1994). As the plumes rise the concentration of the particulate Fe decreases consistently (Fig. 4.9).

The concentrations of the particulate Fe can be significantly different in buoyant plumes in different hydrothermal fields where the concentrations of Fe in the hydrothermal fluids and the dilution factors are similar. As an example, the concentrations of the particulate Fe in the buoyant plumes above the TAG field and Monolith vent are shown in Fig. 4.10. With similar dilution factors the concentration of the particulate Fe in the buoyant plume above the TAG field is several times to an order of magnitude higher than in the buoyant plume above the Monolith vent.

It was mentioned previously that in the TAG hydrothermal field the process of oxidation of Fe(II) ions and precipitation of Fe(III)-hydroxides is completed in the buoyant plume within ~150 m above the vents where the dilution factor is about 3000 (Rudnicki and Elderfield 1993). And according to measurements, even in the nonbuoyant plumes above the Juan de Fuca Ridge, the average concentration of dissolved Fe accounts for 29% of the total concentration of Fe (Massoth et al. 1994). There the process of formation of particulate Fe(III)-hydroxides is much slower than above the TAG hydrothermal field and takes place in a period of time lasting from several hours to several days (Chin et al. 1994).

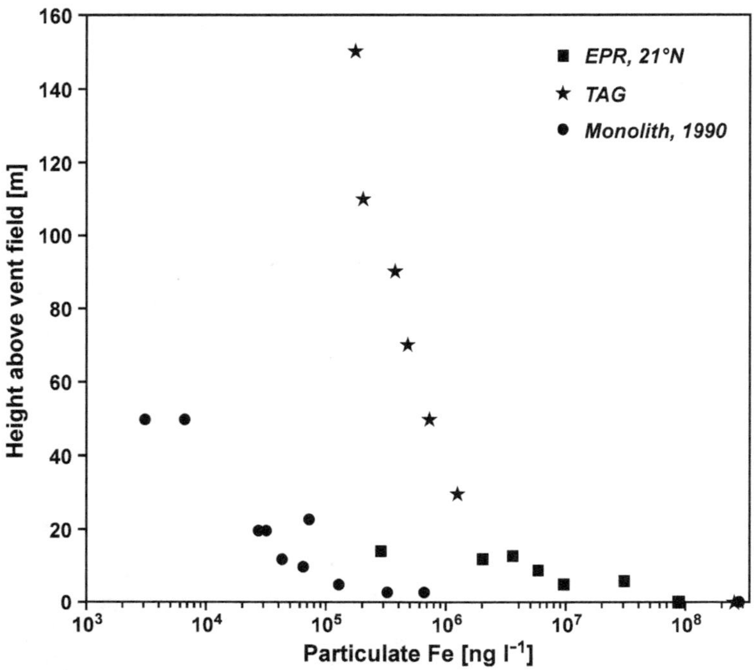

Fig. 4.9. Concentrations of particulate Fe in relation to the height above the vent fields in buoyant hydrothermal plumes. Based on data from Mottl and McConachy (1990), Rudnicki and Elderfield (1993), Feely et al. (1994). At height 0 m concentrations in hydrothermal fluids are shown.

Field and Sherrell (2000) studied the kinetics of the oxidation of Fe(II) in hydrothermal plumes and have shown that it depends mainly on the chemistry of the ambient deep-water, primarily on the pH and content of dissolved O_2, and to a much lesser extent on variations in the composition of the primary vent fluids. They have calculated the rates of oxidation of Fe(II) in ambient ocean water at different hydrothermal sites (Table 4.3). The fastest rates of oxidation of Fe(II) take place at the Rainbow and TAG hydrothermal fields on the Mid-Atlantic Ridge, they are intermediate at the sites on the South East Pacific Rise and Southwest Indian Ridge, and they are the slowest at the sites in the Northeast Pacific.

The behavior of Mn in the buoyant plumes differs from that of Fe. Only an insignificant part of the total concentration of this element is present in the particulate matter in the lower parts of the plumes (Mottl and McConachy 1990), and the Mn/Fe ratios in the particulate matter vary considerably and usually range from 0.0001 to 0.03. These values are much lower than those reported in primary hydrothermal fluids (Trocine and Trefry 1988; Mottl and McConachy 1990; Feely et al. 1994; Gurvich 1998). The Mn/Fe ratio increases somewhat in the buoyant plume of the Monolith vent, where the dilution factor is $(1 \div 2) \cdot 10^3$, but it remains very low; and where the dilution factor is $(12 \div 14) \cdot 10^3$ it is ~0.02 (Fig. 4.11).

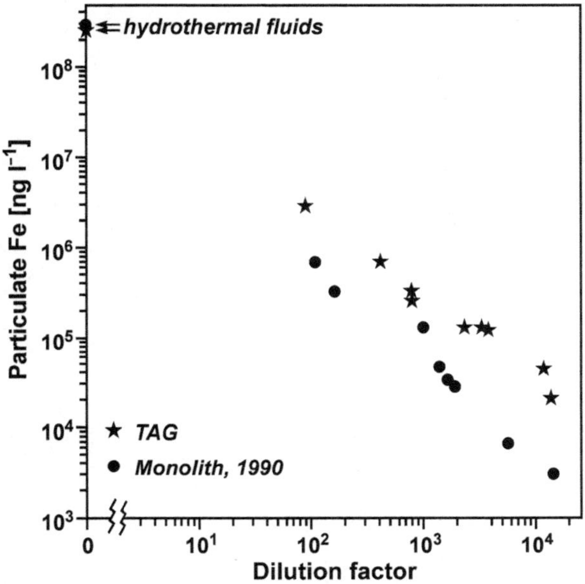

Fig. 4.10. Concentrations of particulate Fe in relation to dilution factors in buoyant hydrothermal plumes above the TAG hydrothermal field and Monolith vent. Based on data from Trocine and Trefry (1988), Feely et al. (1994), Mitra et al (1994).

Table 4.3. Oxidation rates of Fe(II) in ambient ocean water at hydrothermal vent sites. According to Field and Sherrell (2000).

Site	Fe(II) half-life hours	Site	Fe(II) half-life hours
Juan de Fuca Ridge	6.38	South EPR	1.83
Gorda Ridge	3.15	SW Indian Ridge	1.31
21°N EPR	3.7	TAG, MAR	0.45
9°45'N EPR	3.33	Rainbow, MAR	0.29

The Mn/Fe ratio is even lower, about ~0.0002, in the buoyant plume above the TAG hydrothermal field and about 0.1% of the total concentration of Mn is present in particulate matter (Trocine and Trefry 1988; Klinkhammer et al. 1986). In the lower 20 m of the buoyant plume above the 21°N EPR hydrothermal field the Mn/Fe ratio varies from ~0.0009 to 0.011 (Mottl and McConachy 1990), and there is an indication of a decrease in the ratio where an increase in the dilution factor occurs (Fig. 4.11). A relatively high Mn/Fe ratio (0.052) for buoyant plumes has been measured in material that settled 2.5 m from a high-temperature vent at the 13°N EPR hydrothermal field. Probably the increase in the Mn/Fe ratio near the hydrothermal vents may be caused by the formation of Mn(IV)-hydroxides under conditions where the temperatures near the vents are still rather high and the aeration of the water is sufficient to cause oxidation of Mn(II) to Mn(IV) and precipitation, recognizing that the rate of formation of the Mn(IV)-hydroxides in aerated environment increases sharply with a rise in the temperature. Nevertheless the role of this process is minor.

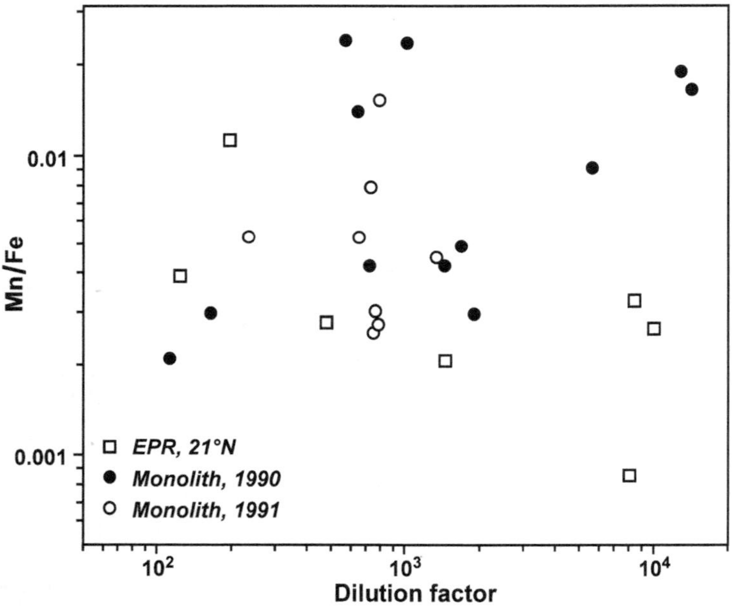

Fig. 4.11. Mn/Fe ratio in particulate matter in relation to the dilution factor in buoyant hydrothermal plumes. Based on the data of Mottl and McConachy (1990) and Feely et al. (1994).

In general, practically all of the dissolved hydrothermal Mn is carried in the buoyant plumes and enters the nonbuoyant plumes. In the buoyant plumes the Mn exhibits almost conservative behavior. For this reason it is used in the calculation of the dilution factors[31] (Mottl and McConachy

[31] By the formula $D = (Mn_h - Mn_{tot}) / (Mn_{tot} - Mn_{sw})$, where Mn_h – concentration of Mn in primary hydrothermal fluid, Mn_{tot} – total (dissolved + particulate) concentration of Mn in a sample, Mn_{sw} – concentration of Mn in ambient ocean water. In buoyant plumes $Mn_{tot} >> Mn_{sw}$ and Mn_{sw} can be neglected.

1990; Rudnicki and Elderfield 1993; Feely et al. 1994; Gurvich 1998; Field and Sherrell 2000, et al.). The Fe_{tot}/Mn_{tot} ratios in the uppermost parts of buoyant plumes together with the Fe_h/Mn_h ratio are used for estimating the loss of Fe from the buoyant plumes[32] and the transfer of Fe to the non-buoyant plumes (Mottl and McConachy 1990; Rudnicki and Elderfield 1993; Field and Sherrell 2000).

The time measured for dissolved Mn(II) to pass into particulate matter[33], including its sorption, oxidation, co-precipitation, etc., in hydrothermal fields and their associated buoyant plumes in the Guaymas Basin, Galapagos Spreading Center, and Endeavor Ridge is respectively, 26 days based on one measurement, 28 to 800 days based on four measurements, and 510 to 1200 days based on four measurements. The rates for the dissolved Mn(II) pass into the particulate matter in the Galapagos Spreading Center and the Endeavor Ridge are respectively from 8 to 130 ng l^{-1} day^{-1} (aver. 62 ng l^{-1} day^{-1}), and from 20 to 140 ng l^{-1} day^{-1} (aver. 69 ng l^{-1} day^{-1}) (Mandernack and Tebo 1993). The measured times are much shorter than the time required for dissolved Mn(II) to pass into particulate matter by chemical processes only (Emerson et al. 1979, 1982). For example, it would require about 50 years (Weiss 1977) instead of 28 to 800 days for this to take place by chemical processes only in the plume above the Galapagos Spreading Center. It has been shown in numerous publications that in an ocean environment, including hydrothermal plumes, the oxidation and sorption of Mn(II) takes place with the participation of bacteria that oxidize manganese (Bromfield and David 1976; Rosson and Nealson 1982; Ehrlich 1983, 1985; Edenborn et al. 1985; Cowen et al. 1986, 1990; Tebo and Emerson 1986; Nealson et al. 1988; Durant et al. 1990; Tambiev and Demina 1992; Mandernack and Tebo 1993; Lilley et al. 1995; Sudarikov et al. 1995, et al.). According to experiments conducted in the hydrothermal fields and the buoyant plumes in the Galapagos Spreading Center and Endeavor Ridge, bacterial activity causes respectively 29 to 70% (aver.

[32] By the formula $Fe_{loss} = 100\% \cdot [1 - (Fe_{tub} / Mn_{tub}) / (Fe_h - Mn_h)]$, where Fe_{tub} and Mn_{tub} – total concentrations of Fe and Mn in the uppermost part of a buoyant plume, Fe_h and Mn_h – concentrations of Fe and Mn in primary hydrothermal fluid. The formula can be used for estimating Fe loss at any height in the buoyant plume if the total concentrations of Fe and Mn at this height are known. Apparently contributions of Fe and Mn delivered by diffusive flows to buoyant plumes should be taken into account in estimations of the Fe losses and in calculation of the dilution factors; otherwise, due to lower Mn/Fe ratio in high-temperature fluids than in diffusive flows (Sect. 4.1), the calculated Fe losses will be overestimated and dilution factors – underestimated.

[33] Ratio of dissolved Mn(II) concentration to experimentally measured rate of dissolved Mn(II) pass into particulate matter.

46%) and 12 to 38% (aver. 24%) of the observed transfer of dissolved Mn(II) to particulate matter (Mandernack and Tebo 1993). However, in buoyant plumes a very small amount of dissolved Mn(II) as well as other elements pass into the particulate matter because of the relatively short time that they are held in these plumes, about 30 to 40 minutes (0.02–0.03 days), that is a period of time 900 to 1300 times shorter than the shortest time measured for dissolved Mn(II) to pass into particulate matter.

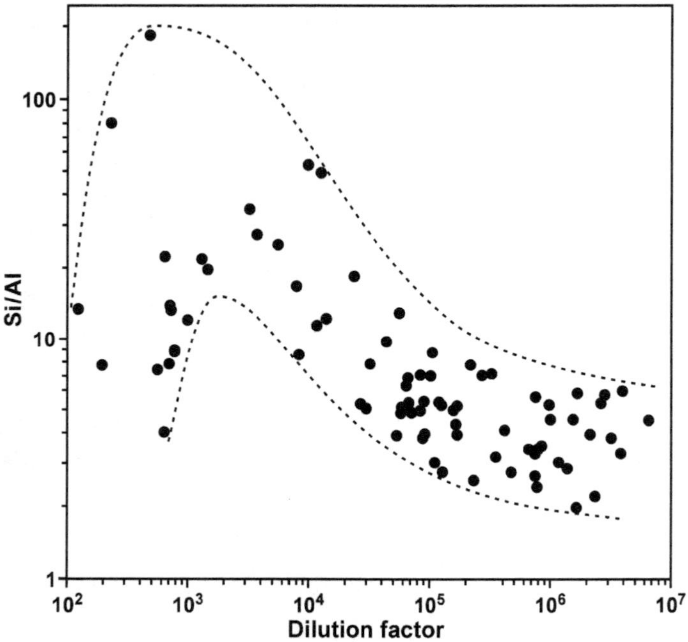

Fig. 4.12. Si/Al ratio in particulate matter in relation to the dilution factor in hydrothermal plumes. Based on the data of Trocine and Trefry (1988), Mottl and McConachy (1990), Feely et al. (1994).

The concentration of Si in the hydrothermal fluids is considerably higher than in ocean water (Fig. 4.1). It exhibits close to conservative behavior when hydrothermal solutions mix with ocean water. However, up to several percent of the total concentration of Si in the buoyant plumes is present in particulate matter (Mottl and McConachy 1990; Feely et al. 1994), and the Si/Al ratio in the particulate matter is relatively high (Fig. 4.12) that apparently indicates the formation of hydrothermal silicate minerals (Table 4.1). Presumably this process is most intensive in the middle parts of the buoyant plumes where the dilution factor values are from ~300 to ~3000. With a further increase in the dilution of hydrothermal solutions by ocean water the Si/Al ratio decreases to the background level.

The concentrations of particulate Cu, Zn, Pb, Cd, Ag, and Co in buoyant plumes vary. However, in the particulate matter in the lower parts of buoyant plumes the Cu/Fe, Zn/Fe, Pb/Fe, Cd/Fe, Ag/Fe, and Co/Fe ratios are higher than in the primary hydrothermal fluids (Mottl and McConachy 1990; Feely et al. 1994; Edmonds and German 2004). In the particulate matter in the lower 14 m of the buoyant plume above the hydrothermal field at 21°N EPR the average Cu/Fe ratio is 20 times higher, the Zn/Fe and Pb/Fe ratios are 4 times higher, the Cd/Fe and Ag/Fe ratios are 10 times higher, and the Co/Fe ratio is 12 times higher than in the primary hydrothermal fluids (Mottl and McConachy 1990).

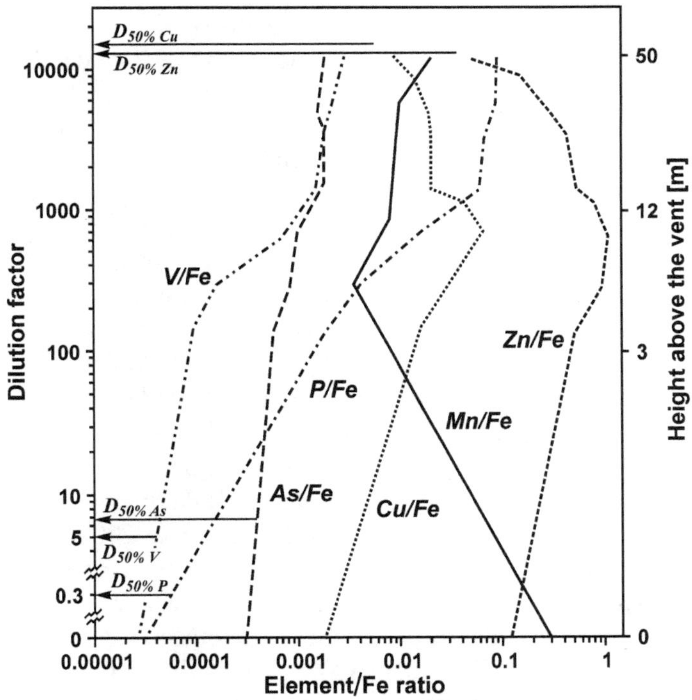

Fig. 4.13. Element/Fe ratios in particulate matter of the buoyant hydrothermal plume in relation to the height above the Monolith vent and the dilution factor. Element/Fe ratios in primary hydrothermal fluids are shown at zero height and dilution factor. Calculated and based on the data of Feely et al. (1994).

Variations in the Cu/Fe and Zn/Fe ratios in particulate matter in the buoyant plume over the Monolith vent are shown in Fig. 4.13. In the lowermost part of the plume these ratios are higher than in the hydrothermal fluids. As the plume rises the ratios increase and reach their maxima at ~10 m above the orifice where the dilution factor is 500 to 1000. As they rise higher and the dilution factor increases the Cu/Fe and Zn/Fe ratios decrease. Patterns like this are interpreted as evidence for relatively more extensive formation of Cu- and Zn-sulfide minerals than Fe-minerals during the early stages of mineral precipitation in the buoyant plume when the hydrothermal fluids mix with ocean water. The Cu-sulfide minerals are usually enriched in Co and the Zn-sulfide minerals in Pb, Cd, and Ag. Apparently a part of the Pb content is present in galena particles.

Usually the size of sulfide mineral particles ranges up to 70 μm. These particles remain in the buoyant plumes as long as they can float in the rising flow of the plume. When there is a decrease in the velocity of the flow the particles settle to the floor in the vicinity of the vent (Feely et al. 1990, 1992, 1994). The maximum Cu/Fe and Zn/Fe ratios mentioned presumably correspond to the levels of neutral buoyancy for the Cu- and Zn-sulfide mineral particles where the concentrations of these particles increase.

The V, As, and P exhibit analogous behavior in the buoyant plumes. The V/Fe, As/Fe, and P/Fe ratios in the particulate matter in the lowermost levels of the buoyant plumes are at their minima. As the plume rises and the dilution factor increases the ratios increase. The maximum gradients in the increase take place when the dilution factor is from 100–200 to 1000–2000. When the dilution factors are higher the ratios increase very slowly or do not vary significantly (Fig. 4.13). This pattern is interpreted as evidence of the co-precipitation of the oxyanions of V, As, and P, and apparently of Cr, U, B, Mo, and W from ocean water by the Fe-hydroxides that formed as the plume is rising. This process is rapid as it usually takes place in the lower parts and ceases in the upper parts of the buoyant plumes (Trocine and Trefry 1988; Feely et al. 1990, 1991a,b, 1992, 1994; German et al. 1991; Rudnicki and Elderfield 1993; Lilley et al. 1995; Savenko 2000; Edmonds and German 2004; German and Von Damm 2004). Even in the buoyant plume of the Monolith vent from the Juan de Fuca Ridge where the process of Fe precipitation is slower than in the plumes over other vent sites (Table 4.3) most of the P/Fe, Cr/Fe, V/Fe, and As/Fe ratios indicate that enrichment of the elements in the hydrothermal particles takes place within the first 50 to 100 m of the rising plume (Feely et al. 1994; Lilley et al. 1995).

Fig. 4.13 shows good evidence that ocean water is the main source of V, As, and P, even in the particulate matter of the buoyant plumes. The maximum gradients of the V/Fe-, As/Fe-, and P/Fe ratios occur as the dilution factors exceed in the values of $D_{50\%}$ for these elements. Apparently some part of the As in particles in the buoyant plumes may be bound in sulfide minerals. The role of the sulfide minerals as carriers of the As is greater in back-arc basins where its concentration in the hydrothermal fluids (Fig. 4.1) and massive sulfides (Gurvich 1990; Lisitzin et al. 1992b,c; Shadlun et al. 1992) is very high.

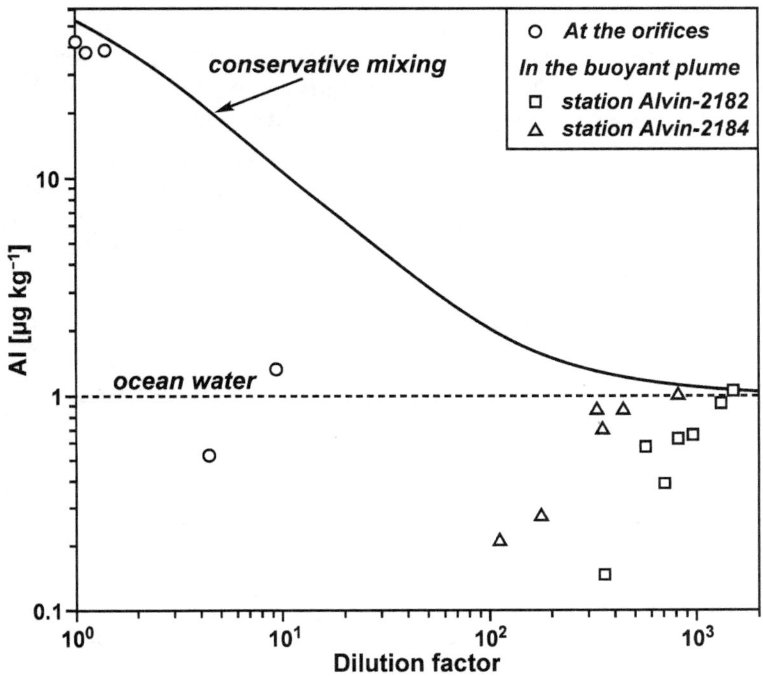

Fig. 4.14. Concentrations of dissolved Al at hydrothermal orifices and in the buoyant hydrothermal plume in the TAG hydrothermal field in relation to the dilution factor. Calculated and based on data from Elderfield et al. (1993a).

The properties of Al in buoyant plumes are similar in many respects to those of V, As, and P. When the Fe-hydroxides are precipitated this element is extensively scavenged from the solution, including the Al carried in the hydrothermal fluids and also the Al that is dissolved in the ocean water. As a result the concentration of dissolved Al in the buoyant plumes becomes lower than that would be at conservative mixing of the hydrothermal fluids and ocean water, but also lower than its background concentration in the ambient ocean water (Fig. 4.14). According to data shown in Fig. 4.14, the Al passes into the particulate matter when the dilution factor is from about 10 to 1000, and especially in the lower parts of buoyant plumes. With a greater dilution factor this process ceases and the concentration of the dissolved Al in the plume water becomes indistinguishable from the concentration in ocean water.

Despite the intensive scavenging of hydrothermal Al by the Fe-oxyhydroxides, the effect of this process on the composition of the metalliferous sediments and the composition of the particulate matter in the buoyant plumes is not significant. In the hydrothermal fluids of the East Pacific Rise and Mid-Atlantic Ridge the average Al/Fe ratio is ~0.0013, in hydrothermal sulfide minerals it is 0.0032 to 0.0069 (Stepanova and Krasnov 1992), when the dilution factor is from 100 to 1000 in particulate matter in the buoyant plume at 21°N EPR it is 0.008 (Mottl and McConachy 1990), above the Monolith vent it is 0.028 (Feely et al 1994), in the TAG hydrothermal field it is 0.001–0.002, as based on estimates from the data of Elderfield et al. (1993a). In the metalliferous sediments that accumulated near the hydrothermal vents the Al/Fe ratio is 0.011 at 21°N EPR (Mottl and McConachy 1990) and 0.012 to 0.027 at the TAG field (German et al. 1993).

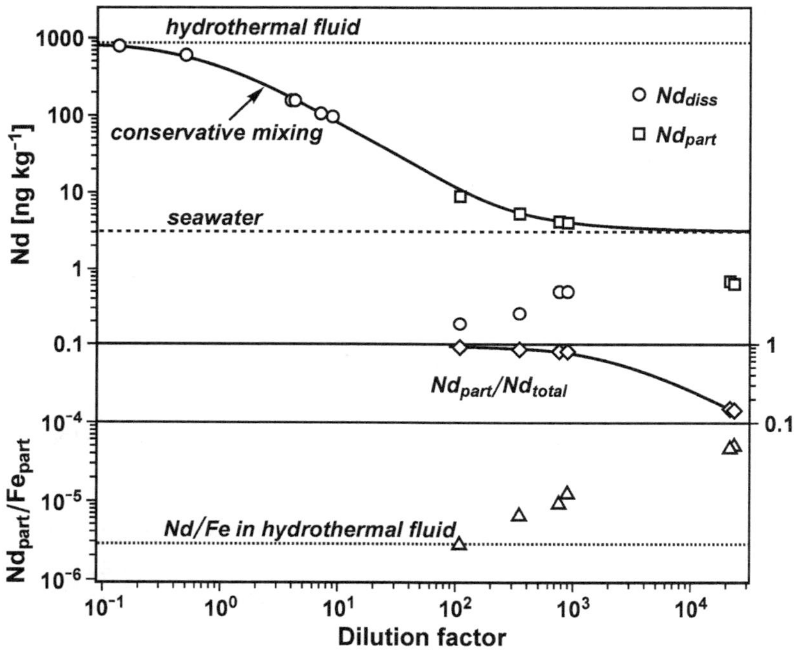

Fig. 4.15. Concentrations of dissolved and particulate Nd, relative part of particulate Nd in the total concentration, and Nd/Fe ratio in particulate matter in relation to the dilution factor in the buoyant hydrothermal plume and transition to the non-buoyant plume in the TAG vent field. Calculated and based on the data of German et al. (1990) and Mitra et al. (1994).

It follows from these data that even if the Al/Fe ratio in the hydrothermal particulate matter of buoyant plumes and in the hydrothermal matter of sediments accumulated in close vicinity to hydrothermal vents are equal to the Al/Fe ratio in the hydrothermal fluids the part of the Al content bound with the hydrothermal matter in these sediments does not exceed 5 to 22% of its total content. At greater distances from the hydrothermal vents the role of hydrothermal Al decreases sharply and in an overwhelming part of the area covered by metalliferous sediments its content is insignificant.

In metalliferous sediments from the TAG field the average Al/Fe ratio is 0.12 (Table 1.10) and the estimated average content of the hydrothermal Al does not exceed 1% of total amount of the Al in the sediments.

The rare earth elements in the early stages of rising of a buoyant plume, and while the sulfide minerals are forming, are mostly present in solution and exhibit a conservative behavior, as shown in the example for Nd in Fig. 4.15. As the plume continues to rise and the Fe-hydroxides precipitate they extensively co-precipitate and sorb both the hydrothermal and the non-hydrothermal REE that are in solution. According to the estimates of Rudnicki and Elderfield (1993), in the uppermost parts of the buoyant plume above the TAG hydrothermal field, at the level of neutral buoyancy, the average contribution of co-precipitation to the amount of particulate REE is 55±20%. There is a minimum of co-precipitation for La (42%), and for the middle REE (26÷45%), and a maximum for Ce (85%) and the heavy REE (62÷85%).

Because of the co-precipitation and sorption when the dilution factor is from 100 to 1000, the particulate REE become predominant, and the concentrations of the dissolved REE become not only lower than their concentrations at conservative mixing of the hydrothermal fluids and ocean water, but lower than their background concentrations in ambient ocean water. Where there is greater dilution the role of the particulate REE decreases and in the upper parts of the plume the dissolved REE become predominant. When the dilution factor becomes less than 100 the REE/Fe ratio in the particulate matter differs very little from the ratios in the hydrothermal fluids (Fig. 1.63); the Nd/Fe ratio is about $(2\div3)\cdot10^{-6}$ (Fig. 4.15). When there is greater dilution the REE/Fe ratio increases owing to the co-precipitation and sorption of the REE by the Fe-oxyhydroxides (Fig. 1.63). The increase in the REE/Fe ratio with dilution has been observed not only in the buoyant plume of the TAG field, but also in other plumes in the 21°N EPR, 9°45'N EPR, and Rainbow hydrothermal fields (Mottl and McConachy 1990; Sherrell et al. 2000; Edmonds and German 2004).

4.3.2. Processes in nonbuoyant hydrothermal plumes

There are no dramatic variations in the physical and chemical conditions in nonbuoyant plumes like those observed in the lower parts of the buoyant plumes. The residence time of fine particulate matter in nonbuoyant plumes is considerably longer than in buoyant plumes. The residence time of particles in the buoyant plumes usually does not exceed 30 to 40 minutes (rising time) but in nonbuoyant plumes even in those that are not far from hydrothermal vents it can be from several days to tens of days or more (Rudnicki and Elderfield 1993). Because of the absence of rising flow fine particles, less than 2 μm in size, are predominant among the particles of hydrothermal origin in the nonbuoyant plumes (Nelsen et al. 1986/1987; Walker and Baker 1988). The particles of hydrothermal origin consist mainly of particulate and colloidal Fe(III)-oxyhydroxides and practically all of the Fe is present in this form. Aggregates of extremely fine-grained particles of Fe-oxyhydroxides are present also in the majority of the biogenic particles sampled in the plumes (Godfrey et al. 1994). Coarser particles of sulfide and sulfate minerals are rarely carried in the nonbuoyant plumes because of their fast settling and dissolution and usually they are observed only in close proximity to the hydrothermal vents (Walker and Baker 1988; Lisitzin 1993; Godfrey et al. 1994).

The concentration of particulate matter of hydrothermal origin gradually decreases in the nonbuoyant plumes where there is an increase in the dilution of the hydrothermal fluids by ocean water as the distances increase from the hydrothermal vents. This decrease in the concentration of particulate Fe, the main hydrothermal element in the particulate matter, can be seen in the nonbuoyant plume over the TAG hydrothermal field (Fig. 4.16). The concentration of particulate Al, which is a proxy of the concentration of background pelagic abiogenic particulate matter, has no determinable variation as the dilution increases. The concentration of particulate Fe (Fe_{part}) also decreases gradually with an increase in the passing of time after the hydrothermal Fe enters into the ocean (Rudnicki and Elderfield 1993).

Concentrations of Fe_{part} can be different in nonbuoyant plumes over different hydrothermal fields even when there are similar concentrations of Fe in the hydrothermal fluids and similar dilution factors. For example, the concentrations of Fe_{part} in the nonbuoyant plume over the TAG hydrothermal field are much higher than those over the nonbuoyant plume at 9°45'N EPR. In the latter case the concentration of Fe_{part} [34] when the dilution factor increases from $7.2 \cdot 10^3$ to $90 \cdot 10^3$ changes from 12% to 53% of the total concentration of Fe (Field and Sherrell 2000; Fig. 4.16). Lower concentrations of Fe_{part} apparently are present where there are not only relatively lower total concentrations of Fe but where the process of oxidation of the Fe(II) is slower (Table 4.3). Strong correlation of the total concentration of Fe with the concentration of dissolved Mn (and with the dilution factor that calculated from the concentration of dissolved Mn or total dissolvable Mn) in the nonbuoyant plumes suggests that the total Fe exhibits close to conservative behavior (Field and Sherrell 2000).

[34] Fe bound in the particulate matter collected on filters with 0.45 μm pore size. The Fe considered as dissolved was that in the sum of dissolved Fe(II) and Fe(III) in the Fe-hydroxide particles <0.45 μm in size (Field and Sherrell 2000).

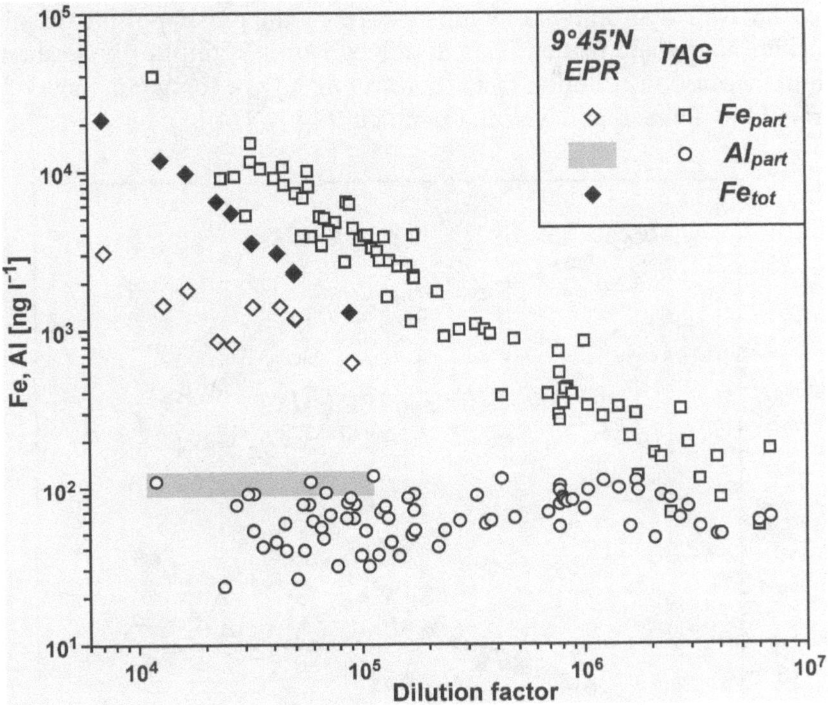

Fig. 4.16. Concentrations of particulate Fe and Al, and of total Fe in relation to the dilution factor in the nonbuoyant hydrothermal plumes over the TAG and 9°45'N EPR hydrothermal fields. Calculated and based on data from Klinkhammer et al. (1986), Trocine and Trefry (1988), German et al. (1991), Sherrell et al. (1999), Field and Sherrell (2000).

Unlike the decrease in the concentration of the dissolved Mn with an increase in the dilution of the hydrothermal fluids with ocean water, the concentration of particulate Mn, Mn_{part}, varies insignificantly within the nonbuoyant plume over the TAG hydrothermal field (Fig. 4.17, Table 4.4). The behavior of the Mn_{part} in the nonbuoyant plume over the TAG hydrothermal field is similar to that in the nonbuoyant plumes over other hydrothermal fields (Fig. 4.18; Feely et al. 1994; James et al. 1995b; Ludford et al. 1996; Edmonds and German 2004 et al.).

Small variations in the concentration of the Mn_{part} in the nonbuoyant plumes over individual hydrothermal fields are caused by the gradual increase in the role of the Mn_{part} in the total concentration of Mn as the dilution factor increases and the total concentration of Mn decreases. Because of the decrease in the concentration of Fe_{part}, the Mn/Fe ratio in the particulate matter increases with an increase in the dilution factor. The measured increase in Mn_{part}/Mn_{tot} and Mn_{part}/Fe_{part} ratios in the nonbuoyant plume over the TAG hydrothermal field is caused by the passing of the dissolved Mn into particulate matter. This dissolved Mn is mainly of hydrothermal origin, because the dilution factor is lower than $D_{50\%}$ for Mn in most of the parts of the plume that have been sampled (Fig. 4.8, 4.17).

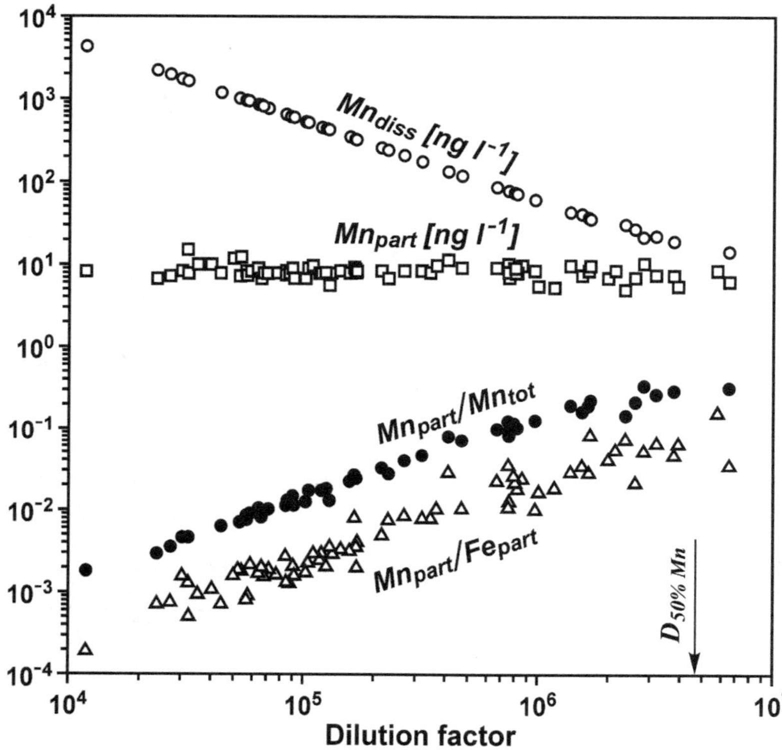

Fig. 4.17. Concentrations of dissolved and particulate Mn, and Mn_{part}/Mn_{tot} and Mn_{part}/Fe_{part} ratios in relation to the dilution factor in the nonbuoyant plume over the TAG hydrothermal field. Calculated and based on the data from Klinkhammer (1986), Trocine and Trefry (1988), and German et al. (1991).

The patterns are similar in the nonbuoyant plumes over other hydrothermal fields. But where the concentrations of Fe_{part} are similar in the plume over the Rainbow field the concentrations of Mn_{part} and the Mn_{part}/Fe_{part} ratios are similar to those in the plume over the TAG field, and are much lower than those in the plumes over the North Cleft segment and at 9°45'N EPR (Fig. 4.18). This is evidence of more active passage of the Mn into particulate matter within the nonbuoyant plumes over the North Cleft segment and at 9°45'N EPR in comparison to that in the plumes over the TAG and Rainbow hydrothermal fields.

Table 4.4. Average particulate concentrations of chemical elements related to increases in the dilution factor in the hydrothermal plume over the TAG hydrothermal field. Calculations based on the data from Klinkhammer (1986), Trocine and Trefry (1988), and German et al. (1991).

Dilution factor	Fe $\mu g\ l^{-1}$	Al	Mn	Cu	Zn	Cd	Pb ng l^{-1}	V	As	Co	Ni	La
$(2\div6)\cdot10^3$*	127		20	2100	5100	39	29	450	230		0.59	2.0
$(0.6\div2)\cdot10^4$	20.3	72	7.3	310	310	6.3	7.2	76			0.67	
$(2\div6)\cdot10^4$	6.8	64	8.6	130	36	0.61	3.8	29	20	0.82	0.47	0.47
$(0.6\div2)\cdot10^5$	2.6	62	7.8	24	7.8	0.039	2.2	12	7.4	0.42	0.45	0.25
$(2\div6)\cdot10^5$	0.6	80	8.7	4.7	3.9	0.033	1.3	2.9	2.3	0.24	0.23	0.14
$(0.6\div2)\cdot10^6$	0.23	86	7.3	1.7	3.2	0.039	0.91	1.2	0.37	0.18	0.22	0.100
$(2\div4)\cdot10^6$	0.12	57	6.8	2.8	2.5	0.029	0.51	0.59		0.12	0.29	0.097

* buoyant plume

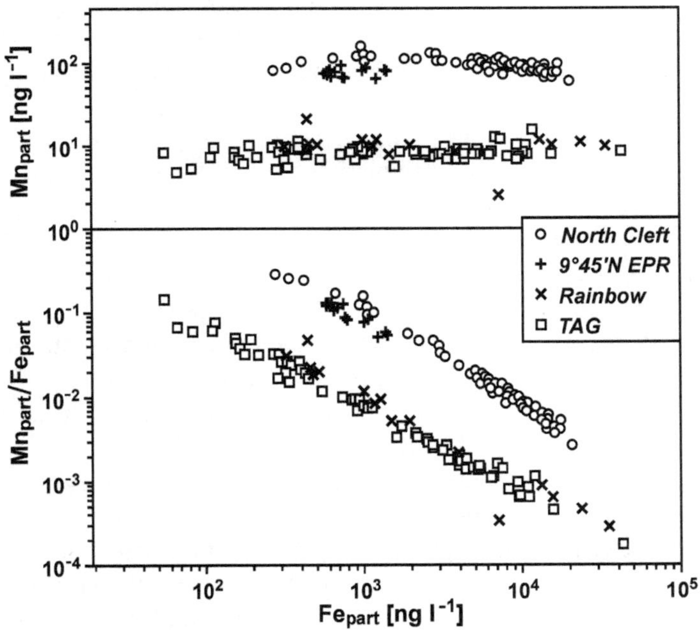

Fig. 4.18. Concentrations of particulate Mn and Mn_{part}/Fe_{part} ratios in relation to the concentrations of particulate Fe in nonbuoyant plumes over the TAG, Rainbow, North Cleft segment, and 9°45'N EPR hydrothermal fields. Based on the data from Trocine and Trefry (1988), German et al. (1991), Feely et al. (1994), Field and Sherrell (2000), and Edmonds and German (2004).

Within the nonbuoyant plumes over the Broken Spur and 24°30'N MAR hydrothermal fields, as over the TAG field, the Mn_{part}/Fe_{part} ratios, even where there are rather low concentrations of Fe_{part} (<200 ng l^{-1}), do not exceed 0.1 (Krasnov et al. 1992; Sudarikov et al. 1992; James et al. 1995b). Within the nonbuoyant plumes over the Juan de Fuca Ridge, Galapagos Spreading Center, Guaymas Basin, North Lau Basin, and East Manus Basin the Mn_{part}/Fe_{part} ratios are ≥0.3 at relatively low concentrations of Fe_{part} (Bolger et al. 1978; Gordeev and Demina 1979; Gordeev et al. 1979; Tambiev and Demina 1992; Massoth et al. 1994; Lisitzin et al. 1997). The reason for this probably can be attributed to the transfer of dissolved Mn into particulate matter by the action of bacteria that oxidize manganese.

The Broken Spur, 24°30'N MAR, and TAG hydrothermal fields are located in low-productive areas of the Atlantic Ocean, and the hydrothermal fields of the Juan de Fuca Ridge, Galapagos Spreading Center, Guaymas Basin, North Lau Basin, and East Manus Basin are located in mid-productive and high-productive areas (Koblents-Mishke 1977; Vinogradov et al. 1996). The production of bacteria and biomass of bacteria are greater in those areas of the ocean where the production of phytoplankton is greater (Sorokin 1977). As for the action of chemosynthesis, it is somewhat greater only in the close proximity to hydrothermal vents. In a trap positioned 21 m over the ocean floor and 4 m away from a high-temperature vent in the hydrothermal field at 47°57.05'N, 129°06.30'W on the Endeavor Ridge, that was located in a mid-productive area, 95% of the organic flux was developed by chemosynthesis, and in traps positioned within a nonbuoyant plume over this field, 26–36% of the organic flux was developed by chemosynthesis (Roth and Dymond 1989). In a trap positioned 1.5 m above the ocean floor 80 m from of a group of three high-temperature hydrothermal vents in the 13°N EPR hydrothermal field within a mid-productive area, one third of the organic flux was developed by chemosynthesis (Khripounoff and Alberic 1991). According to these data, in the mid-productive areas of the ocean the role of chemosynthetic organic matter is significantly greater only in close proximity to the high-temperature hydrothermal vents, and in nonbuoyant plumes in the mid-productive and high productive areas the photosynthetic organic matter is prevalent. One can deduce from this that the production of the biomass of manganese oxidizing bacteria in nonbuoyant plumes in mid- and high-productive areas of the ocean, with other factors being equal, is greater than in low-productive areas.

The periods of time that have been measured for the passage of dissolved Mn into the particulate matter in nonbuoyant plumes over the Galapagos Spreading Center are from 34 to 1200 days (aver. 390 days, in 8 measurements), over the Endeavor Ridge they are from 380 to 5700 days (aver. 1600 days, in 16 measurements) (Mandernack and Tebo 1993), over the Guaymas Basin from 4 to 14 days (Campbell et al. 1988a), and over the southern part of the Juan de Fuca Ridge the period is 475 days (Cowen et al. 1990). These times are comparable to the time that hydrothermal matter remains in the nonbuoyant plumes, which may be from tens to hundreds of days, even within a distance of a kilometer from the hydrothermal vents (Rudnicki and Elderfield 1993). The rates at which the dissolved Mn passes into particulate matter in the nonbuoyant plumes over the Galapagos Spreading Center and Endeavor Ridge are from 0.5 to 34 (aver. 8.4) and from 0.3 to 11 (aver. 1.7) ng l^{-1} day^{-1} (Mandernack and Tebo 1993). In the nonbuoyant plume over the TAG hydrothermal field the average rate for the dissolved Mn to pass into particulate matter is about 0.11 ng l^{-1} day^{-1} (my estimate), which is, respectively, 70 and 14 times less than in the nonbuoyant plumes over the Galapagos Spreading Center and Endeavor Ridge. This coincides with the presumed rate for the dissolved Mn to pass into particulate matter by biological processes. In the nonbuoyant plumes over the Galapagos Spreading Center and Endeavor Ridge bacterial activity causes 25 to 68% (aver. 50%) and 0 to 70% (aver. 38%) respectively, of the dissolved Mn to pass into particulate matter.

Over the Endeavor Ridge the influence of bacteria in the transfer of the dissolved Mn into particulate matter increases over distances as far as 17 km from the hydrothermal vents. At distances from 0 to 2 km the bacterial influence averages 28%, at 4 to 10 km it is 37%, at 17 km it is 68% and its influence correlates positively with the increase in the number of bacteria covered by manganese bearing film. Under the nonbuoyant plumes bacteria greatly influence the passing of the dissolved Mn into particulate matter. Over the Galapagos Spreading Center 600 m from the hydrothermal vents the influence of bacteria as measured was 58% (one measurement), over the Endeavor Ridge 4 km from hydrothermal vents it was 54–76% (two measurements); the rates of passing of the dissolved Mn into particulate matter was respectively 2.5 and 1.1–3.3 ng l^{-1} day^{-1}. These values are higher than the rates measured within nonbuoyant plumes at the same stations. The times required for the dissolved Mn to pass into particulate matter below the nonbuoyant plumes over the Galapagos Spreading Center and Endeavor Ridge are 90 and 118–450 days respectively; they are shorter than those within the plumes where the average values are 390 and 1600 days (Mandernack and Tebo 1993).

In the opinion of Mandernack and Tebo (1993), the greater role of bacteria in the transfer of the dissolved Mn into particulate matter under the nonbuoyant plumes is caused by the settling from the nonbuoyant plumes of bacteria that are covered with the manganese films, because the bacteria act as additional centers for the capture of the dissolved Mn. Under the nonbuoyant plume in the Guaymas Basin very large colonies (tens of microns and on the average 10 to 30 μm in size) of such bacteria have been found; they consist mainly of organic matter and Mn-oxyhydroxides (Tambiev 1988, 1989; Tambiev and Demina 1992).

The increase in the role of bacteria under the nonbuoyant plumes results in an increase in the concentration of particulate Mn and of the Mn_{part}/Fe_{part} ratio with depth (Bolger et al. 1978; Campbell et al. 1988a; Cowen et al. 1990; Cowen and Li 1991; Mandernack and Tebo 1993). A section across the East Pacific Rise at 15°S that extends as far as 80 km to the west of the axis (Feely et al. 1996) is a good example. Under the nonbuoyant plume, where the concentrations of the total particulate matter and Fe_{part} are considerably lower than within the plume, the Mn_{part} concentration is similar to that within the nonbuoyant plume and the Mn_{part}/Fe_{part} ratio is considerably higher (Fig. 4.19).

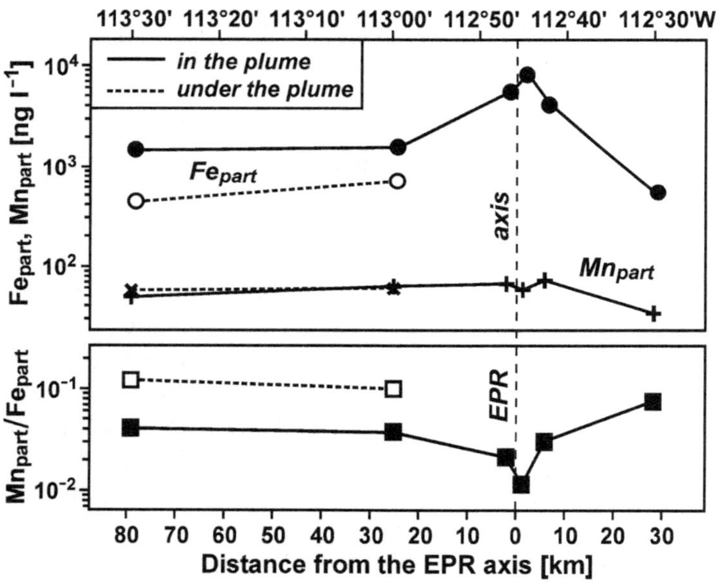

Fig. 4.19. Average concentrations of particulate Fe and Mn and Mn_{part}/Fe_{part} ratios in the nonbuoyant plume and under the nonbuoyant plume along the section across the East Pacific Rise at 15°S. Calculated and based on data from Feely et al. (1996).

In nonbuoyant plumes over the East Pacific Rise between 13°33'S and 18°40'S the Mn_{part} concentrations vary from 10 to 40 ng l^{-1} in the northern part and from 400 to 600 ng l^{-1} in the southern part of the studied area (Feely et al. 1996; Fig 4.20). This area is located in the low-productive zone, and if the Mn_{part} concentrations from 10 to 40 ng l^{-1} are close to those observed in the nonbuoyant plumes from other low-productive areas, the concentrations of 400 to 600 ng l^{-1} not only exceed these but also the concentrations observed in the nonbuoyant plumes from the mid- and high-productive areas. In the southern part the ratios of Mn_{part}/Fe_{part} greatly exceed those in areas where similar concentrations of Fe_{part} are present in the nonbuoyant plumes in other low-productive areas (Fig. 4.20, 4.18). Over the axial part of the EPR between 17°35'S and 18°40'S both dissolved gases and the bacterial counts are significantly higher than in the area north of 17°20'S. Apparently higher concentrations of volatile gases fertilize the local microbial population, which enhances the oxidation of the dissolved Mn and increases the concentration of Mn_{part} (Urabe et al. 1995; Feely et al. 1996).

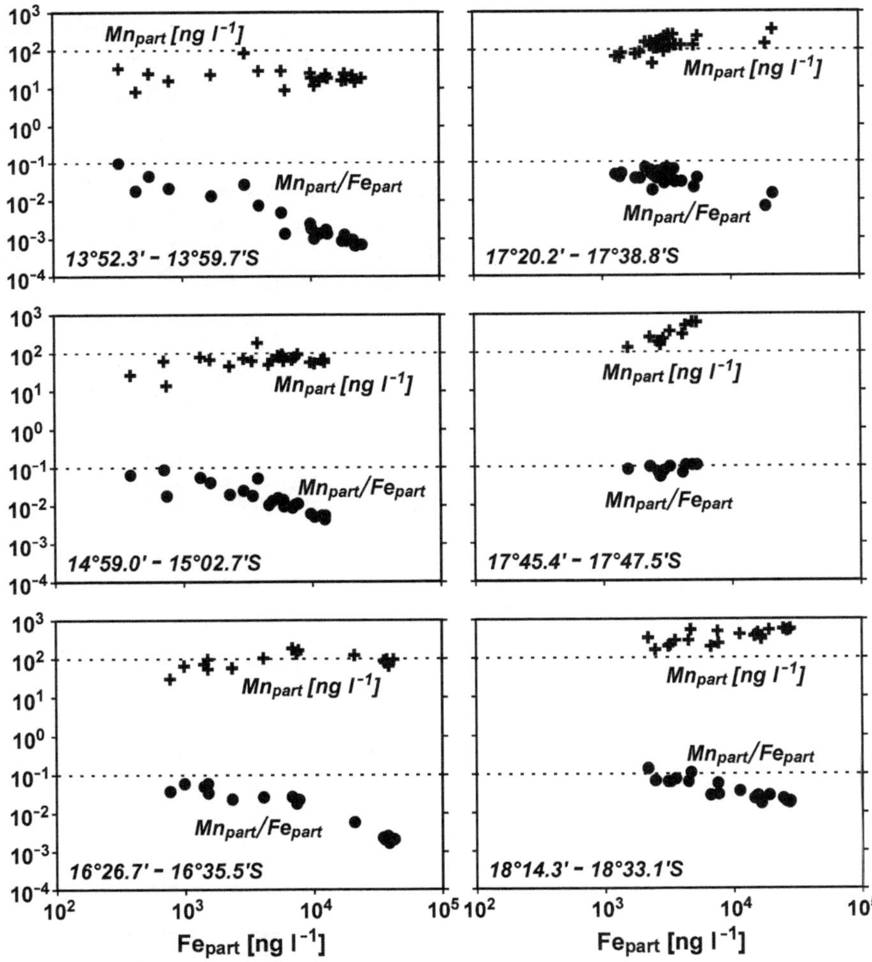

Fig. 4.20. Concentrations of particulate Mn and Mn_{part}/Fe_{part} ratios in relation to the concentrations of particulate Fe in the nonbuoyant plumes over the axial part of the East Pacific Rise. Based on the data from Feely et al. (1996).

With an increase in the dilution of hydrothermal fluids by ocean water (and a decrease in the concentration of Fe_{part}) in the transition from the buoyant to the nonbuoyant plume and subsequently in the nonbuoyant plume the concentrations of particulate Cu, Zn, and Cd decrease first and then change somewhat. In the nonbuoyant plume over the TAG hydrothermal field the decrease in the concentrations of Cu_{part} and Zn_{part} is evident when the dilution factor is about $(2\div6)\cdot10^5$ and the concentration of Fe_{part} is about 0.6 µg l^{-1}, and the decrease in the concentration of Cd_{part} ceases when the dilution factor is about $(0.6\div2)\cdot10^5$ and the concentration of Fe_{part} is about 2.6 µg l^{-1} (Table 4.4). In the nonbuoyant plume over the North Cleft segment the decrease in the concentration of Cu_{part} and Zn_{part} ceases when the concentration of Fe_{part} is about 2.5 µg l^{-1}, and in the nonbuoyant plume over the axial part of the East Pacific Rise from 13°33'S to 18°40'S when the concentration of Fe_{part} is about 1 µg l^{-1} (Table 4.5).

Table 4.5. Average particulate concentrations of chemical elements with decreases in the concentrations of particulate Fe within the nonbuoyant hydrothermal plumes over the North Cleft segment, and the axial part of the East Pacific Rise from 13°33'S to 18°40'S. Calculated on the data from Feely et al. (1994, 1996).

Fe_{part}	Al	Mn	Cu	Zn	V	As	P	Ni
µg l^{-1}	ng l^{-1}							
				North Cleft				
10÷25	120	75	20	43	35	29	1510	
5÷10	170	90	11	15	20	15	770	
2.5÷5.0	260	100	12	10	11	7.2	370	
1.0÷2.5	190	110	4.5	3.4	4.7	2.9	140	
<1.0	190	110	4.4	3.7	2.3	1.1	49	
				East Pacific Rise				
100÷150			2080	2500	290	313	11700	
30÷100			215	160	120	105	4250	
10÷30			130	35	41	38	1560	0.12
3.0÷10			24	8.8	13	7.0	440	0.056
1.0÷3.0			14	6.8	6.2	3.9	190	0.055
<1.0					1.9	0.73		0.063
0.3÷1.0			4.6	4.2			54	
0.1÷0.3			3.4	4.7			29	
Average	15	144						

As to the Cu/Fe, Zn/Fe, and Cd/Fe ratios in the particulate matter, in the transition from the buoyant to the nonbuoyant plume and subsequently in the nonbuoyant plume they decrease at first to a minimum and then increase (Fig. 4.21, 4.22). The decrease in the Cu/Fe, Zn/Fe, and Cd/Fe ratios in the particulate matter to minimum values as the dilution factor increases with the distance from the hydrothermal vents, apparently is evidence of a decrease in the role and then the disappearance of sulfide minerals enriched in Cu, Zn, and Cd in the particulate matter. The subsequent increase in these ratios obviously is caused by an increase in the role of the Cu, Zn, and Cd passed from solution into the particulate matter because of sorption by the Fe-oxyhydroxides. This increase occurs for these three elements when the dilution factors exceed the values of $D_{50\%}$ (Fig. 4.21, 4.8). It is evident that ocean water is the main source of the Cu, Zn, and Cd that pass into the particulate matter after the sulfide minerals settle.

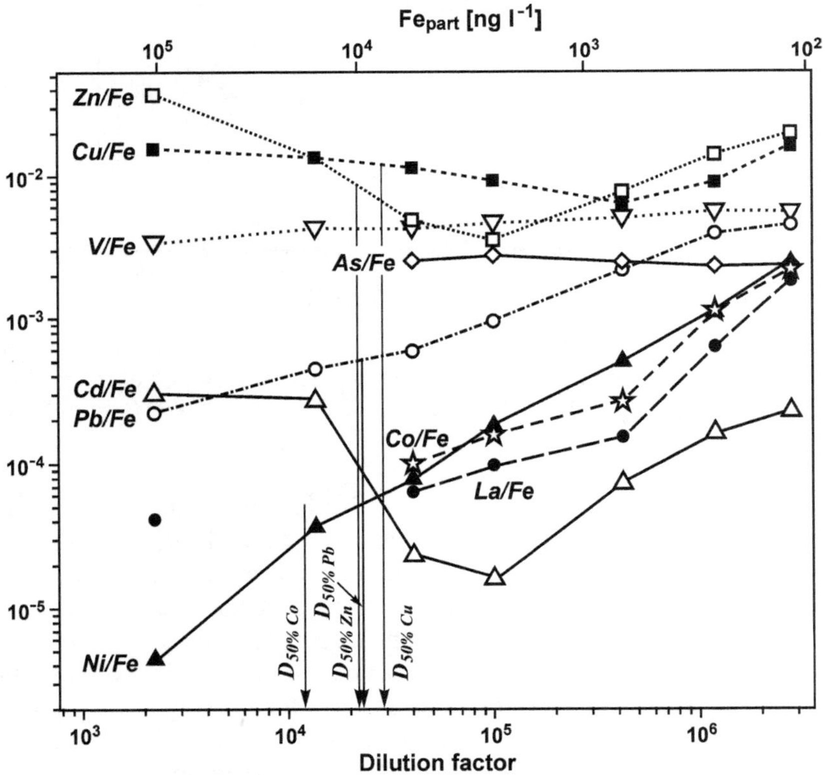

Fig. 4.21. Element/Fe ratios in particulate matter of the nonbuoyant plume over the TAG hydrothermal field in relation to the dilution factor. Calculated and based on the data from Klinkhammer et al. (1986), Trocine and Trefry (1988), German et al. (1990, 1991).

Where the dilution factor increases to $(2{\div}6)\cdot 10^5$ the Cu/Fe ratio is at a minimum, and with an increase to $(2{\div}4)\cdot 10^6$ the Cu/Fe ratio increases threefold, and as the dilution factor increases from $(0.6÷2)·10^5$, where the Zn/Fe and Cd/Fe ratios are minimum, to $(2÷4)·10^6$, the Zn/Fe and Cd/Fe ratios increase sixfold and twentyfold respectively (Fig. 4.21). This indicates that when the dilution factor is $(2÷4)·10^6$, the ocean water is the source of about ≥75%, ≥85%, and ≥95% of the Cu, Zn, and Cd respectively in the particulate matter that is bound with the hydrothermal material.

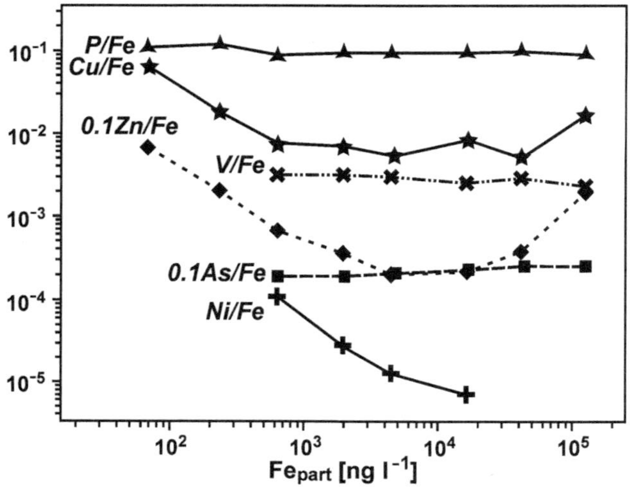

Fig. 4.22. Element/Fe ratios in particulate matter in relation to the concentrations of the particulate Fe in the nonbuoyant plumes over the axial part of the East Pacific Rise from 13°33'S to 18°40'S. Calculated and based on the data from Feely et al. (1996).

As the dilution factor increases from $2·10^3$ to $2·10^6$ in the transition from the buoyant to the nonbuoyant plume and subsequently in the nonbuoyant plume over the TAG hydrothermal field the concentration of particulate Pb decreases steadily (Table 4.4), and the Pb/Fe ratio increases steadily (Fig. 4.21). This is an indication of the continual passage of dissolved Pb into the particulate matter. In the nonbuoyant plume the increase in the Pb/Fe ratio occurs when the dilution factor exceeds $D_{50\%}$ for Pb (Fig. 4.21), which is evidence indicating that ocean water is the main source of the particulate Pb bound in the hydrothermal matter in the nonbuoyant plume. When the dilution factor reaches $(2÷4)·10^6$ the ocean water is the source of >90% of the Pb in the particulate matter that is bound with the hydrothermal material.

In the nonbuoyant plume over the TAG hydrothermal field at increase in the dilution factor up to $(2÷6)·10^5$ the concentrations of particulate Co, Ni, and La decrease. Where the dilution factors have higher values the concentration of Co_{part} continues to decrease, the La_{part} concentration does not vary significantly, and the concentration of Ni_{part} does not vary until the dilution factor exceeds ~$2 \cdot 10^6$ when it starts to increase (Table 4.4). In the nonbuoyant plumes over the southern East Pacific Rise the Ni_{part} concentration exhibits a similar pattern (Table 4.5) at decrease in the Fe_{part} concentration.

The Co/Fe, Ni/Fe, and La/Fe ratios in the particulate matter in the nonbuoyant plume over the TAG hydrothermal field increase continuously as the dilution factor increases along with the decrease in the concentration of Fe_{part} (Fig. 4.21). These increases are evidence of the continuous passage of Co, Ni, and La from solution into the particulate matter, and the Ni/Fe ratio exhibits a similar pattern in the nonbuoyant plumes over the southern East Pacific Rise. The Co/Fe, Ni/Fe, and La/Fe ratios increase and become greater by one and a half orders of magnitude or more when the dilution factors are much higher than $D_{50\%}$ for Co and especially for Ni and La (Fig. 4.8, 4.21). This is evidence, which shows that ocean water is the main source of these elements in the particulate matter of the nonbuoyant plume. When the measured dilution factors in the nonbuoyant plume reach a maximum the ocean water becomes the source of 90–95% or more of the particulate Co, Ni, and La bound with the hydrothermal matter.

With an increase in the dilution factor and a decrease in the concentration of Fe_{part} that take place in the transition from buoyant to nonbuoyant plumes and subsequently in the nonbuoyant plumes, the concentrations of particulate P, V, and As decrease continuously (Tables 4.4, 4.5). The P/Fe, V/Fe, and As/Fe ratios vary little within the nonbuoyant plumes; the V/Fe ratio shows a slight increase and the As/Fe ratio decreases very little (Fig. 4.21, 4.22). The very low variation in these ratios is evidence showing that there is very little transfer of dissolved P, V, and As into the particulate matter in the nonbuoyant plumes. The rapid process within the buoyant plume of co-precipitation of the oxyanions of P, V, and As has ceased (Sect. 4.3.1), and the subsequent sorption of these elements from the ocean water in the nonbuoyant plume is insignificant. Presumably this process plays some role for V only, which is indicated by the small increase in the V/Fe ratio in the particulate matter as the dilution factor increases and the concentration of Fe_{part} decreases in the nonbuoyant plumes over both the TAG hydrothermal field and the axial part of the East Pacific Rise (Fig. 4.21, 4.22). The V/Fe ratio stabilizes when the dilution factors measured are highest. This is evidence indicating that scavenging of V from the ocean water by Fe-oxyhydroxides has ceased or it is imperceptible. The slight decrease in the As/Fe ratio associated with an increase in the dilution factor and the decrease in the concentration of Fe_{part} in the nonbuoyant plumes both over the TAG hydrothermal field and the southern East Pacific Rise (Fig. 4.21, 4.22) apparently is evidence of the settling of small amounts of remnant sulfide minerals that entered the nonbuoyant plumes from the buoyant plumes.

According to laboratory experiments on the co-precipitation and sorption of P from ocean water by hydrothermal Fe-hydroxides (Berner 1973), the first process ceases quickly within a few hours, and the second process is much slower as equilibrium is not indicated even in 210 days in the experiments. The final P/Fe ratio measured in the experiments varied from 0.08 to 0.12 and coincided with the P/Fe ratio in the particulate matter of the nonbuoyant plumes over the North Cleft segment which varies from 0.08 to 0.11 (Feely at al. 1994), and over the axial part of the southern East Pacific Rise where it varies from 0.085 to 0.12 (Fig. 4.22; Feely et al. 1996).

The main features in the chemical compositions and the physical characteristics of the high-temperature hydrothermal fluids in the open rifts (that are not covered with sedimentary strata) in different areas of the ocean are similar (Sects. 4.1 and 4.2). In these areas the main characteristic physical and chemical features of the bottom ocean water are also similar. It is evident that in different areas of the ocean the general development of the processes that are caused by the mixing of hydrothermal fluids and bottom ocean water, including the processes in the buoyant and nonbuoyant plumes, are similar. Minerals formed in the buoyant plumes either settle in the vicinity of the hydrothermal vents or migrate into the nonbuoyant plumes. The dispersion of the hydrothermal material and its settling to form the main fields of metalliferous sediment occur mainly from the nonbuoyant plumes. The extent to which the processes take place in the nonbuoyant plumes as well as the size of the plumes depend on the volume and intensity of the contribution of high-temperature hydrothermal fluids to the ocean. In general with an increase in the volume and contribution of hydrothermal fluids the amount of ocean water needed to achieve the same dilution of the hydrothermal fluids and the area of nonbuoyant plumes must have increased.

With other factors being equal, the intensity of the contribution of hydrothermal matter into the ocean in any area is reflected by the amount of Fe_{part} and the total amount of dissolvable manganese, TDM, within a nonbuoyant plume (Baker and Massoth 1987; Nelsen et al. 1986/1987; Fig. 4.23). Therefore comparisons of areas with "hydrothermal" concentrations of Fe_{part} (>3000 ng l^{-1}) and TDM (>400 ng l^{-1}) that occur where the dilution factor is $<(0.6\div2)\cdot10^5$ in the nonbuoyant plumes in three areas of modern hydrothermal activity are illustrative: a) within the TAG hydrothermal field in the slow-spreading Mid-Atlantic Ridge there is one active hydrothermal mound that has several hydrothermal vents; the nearest hydrothermal vents known are ~180 km to the south and ~330 km to the north of the field; b) over the North Cleft segment of the medium-spreading Juan de Fuca Ridge there are four hydrothermal vents within an area that is about 3 km long, the nearest hydrothermal vents known occur ~35 km to the south; c) 10 km to the east and to the west of the axis of the very fast-spreading East Pacific Rise between 21°20'S and 22°40'S, the exact number of hydrothermal vents is not yet known within this area that is about ~150 km long; however, according to indirect evidence, there are a great number.

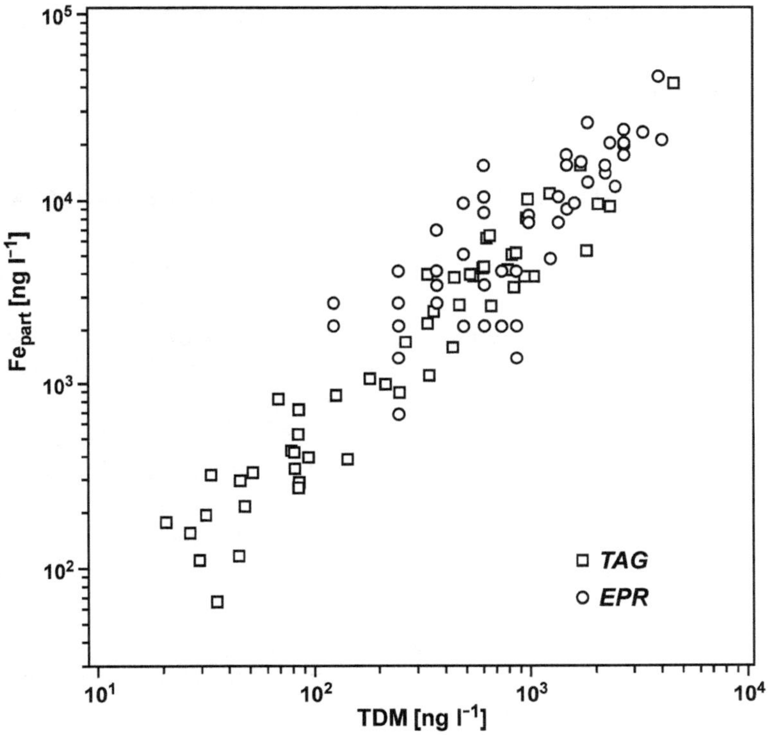

Fig. 4.23. Concentration of Fe_{part} in relation to the concentration of TDM in the nonbuoyant plumes over the TAG hydrothermal field and the axial part of the East Pacific Rise from 21°20'S to 22°40'S. Based on the data from Klinkhammer et al. (1986), Trocine and Trefry (1988), German et al. (1991), Krasnov et al. (1992b).

The area with the "hydrothermal" concentrations of Fe_{part} and TDM in the nonbuoyant plume over the TAG field is about 15 to 30 km^2 (Sudarikov et al. 1992; Rudnicki and Elderfield 1993), over the North Cleft segment the area is about 250 km^2 (Feely et al. 1994), and over the East Pacific Rise between 21°20'S and 22°40'S it is about 3000 km^2 (Krasnov et al. 1992b). This is in agreement with: a) the relationships of hydrothermal venting probability with spreading rate; b) the relationships of plume incidence with spreading rate (Sect. 4.1); c) the increase in the accumulation of hydrothermal metal-bearing matter in the bottom sediments in a sequence: Mid-Atlantic Ridge → Juan de Fuca Ridge → East Pacific Rise.

Studies of the TDM, $^3He/^4He$, and temperature anomalies in ocean water at depth of about 2500 m have shown that in the Southeast Pacific nonbuoyant plumes or their fragments occur at distances up to 2000 km to the west of the EPR axis (Klinkhammer and Bender 1980; Klinkhammer and Hudson 1986; Lupton and Craig 1981; Lupton 1995; Lupton et al. 2004). This is comparable to the distance from the EPR axis where recent metalliferous sediments occur in the Southeast Pacific (Sect. 1.1.1).

4.4. Settling fluxes of hydrothermal particulate matter

Not only the concentrations of particulate hydrothermal matter, the contents of chemical elements bound in it but also their settling fluxes vary with the distance from hydrothermal vents. It is important in this study to trace the variations in the settling fluxes of the hydrothermal matter and the components that are bound in this matter as distances increase from the hydrothermal vents. Up to now only a few measurements have been made of these variable factors in areas of modern hydrothermal activity by using sediment traps that were located at measured distances from the hydrothermal vents and fields. The data for several areas are shown in Table 4.6. They include measurements of fluxes near the hydrothermal vents and at different distances from the vents and hydrothermal fields.

The fluxes of almost all chemical elements are higher near the hydrothermal vents than at distance away from them. The fluxes are higher within the nonbuoyant plumes and below them than above the plumes (Dymond and Roth 1988; Lukashin et al. 1999). Variations in the fluxes of Fe, Mn, Cu, and Zn bound in the hydrothermal matter with increasing distance from the hydrothermal vents in the areas of hydrothermal activity on the Endeavor Ridge and at 13°N EPR are shown in Fig. 4.24. The fluxes of all the elements tend to decrease with distance from the hydrothermal vents, but the relative decreases in the fluxes of individual elements are different. This is evidence of the differentiation of the elements with distance from the hydrothermal vents. The differentiation of the elements can be seen in a graph of the variations in the ratios of the fluxes of Mn, Cu, and Zn to the flux of Fe in relation to the increasing distance from the hydrothermal vents (Fig. 4.25).

Close to the hydrothermal vents, <0.5 to 4 m, the Cu/Fe ratio is relatively high; 5 to 20 m away from the vents it is lower, and further away the ratio increases gradually. The Zn/Fe ratio close to the hydrothermal vents is lower than the ratio in matter collected 5 to 20 m away from the vents. The Zn/Fe ratio decreases and has minimum values at locations about 80 to 300 m from the vents and at locations further distant the ratio increases gradually. The Mn/Fe ratio increases gradually with distance from the hydrothermal vents.

Table 4.6. Settling fluxes of chemical elements at different distances from hydrothermal vents and fields.

Distance, m	Fe ($mg\ cm^{-2}\ ka^{-1}$)	Mn ($mg\ cm^{-2}\ ka^{-1}$)	Cu ($mg\ cm^{-2}\ ka^{-1}$)	Zn ($mg\ cm^{-2}\ ka^{-1}$)	Ref.
Endeavor Ridge					
21 m above the bottom					
4	4367	7.2	332	285	1
lower boundary of the nonbuoyant plume					
2000	218	3.4	14.6	44.9	1
upper part of the nonbuoyant plume					
2000	74.5	0.79	4.3	11.7	1
13°N EPR					
1.5 m above the bottom					
2.5	3836	198	94	495	2
80	836	16	9.4	34	2
<0.5	84600	63	317	389	7
1.5	3210	10	22	1020	7
5	2050	6	14	539	7
12	739	20	5	169	7
20	3400	31	29	567	7
5 m above the bottom					
300	79	9.2	1.55	2.2	7
50 m above the bottom					
15000	66	1.8	1.2	6.2	2
North Cleft segment, JDF					
9 m above the bottom					
10	113	36			4
18–150 m above the bottom					
~500	28	3.2			4
Rainbow field, MAR					
1.5 m above the bottom					
2	18100		12600		
150 m above the bottom					
500	93		38		5
1000	5.3		2.9		5
TAG field, MAR					
2400	47.1				3
Broken Spur field, MAR					
0.9 m above the bottom					
3	24300		460	460	6
38 m above the bottom					
300	2.4			0.22	6
33 m above the bottom					
2500	6.9			1.3	6

References: 1 – Dymond and Roth (1988), 2 – Khripounoff and Alberic (1991), 3 – Demina et al. (1992), 4 – Feely et al. (1994), 5 – Khripounoff et al. (2001), 6 – Lukashin et al. (1999), 7 – German et al. (2002).

The increase in the Cu/Fe ratio close to the vents apparently is evidence of the settling of Cu-bearing sulfide particles from the buoyant plumes. This settling ceases within 5 to 20 m from the vents. The consequent increase in the Cu/Fe ratio results from scavenging of the dissolved Cu by the Fe-oxyhydroxides. Maximum settling of the Zn-bearing sulfide minerals obviously takes place close to the vents, but some further away from the vents than the Cu-bearing sulfide minerals settle. The settling of Zn-bearing sulfide minerals ceases at further distances, 20 to 80 m, from the vents. The increase in the Zn/Fe ratios from 300 to 2000 m and at least as far as 15 km from the vents apparently is evidence of the scavenging of the dissolved Zn by the Fe-oxyhydroxides. The gradual increase in the Mn/Fe ratio apparently results from the gradual oxidation and scavenging of Mn by the Fe-oxyhydroxides. The role of oxidized and scavenged Mn increases with time (Sects. 4.3.1, 4.3.2), and the older the hydrothermally derived settling particulate matter is the richer it is in Mn. In general the greater distance is from the vents the older it is.

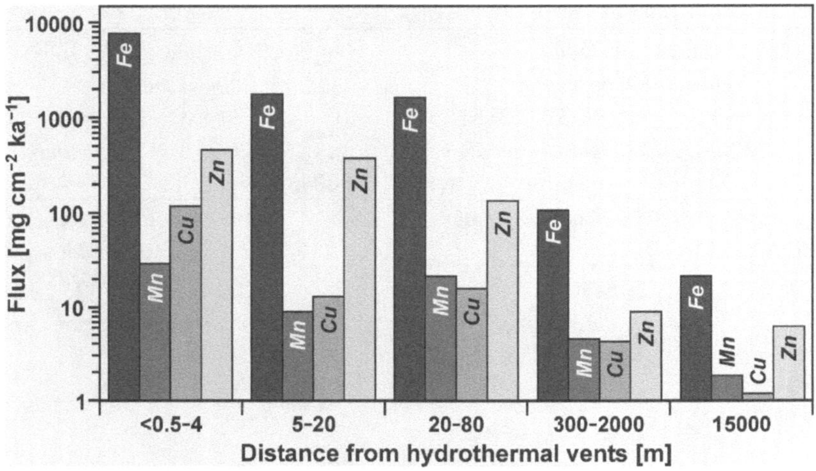

Fig. 4.24. Average settling fluxes of Fe, Mn, Cu, and Zn that are bound in the hydrothermal matter in relation to increasing distances from the hydrothermal vents as shown by the average for data from areas of hydrothermal activity on the Endeavor Ridge and at 13°N EPR. Calculated and based on data from Dymond and Roth (1988), Khripounoff and Alberic (1991), German et al. (2002).

The data from Table 4.6 enable quantitative estimates to be made of the variations in the settling fluxes of hydrothermal matter in relation to the increasing distances from hydrothermal vents. Plots of the settling fluxes of the hydrothermal Fe (Fe_h*) in relation to the distances from the hydrothermal vents are shown in Fig. 4.26 for the hydrothermal fields and adjacent areas of the Endeavor Ridge, 13°N EPR[35], North Cleft segment of the Juan de Fuca Ridge, as well as for the Broken Spur and Rainbow hydrothermal fields of the Mid-Atlantic Ridge.

In spite of the different settling fluxes of the hydrothermal Fe at similar distances from the vents in different hydrothermal fields the characteristics

[35] The data for the flux measured at distance <0.5 m from the vent (Table 4.6) have not been taken into consideration because the measurement of the distance was not precise.

of the relative variations in the fluxes with increasing distances from vents are similar in the data for the Pacific Ocean and in the data for the Atlantic Ocean. The slopes of the regression lines for the data from the Pacific Ocean (the Endeavor Ridge, 13°N EPR, and North Cleft segment) vary from –0.45 to –0.59 and for the data from the Atlantic Ocean (the Rainbow and Broken Spur hydrothermal fields) from –1.23 to –1.43 (Fig. 4.26).

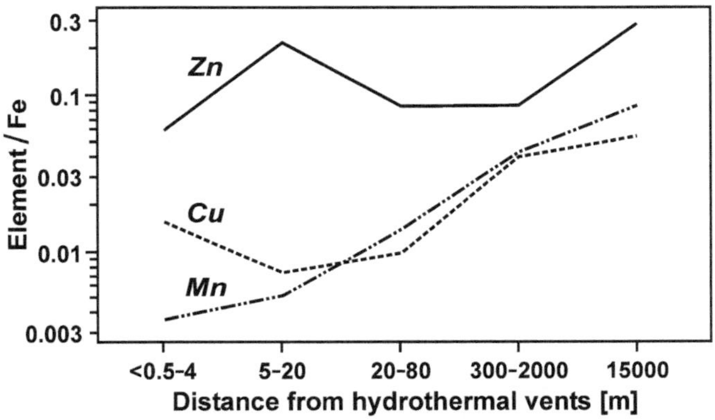

Fig. 4.25. Ratios of the settling fluxes of Mn, Cu, and Zn bound within the hydrothermal matter to the settling flux of hydrothermal Fe related to increasing distance from hydrothermal vents, using the average for data from the areas of hydrothermal activity on the Endeavor Ridge and at 13°N EPR. Calculated and based on the data from Dymond and Roth (1988), Khripounoff and Alberic (1991), German et al. (2002).

Taking notice of this, the average relative variations have been calculated for the settling fluxes of Fe_h in relation to the distances from the hydrothermal vents for the hydrothermal fields in the Pacific Ocean and for the hydrothermal fields in the Atlantic Oceans. The closest measured distances from the vents for the settling fluxes of Fe_h that have been measured vary from 1.5 m to 10 m (Table 4.6) and a distance of 2.5 m has been used as a base distance for further calculations. The settling fluxes of Fe_h at distances of 2.5 m from the vents ($Fe_h{}^*{}_{2.5}$) have been calculated from the regression equations for each area (Fig 4.26). $Fe_h{}^*/Fe_h{}^*{}_{2.5}$ ratios have been calculated for each area. The results of these calculations are plotted on Fig. 4.27. The relative variations of the settling flux of Fe_h in relation to the distance from the hydrothermal vents for 12 points from the Pacific Ocean are described by the regression equation (R=0.95):

$$\lg(Fe_h{}^*/Fe_h{}^*{}_{2.5}) = 0.23 - 0.56\lg D, \qquad (4.1)$$

and for 6 points from the Atlantic Ocean, by the regression equation (R= 0.94):

$$\lg(Fe_h^*/Fe_{h\ 2.5}^*) = 0.54 - 1.34\lg D, \quad (4.2)$$

where D is distance from hydrothermal vents, m.

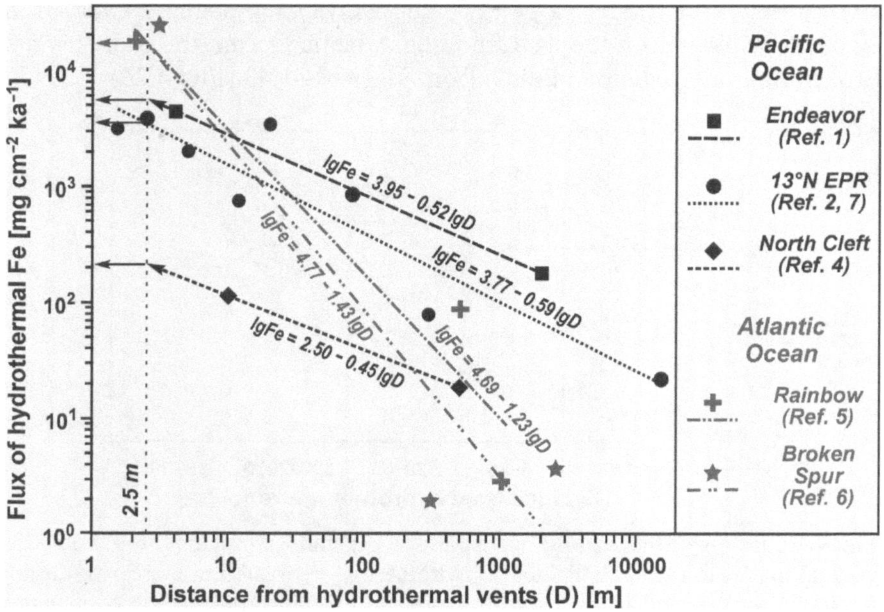

Fig. 4.26. Plots of the settling fluxes of hydrothermal Fe in relation to the distance from the hydrothermal vents for the hydrothermal fields and adjacent areas of the Pacific and Atlantic Oceans. Straight lines are regression lines for groups of data named in the legend; references are the same as for Table 4.6. Based on the data from Dymond and Roth (1988), Khripounoff and Alberic (1991), Feely et al. (1994), Lukashin et al. (1999), Khripounoff et al. (2001), German et al. (2002).

In spite of the few points where data are available, the evidence shows that the settling flux of the Fe_h derived from the Broken Spur and Rainbow hydrothermal fields in the Atlantic Ocean decreases as the distance from the vents increases to a much greater extent than in the Pacific Ocean (Fig. 4.26, 4.27). According to calculation from the regression equations (4.1) and (4.2), at 100 m distance from vents the difference is ~20-fold, at 500 m ~60-fold, at 1000 m ~110-fold, at 2000 m ~180-fold. Differences such as these cannot be explained by the additional fluxes of Fe_h derived from vents in the neighboring hydrothermal fields. They can increase the flux, but not to the extent indicated, because there are not enough additional sources of Fe_h located near the hydrothermal fields in the Pacific Ocean where the settling fluxes were measured and discussed here. Presumably the much higher rates of oxidation of the Fe(II) in the hydrothermal fields in the Atlantic Ocean (Table 4.3) could be a reason for this difference. Further studies are needed to solve this problem. If it can be confirmed that there is a much greater decrease in the settling flux of Fe_h as the distances increase from the hydrothermal vents in the Atlantic Ocean, one can feel assured that one more factor influences the dispersion of hydrothermal matter and the size of the fields of metalliferous sediments.

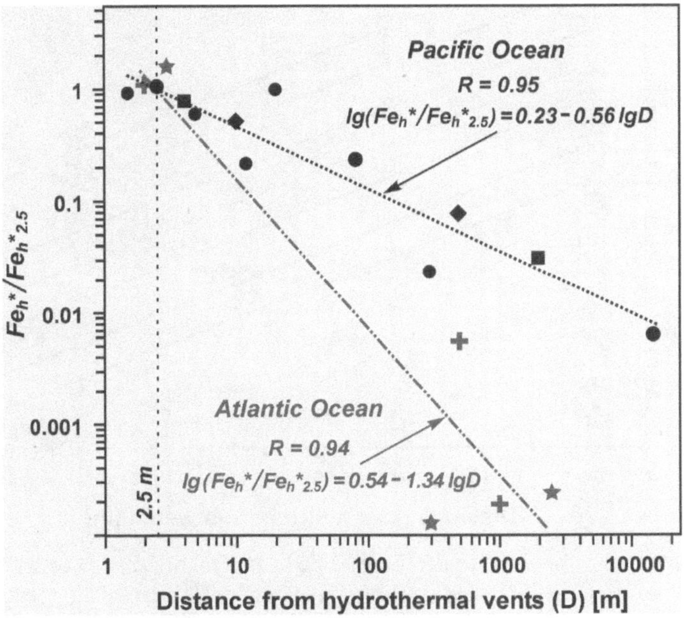

Fig. 4.27. Plots of calculated $Fe_h^*/Fe_h^*{}_{2.5}$ ratios for settling fluxes of hydrothermal Fe in relation to distance from hydrothermal vents for the hydrothermal fields and adjacent areas of the Pacific and Atlantic Oceans. The regression line for the Pacific Ocean is based on the data for the Endeavor Ridge, North Cleft segment, and 13°N EPR; the regression line for the Atlantic Ocean is based on the data for the Broken Spur and Rainbow hydrothermal fields. The symbols are the same as in Fig. 4.26.

The regression equation (4.1) for the data from the Pacific Ocean enables estimates to be made of the settling fluxes of Fe_h as distances increase from the vents when the different settling flux values for Fe_h were measured at a distance of 2.5 m from the vents (Fig. 4.28).

It should be emphasized that the settling flux of the Fe_h at some distance from a vent can result from this vent as well as from other vents in the area, if they exist. An increase in the Fe_h settling flux can result from both an increase in the hydrothermal contribution from an individual vent and from the increase in flux from a number of vents. The settling flux of the Fe_h at a long distance from points of hydrothermal contribution apparently has characteristics that are an integration of the intensity of the hydrothermal activity in an area or a large segment of a spreading ridge and is not representative of the hydrothermal contribution of individual vents.

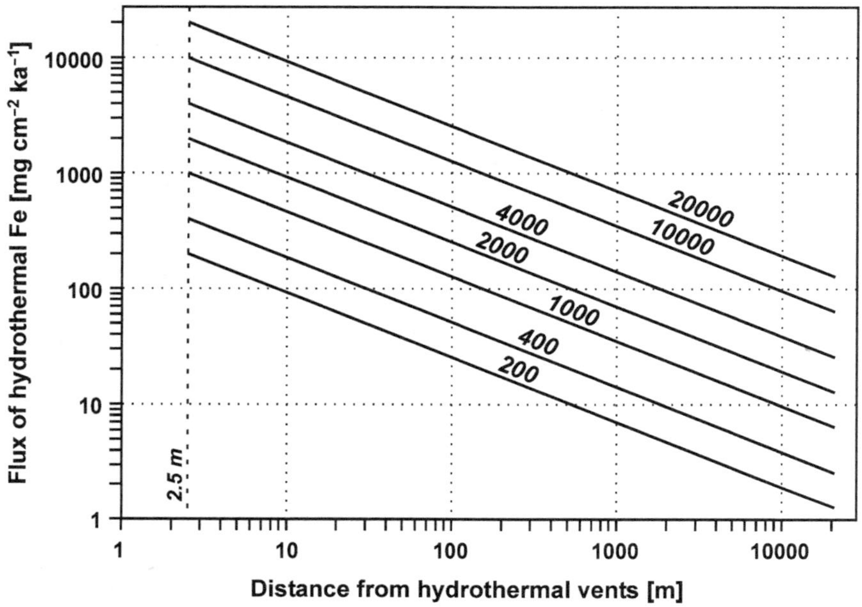

Fig. 4.28. Settling fluxes of hydrothermal Fe at increasing distances from the hydrothermal vents when different settling fluxes of hydrothermal Fe were measured at 2.5 m from the vents. Estimations for Pacific Ocean areas.

4.5. Differentiation of chemical elements during formation of metalliferous sediments in the ocean

At different stages of hydrothermal mineral formation and sedimentation as the distance increases away from the hydrothermal vents the chemical elements bound in the hydrothermal metal-bearing matter exhibit varied behavior that results to their differentiation at these stages.

The differentiation of the elements is reflected in the composition of the metal-bearing matter in the sequence that extends from the high-temperature hydrothermal fluids to distal metalliferous sediments. Members of this sequence include: primary hydrothermal fluids, metalliferous sediments on and near massive hydrothermal edifices, particulate matter of buoyant plumes, proximal metalliferous sediments including metalliferous sediments accumulated in hydrothermal fields up to ≤1–2 km from hydrothermal vents and massive hydrothermal accumulations, particulate matter of nonbuoyant plumes within and in the vicinity of hydrothermal fields (recognizing that there are very few data for the nonbuoyant plumes that are far away from the hydrothermal fields), and distal metalliferous sediments. The differentiation can be demonstrated by using a geochemical reference element. In this case Fe appears to be the best reference element. It is the main chemical component in the metal-bearing matter in all of the members in the sequence from primary hydrothermal fluids to the distal metalliferous sediments that accumulated at distances up to many hundreds of kilometers from the hydrothermal fields. In the first step in the calculation of factors for the differentiation of chemical elements in relation to Fe, the average contents of the elements in the metal-bearing matter in all of the members of the sequence were divided by the respective contents of Fe. Then the quotients obtained were normalized in relation to the respective quotients for the primary hydrothermal fluids. Where chemical elements have differentiation factors in some members of the sequence that are >1, it is evident that these elements are enriched in comparison to the Fe relative to composition of the primary hydrothermal fluids, and if they are <1 that the elements are depleted. The differentiation factors are equal to 1 for all of the chemical elements in the primary hydrothermal fluids and for Fe in all the members of the sequence. Calculations for the differentiation factors for Mn, Cu, Zn, Co, Pb, Ni, La, and V in the sequence of members are shown in graphic form in Fig. 4.29.

There are no data at the present time that can enable the calculations of the differentiation factors for these chemical elements in all of the members in a sequence in a single separate area of hydrothermal contribution to the ocean. For this reason, the composition of the primary hydrothermal fluids has been calculated from data for the hydrothermal fields on the East Pacific Rise and for the basalt hosted fields on the Mid-Atlantic Ridge, the composition of the metalliferous sediments on or near hydrothermal edifices from the data for the 21°N EPR and TAG hydrothermal fields, the composition of particulate matter in the buoyant plumes from the data for the 21°N EPR, 13°N EPR, and TAG hydrothermal fields as well as for the Monolith vent, the composition of proximal metalliferous sediments from the data for the 21°N EPR and TAG hydrothermal fields, the composition of particulate matter in nonbuoyant plumes from the data for the EPR, North Cleft segment, TAG, Broken Spur, and MARK hydrothermal fields, and the composition of distal metalliferous sediments from the data for the field with an area of $\sim 2 \cdot 10^6$ km^2 in the Indian Ocean around the Rodriguez Triple Junction and the field with an area of $10 \cdot 10^6$ km^2 in the Southeast Pacific.

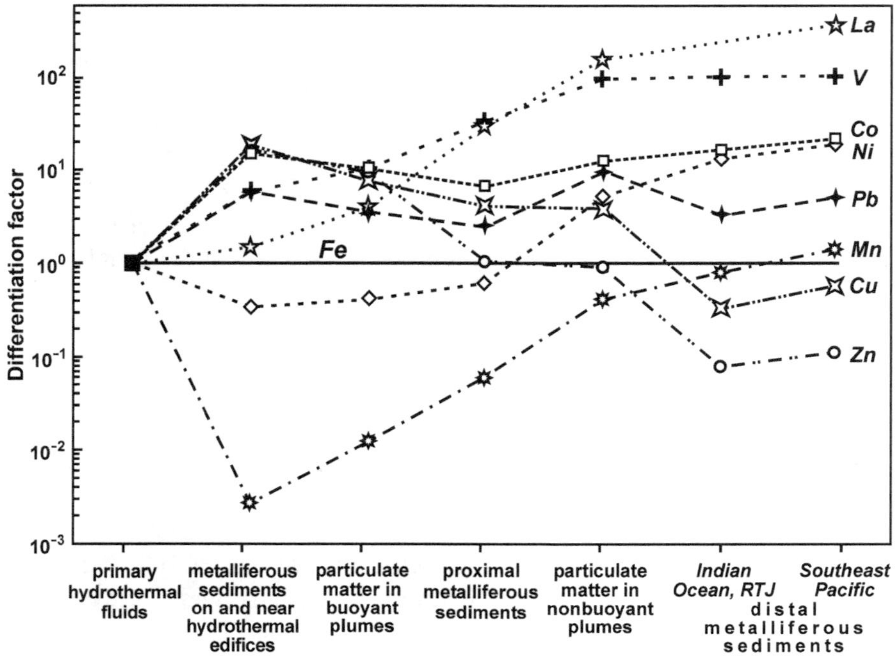

Fig. 4.29. Differentiation factors for chemical elements in the sequence that extends from primary hydrothermal fluids to distal metalliferous sediments.

The calculated differentiation factors should be considered as preliminary only. Nevertheless they reflect concisely the variations in the composition of the hydrothermal metal-bearing matter in the main stages from its entering the ocean to its accumulation in the distal metalliferous sediments.

The differentiation factor for manganese (K_{Mn}) in the metal-bearing matter increases from the metalliferous sediments on or near hydrothermal edifices to the end of the sequence in the distal metalliferous sediments. However, it is considerably less than 1 in the stages of the sequence before the particulate matter in the nonbuoyant plumes (Fig. 4.29). It is evident that there is much less fixation of Mn in comparison to that of Fe as a component of the metal-bearing matter, and a prevalent part of the Mn is transported outside the hydrothermal fields in solution. The K_{Mn} is somewhat lower than 1 in the distal metalliferous sediments from the Indian Ocean and it is considerably higher than in other members of the sequences located within hydrothermal fields and in the particulate matter from the nonbuoyant plumes. But it is lower than the K_{Mn} in the distal metalliferous sediments from the Southeast Pacific, which is the only area where the K_{Mn} exceeds 1. This evidence indicates that the dissolved Mn passes into solid phases of the hydrothermal metal-bearing matter mainly outside the hydrothermal fields and that there is continuous enrichment of Mn in the metal-bearing matter as the distance increases from the hydrothermal fields.

The question emerges as to how the dissolved Mn passes into a solid phase outside the hydrothermal fields. Does the dissolved Mn pass mainly into the particulate matter in the nonbuoyant plumes, or into the particulate matter under the nonbuoyant plumes, or after the metal-bearing matter is deposited in the metalliferous sediments on the seafloor? Studies of the Mn in nonbuoyant plumes (Sect. 4.3.2) provide evidence which shows that only in the high productive zones or locally in areas where there are high concentrations of volatile gases are the Mn_{part}/Fe_{part} ratios comparable with the Mn/Fe ratios in the metal-bearing matter in the distal metalliferous sediments, 0.25 to 0.35 (Fig 4.18, 4.20), and usually they are lower. The Mn_{part}/Fe_{part} ratio in particulate matter under the nonbuoyant plumes is considerably higher than inside the plumes. The measurements in the low-productive zone as far as 80 km to the west of the East Pacific Rise axis at 15°S have shown that the Mn_{part}/Fe_{part} ratio under the nonbuoyant plume, 0.1–0.12, is about three times higher than within the plume (Fig. 4.19). But this ratio is from 2 to 3 times lower than the Mn/Fe ratio that is about 0.3 in the distal metalliferous sediments that accumulated some tens of kilometers to the west of the East Pacific Rise (Fig. 1.5, 1.7; Boström and Peterson 1969; Dymond 1981; Kunzendorf et al. 1984/1985; Müller et al. 1985; Walter and Stoffers 1985; Gurvich 1988; Dekov 1994; Sudarikov et al. 1995; Renner et al. 1997). Presumably this evidence indicates that within some tens of kilometers from the EPR axis a considerable part of the Mn content bound in the metal-bearing matter has passed from solution into the distal metalliferous sediments after their deposition.

Unfortunately, no data are available to the writer on the concentrations of Mn_{part} and Fe_{part} in single samples from the nonbuoyant hydrothermal plumes located at distances far from the EPR axis. However there are data on the concentrations of Mn_{part} and of Al_{part} (which is considered as a proxy of the background non-biogenic particulate material) in fragments of nonbuoyant hydrothermal plumes located at distances far from the EPR axis in the low-productive zone near 20°S (Fig. 4.30). As the distance increases to the west of the EPR axis there is little variation in the concentration of Al_{part}, but the concentration of Mn_{part} and the Mn_{part}/Al_{part} ratio decrease considerably. The Mn/Al ratio in the distal metalliferous sediments decreases with the distance from the EPR axis, but at similar distances from the EPR axis it is considerably higher than the ratio for Mn_{part}/Al_{part}.

There are few data for the composition of the particulate matter under the nonbuoyant plumes at long distances from the EPR axis. In one sample, collected 40 m above the bottom and at a distance of 1000 km from the EPR axis, in which an admixture of re-suspended material from the bottom is absent, the Mn_{part}/Al_{part} ratio is 0.48, in another sample collected 73 m above the bottom at a distance of 260 km from the EPR axis the Mn_{part}/Al_{part} ratio is 0.058 (Klinkhammer and Hudson 1986). The ratio for the first sample is comparable to the Mn/Al ratio in sediments, but in the second sample it is much lower. The available data provide evidence to show that in the low-productive areas of the ocean a small part of the Mn accumulated in the metal-bearing matter of the distal metalliferous sediments passes from solution into solid phases of the particulate matter in the nonbuoyant plumes. The data available do not provide an answer to question where the Mn mostly passes into the solid phase from solution: under the plume into particulate matter or into the sediments after they are deposited. Additional measurements are needed from samples of the particulate matter under the plumes and from samples of the distal metalliferous sediments collected at the same locations at distances far from the hydrothermal fields.

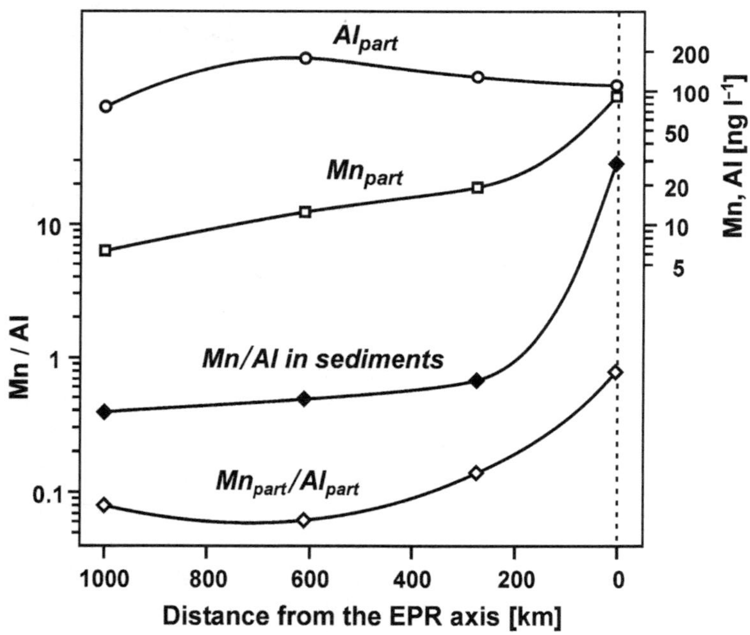

Fig. 4.30. Concentrations of particulate Mn and Al and their ratio in nonbuoyant hydrothermal plumes, as well as Mn/Al ratio in metalliferous sediments at increasing distances to the west of the East Pacific Rise near 20°S. Based on information from Klinkhammer and Hudson (1986) and original data.

Because of the bacterial activity in mid- and high-productive areas of the ocean the intensity to which the dissolved Mn passes into solid phases of the particulate matter in the nonbuoyant plumes is much higher than in the low-productive areas (Sect. 4.3.2). The manganese content in metalliferous sediments in these areas is influenced further by the Eh conditions.

In the metalliferous sediments on or near the hydrothermal edifices the differentiation factors for Cu, Zn, Pb, and Co (K_{Cu}, K_{Zn}, K_{Pb}, and K_{Co}) are much higher than 1 and the factors for Cu and Zn have maximum values for the whole sequence, and they decrease in the particulate matter of the buoyant plumes and in the proximal metalliferous sediments. These high values result from the extensive formation of sulfide minerals containing Cu, Zn, Pb, and Co rather than the development of Fe-minerals in the lower parts of the buoyant plumes (Sect. 4.3.1). The material in the metalliferous sediments on or near the hydrothermal edifices settled from the lower parts of the buoyant plumes, and the average particulate matter in the buoyant plumes is lower in Cu, Zn, Pb, and Co than the particulate matter in the lower parts of these plumes. The K_{Cu}, K_{Zn}, K_{Pb}, and K_{Co} factors in the proximal metalliferous sediments are lower than in the particulate matter in the buoyant plumes. This results from the settling of a considerable amount of sulfide minerals near the hydrothermal vents and by an increase in the amount of Fe-oxyhydroxides that have been precipitated and have not scavenged any significant amounts of Cu, Zn, Pb, and Co from solution at this stage. The average K_{Cu} and K_{Zn} factors are lower and K_{Pb} and K_{Co} factors are higher in the particulate matter of the nonbuoyant plumes than in the particulate matter of the buoyant plumes. The K_{Cu} and K_{Zn} factors are influenced mainly by the settling of particles of Cu- and Zn-sulfide minerals from the buoyant plumes and from proximal parts of the nonbuoyant plumes. Oxidative dissolution of sulfide minerals is of secondary importance (Rudnicki and Elderfield 1993). The settling of particles of Cu- and Zn-sulfide minerals ceases when the dilution factor is about $(0.6\div6)\cdot10^5$. When the dilution factors are higher in the nonbuoyant plumes the scavenging of Cu and Zn from solution by Fe-oxyhydroxides is prevalent. The scavenging of Pb and Co from solution by the Fe-oxyhydroxides is prevalent when the dilution factors are $>10^3$ (Sect. 4.3.2, Fig. 4.21). In the distal metalliferous sediments the K_{Co} factor is higher than in particulate matter from the nonbuoyant plumes and the K_{Zn}, K_{Cu}, and K_{Pb} factors are lower. The increase in the K_{Co} factor in the distal metalliferous sediments probably is caused by the presence of Mn(IV)-oxyhydroxides, which carry a considerable amount of Co (Sects 1.1.5, 1.4). Probably the K_{Cu}, K_{Zn}, and K_{Pb} factors in the transition from the particulate matter of the nonbuoyant plumes to the distal metalliferous sediments are influenced by the partial desorption of Cu, Zn, and Pb from the metal-bearing matter. This is in agreement with the experimentally demonstrated sequence of ion-exchange mobility of metals:

Mn(IV) << [Co(III), Ce(III)] << La < [Ni, Co(II)] < [Pb, Cu, Cd, Zn]

(Chelishchev et al. 1992; Novikov 1994). The distal metalliferous sediments from the Southeast Pacific have higher values for the K_{Co}, K_{Cu}, K_{Zn}, and K_{Pb} factors than those from the Indian Ocean (Fig. 4.29). Apparently this is because the width of the metalliferous sediment field in the Southeast Pacific is greater than that in the Indian Ocean and as the distances from ridge axes increase the metal-bearing matter is continuously enriched in elements scavenged from ocean water during migration from the axes to places of deposition.

Where V enters the particulate metal-bearing matter mainly by the co-precipitation of Fe-hydroxides from the ocean water (Sect. 4.3.1) an increase in the differentiation factor (K_V) from the hydrothermal fluids to that in the particulate matter of the nonbuoyant plumes is observed (Fig. 4.29). The particulate matter of the nonbuoyant plumes scavenges scarcely any of the V from ocean water (Sect. 4.3.2, Fig. 4.21, 4.22), however it contains particles of Fe-oxyhydroxides that have scavenged V during their formation in the buoyant plumes. Further, the K_V factor values are almost invariable and do not refute the evidence for joint migration of particulate Fe and V in the components of the metal-bearing matter outside of the hydrothermal fields prior to their deposition on the ocean bottom.

The differentiation factor for Ni (K_{Ni}), which is not accumulated extensively in hydrothermal sulfide minerals (Stepanova and Krasnov 1992), is <1 in the metalliferous sediments on and near the hydrothermal edifices, in the particulate matter of the buoyant plumes, and in the proximal metalliferous sediments. Because of the sorption of Ni from the ocean water by the Fe-oxyhydroxides, the intensive incorporation of it into the particulate hydrothermal metal-bearing matter begins in the upper horizons of the buoyant plumes and continues to be very active in the nonbuoyant plumes (Fig. 4.21, 4.22), not only within the hydrothermal fields but also outside of them. Apparently this results in a steady increase in the K_{Ni} factor from the particulate matter of the buoyant plumes to the distal metalliferous sediments of the Indian Ocean and then of the Southeast Pacific. The Mn-oxyhydroxides, which contain a considerable amount of the Ni in the distal metalliferous sediments (Fig. 1.49), obviously play an important role in its accumulation as well.

The differentiation factor for La (K_{La}) increases steadily from the hydrothermal fluids to the distal metalliferous sediments. This increase begins in the lower parts of the buoyant plumes at the moment when the Fe-hydroxides begin to form, and the co-precipitation of the REE from the hydrothermal fluids and the ambient ocean water takes place. Then their sorption begins. It was mentioned in Sect. 4.3.1 that at the level of neutral buoyancy the average contribution of co-precipitation to the amount of particulate REE is estimated to be 55±20%, and for the La 42% (Rudnicki and Elderfield 1993). The occurrence of some REE contents from hydrothermal sources (indicated in by the positive Eu anomalies) is observed in the metalliferous sediments on and near the hydrothermal edifices, in the particulate matter of the buoyant plumes, and in the proximal metalliferous sediments (Fig. 1.62). However, in most cases the metal-bearing matter in them contains REE that are derived mainly from the ocean water. In the nonbuoyant plumes the sorption of the REE by particles of Fe-oxyhydroxides is predominant. The enrichment of the metal-bearing matter in the REE increases with the time of its contact with ocean water and the K_{La} factor increases likewise. In the distal metalliferous sediments the K_{La} factor is higher than in the particulate matter of the nonbuoyant plumes. There are no data on the average content of La in the distal metalliferous sediments from the Indian Ocean. Presumably, the value of the K_{La} factor in these sediments is somewhere between the values for the particulate matter of the nonbuoyant plumes and those for the distal metalliferous sediments from the Southeast Pacific. The continuous increase in the REE/Fe and La/Fe ratios in the metal-bearing matter with increasing distance from the ridge axes is known from studies of metalliferous sediments from the East Pacific Rise and the Juan de Fuca Ridge (Olivarez and Owen 1989; German et al. 1997).

4.6. Reflection of metallogenic specialization of hydrothermal fluids and near-vent massive accumulations of sulfide minerals in the composition of metalliferous sediments

Studies of the changes in composition of the hydrothermal metal-bearing matter at different stages of hydrothermal mineral formation and sedimentation with increasing distance from the hydrothermal vents have shown significant differentiation of chemical elements including Cu and Zn relative to Fe. Cu and Zn are the main metallic elements of interest in the ocean submarine hydrothermal systems and indicate metallogenic specialization of the hydrothermal fluids and the massive accumulation of sulfide minerals.

Some of the chemical elements contained in the metal-bearing matter of the metalliferous sediments in the proximity of the hydrothermal fields are mainly of hydrothermal origin including Fe, Cu, and Zn. However, during the formation and development of the composition of the metal-bearing matter in the proximal and distal metalliferous sediments some differentiation of Fe, Cu, and Zn takes place. It is of interest to trace the differences in the average ratios of Fe, Cu, and Zn in hydrothermal fluids, in massive accumulations of sulfide minerals, and in proximal and distal metalliferous sediments for specific hydrothermal fields. These comparisons can be made by using the triangular diagram shown in Fig. 4.31.

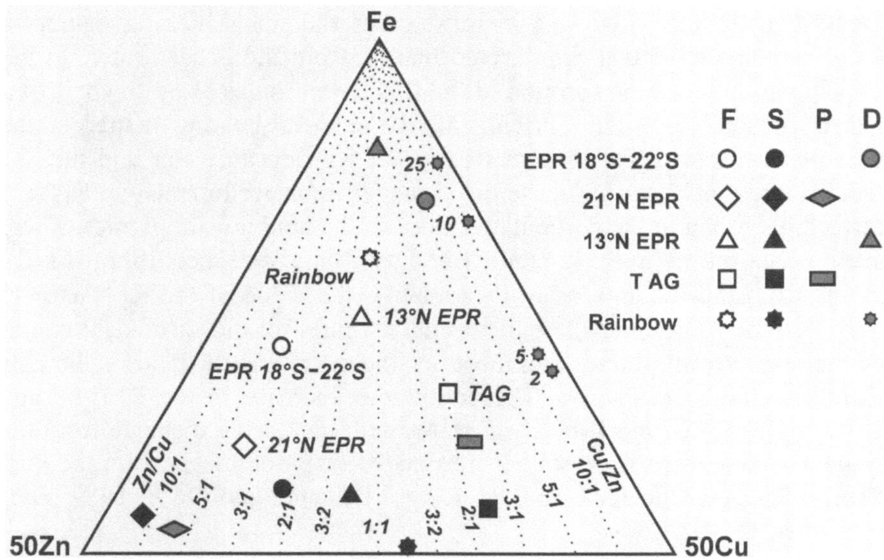

Fig. 4.31. Average ratios of Fe, Cu, and Zn in hydrothermal fluids (F), massive sulfide accumulations (S), and metal-bearing matter of proximal (P) and distal (D) metalliferous sediments. Figures near the points for the Rainbow hydrothermal field show the distances that sediments are from the field. Based on data from Mottl and McConachy (1990), Stepanova and Krasnov (1992), Cherkashev (1992), Dekov (1994), Gurvich (1998), Bogdanov et al. (2002), Cave et al. (2002), Douville et al. (2002).

In the high-temperature hydrothermal fluids from all of the hydrothermal fields relative roles of Fe are higher than in the massive accumulations of sulfide minerals and metal-bearing matter of the proximal metalliferous sediments but lower than in the metal-bearing matter of the distal metalliferous sediments. This is the result of the predominant accumulation of hydrothermal Cu and Zn in the solid phases of the metal-bearing matter in the hydrothermal fields, and a major part of the hydrothermal Fe accumulates outside of the hydrothermal fields. Considering the ratios of Cu and Zn, in spite of their considerable differentiation in specific hydrothermal fields, the hydrothermal fluids, massive accumulations of sulfide minerals, and the metal-bearing matter of the proximal metalliferous sediments show uniform metallogenic specializations: in the TAG hydrothermal field copper is prevalent, in the 21°N EPR hydrothermal field zinc is prevalent, and in the 13°N EPR and Rainbow hydrothermal fields concentrations of copper and zinc are comparable (Fig. 4.31). The data for the TAG and 21°N EPR hydrothermal fields are evidence, which shows that the composition of the metal-bearing matter of the proximal metalliferous sediments reflects metallogenic specialization in the hydrothermal fluids and the massive accumulations of sulfide minerals.

The ratios of Cu and Zn in the metal-bearing matter of the distal metalliferous sediments can be similar (13°N EPR) or very different (EPR 18°S–22°S, Rainbow) to the ratios in the hydrothermal fluids and massive accumulations of sulfide minerals (Fig. 4.31). However, the similarities in the distal metalliferous sediments in the area of 13°N EPR result not from the inheritance of the composition of the hydrothermal fluids and massive accumulations of sulfide minerals, but are related to the similar or close values of the ratios for the Zn and Cu entered into in the metal-bearing matter of the distal metalliferous sediments from seawater. There are no such similarities found in other hydrothermal fields and the associated distal metalliferous sediments. The hydrothermal fluids and massive accumulations of sulfide minerals at 18°S–22°S on the EPR are richer in Zn than in Cu, but the associated distal metalliferous sediments are richer in Cu. Distal metalliferous sediments and even the sediments that are very close to the Rainbow hydrothermal field do not have the same average composition as those in the hydrothermal fluids and the massive accumulations of sulfide minerals in the field (Fig. 4.31). These data provide evidence, which shows that the composition of the metal-bearing matter in the distal metalliferous sediments does not reflect the metallogenic specialization in the hydrothermal fluids and massive accumulations of sulfide minerals as that found in the combined hydrothermal fields.

Iron is the only chemical element of direct hydrothermal origin in the metal-bearing matter of the distal metalliferous sediments. Although the Mn in the metal-bearing matter is mainly of hydrothermal origin, it had been largely assimilated in the ocean water before its accumulation in the distal metalliferous sediments. As for other chemical elements including Cu and Zn, they are predominantly of exogenic origin and accumulate in the metal-bearing matter of the distal metalliferous sediments mainly by processes of sorption and co-precipitation from ocean water. The composition of the distal metalliferous sediments does not reflect an average metallogenic specialization in the hydrothermal fields. Nevertheless, during the formation of the distal metalliferous sediments and metalliferous strata the accumulation rates of the metal-bearing matter and, with other factors being equal or constant, its contents reflect the distances from the hydrothermal fields and the intensity of the hydrothermal activity. This is shown in the next section in the example of the field of the distal metalliferous sediments in the Southeast Pacific, the largest field in the World Ocean.

4.7. Formation of distal metalliferous strata

The largest field of recent distal metalliferous sediments is in the Southeast Pacific. It is located entirely within the southern arid climatic zone where the processes are fairly uniform for the deposition of pelagic sediments. The part of the East Pacific Rise, around which the field is located, has the whole range of spreading rates, from ~8 cm a^{-1} to ~20 cm a^{-1}, at which the large fields of metalliferous sediments occur in the Pacific Ocean. These features of the field in the Southeast Pacific enable an attempt to be made to show general regularities in distribution of different parameters in the field and to consider variations in these parameters that depend on conditions that are either conducive to or inhibit the formation of metalliferous sediments.

The spreading rate is the main characteristic feature that has been quantified among the major factors that influence and control the formation of metalliferous sediments. A quantitative correlation of the (Fe+Mn)/Al ratio in the metalliferous sediments from the Southeast Pacific with the spreading rate of the EPR was found for the first time by Boström (1973). Later Lisitzin (1978) found a correlation of the width of the field of metalliferous sediments in the Southeast Pacific with the spreading rate of the EPR. These researches have demonstrated that, with other factors being equal, the size dimensions of the fields of metalliferous sediments and the contents of the metal-bearing matter in these sediments quantitatively depend on the spreading rate of the active ridge that contributed the metal-bearing matter in these sediments. Since these fundamental relationships were found a large number of data has been obtained in the distribution, composition and the accumulation rates of recent and ancient ocean metalliferous sediments. The relationships have been shown as well between the probability of hydrothermal venting and the spreading rate and the relationship of the incidence of hydrothermal plumes and the spreading rate (Sect. 4.1). All of these data enable the creation of an average quantitative model for the formation of distal metalliferous strata on the East Pacific Rise in the Southeast Pacific.

According to the concept of sedimentary formations in a spreading ocean (Lisitzin et al. 1973; Lisitzin 2001, 2004), the formation of metalliferous sediment strata begins on either side of the axis of a spreading ridge in close proximity to the axis whence the hydrothermal metal-bearing matter contained in the particulate matter of the hydrothermal plumes and in solution is contributed to the ocean. This matter is carried by bottom currents on either side of the ridge and settles to the bottom where it mixes with the background sedimentary material and forms metalliferous sediments and strata. As the distance increases from the ridge axis the ratio in the content of the metal-bearing matter to the content of non-biogenic background material gradually decreases. According to the definition of metalliferous sediments given in Introduction, biogenic material is not considered as a dilutant of the metal-bearing matter. The content of the metal-bearing matter in the metalliferous sediments also decreases and at some distance from the ridge axis the content of Fe in the abiogenic (carbonate- and silica free) material (Fe_{CSFB}) becomes <10%, and the sediments are considered to be non-metalliferous.

If the accumulation of a metalliferous stratum took place on an ocean floor that is static (non-moving) and the intensity of the hydrothermal activity, the pattern in the dispersion of the metal-bearing matter, and the rate of accumulation of the background abiogenic matter were constant over a period of time the content in the abiogenic matter and the accumulation rate of the metal-bearing matter in the metalliferous sediments would vary only in horizontal directions, and would decrease with increasing distance from a ridge axis, and would be constant in vertical sections. The thickness of the stratum would decrease as the distance increases from the axis of the ridge. The stratum would accumulate during the entire time of the contribution of the hydrothermal metal-bearing matter. If the contribution of the metal-bearing matter were to continue over an indefinitely long period of time the thickness of the stratum could increase to infinity.

In the existing ocean where the ocean floor is mobile and moves on both sides of the ridge axis as a result of the spreading, the pattern in the accumulation of a single metalliferous stratum is very different. Under these conditions a metalliferous stratum represents a bedding of laterally displaced elementary isochronal layers. If the ocean floor were not-moving this displacement would be absent. The extent of the displacement between any two layers is equal to the product of multiplying the average half-rate of spreading by the time that passes between the deposition of these layers. Because of the displacement of the elementary layers during the accumulation of the metalliferous strata on the mobile ocean floor the content and the accumulation rate of the metal-bearing matter may vary in the horizontal and vertical directions. In this case, with constant intensity of the hydrothermal activity, the pattern in the dispersion of the metal-bearing matter, and the accumulation rate of the background abiogenic matter through time, the content of the abiogenic matter and the accumulation rate of the metal-bearing matter and of any other components and chemical elements vary upwards in vertical sections at locations some distance from the ridge axis and correspond to variations with increasing distance from the ridge axis to the point of the profile (Fig. 4.32). The modern boundary of the field of distal metalliferous sediments is also the top boundary of a metalliferous stratum that is completely formed.

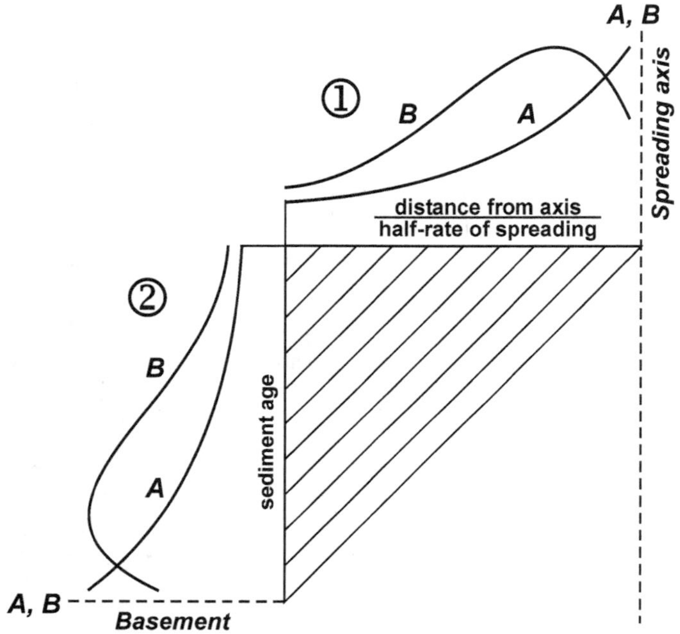

Fig. 4.32. Correspondence in the distribution of the contents or accumulation rates of components A and B at constant hydrothermal activity and conditions of sedimentation through time: ① in surface distal metalliferous sediments along a section perpendicular to a spreading axis, and ② in distal metalliferous sediments in the vertical profile at the end of this section.

The boundaries of the ancient fields of metalliferous sediment have now moved away from the ridge axes to distances greater than the width of the present field and they are overlain at present by strata of non-metalliferous sediments. Because of the existence of the moving ocean floor, the thickness of the metalliferous stratum cannot be infinite even in an infinitely long period of time in the contribution of the hydrothermal metal-bearing matter to the ocean floor. The stratum accumulates only during the time that elapses while the ocean floor moves from the ridge axis to the external boundary of the field of distal metalliferous sediments and the thickness of the stratum is determined by the time required for this movement and the average sedimentation rate during this time.

An ideal situation is presented in which there is a process that is constant in time for the accumulation of metalliferous sediment strata during the movement of the ocean floor and it does not take into account the spatial and temporal variations in parameters that control and influence the real existing process. Nevertheless, even considering an idealized pattern in the accumulation of the metalliferous strata at various spreading rates and with various dilution of the metal-bearing matter by background abiogenic material, this ideal concept can be useful for understanding the actual course of this process and the influence of different factors in it. Knowledge of the idealized conception of the process can be helpful in the understanding actual patterns in distribution of different parameters in ancient metalliferous strata and in the solution of reconstructive problems.

After a qualitative description of the process of formation and deposition of a metalliferous stratum on the moving ocean floor, an attempt is made to show an average quantitative pattern in the course of this process on the East Pacific Rise. Recognizing that Fe is the main component in the metal-bearing matter of metalliferous sediments and its content is used for distinguishing them from non-metalliferous sediments, the distribution of the contents of Fe_{CSFB} and the distribution of the accumulation rates of Fe in the metalliferous sediments in the Southeast Pacific have been used for the construction of this pattern. The maps showing the distribution of these parameters published by Bogdanov and Chekhovskikh (1979) and Bogdanov et al. (1979c) and data from subsequent works have been the main sources of data and materials used in the construction of this pattern. The average accumulation rate of the background abiogenic material in the Southeast Pacific is about 20 mg cm^{-2} ka^{-1} (Sect. 1.1.1). Considering that within the Southeast Pacific the accumulation rate of Al, the proxy of the background abiogenic material, in recent metalliferous has little variation (Lisitzin et al. 1980b), a value of 20 mg cm^{-2} ka^{-1} will be assumed to be a constant that characterizes the accumulation of the background abiogenic material in the recent metalliferous sediments on the East Pacific Rise.

There is a distinct quantitative relationship in the Southeast Pacific between the half-rate of spreading and a half-width of the field of metalliferous sediments[36] when the boundary is 10% of the Fe_{CSFB}. There are similar relationships when the values of Fe_{CSFB} are 15, 20, and 25%. This enables the plotting of the family of curves that shows a relationship of the Fe_{CSFB} contents in the metalliferous sediments with the half-rate of spreading and with the distance from the EPR axis (Fig. 4.33).

[36] The use of the half-rate of spreading and the half-width of the field of metalliferous sediments goes beyond consideration of bottom- and deep currents carrying hydrothermal matter in the directions perpendicular to the ridge axis.

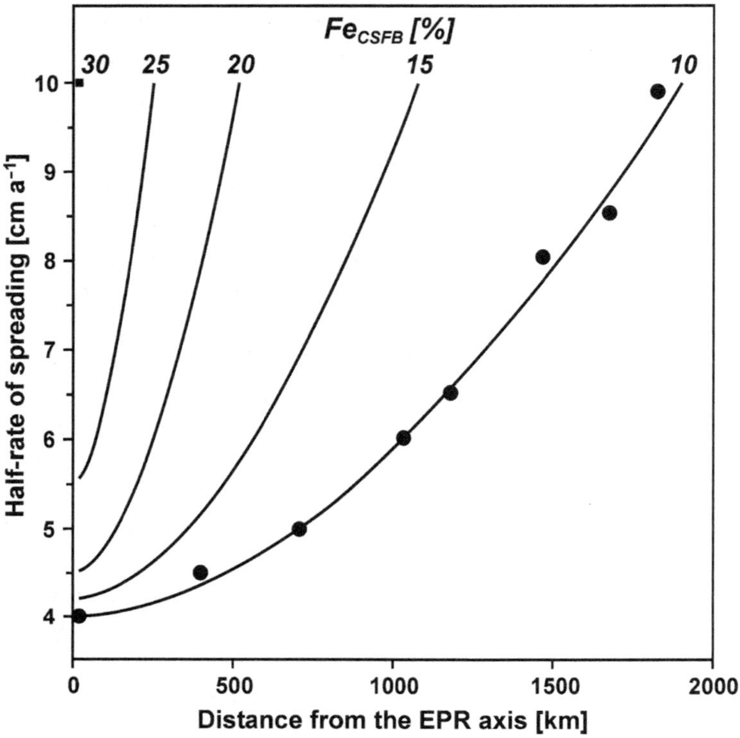

Fig. 4.33. Averaged relationship of Fe_{CSFB} content in distal metalliferous sediments from the Southeast Pacific with distance from the EPR axis and half-rate of spreading.

The average distribution of the Fe_{CSFB} contents in metalliferous sediments along a generalized section perpendicular to the EPR axis related to different half-rates of spreading can be shown by changing the coordinates of this relationship (Fig. 4.34).

On the basis of Fig. 4.34, of the average composition of the background material and of its accumulation rate, and of the data for the distribution of the Fe accumulation rates in metalliferous sediments, an average distribution of the accumulation rate of the hydrothermal Fe in metalliferous sediments along a generalized section perpendicular to the EPR axis associated with the different half-rates of spreading has been calculated (Fig. 4.35). For distances 20 to 2000 km from the EPR axis the relationship or dependence of a decrease in the rate of accumulation of the hydrothermal Fe (Fe_h*) with distance D (km) in comparison with Fe_h* at 20 km distance from the EPR axis is approximated by the best fit equation:

$$Fe_h^*/Fe_{h20}^* = (1.58 - 0.446 \lg D)^2 \quad (4.3)$$

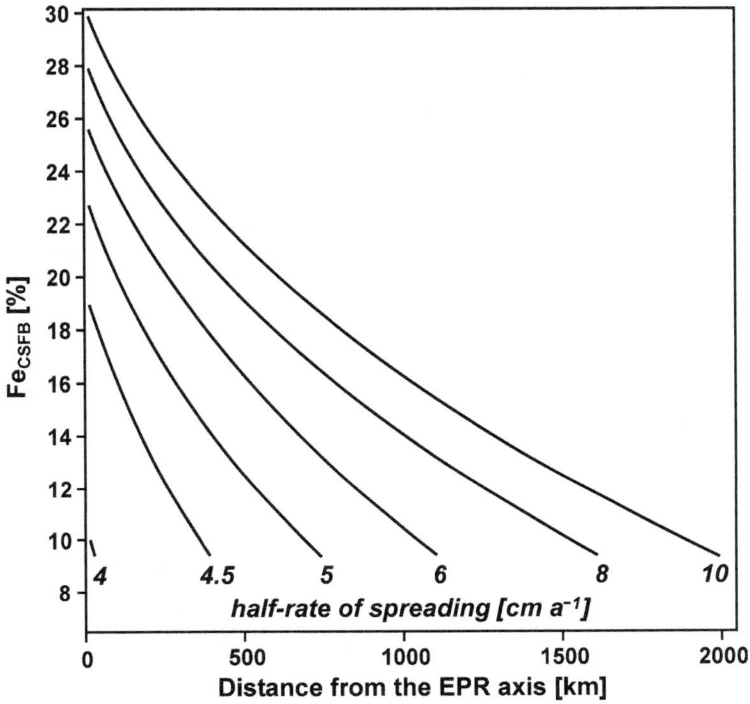

Fig. 4.34. Averaged relationship of the Fe_{CSFB} content in distal metalliferous sediments from the Southeast Pacific with distance from the EPR axis and half-rate of spreading.

The graph in Fig. 4.33 demonstrates the relationship of the contents of the Fe_{CSFB} and of the half-width of the field of metalliferous sediments (isoline 10%) with the half-rate of spreading when the accumulation rate of the background abiogenic material is 20 mg cm^{-2} ka^{-1}. Variations in this parameter change the half-width of the field and the pattern of distribution of the Fe_{CSFB}. The changes with an increase in the rate of accumulation of the background abiogenic material can be estimated with some accuracy; but changes with a decrease in the rate of accumulation can be made only by extrapolation and are less accurate. The half-widths of the field of metalliferous sediments, when accumulation rate of the background abiogenic material has values different than 20 mg cm^{-2} ka^{-1}, are calculated from the data of Fig. 4.33 and Table 1.1 and are shown in Fig. 4.36.

It is evident from Fig. 4.36 that the higher accumulation rate of the background abiogenic material the higher spreading rate is a boundary value for formation of metalliferous sediments, and with an increase in the accumulation rate of the background abiogenic material the half-width of the field of metalliferous sediments decreases.

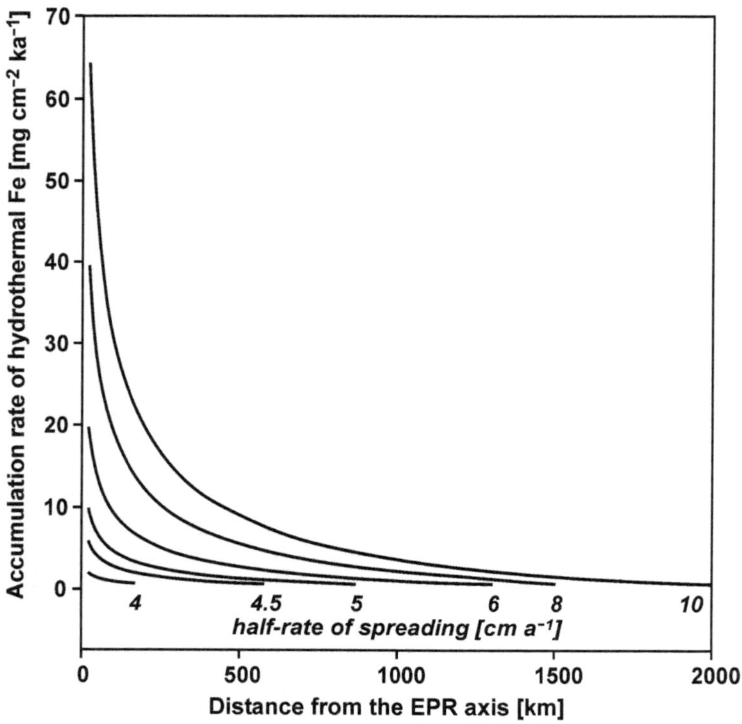

Fig. 4.35. Averaged relationship of the accumulation rate of hydrothermal Fe in distal metalliferous sediments from the Southeast Pacific with distance from the EPR axis and half-rate of spreading.

The data from Fig. 4.36 enable the calculation of the time required for the formation of metalliferous strata at different spreading rates and different accumulation rates of the background abiogenic material (Fig. 4.37). As the duration of time required for completion of the formation of a metalliferous stratum is equal to the time of movement with the half-rate of spreading from the ridge axis to the external boundary of the field of surface metalliferous sediments, this period of time can be calculated as a quotient in the division of the half-width of the field by the half-rate of spreading. Increase in the accumulation rate of the background abiogenic material results in decrease in the duration of time required for formation of the stratum (Fig. 4.37).

The graph in Fig. 4.37 shows an interesting feature; with a constant rate of accumulation of the non-metalliferous abiogenic material the duration of time required for the formation of the stratum varies little over a wide range of the half-rates of spreading. For example, when the average accumulation rate of this material is 20 mg cm^{-2} ka^{-1} there is only a narrow range in the half-rate of spreading, from 4 to ~6 cm a^{-1}, and the duration of time for the formation of the stratum varies greatly, from 0 to ~18 Ma, and in the range from 6 to 10 cm a^{-1} it changes a little and is about 18–19 Ma. This duration of time is similar to maximal actual values (Fig. 3.3). Decreases in the duration of time required for the formation of strata are the greatest when the values of the half-rate of spreading are minimum or close to minimum and the accumulation rates of the background abiogenic material are low (Fig. 4.37).

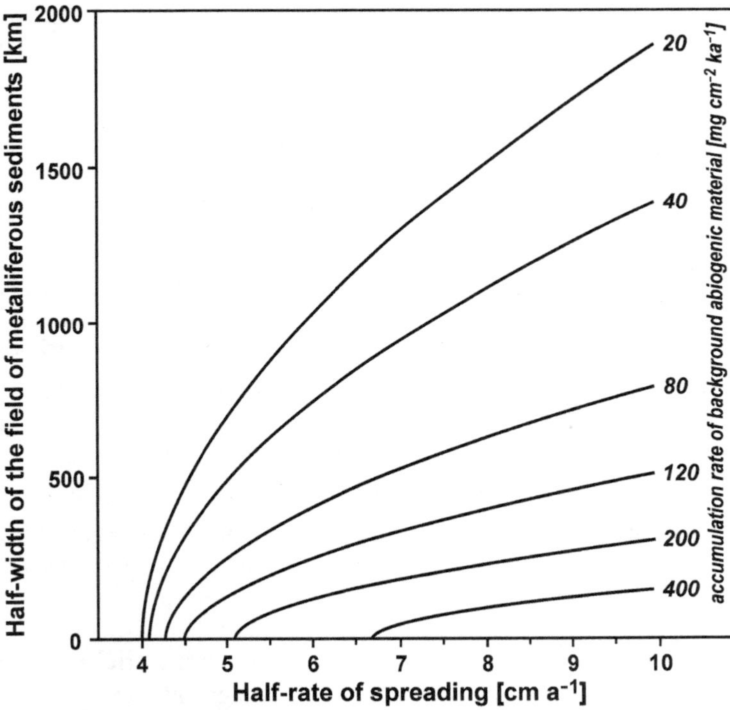

Fig. 4.36. Dependence of half-width of the field of metalliferous sediments on half-rate of spreading and accumulation rate of the background abiogenic material.

With changes in the spreading rate and the accumulation rate of the background abiogenic material the total amount of abiogenic material including the metal-bearing and background material in the metalliferous stratum changes as well. These two parameters have multiple values depending on the total amount of abiogenic material in the stratum (Fig. 4.38). If the accumulation rate of the background abiogenic material is constant the total amount of the abiogenic material in the stratum increases with an increase in the half-rate of spreading, and the average content of the metal-bearing matter increases as well. This reflects an increase in the intensity of hydrothermal activity with an increase in the spreading rate, which results in an increase in the content of the metal-bearing matter in the stratum.

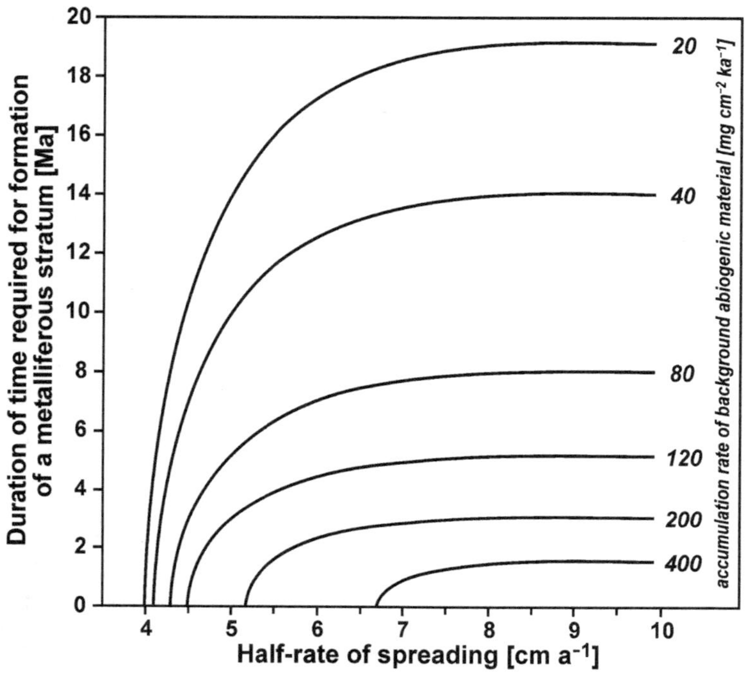

Fig. 4.37. Dependence of duration of time required for the formation of metalliferous strata on half-rate of spreading and accumulation rate of the background abiogenic material.

Dependence of the total amount of the abiogenic material in the stratum on variations in the accumulation rate of the background abiogenic material is more complex. If the half-rate of spreading remains constant and there is an increase in the accumulation rate of the background abiogenic material the total amount increases at first, reaches a maximum, and then decreases. With an increase in the half-rate of spreading, the position of the maximum is transformed to higher accumulation rates of the background abiogenic material. Zero values of the total amount of the abiogenic material in the stratum correspond to conditions when the average content of the Fe in sediments becomes 10%. It must be noted that the maxima for the total amount of abiogenic material in the stratum occur when there are optimum proportions for the accumulation rate of background abiogenic material and the duration of time required for the formation of the stratum.

The described regularities concern the abiogenic material of metalliferous strata. Transition from gross amount of the abiogenic material in the stratum (per unit of area) to gross amount of the abiogenic plus biogenic material can be carried out by multiplication of the amount of the abiogenic material by a factor (K) depending on the average content of the biogenic material (B) in the stratum. K = 100/(100–B), where B is in %. Calculation of the thickness of the stratum can be carried out by division of the gross amount of the abiogenic plus biogenic material in the stratum per unit of area by average bulk dry density for salt-free matter of sediments.

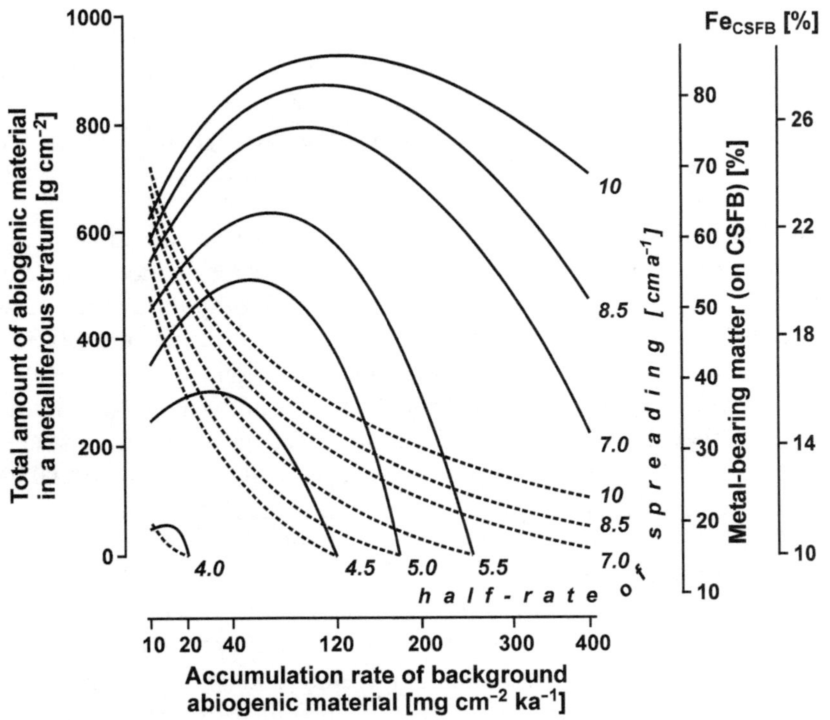

Fig. 4.38. Variations in total amount of abiogenic (metal-bearing + background) material in metalliferous strata (full lines) and average content of metal-bearing matter and Fe (hachure lines) with changes in accumulation rate of the background abiogenic material and half-rate of spreading.

Even highly simplified quantitative consideration of the formative process for metalliferous strata shows that this process depends on a number of factors and their influence is not always shown by a single value. Nevertheless, the influence of the factors considered in the model should be reflected in the metalliferous strata in their specific properties including thickness and chemical composition and the factors also influence the period of time required for the formation of the strata.

4.8. Formation of composition of metalliferous sediments in the Red Sea

In the Red Sea rift and in the rift systems of the open ocean the mobilization of the metal-bearing matter and its passing into hydrothermal fluids take place when seawater and/or brines interact with tholeiitic basalts under the PT-conditions, which apparently vary not much along rift systems (Lisitzin et al. 1990). Mainly the temperatures of the effluent hydrothermal solutions in hydrothermal orifices and conditions near the bottom cause the observed differences in the composition of the metalliferous sediments in the Red Sea. Many deeps in the Red Sea rift zone are filled with heavy brines that differ considerably in temperature and salinity, and the concentrations of the major ions and the dimensions of the brine pools are very different (Tables 2.1, 2.2).

The Late Miocene salt-bearing strata are a source of the salts in the brines. There are three main ways that the salts can enter the deeps: by exposure of the salt-bearing strata on their walls, by the migration of the seawater and its hydrothermal circulation through the salt-bearing strata as it descends in the circulation system, and by the migration of the hydrothermal fluids through the salt-bearing strata as they ascend in the circulation system (Fig. 4.39). In the deeps with the exposure of the salt-bearing strata on the walls (Fig. 4.39a, d) the brines would exist without hydrothermal activity; and in the deeps, where there is no exposure of salt-bearing strata (Fig. 4.39b, c), the brines appear only in the periods of time when there is intensive and sometimes moderate hydrothermal activity. The brine layers do not form when there is low hydrothermal activity and the circulating bottom seawater dilutes the highly saline hydrothermal solutions contributed to a deep. In such cases the temperature and salinity of the bottom water in a deep are slightly higher than in ambient seawater and these conditions exist in the NE Thetis, Hadarba[37], Gypsum, and Vema Deeps at the present time (Tables 2.1, 2.2).

The salinity of the brines and the thickness of the brine layers depend on a number of factors including the following: morphology of the deeps, size and dimensions of the exposures of the salt-bearing strata, intensity of hydrothermal activity, bottom circulation, etc. The brine pools act as traps of heat and matter that are contributed from the crust. Because of the weak exchange of the brines with seawater the dispersion of the chemical elements contributed by the hydrothermal solutions is limited.

[37] According to Bäcker et al. (1975), the salinity of pore water in sediments from the Hadarba Deep is somewhat higher than the salinity of ambient seawater.

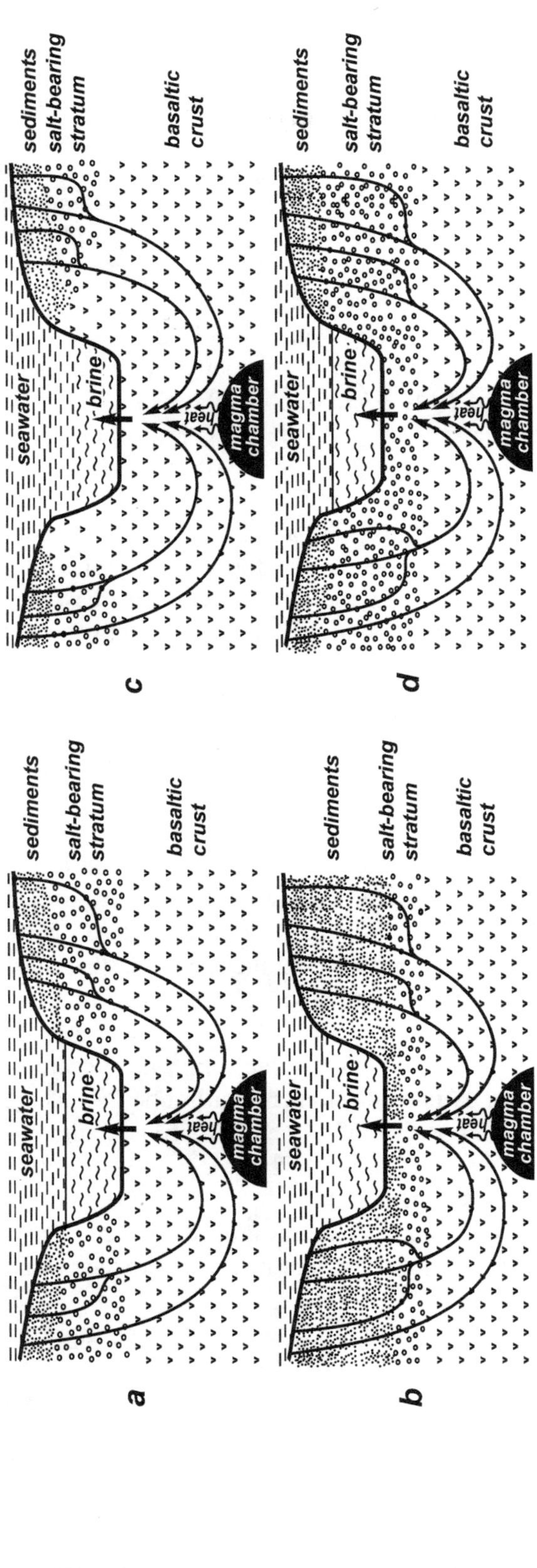

Fig. 4.39. Simplified schemes of the hydrothermal circulation systems and the formation of brines in different deeps of the Red Sea rift zone, related to their geological position (without scale). ***a*** – the sediment and salt-bearing strata have been faulted during rifting, the salt-bearing stratum is exposed on the walls of a deep, hydrothermal solutions are ejected from the basaltic crust (as this occurs in the Atlantis II, Erba, Shagara, Suakin, Port Sudan, Albatross Deeps); ***b*** – the sediment and salt-bearing strata have not been faulted during rifting, there is no exposure of the salt-bearing stratum on the walls of the deep, before the ejected hydrothermal solutions migrate through the salt-bearing and other sediment strata (as this occurs in the Vema, Gypsum, Kebrit Deeps); ***c*** – the sediment and salt-bearing strata have been faulted during rifting, there is no exposure of the salt-bearing stratum on the walls of the deep, hydrothermal solutions are ejected from the basaltic crust (as this occurs in the Thetis and Hadarba Deeps); ***d*** – the sediment and salt-bearing strata have not been faulted during rifting, the salt-bearing stratum is exposed on the walls of the deep, before the ejected hydrothermal solutions migrate through the salt-bearing stratum (as this occurs in the Valdivia Deep).

During circulation through the crust the brines undergo considerable metamorphism. During their metamorphism and when the temperatures in the crust are >200°C the brines lose SO_4^{2-} and Mg^{2+} ions. Also the brines lose these ions in the deeps when they come into contact with high-temperature hydrothermal fluids, which are metamorphosed brines with low concentrations of the SO_4^{2-} and Mg^{2+} ions. For example, in the Atlantis II Deep the brines are highly depleted in these ions in comparison with seawater (Table 2.2, Fig. 2.4), and these ions are lost in the deep near the orifices of hydrothermal vents. Bottom sediments near the orifices are highly enriched in sulfate minerals, anhydrite and gypsum as well as in Mg-minerals including saponite, chlorite, talc, amphiboles, serpentine, vermiculite, cerolite, and stevensite (Fig. 2.10, 2.11; Bäcker and Richter 1973; Schneider und Schumann 1979; Singer and Stoffers 1981, 1987; Pottorf and Barnes 1983; Oudin et al. 1984; Cole 1988; Ramboz et al. 1988; Butuzova 1998).

The absence or presence of dissolved oxygen, and the existence of reducing or oxidizing conditions greatly influence the formation of minerals, which is very different in the brines and seawater. The entry of dissolved oxygen into the brines depends on the amount of exchange between the brines and seawater. With greater thickness of brine layers and greater differences in the densities of the brines and seawater the brines become more effective as traps for matter and heat, and conditions in the brines become more reducing. However, at times when the intensity of the entry of hydrothermal solutions into a deep is very high, they can disturb the continuity of the brines, which results in partial or complete mixing of the brines with seawater. When the interaction of the hydrothermal fluids with the brines takes place additional amounts of dissolved oxygen enter the brines and the conditions become favorable for the formation of oxidized sediments. The higher the extent of the hydrothermal flow and the lower the thickness of the brine layers the more favorable conditions develop for the formation of oxidized sediments.

There is a wide spectrum of conditions for the deposition of hydrothermal sediments in many different parts of the Red Sea rift zone and where considerable variation in the composition of the metalliferous sediments can occur. In spite of the significant variation in the mineral composition of the metal-bearing matter in the sediments of the Red Sea deeps, this mineral matter is composed mainly of several mineral groups: sulfides, ferriferous silicates, oxyhydroxides of Fe and Mn, and rarely carbonates of Fe and Mn.

If it is assumed that the composition of the primary metal-bearing hydrothermal fluids varies little along the rift and that the fluids unload their content of metals mainly in zones where they have contact with bottom water or brines, it is possible to present a general scheme for the formation of the composition of the metal-bearing matter in the metalliferous sediments in the deeps of the Red Sea rift zone. In the creation of this scheme it was supposed that with increasing distance from the zone of contact of the metal-bearing hydrothermal fluids with the bottom water or brines the differentiation of the chemical elements and the occurrence of the main minerals in the metal-bearing matter follow the general distribution patterns demonstrated by numerous data obtained in the research on the evolution of the hydrothermal systems in the ocean rift zones (Lisitzin et al. 1990). The mineral sequence is as follows: *hydrothermal sulfide minerals, magnetite → Fe-oxyhydroxide minerals, nontronite → Mn-oxyhydroxide minerals.* Depending on the position of the zone of contact the metalliferous sediments can be divided into three groups (Gurvich and Bogdanov 1986c).

Metalliferous sediments of the **first group** form under conditions where hydrothermal solutions migrate from the zone of mobilization to the surface of the sea floor without essential changes taking place. The temperature of the solutions exceeds 300°C, and the concentrations of metals are close to or about the same as those in the primary hydrothermal fluids.

When the high-temperature hydrothermal solutions come into contact with bottom water or brines there is intensive unloading of the metals carried by these solutions, except for Mn. Under the conditions of the deeps filled with stratified brines, Fe-, Cu- and Zn- sulfide minerals precipitate at or near the orifices of the high-temperature vents.

The formation of minerals located at some distance from the hydrothermal orifices is dependent on the physical and chemical conditions that prevail near the bottom, and first of all on the Eh. Under the conditions that exist in well-stratified bottom brines the ingress of dissolved oxygen from the overlying seawater is limited. As a result unusual conditions for the precipitation of the main hydrothermal elements, Fe and Mn, are present. Fe- and Mn-oxyhydroxide minerals form in the upper parts of the brine layers and in the transition zone from the brines to seawater respectively. A differentiation of the Fe and Mn occurs during the sedimentation of these oxyhydroxide minerals. Most of the Fe-oxyhydroxides reaches the bottom and accumulates in the bottom sediments, and the Mn-oxyhydroxides are dissolved during their settling and return into solution. The hydrothermal Mn accumulates on the floor mainly in areas outside of the brine deeps.

The processes of co-precipitation and sorption of trace elements are specific. On the one hand subacid low-reducing conditions in the brines are less favorable for the co-precipitation and sorption scavenging of chemical elements compared with those in low-alkaline oxidizing conditions in seawater. On the other hand the brines, acting as traps, are highly enriched in many chemical elements and this promotes their accumulation in the bottom sediments.

Similar near-bottom conditions are developed in the deeps in two cases: a) when there is intensive discharge of high-temperature hydrothermal effluent into a deep that is filled with thick layer of dense brines and the stratification of the brines is disturbed; b) a brine layer is formed when there is intensive discharge of highly saline high-temperature hydrothermal solutions into a deep where brines are absent. In both cases at and near the hydrothermal orifices hydrothermal sulfide minerals and/or magnetite are precipitated along with Fe-oxyhydroxide minerals. Fe-oxyhydroxide minerals are precipitated mainly at some distance from the orifices.

Metalliferous sediments of the **second group** form under conditions where the hydrothermal solutions that enter a deep have been altered in composition in comparison with the primary hydrothermal fluids. The zone of contact of the metal-bearing hydrothermal fluids with the bottom water or brines apparently is located below the bottom surface at near-surface levels of the crust. In this zone partial unloading of the metals and the formation of sulfide minerals takes place. The residual hydrothermal solutions are depleted considerably in Fe, Cu, Zn, and S. Their temperatures are presumed to be from 30° to 300°C. In the metalliferous sediments formed on the bottom near the orifices of the mid-temperature hydrothermal vents the metal-bearing matter is composed mainly of Fe-oxyhydroxide minerals and ferriferous silicate minerals. Hydrothermal sulfide minerals are absent or present only in small amounts. In environments, where the conditions are reducing, Fe-sulfide minerals are formed by the sulfate-reduction. As the distances increase from the orifices the metal-bearing matter is enriched gradually in Mn. In the brine-filled deeps the sediments accumulated near the brine-seawater boundary are highly enriched in Mn.

Metalliferous sediments of the **third group** form under conditions where the hydrothermal solutions that enter a deep have been highly altered in composition in comparison with the primary hydrothermal fluids. The unloading of the major contents of the hydrothermal metals in sulfide and probably in silicate mineral matter takes place in the subsurface levels of the crust. The residual hydrothermal solutions are highly depleted in Fe, Cu, and Zn but the concentrations of Mn do not change significantly. On the sea floor near the orifices of the low-temperature (<30°C) hydrothermal springs, under favorable redox conditions, sediments accumulate in which the Mn-oxyhydroxide minerals are of significance in the metal-bearing matter. As the distance from the orifices increases the amount of Mn in the sediments increases and may become prevalent. In the deeps filled with thick layers of brines the existing redox conditions are not favorable for the precipitation and sedimentation of Mn-oxyhydroxide minerals. In the environmental conditions near to the orifices of hydrothermal springs and at some distance from them the metal-bearing matter, if it accumulates at all, is composed of Fe-oxyhydroxide and sulfide minerals and the latter are formed by the sulfate-reduction. The Mn-bearing minerals apparently precipitate and settle to the floor near the boundary between the brines and seawater.

The **fourth group** of metalliferous sediments was recognized in studies of the formation of metalliferous sediments in the Red Sea rift zone. The metal-bearing matter of these sediments is formed in environments at or near the floor and deposited during the migration of heavy brines that are enriched in metals. The migration involves a) the flow of the brines enriched in metals from an orifice or place of hydrothermal effluence down slope toward a deep where there is no hydrothermal effluent, and b) the overflow of the brines from a deep with hydrothermal effluent to another deep located at lower level, and then from the second deep to a deep located at still lower level. There is no hydrothermal effluent in the second and third deeps. Where the bottom relief is favorable, metalliferous sediments of the fourth group can accumulate from hydrothermal metal-bearing solutions of any composition that enter the first deep. The composition of the metal-bearing matter that accumulates in the metalliferous sediments of the fourth group and the volume of sediments depend not only on the place of deposition, near to hydrothermal orifices, on slopes, within deeps etc., but also on the degree of unloading of metals from the hydrothermal solutions when they enter the sea. General consideration of all possible variants is beyond our purpose here since the only metalliferous sediments known for certain to belong to the fourth group are in the Nereus and Chain Deeps.

Fe and Mn are not the only elements that undergo differentiation related to the intensity of hydrothermal effluence in zones located in the contact areas of the metal-bearing hydrothermal fluids and the bottom water or brines. Other chemical elements carried by the hydrothermal solutions are differentiated also. For example, Cu is precipitated in sulfide minerals in the early stages of hydrothermal unloading of metals into the metalliferous sediments of the first group, and Zn accumulates also in the metalliferous sediments of the other groups.

Metalliferous sediments that occur in separate deeps of the Red Sea rift zone may belong not only to one of the named groups but also alternate in a vertical section from the first to the third group depending on the variations in the intensity of the hydrothermal activity through time. The regularities described in the distribution of the major groups of minerals in the metal-bearing matter in the metalliferous sediments of the Red Sea rift zone are of a general character. They do not include some of the features that are related to local conditions and the processes that bind the metals in carbonate minerals. The possible redistribution of chemical elements in the metalliferous sediments during their alteration after sedimentation has not been studied in detail.

Three major and four additional facies zones can be considered that are related to the distance from a hydrothermal orifice and their location relative to the boundary between brines and seawater. These major zones are: I – the zone in the vicinity of a hydrothermal orifice, II – the zone of the deep, III – the zone near to the brine-seawater boundary in the brine-filled deeps, or the distal zone in the deeps where brines are absent. The additional zones are: II' – the zone of the slope, and II" – the zone of the deep in the case of run-off of the metal-bearing brines from the place of hydrothermal effluence down a slope toward a deep where hydrothermal vents are absent; II_A – the zone of the second deep, and II_B – the zone of the third deep for the case where there is an overflow of brines from a deep with hydrothermal effluence into the second deep (A), and then from the second deep into the third deep (B).

The chemical and mineral compositions of the metal-bearing matter in metalliferous sediments depend on the following major factors:

a) temperature and degree of unloading of hydrothermal solutions that enter a deep;

b) intensity of hydrothermal effluence;

c) presence or absence of brines in a deep;

d) origin of brines, from the exposures of salt-bearing strata on the walls of a deep or from the migration of seawater that is associated with hydrothermal circulation through salt-bearing strata;

e) thickness of a brine layer;

f) presence or absence of disturbances in the continuity of the brines by effluent hydrothermal solutions;

g) distance of a point of deposition from a hydrothermal orifice;

h) location of effluence of hydrothermal solutions in a deep or on a slope;

i) occurrence of overflow of brines from a deep with hydrothermal effluence to a shallower deep where hydrothermal effluence is absent.

Three major and four additional facies analogous to the facies zones of metalliferous sediments can be considered by relating the distance that a point of deposition of sediments is located from a hydrothermal orifice. The major facies are: I – the near orifice facies, II – the facies of the deep, III – the facies of the brine-seawater boundary (distal facies if the brines are absent). The additional facies are: II' – the facies of the slope, II" – the facies of the deep, II_A – the facies of the second deep, and II_B – the facies of the third deep.

In general the factors named above cover practically the whole range of conditions for the formation of the metal-bearing matter in the metalliferous sediments in the deeps in the Red Sea rift zone. The various combinations of the factors can account for the environment and conditions in which the sediments or individual interlayers form in specific deeps in specific intervals of geologic time. Variations in the conditions for the formation of the metal-bearing matter in the metalliferous sediments in the deeps of the Red Sea rift zone and the manner in which they influence the chemical composition of this matter are shown in Fig. 4.40–4.43. Specific and in some instances somewhat limited data for some of the specific intervals or parts of the sedimentary layers in separate deeps that presumably correspond to variations in the sedimentary environment have been used in defining the characteristics of the chemical composition of the metal-bearing matter in the sediments. These data have been taken mainly from Baturin et al. (1969), Hot brines … (1969), Baumann et al. (1973), Bäcker (1976, 1982), Bignell and Ali (1976), Bignell et al. (1976a), Butuzova et al. (1980, 1981), Cronan (1980), Shanks and Bischoff (1980), Pottorf and Barnes (1983), Zierenberg and Shanks (1983), Bogdanov et al. (1986), Scholten et al. (1991, 2000), Anschutz and Blanc (1995). As an illustration the ratios of the contents of some chemical elements, Fe, Mn, Zn, Cu, Co, Ni, that are bound in the metal-bearing matter, to the sum of the contents of Fe and Mn in the metal-bearing matter, are shown in the diagrams in Fig. 4.40–4.43. These ratios show characteristic features in the contents of the elements in the metal-bearing matter. These elements were not chosen at random because the Fe, Mn, Zn, and Cu in the metalliferous sediments of the Red Sea are mainly of hydrothermal origin. Cobalt is commonly of hydrothermal origin but it does not form separate minerals. The content of Ni contributed from hydrothermal sources to the metalliferous sediments is insignificant, and any excessive accumulation of it in the metalliferous sediments, if it exists, results from the scavenging of this element from the seawater by dispersed phases of the metal-bearing matter. The diagrams in Fig. 4.40–4.43 show only the characteristic trends in the lateral variations in the chemical composition of the metal-bearing matter, since in the majority of the deeps the locations of active hydrothermal vents, or of the extinct ones, are not known or have only been assumed. In the classification given for conditions for the formation of the metal-bearing matter in the metalliferous sediments in the Red Sea rift zone the levels of the location of the zones of interaction of the metal-bearing hydrothermal fluids with brines or seawater are considered first. These levels are: at the surface of the sea bottom, below the bottom surface in near-surface levels of the crust, and subsurface levels below the surface of the crust. At these locations of the zones of interaction the temperatures of hydrothermal solutions discharged into the deeps are: >300°C, 30÷300°C, and <30°C respectively.

The different kinds of geological settings in the deeps and the locations of the effluence of the hydrothermal solutions are considered in Fig. 4.39.

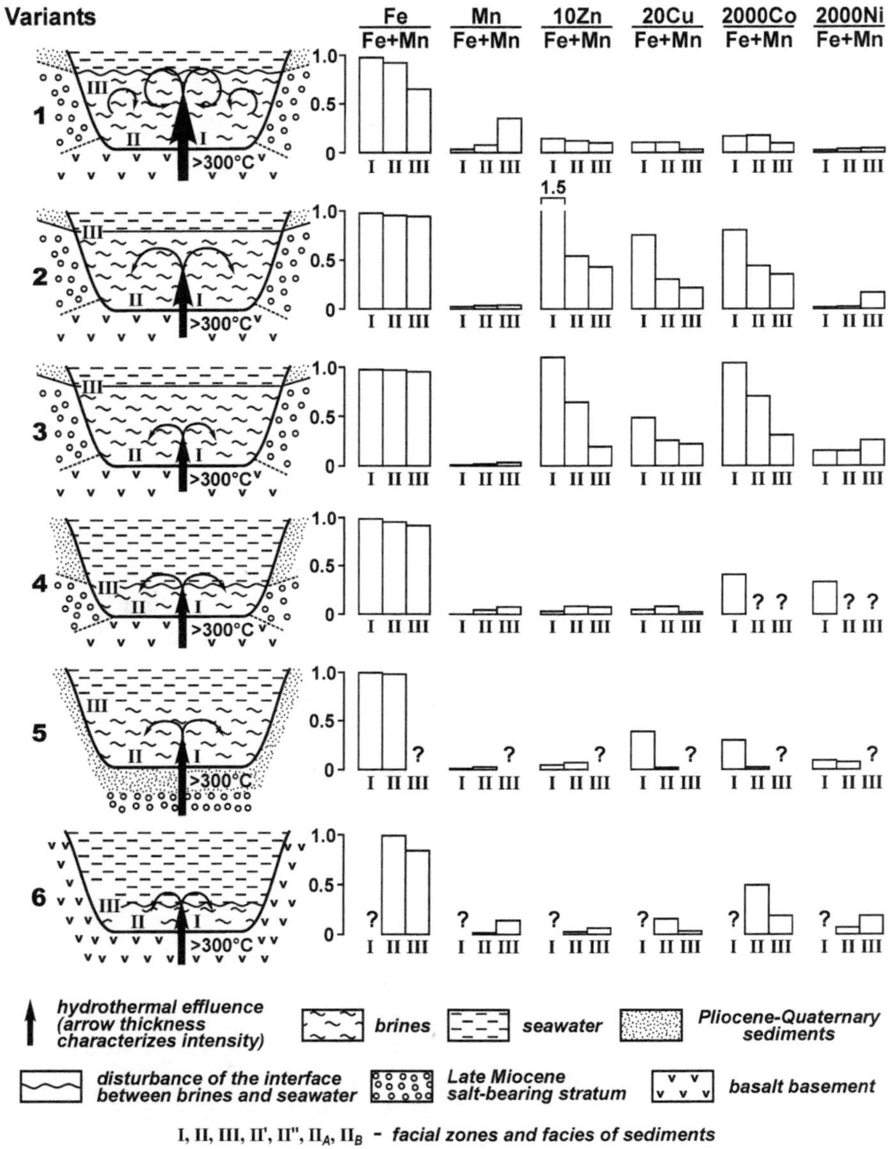

Fig. 4.40. The first group variants of conditions for formation of metal-bearing matter of metalliferous sediments in the deeps of the Red Sea rift zone.

In **the first group** the conditions for the formation of the metal-bearing matter in deeps where there is effluence of hydrothermal solutions that have temperatures >300°C are shown in Fig. 4.40. The zone of contact of the primary hydrothermal fluids with brines or seawater is located at the level of the sea floor.

Variant 1. An example is the conditions for the formation of the sediments of the CO Zone in the Atlantis II Deep. There is very high-intensive effluence of high-temperature hydrothermal solutions into a deep where a thick brine layer exists. A salt-bearing stratum exposed on the walls of the deep and hydrothermal solutions are sources of the salts in the brines, and the hydrothermal solutions are ejected from basaltic crust (Fig. 4.39a). Because of the intensive effluence of the solutions the continuity of the brine layer is disturbed, dissolved oxygen can enter the brines from the overlying seawater, and some amounts of chemical elements that are dissolved and in particulate matter can leave the deep. The near orifice facies of metalliferous sediments is rich in Fe, that is present in hydrothermal sulfide minerals, well-crystallized hematite, and magnetite; the content of Mn is insignificant. At a distance from the orifices the Mn content increases considerably and reaches 10 to 50% of the sum of the Fe and Mn contents. Apparently the high Mn content results from the entry of dissolved oxygen into the brines and the oxidation of hydrothermal Mn(II), at least in part, to Mn(IV), which forms Mn-oxyhydroxide minerals and settles to the bottom, especially in areas near to the brine-seawater boundary. The contents of Zn, Cu, and Co in the metal-bearing matter of the sediments are generally low. This results from the significant dissipation of these elements outside of the deep and the lack of precipitation of Zn- and Cu- sulfide minerals in most parts of the deep where oxidizing conditions are predominant. Because of the very high accumulation rate of the metal-bearing matter the content of Ni that is scavenged from the seawater is insignificant.

Variant 2. An example is the conditions for the formation of the sediments of the SU_1 and SU_2 Zones in the Atlantis II Deep. There is high-intensive effluence of high-temperature hydrothermal solutions into a deep where a thick brine layer exists. The salt-bearing stratum exposed on the walls of the deep and hydrothermal solutions are the sources of the salts in the brines, and the hydrothermal solutions are ejected from basaltic crust (Fig. 4.39a). The continuity of the brine layer is not disturbed and almost no dissolved oxygen enters into the brines. Fe in the hydrothermal sulfide- and oxyhydroxide minerals is prevalent in the metal-bearing matter of sediments in the near orifice facies and in the facies of the deep. The content of sulfide minerals is higher in the near orifice facies. Fe-oxyhydroxide minerals form in the upper part of the brine layer. Mn precipitates in particulate Mn-hydroxide minerals in the brine-seawater boundary zone. The Mn-hydroxide minerals are dissolved during their settling in the brine layer. The process of Mn precipitation and dissolution is cyclic. The contents of Mn are low in the metal-bearing matter in the sediment of the facies of the deep and especially in the near orifice facies. The content of Mn in the metal-bearing matter is higher in sediments of the facies of the brine-seawater boundary. Only a very small part of the Mn contributed by the hydrothermal solutions accumulates in the sediments within the deep. A predominant part of it is dissipated outside of the deep. The contents of Zn, Cu, and Co in the metal-bearing matter of sediments are high. They gradually decrease as the distance increases from the hydrothermal orifices; near the orifices the conditions are favorable for the accumulation of Zn and Cu in sulfide minerals. Anhydrite can accumulate in sediment of the near orifice facies. Some accumulation of Ni derived from the seawater takes place in sediments of the facies of the brine-seawater boundary.

Variant 3. An example is the conditions for the formation of the sediments of the AM Zone in the Atlantis II Deep. Intensive effluence of high-temperature hydrothermal solutions that is less intensive than in Variant 2 takes place in a deep where a thick brine layer exists. The salt-bearing stratum exposed on the walls of the deep and hydrothermal solutions are the sources of the salts in the brines and the hydrothermal solutions are ejected from the basaltic crust (Fig. 4.39a). The continuity of the brine layer is not disturbed and very little amount of dissolved oxygen enters the brines. The pattern for sedimentation of the metal-bearing matter is similar to that in Variant 2, however hydrothermal sulfide minerals and anhydrite accumulate only in the sediments of the near orifice facies. Fe-oxyhydroxide minerals are prevalent in the sediments of the facies of the deep. The contents of Zn, Cu, and Co in the metal-bearing matter decrease gradually as the distance increases from the hydrothermal orifices. In general the contents of Zn and Cu are somewhat lower, and the contents of Co are a little higher than in Variant 2, and all of the metal contents are considerably higher than those in Variant 1. Ni derived from seawater accumulates in sediments of all the facies, but its content is higher in the metal-bearing matter in the sediment of the facies of the brine-seawater boundary.

Variant 4. An example is the conditions for the formation of sediments in the Erba Deep during periods of intense hydrothermal activity. The effluence of high-temperature hydrothermal solutions takes place in a deep where there is a thin brine layer. The salt-bearing stratum exposed on the walls of the deep and hydrothermal solutions ejected from the basaltic crust are the sources of the salts in the brines (Fig. 4.39a). Because of the effluence of the hydrothermal solutions the continuity of the brine layer is disturbed and dissolved oxygen enters into the brines from the overlying seawater. Oxidizing conditions are predominant within the deep. A considerable part of the content of the chemical elements contributed by the hydrothermal solutions leaves the deep, obviously because of the intensive hydrothermal flow. Minerals bearing Fe dominate in the metal-bearing matter of the sediments, especially in the sediments of the near orifice facies. Fe-oxyhydroxide minerals are prevalent. The magnetite content is significant in the sediments of the near orifice facies along with hydrothermal sulfide minerals and gypsum. The Mn-oxyhydroxide mineral contents are notable only in the sediments of the facies of the brine-seawater boundary.

Variant 5. An example is the conditions for the formation of sediments in the Gypsum Deep during periods of intense hydrothermal activity. There is no exposure of salt-bearing strata on the walls of the deep. Before entering the deep the hydrothermal solutions migrate through salt-bearing strata and transport salts into the deep (Fig. 4.39b). Presumably some disturbance in the continuity of the brine layer takes place. Probably dissolved oxygen enters the deep, but in limited amounts. The redox conditions and conditions for the formation of mineral components in the metal-bearing matter apparently are intermediate between those for Variants 2 and 4. Hydrothermal sulfide minerals of Fe are predominant in the metal-bearing matter in the sediments of the near orifice facies, and magnetite is present as well. The metal-bearing matter has notable contents of Cu and Co. Fe-oxyhydroxide minerals are prevalent in the metal-bearing matter in the sediments of the facies of the deep, and hydrothermal sulfide minerals and magnetite are present in small amounts. Here, as well as in the metal-bearing matter in the sediments of the near orifice facies, Mn minerals are absent and the contents of Mn are low. The metal-bearing matter in the sediments of both facies has a low content of Zn. The accumulation of Ni in the metal-bearing matter of the sediments is insignificant.

Variant 6. An example is the conditions for the formation of sediments in the Thetis Deep during periods of intensive hydrothermal activity. The effluence of high-temperature hydrothermal solutions takes place in a deep without the exposure of salt-bearing strata on the walls of the deep. Hydrothermal solutions entering the deep are highly saline because of the migration of the seawater that is involved in the hydrothermal circulation system through the salt-bearing strata when it descends through them in the circulation system (Fig. 4.39c). Hydrothermal solutions contribute sufficient salts into the deep to maintain the existence of a brine layer, the continuity of which is disturbed. Dissolved oxygen exists in the brines, but in concentrations lower than in seawater. Redox conditions in the brines are more reducing than those in seawater. Under such conditions magnetite, goethite, and lepidocrocite can form, Mn(II) is not oxidized and Mn(IV)-hydroxide minerals do not precipitate. Mainly Fe-oxyhydroxide minerals and magnetite, and to a lesser extent, hydrothermal sulfide minerals accumulate in the metal-bearing matter in the sediments of the facies of the deep. The metal-bearing matter in the sediments of the facies of the brine-seawater boundary is composed mainly of Fe- and Mn-oxyhydroxide minerals; and the Fe-minerals are prevalent. The contents of Zn, Cu, and Co in the metal-bearing matter are comparable to those in Variants 1 and 4. The contents of Cu and Co in the metal-bearing matter in the sediments of the facies of the deep are higher than in the metal-bearing matter in the sediments of the facies of the brine-seawater boundary. The latter have a higher content of Ni that is derived from seawater.

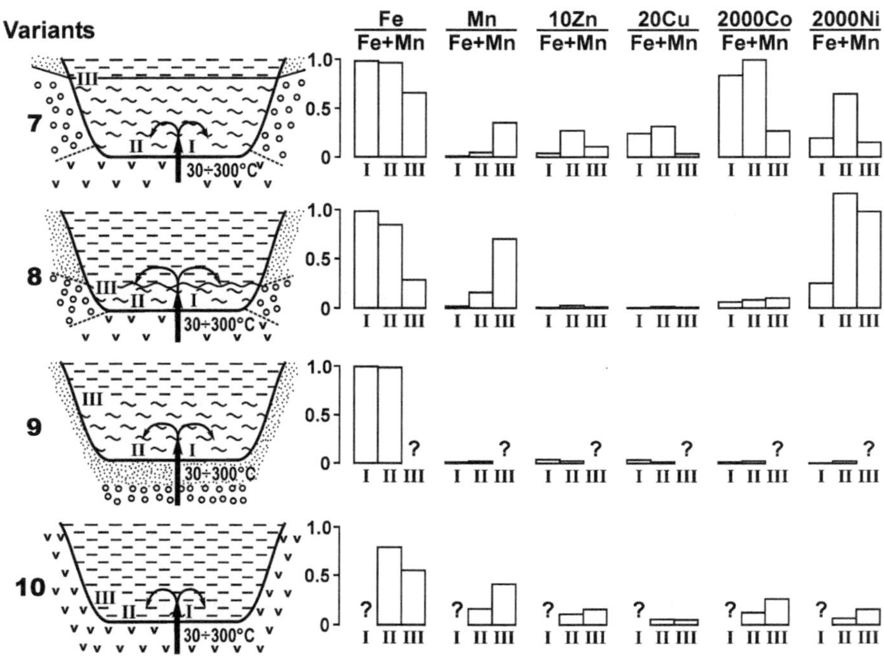

Fig. 4.41. The second group variants of conditions for formation of metal-bearing matter of metalliferous sediments in the deeps of the Red Sea rift zone. Legend see in Fig. 4.40.

The second group has the characteristic environmental conditions for the formation of the metal-bearing matter that exist in deeps where there is effluence of hydrothermal solutions that have temperatures in the range from 30° to 300°C (Fig. 4.41). The zone of contact of the primary metal-bearing hydrothermal fluids with the bottom water or brines is located near the surface of the floor. Partial unloading of the metals and the formation of sulfide minerals takes place in this zone. The residual solutions that reach the deeps are depleted considerably in Fe, Cu, Zn, and S. The effluence of the hydrothermal solutions is not intense and they can disturb the continuity of only a thin brine layer. Variations in the intensity of the effluence apparently can cause a shift in the boundary of the zone of the near orifice facies. The more intensive the effluence the larger the area of sediments of the near orifice facies.

Variant 7. An example is the conditions for the formation of interlayered metalliferous sediments in the Suakin, Port Sudan, and Albatross Deeps. The effluence of mid-temperature hydrothermal solutions takes place in a deep where there is a thick brine layer. The salt-bearing stratum exposed on the walls of the deep and hydrothermal solutions are the sources of the salts in the brines and the hydrothermal solutions are ejected from the basaltic crust (Fig. 4.39a). The continuity of the brine layer is not disturbed. Very little dissolved oxygen enters the brines. The metal-bearing matter in the sediments of the near orifice facies and the facies of the deep is mainly composed of Fe-oxyhydroxide minerals and/or sulfide minerals and the latter result from the sulfate-reduction. The content of Mn is low. Minerals containing Mn(IV) are absent. In the metal-bearing matter in the sediments of the facies of the deep the Mn content increases slightly because of the occurrence of manganosiderite minerals. The metal-bearing matter of the facies of the brine-seawater boundary is enriched considerably in Mn. The contents of Zn and Cu in the metal-bearing matter of the sediments are less than in that of the sediments in Variants 2 and 3. The content of Cu is higher in the metal-bearing matter in the sediments of the near orifice facies and facies of the deep, and the content of Zn is higher in the metal-bearing matter in the sediments of the facies of the deep. The content of Co is significantly high and its distribution is similar to that of the content of Cu. The content of Ni is high, especially in the metal-bearing matter in the sediments of the facies of the deep; this is evidence of the contribution of the trace elements from seawater.

Variant 8. An example is the conditions for the formation of interlayered metalliferous sediments in the Shagara and Erba Deeps, except for the sediments in the Erba Deep, which correspond to the conditions considered in Variant 4. The effluence of mid-temperature hydrothermal solutions takes place in a deep where there is a thin layer of brines. The salt-bearing stratum exposed on the walls of the deep and hydrothermal solutions are the sources of the salts in the brines and the hydrothermal solutions are ejected from the basaltic crust (Fig. 4.39a). Because of the effluence of the hydrothermal solutions the continuity of the brine layer is disturbed, and dissolved oxygen enters the brines from the overlying seawater. Oxidizing conditions are predominant throughout the deep. On the effluence of the hydrothermal solutions a considerable amount of chemical elements contributed by them rises to the upper boundary of the brine layer or higher. The elements that pass into the particulate matter, primarily Fe and then Mn, settle on the floor. Other elements, such as Zn, Cu, and Co, are carried out of the deep. The differentiation of Fe and Mn occurs within the deep: Fe enriches the metal-bearing matter in the sediments of the near orifice facies most of all, and Mn enriches the sediments of the facies of the brine-seawater boundary. The high content of Ni, especially in the metal-bearing matter in the sediments of the facies of the deep and in the sediments of the facies of the brine-seawater boundary, is the highest in all of the variant in the second group, which is evidence that seawater is an essential source of the trace elements.

Variant 9. An example is the conditions for the formation of the interlayered metalliferous sediments in the Vema, Gypsum, Kebrit, and Valdivia Deeps, except for the sediments of the Gypsum Deep, which correspond to those described for Variant 5. There is no exposure of salt-bearing strata on the walls of the deeps, except in the Valdivia Deep. Before entering the deeps the hydrothermal solutions migrate through salt-bearing strata and contribute salts to the deeps (Fig. 4.39b, d). The thickness of the brine layer and disturbances in its continuity may vary depending on the morphology of a deep, the intensity of the hydrothermal effluence, the density of the hydrothermal solutions and brines, and dynamic of bottom water. The redox conditions may vary from low-oxidizing to low-reducing. When low-oxidizing conditions exist the metal-bearing matter in sediments of the near orifice facies and facies of the deep is composed mainly of Fe-oxyhydroxide minerals, and when low-reducing conditions exist the contents of the sulfide minerals that formed by the sulfate-reduction are significant. Mn-oxyhydroxide minerals are absent in both of these sediment facies. The contribution and accumulation of Cu and Co from hydrothermal sources is practically absent because of their significant unloading from the hydrothermal solutions before their entering the deep. The content of Zn, although presents in small amounts, is higher[38] in the metal-bearing matter in the sediments of the near orifice facies. The content of Ni in the metal-bearing matter in the sediments is very low.

Variant 10. An example is the conditions for the formation of metalliferous sediments in the Thetis and Hadarba Deeps with the exception of the sediments in the Thetis Deep where the conditions correspond to those considered in Variant 6. The effluence of mid-temperature hydrothermal so-

---

[38] Hydrothermal sulfide mineral chimneys have been dredged in the Kebrit Deep (Puchelt and Laschek 1984; Blum and Puchelt 1991; Hembelen et al. 1996; Stoffers et al. 1998). These sulfide chimneys are rich in Fe and Zn and have very low concentrations of Cu. This is interpreted as a result of the relatively low (120°C to 300°C) temperatures of the hydrothermal solutions (Scholten et al. 2000). Cu-bearing phases may be found in the deeper stockworks or fractures under the deep (Blum and Puchelt 1991).

lutions occurs in a deep without brines. There is no exposure of salt-bearing strata on the walls of the deep. Hydrothermal solutions entering the deep are highly saline because of the migration of the seawater that is involved in the hydrothermal circulation system through the salt-bearing strata when it descends through them in the circulation system (Fig. 4.39c). The solutions contribute the salts to the deep, but the contribution is not high enough to maintain a significant brine layer; salinity and temperature of the water at the seafloor are some higher than in normal seawater at these depths. The redox conditions near the floor are oxidizing. The metal-bearing matter in the sediments of the facies of the deep and the distal facies is composed of Fe- and Mn-oxyhydroxide minerals. The amounts of Mn-oxyhydroxide minerals increases with the distance from the hydrothermal orifices. The content of Cu in the metal-bearing matter is low. The contents of Zn, Co, and Ni are higher in the metal-bearing matter in the sediments of the distal facies. The accumulation of Ni in the metal-bearing matter is evidence of the contribution of the trace elements from seawater.

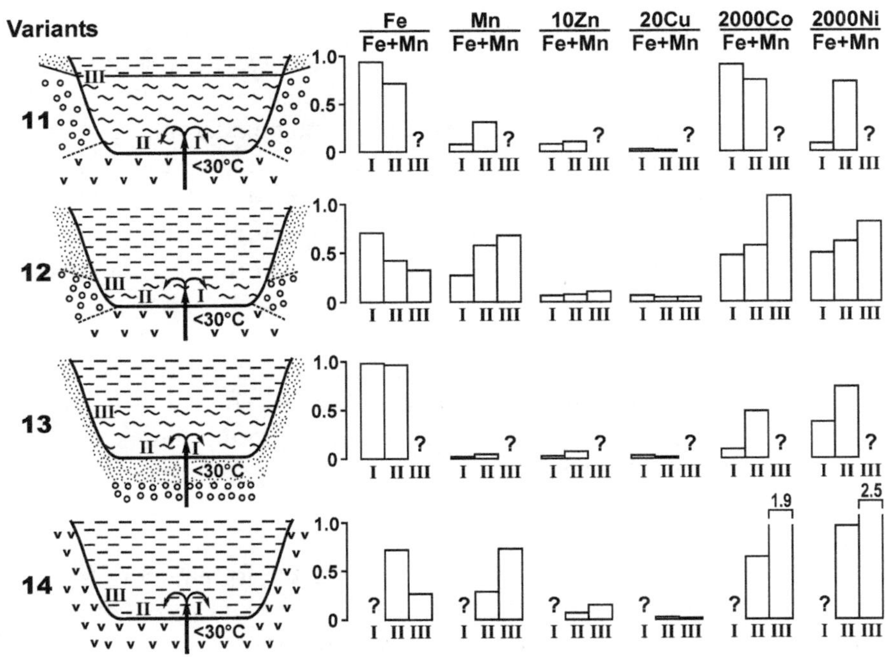

Fig. 4.42. The third group variants of conditions for formation of metal-bearing matter of metalliferous sediments in the deeps of the Red Sea rift zone. Legend see in Fig. 4.40.

The third group has the characteristic environmental conditions for the formation of metalliferous sediments that have a low content of the metal-bearing matter in deeps where there is effluence of hydrothermal solutions that have temperatures <30°C (Fig. 4.42).

The zone of contact of the primary metal-bearing hydrothermal fluids with seawater or with brines is located in the subsurface horizons of the crust. The unloading of most of the metals from hydrothermal fluids in the components of sulfide minerals and probably in the silicate minerals occurs in this zone. The residual hydrothermal solutions are highly depleted in Fe, Cu, and Zn, but not in Mn. The intensity of the effluence of the hydrothermal solutions is generally low and they do not disturb the continuity of the thick or sometimes thin brine layers. The content of the metal-bearing matter in the sediments is usually not high, <10%. Under favorable conditions it can accumulate in significant amounts only in areas close to the hydrothermal orifices.

Variant 11. An example is the conditions for the formation of interlayered low-metalliferous sediments, as found in the Suakin, Port Sudan, and Albatross Deeps. The effluence of low-temperature hydrothermal solutions occurs in a deep where there is a thick layer of brines. The salt-bearing stratum exposed on the walls is the main source of the salts in the brines and the hydrothermal solutions are ejected from the basaltic crust (Fig. 4.39a). The continuity of the brine layer is not disturbed. Little or no dissolved oxygen enters the brines. The metal-bearing matter in the sediments of the near orifice facies and facies of the deep is composed mainly of Fe-oxyhydroxide and/or sulfide minerals that are formed by the sulfate-reduction. The content of Mn is low. Minerals bearing Mn(IV) are absent. In the metal-bearing matter in the sediments of the facies of the deep the Mn content increases because of the presence of manganosiderites. Mn(IV)-oxyhydroxide minerals probably occur in the sediments of the facies of the brine-seawater boundary. The contents of Zn and especially of Cu in the metal-bearing matter of the sediments are low. The content of Co is relatively high, especially in the metal-bearing matter in the sediments of the near orifice facies. The metal-bearing matter in the sediments of the facies of the deep is enriched in Ni, which is evidence of the contribution of the trace elements from the seawater.

Variant 12. An example is the conditions for the formation of interlayered low-metalliferous sediments as in the Shagara and Erba Deeps. The effluence of low-temperature hydrothermal solutions occurs in a deep that has a thin brine layer. The salt-bearing stratum exposed on the walls is the main source of the salts in the brines and the hydrothermal solutions are ejected from the basaltic crust (Fig. 4.39a). Although the continuity of the brine layer is not disturbed significantly by the effluence of the hydrothermal solutions, dissolved oxygen enters the brines from the overlying seawater in amounts that are sufficient to enable the development of oxidizing or low-oxidizing conditions. The metal-bearing matter in the sediments is composed of Fe- and Mn-oxyhydroxide minerals. Fe-oxyhydroxide minerals are prevalent in the sediments of the near orifice facies, Mn-oxyhydroxide minerals are present in the sediments of the facies of the deep and especially in the sediments of the facies of the brine-seawater boundary. The contents of Zn and especially of Cu in the metal-bearing matter of the sediments are low. The content of Cu decreases from that in the sediments of the near orifice facies to that in the sediments of the facies of the brine-seawater boundary, and the content of Zn varies conversely. The contents of Co and Ni are significantly high in the metal-bearing matter in the sediments and maximum values occur in the sediments of the facies of the brine-seawater boundary. This evidence indicates that seawater is an important source of the trace elements.

Variant 13. An example is the conditions for the formation of interlayered low-metalliferous and non-metalliferous sediments as in the Vema, Gypsum, and Kebrit Deeps. The hydrothermal solutions migrate through salt-bearing strata before their effluence and contribute the salts to a deep (Fig. 4.39b, d). A brine layer may be present or absent depending on the morphology of a deep, the intensity of hydrothermal effluence, the density of the hydrothermal solutions and brines, and the dynamics of the bottom water.

When a brine layer is present in a deep it acts as a trap and inhibits the dispersion of the hydrothermal matter. The thickness of the layer and the amount of disturbance in its continuity can vary and the redox conditions may be low-oxidizing or low-reducing. When low-oxidizing conditions exist the metal-bearing matter in the sediments of the near orifice facies and facies of the deep is composed mainly of Fe-oxyhydroxide minerals and when conditions are reducing the sulfide mineral content formed by the sulfate-reduction becomes significant. Mn-oxyhydroxide minerals are absent, at least in the sediments of these two facies. Because of the considerable unloading of the hydrothermal solutions before entering the deep the contents of Zn and especially of Cu in the metal-bearing matter of the sediments are low. The content of Cu is higher in the sediments of the near orifice facies and the content of Zn is higher in the sediments of the facies of the deep. The contents of Co and Ni are significant in the metal-bearing matter in the sediments of the facies of the deep and this is evidence of the contribution of the trace elements from seawater.

Where a brine layer is absent the hydrothermal matter contributed to the deep is highly dispersed and its content in the sediments is very low. Apparently iron-manganese and/or manganese hydrothermal mineral crusts can form near the orifices of the hydrothermal springs.

Variant 14. An example is the conditions for the formation of low metalliferous as in the Hadarba Deep. The effluence of low-temperature hydrothermal solutions occurs in a deep where brines are absent. There is no exposure of salt-bearing strata on the walls of the deep. The hydrothermal solutions entering the deep are saline because of the migration of the seawater that is involved in the hydrothermal circulation system through the salt-bearing strata when it descends through them in the circulation system (Fig. 4.39c). They contribute the salts to the deep, but in very small amounts and a stable brine layer does not form. The redox conditions near the bottom are oxidizing. The metal-bearing matter in the sediments of the facies of the deep and of the distal facies are composed of Fe- and Mn-oxyhydroxide minerals. Fe-oxyhydroxide minerals are prevalent in sediments of the facies of the deep, and Mn-oxyhydroxide minerals are prevalent in the sediments of the distal facies. The metal-bearing matter has a low content of Zn, Cu is practically absent and the contents of Co and Ni are significant. The metal contents in the metal-bearing matter from the sediments of the distal facies have maximum values for the Co and Ni contents compared to the values found in all of the variants. This is evidence of the importance of seawater as a source of trace elements.

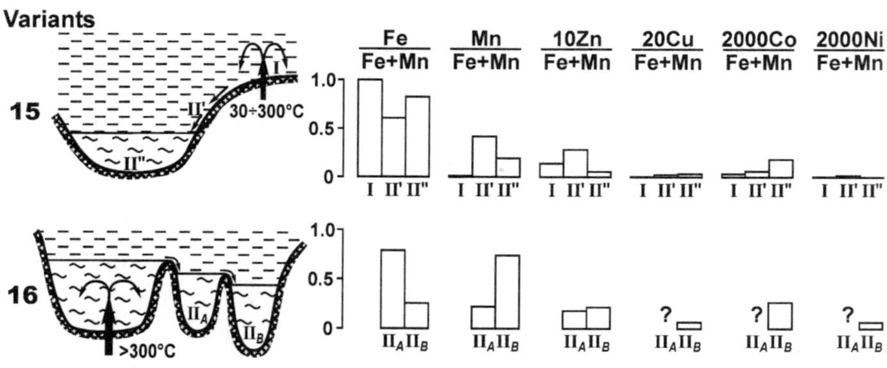

Fig. 4.43. The fourth group variants of conditions for formation of metal-bearing matter in metalliferous sediments in the deeps of the Red Sea rift zone. Legend see in Fig. 4.40.

The fourth group has the characteristic environmental conditions for the formation of metalliferous sediments during the migration of heavy brines that are enriched in metals. As mentioned previously there are many variants for this group that could be considered but only two of them have been explored (Fig. 4.43).

Variant 15. An example is the conditions for the formation of metalliferous sediments in the East Nereus Deep. The effluence of mid-temperature high-saline hydrothermal solutions occurs at a marginal fault scarp of a deep. The brines flow down a slope and accumulate in the deepest part of the deep. The metal-bearing matter in the sediments of the near orifice facies consists mainly of Fe-oxyhydroxide minerals. At a distance from the orifice the metal-bearing matter of the sediments of the slope facies is composed of Fe-oxyhydroxide and significant amount of Mn-oxyhydroxide minerals. The sediments of these two facies form in oxidizing conditions. And the redox conditions for the formation of the sediments of the facies of the deep depend on proportion of the brine inflow and the dimensions of the deep. The conditions can be either low-oxidizing or low-reducing. When there are low-oxidizing conditions and a great inflow of brines the metal-bearing matter is composed of Fe- and Mn-oxyhydroxide minerals. When there are low-reducing conditions and a low inflow of brines Fe-oxyhydroxide minerals form along with the sulfide minerals formed by the sulfate-reduction. The average values for the chemical compositions of this matter are shown in Fig. 4.43.

The metal-bearing matter in the sediments of the near orifice facies, and especially the facies of the slope, are enriched somewhat in Zn. The content of Zn is low in the metal-bearing matter in the sediments of the facies of the deep. The contents of Cu and Co are relatively low in all three facies and increase from the contents in the sediments of the near orifice facies to those in the sediments of the facies of the deep. There is practically no accumulation of Ni, which is considered as evidence that shows that seawater is an insignificant source of the trace elements.

Variant 16. An example is the conditions for the formation of metalliferous sediments in the Chain A and Chain B Deeps. The effluence of high-temperature hydrothermal solutions occurs in a deep-reservoir that is filled with brines. The Atlantis II Deep is a deep-reservoir for the Chain Deeps (Fig. 2.2). The brines overflow from the deep-reservoir into the A deep, and then from the A deep into the B deep. The characteristics of the metal-bearing matter that accumulates in the deep-reservoir were described in Variants 1 to 3. The metal-bearing matter in the deeps A and B is composed mainly of Fe- and Mn-oxyhydroxides. The Fe-oxyhydroxides are prevalent in the A deep, and Mn-oxyhydroxides are prevalent in the B deep, which demonstrates the differentiation of Fe and Mn. The contents of Zn in the metal-bearing matter of the sediments in both of the deeps are similar and are not high. The contents of Cu, Co, and Ni in the metal-bearing matter in the sediments of the B deep are relatively low.

CHAPTER 5 RECONSTRUCTION OF THE HISTORY OF HYDROTHERMAL ACTIVITY AND MINERAL FORMATION IN THE OCEANS BASED ON THE STUDIES OF METALLIFEROUS SEDIMENTS

While direct study of hydrothermal activity and the associated mineral formation as well as study of their scales are possible in the modern ocean, researchers are deprived of such opportunity for past geologic time. Even in cases where hydrothermal deposits that formed in the geological past are preserved they have been covered by sediments or overlain by lava and have become inaccessible after a time for observation and sampling. Rarely are these buried hydrothermal deposits found during deep-sea drilling or coring operations, and their direct study on an ocean scale is impossible at present. Even within hydrothermal fields, without special preliminary detailed surveys, the probability of intersecting hydrothermal edifices that are covered by sediments becomes a rare event. For example, within the TAG hydrothermal field, which has an area of about 25 km^2, after many years of investigations only once has a gravity core intersected a gossan over a sediment-covered hydrothermal mound (Bogdanov et al. 1995b). Studies of the metalliferous sediments sampled in the gravity- and piston cores, and in the cores obtained in the deep-sea drilling program can be very useful for the reconstruction of the history of hydrothermal activity and mineral formation.

Only a small part of the metal-bearing hydrothermal matter contributed to the ocean by high-temperature hydrothermal solutions accumulates in massive hydrothermal bodies. A prevalent part of this matter accumulates in the bottom sediments in the regions adjacent to the centers of hydrothermal activity. It has been shown in the previous chapters that, taking into account the distance from hydrothermal fields and/or vents and in the absence of the redeposition of sediment material, the accumulation rate of the hydrothermal metal-bearing matter in both proximal- and in distal metalliferous and low-metalliferous sediments reflects the intensity of the hydrothermal activity and, as a consequence, the amount of mineral formation and deposition associated with it. As for the proximal metalliferous sediments, their composition also reflects metallogenic specialization of the hydrothermal fluids and the massive accumulation of sulfide minerals. The accumulation of coarse-grained proximal metalliferous sediment reflects stages of active hydrothermal discharge and the partial destruction of massive hydrothermal bodies. These properties of the metalliferous sediments enable their use in reconstructing the location and intensity of hydrothermal activity and mineral formation in past geological time, and in determining areas of present location of buried hydrothermal deposits.

5.1. Reconstruction of the history of hydrothermal activity and mineral formation based on the studies of sediment cores

The Active hydrothermal mound in the **TAG hydrothermal field** was the first massive hydrothermal accumulation where attempts have been made to reconstruct the history of hydrothermal activity and the mineral formation with use of data from studies of proximal metalliferous sediments (Lisitzin et al. 1989, 1990; Bogdanov et al. 1992). Core AMK-1785 (Fig. 5.1, 5.2) was chosen for this reconstruction.

Two layers that have high contents of Fe and Cu, the main hydrothermal metals in the metalliferous sediments from the TAG hydrothermal field Sect. 1.6.1), are noticeable in the core. Mineral indicators of the greenstone alteration of the basalt rocks are present in the sediments. Their total content in the carbonate-free part of the sandy-silt grain size fraction of the sediments rarely exceeds 5%. But increased contents exceeding 10% exist in two horizons at 145–147 cm and 32–35 cm in the core. Their ages as dated are 24–25 ka and 15–17 ka and are located immediately below the layers of metalliferous sediments. There is no correlation between the contents of these minerals and the contents of basalt fragments including the altered ones. The volcanic material in the TAG hydrothermal field has ages as dated of ~27 ka, ~18 ka, and 8–10 ka (Bogdanov et al. 1992, 1994). These data provide evidence of alternative stages of hydrothermal and volcanic activity and exhibit a cyclicity of these processes in periods of about 10 ka.

Following these studies materials were obtained on metalliferous sediments accumulated near the Mir Zone and Alvin Zone. These data are summarized in Fig. 5.2. The results of dating of the massive sulfide samples from the hydrothermal accumulations in the TAG hydrothermal field are shown as well. Synchronism in the stages of intensive accumulation of the metal-bearing matter in the sediments, regardless of the place of their

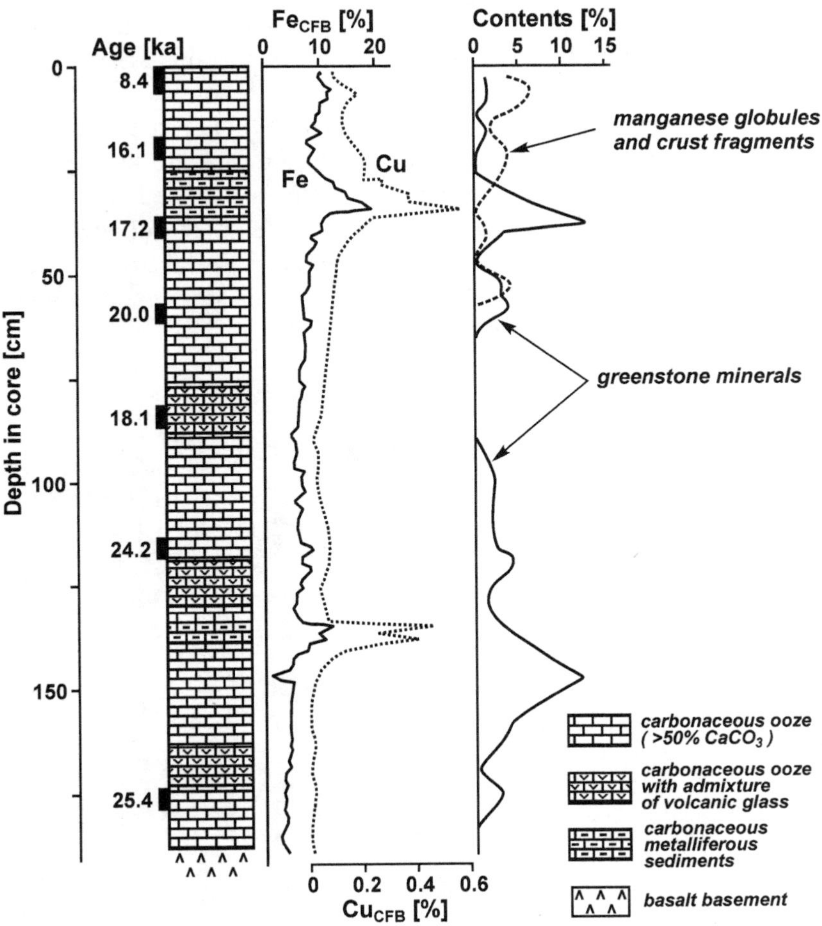

Fig. 5.1. Composition of bottom sediments in Core AMK-1785. Contents of greenstone minerals, manganese globules and crust fragments are shown in the sandy-silt grain size fraction of sediments. After Lisitzin et al. (1990).

deposition in relation to the massive hydrothermal accumulations, is evident at least in the last 25 ka. During this time there are three stages, ≥25–23 ka BP, 17–12 ka BP, and 7–3 ka BP, that demonstrate cyclicity within periods of ~10 ka. A reflection of the present phase of hydrothermal activity, which began ~100 years ago at the Active mound (Lalou et al. 1993), has not been found in the sediments collected nearby[39]. With an average sedimentation rate in the TAG hydrothermal field of ~2 cm ka^{-1}, a layer

[39] Presumably, it exists in sediments sampled in the cores collected directly on the Active mound (German et al. 1993). Unfortunately these sediments have not been dated.

about 2 mm thick would accumulate during this period of time, but with the common sampling methods in use, layers such as this are not detected or lost. On the other hand, within the TAG hydrothermal field the thickness of the layer in which mixing takes place due to bioturbation is about 6 cm (Kuptsov 1993). With the sedimentation rate of 2 cm ka^{-1} this thickness of sediments corresponds to a period of time of ~3 ka. For this reason, together with the less than ideal sampling methods, it has not been possible to identify the sediments accumulated during the final stage of high-temperature hydrothermal activity at the Mir mound. This stage ended about 700 years ago (Lalou et al. 1995; Fig. 5.2).

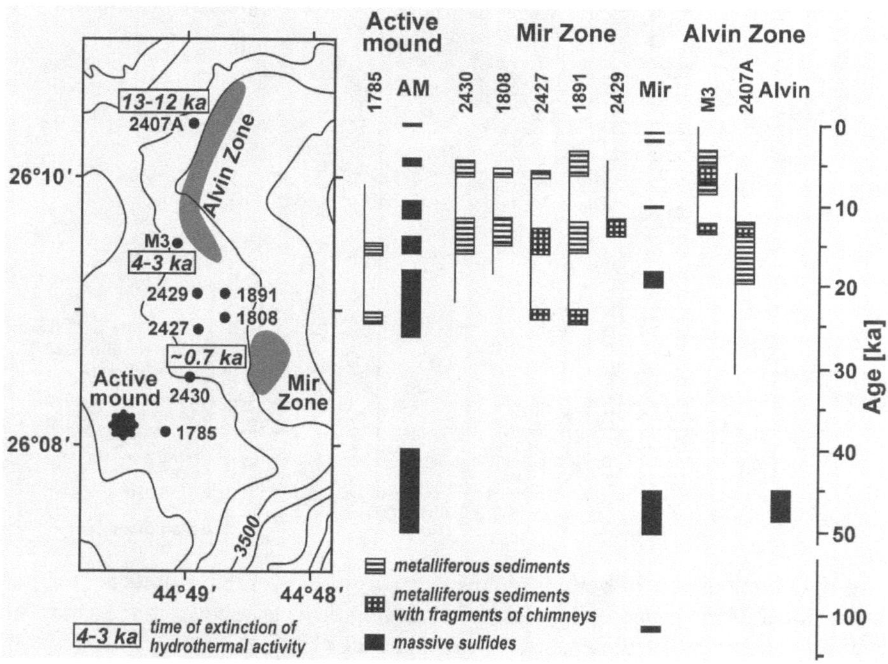

Fig. 5.2. Stages of intensive accumulation of the metal-bearing matter in sediments from cores collected at the TAG hydrothermal field and isotopic ages of samples of massive sulfides collected from hydrothermal accumulations. Cores AMK-1785, 1808, and 1891 were collected on Cruise 15 and Cores AMK-2407A, 2427, 2429, and 2430 on Cruise 23 of R/V *Akademik Mstislav Keldysh*. Data on these cores – materials of P.P. Shirshov Institute of Oceanology, data on Core M3 from Metz et al. (1988), data for the ages of the massive sulfides from Lalou et al. (1993, 1995).

There is another point to be considered. If at the Active mound the cyclicity of hydrothermal activity with a period of about 10 ka really exists, the next stage of activity should be in 3–7 ka. At present hydrothermal activity and formation of mineral matter at the Active mound probably are depressed. The existence of mineral formation during the depressed stages of hydrothermal activity, that is not reflected or poorly reflected in sediments, can explain the existence of hydrothermal sulfide minerals that have ages that do not correspond to the ages of the mentioned stages of increased hydrothermal activity and the accumulation of the metal-bearing matter in the sediments (Fig. 5.2).

Gradual narrowing of the area of hydrothermal activity in the TAG hydrothermal field apparently has been taking place at least during the last 25 ka. The approximate time of extinction of hydrothermal activity in the Alvin Zone can be established from the data shown in Fig. 5.2. The synchronism in the stages of intensive accumulation of the metal-bearing matter in sediments, regardless of the place of their deposition relative to that of the massive hydrothermal deposits, is evidence of synchronism in the stages of hydrothermal activity. In the northern part of the Alvin Zone, based on the data from Core AMK-2407A, hydrothermal activity ceased 13–12 ka BP, and in the southern part, based on the data from Core M3, about 4–3 ka BP. The last stage of hydrothermal activity in the northern part is synchronous to the penultimate stage in the southern part of the zone. At the Mir mound hydrothermal activity became extinct about 700 years ago (Lalou et al. 1995). The gradual narrowing of the area of hydrothermal activity apparently results from the gradual reduction in the size of the magma chamber (Bogdanov 1997).

The history of hydrothermal activity and volcanism in the **Axial Seamount, Juan de Fuca Ridge**, during the last 35 ka has been reconstructed on the basis of data that show temporal variations in the occurrences of the indicators of volcanism and hydrothermal activity in the sediments that accumulated in the area of this seamount (Fig. 5.3) and on temporal variations in the accumulation rate of hydrothermal Fe (Fig. 1.43). This is summarized in Fig. 5.4 and described on the results of the investigations of Bogdanov et al. (1990) and Lisitzin et al. (1990).

The Axial Seamount is located immediately above the Cobb hot spot (Desonie and Duncan 1990). A volcanic body of the Axial Seamount formed when this hot spot was close to or near the spreading axis of the Juan de Fuca Ridge. However, the first complex group of effusive rocks dated >30 ka BP (Lisitzin et al. 1990) presumably formed on the flank of the rift zone; at that time the axis of the Juan de Fuca Ridge was not located above the Cobb hot spot.

In the period *from 30 to 20 ka BP*, following the termination of the first stage of volcanism, the mount located adjacent to the tectonic zone of the spreading ridge was dissected by a system of fissures. Channel ways were formed for the metal-bearing hydrothermal fluids and steady hydrothermal circulation developed. Massive hydrothermal deposits were formed on the surface of the mount at the orifices of hydrothermal discharge and metalliferous sediments accumulated around the hydrothermal field.

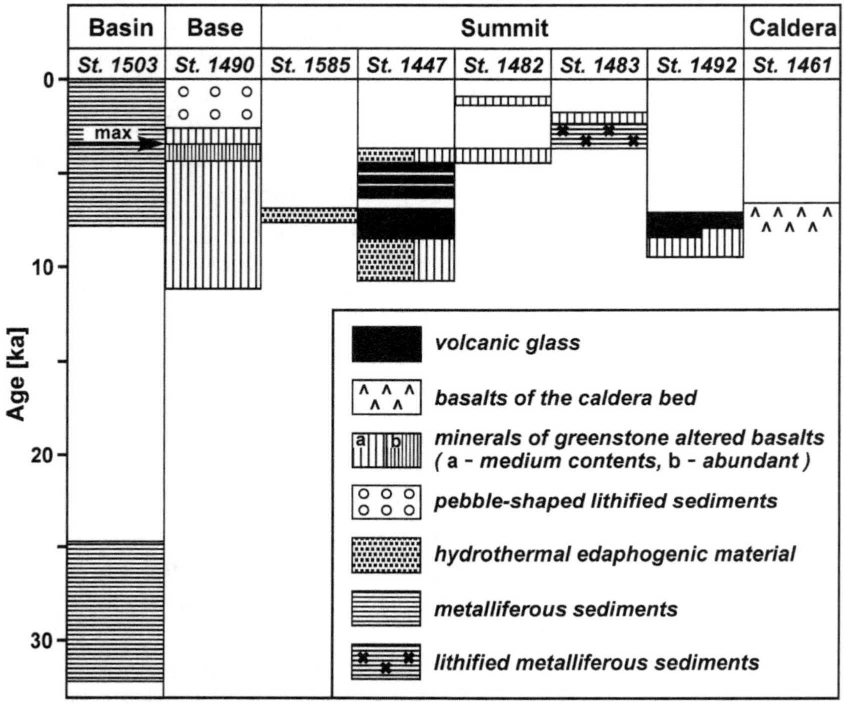

Fig. 5.3. Temporal variations in the occurrence of the indicators of volcanism and hydrothermal activity in sediment from the area of the Axial Seamount. After Bogdanov (1990a).

From 12 to 7 ka BP, hydrothermal activity was terminated almost completely. Basalt rocks blanketed and overlapped the volcanic complex of the mount and the hydrothermal bodies that had formed earlier. After the voiding of a magmatic chamber a collapse caldera formed. This stage of volcanism, presumably, was associated with a new geologic setting when the spreading center of the Juan de Fuca Ridge intersected the Axial Seamount.

From 6 to 3 ka BP the next active stage of hydrothermal activity occurred; but it was less intensive than the previous one.

In the period *from 2 to 0 ka BP*, there was a stage of caldera volcanism that was associated with the spreading of the plates; the spreading axis intersected the central part of the caldera. Hydrothermal activity of low intensity developed at the periphery of the caldera along the fault zones that formed the caldera. There is a decline at the present time in hydrothermal discharge following the eruptions and volcanic activity on the seafloor at the Axial Seamount as observed by Baker et al. (2004).

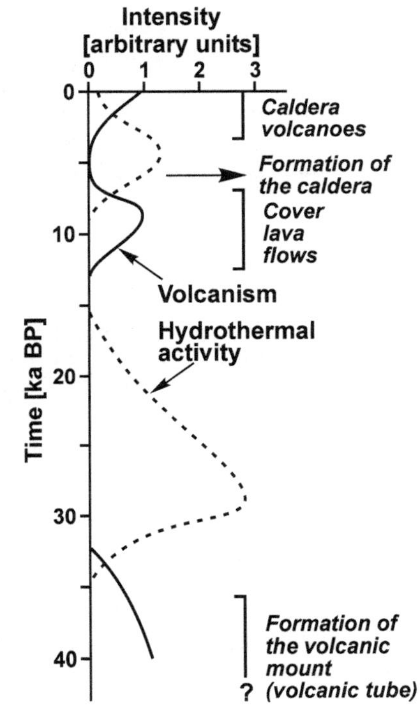

Fig. 5.4. Temporal variations of volcanism and hydrothermal activity at the Axial Seamount. After Lisitzin et al. (1990) with some changes.

The history of hydrothermal activity in the **southern part of the Juan de Fuca Ridge** can be reconstructed on the results of studies of the metalliferous sediments in Core KC-1 (44°44.21'N, 30°33.38'W, depth 2635 m) that accumulated during the last 130 ka (Olivarez and Owen 1989).

The core was collected 15.6 km west of the Juan de Fuca Ridge axis. With a half-rate of spreading of 3 cm a^{-1} during the past 130 ka the distance from the place of accumulation of the oldest sediments in the core to the ridge axis increased by 3.9 km; i.e. the oldest sediments accumulated 11.7 km from the axis. Factor analysis has been carried out for ascertaining the different sources of chemical elements in the sediments (Olivarez and Owen 1989). The relative contents of proximal and distal[40] hydrothermal

[40] These terms used by Olivarez and Owen (1989) have another meaning than the meaning used in this book, because sediments in the core have accumulated far from and outside of hydrothermal fields (or a field).

components in the detrital-free material have been calculated. The proximal component is enriched in Fe, distal component in Ba, Co, Mn, Ni, and REE. The relative contents of the components have varied with time. In general, in the more ancient sediments accumulated closer to the ridge axis the amount of the proximal component is higher (Fig. 5.5).

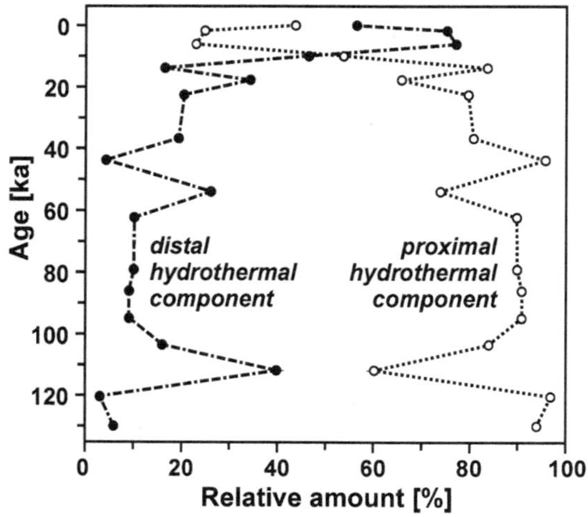

Fig. 5.5. Temporal variations in the relative amounts of proximal and distal hydrothermal components in detrital-free matter of sediments from Core KC-1. Based on data from Olivarez and Owen (1989).

Temporal variations in the accumulation rate of hydrothermal Fe (Fe_h*) in the sediments from Core KC-1 estimated on the data from Olivarez and Owen (1989) are shown in Fig. 5.6. These values are considerably lower than those observed at similar distances from the EPR axis (Boström 1973; Shimmield and Price 1988; Lisitzin et al 1990; Dekov 1994, et al.). In the 130 ka that transpired during the accumulation of the sediments collected in the core the distance of the place of their deposition from the Juan de Fuca Ridge axis increased from 11.7 km to 15.6 km (Fig. 5.6).

A general trend in the temporal variations of Fe_h* exhibits a decrease with decreasing sediment age and an increase in the distance of their deposition place from the ridge axis (Fig. 5.6). If the intensity of hydrothermal activity during the time of sediment deposition and other factors influencing the accumulation of hydrothermal matter, except for the distance of the deposition place from the ridge axis, remained constant and invariable, the value of the Fe_h* would gradually decrease with decreasing age of the sediments because of the increasing distance of the sediment deposition from hydrothermal vents.

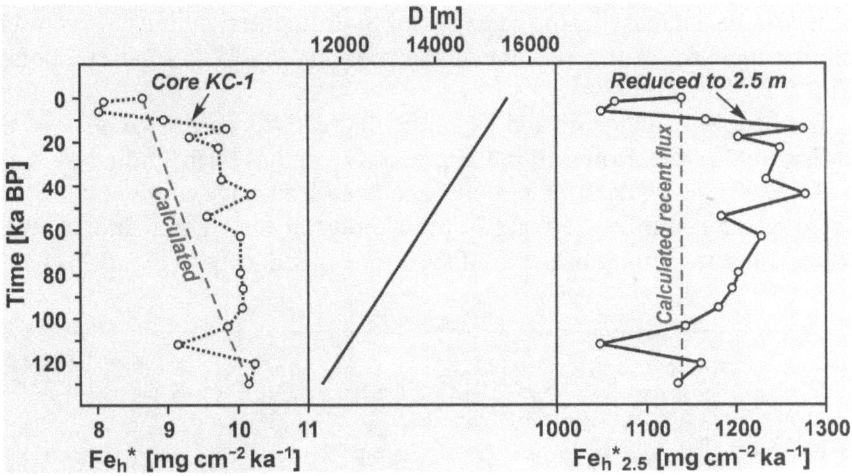

Fig. 5.6. Temporal variations in the distance from the deposition (D) place to the axis of the Juan de Fuca Ridge for sediments collected in Core KC-1, and for Fe_h* estimated from the data of Olivarez and Owen (1989) in comparison with Fe_h* variation calculated from the equation (4.1), $\lg(Fe_h^*/Fe_h^*{}_{2.5}) = 0.23 - 0.56\lg D$, for the value of Fe_h* at the top of the core, as well as calculated values of Fe_h* reduced to the distance of 2.5 m from hydrothermal vents.

Comparison of the data from Core KC-1 is shown in Fig. 5.6 with variation in the settling flux of the hydrothermal Fe that is related to the increasing distance from the axis, as calculated from the empirical equation (4.1) $\lg(Fe_h^*/Fe_h^*{}_{2.5}) = 0.23 - 0.56\lg D$ (see Sect. 4.4), with the value of Fe_h* observed in the upper part of the core or at a distance of 15600 m from the ridge axis. There is good agreement of the general trend of the values for Fe_h* in the core with those in the curve calculated from the equation (4.1). This agreement enables the estimation of the values of Fe_h* near hydrothermal vents during the last 130 ka by using the equation (4.1) and values for Fe_h* that have been calculated for the distance 2.5 m from hydrothermal vents. They are plotted in Fig. 5.6 and show that during this time the settling flux of the hydrothermal Fe near the vents varied little and was from ~1000 to ~1300 mg cm^{-2} ka^{-1}. These values are higher than those calculated for the North Cleft segment of the Juan de Fuca Ridge but lower than those for hydrothermal fields in the Endeavor Ridge and 13°N EPR areas, as well as for the Rainbow and Broken Spur hydrothermal fields (Fig. 4.26). Temporal variations of the $Fe_h^*{}_{2.5}$ values show that ~130, 120, 105, and 10 ka BP the hydrothermal activity at the ridge axis was similar to that in recent times. From 120 to 105 ka BP and from 10 to 1 ka BP it was lower than the recent activity, and in an interval from 105 to 10 ka BP it was higher than the recent activity. During the intervals of time when there was an increase in the intensity of hydrothermal activity there was a relative increase in the amount of the proximal hydrothermal component (Fig. 5.5).

The history of volcanic and hydrothermal activity in **the vicinity of the Rodriguez Triple Junction** in the Indian Ocean has been studied by Kuhn et al. (2000), using data for samples analyzed from the sediment cores collected in the central valley and in the vicinity of the Central Indian Ridge axis and in the vicinity of the Southeast Indian Ridge axis (Fig. 5.7).

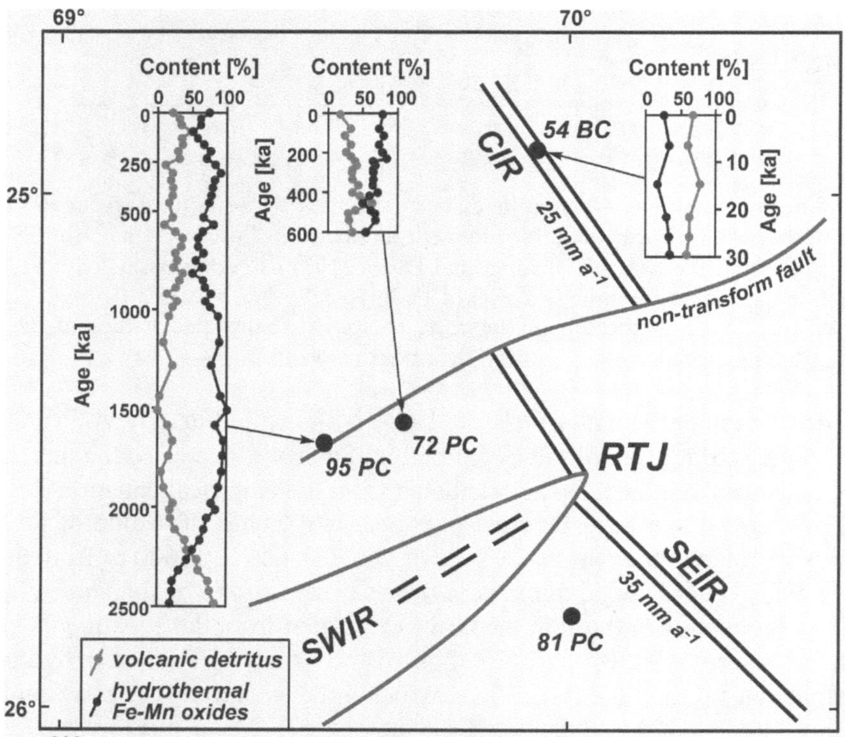

Fig. 5.7. Locations of the cores in the area of the Rodriguez Triple Junction (RTJ) and temporal variations in the relative contents of the volcanic detritus and hydrothermal Fe-Mn oxide end-members. CIR – Central Indian Ridge, SEIR – Southeast Indian Ridge, SWIR – Southwest Indian Ridge. 25 mm a^{-1} and 35 mm a^{-1} – half-rates of spreading for CIR and SEIR respectively. Compiled from Kuhn et al. (2000).

Multivariant analysis of the carbonate-free contents of Si, Al, Ti, Mg, K, Fe, Mn, Cu, Zn, and Ni has shown that the sediments that have accumulated in the vicinity of the Rodriguez Triple Junction consist of two main end-members that are mixed in different proportions in the samples studied. The first end-member consists of volcanic detrital components and the second of hydrothermal Fe-Mn oxides. Temporal variations in the relative contents of these end-members in the sediments that have been dated are shown in Fig. 5.7. In spite of the high relative contents of the hydrothermal Fe-Mn oxide end-member, the accumulation of both end-members in the cores from the axial zone of the Central Indian Ridge (Cores 95 PC and 72 PC) reflect the evolution of a non-transform fault that separates the first and second Central Indian Ridge segments rather than the history of the spreading axis. Data from Core 81 PC reflect magmatic-volcanic activity of the rift axis (Kuhn et al. 2000).

Temporal variations in the Fe_h* in the sediment that has been dated from Core 81 PC estimated on the data from Kuhn et al. (2000) are shown in Fig. 5.8, and data for the background sediments have been taken from Table 1.5. During the last 520–530 ka while these sediments accumulated the distance from the place of their deposition to the Southeast Indian Ridge axis increased from 1.0 km to 19.5 km (Fig. 5.8).

A general trend in the temporal variations of Fe_h* in sediments from Core 81 PC exhibits a decrease with decreasing sediment age and increasing distance of their deposition from the ridge axis. A best fit equation for the relationship of the Fe_h* and the distance data is:

$$\lg Fe_h^* = 2.96 - 0.57 \lg D.$$

The correlation coefficient is high (R=0.88) in spite of the temporal fluctuations in the Fe_h* values relative to the general trend (Fig. 5.8).

It follows from this equation that the value of $Fe_h^*{}_{2.5}$ for the best-fit relationship is 541 mg cm^{-2} ka^{-1}. Using this value the equation can be got:

$$\lg(Fe_h^*/Fe_h^*{}_{2.5}) = \lg Fe_h^* - Fe_h^*{}_{2.5} = 2.96 - 0.57\lg D - \lg 541 =$$

$$= 0.23 - 0.57\lg D.$$

This equation is practically the same as the empirical equation (4.1),

$$\lg(Fe_h^*/Fe_h^*{}_{2.5}) = 0.23 - 0.56\lg D,$$

that shows the relative decrease of Fe_h* with increasing distance from hydrothermal vents in the Pacific Ocean (Sect. 4.4). The equation can be used for estimating Fe_h* values near hydrothermal vents in the past 520–530 ka. The values of Fe_h* reduced to the distance of 2.5 m from the vents have been calculated. They are plotted in Fig. 5.8 and show that during this time the settling flux of the hydrothermal Fe near the vents varied from ~300 to ~900 mg cm^{-2} ka^{-1}. These values are higher than those for the North Cleft segment of the Juan de Fuca Ridge but lower than those for the southern part of the Juan de Fuca Ridge (Fig. 5.6) and for hydrothermal fields of the

Endeavor Ridge and 13°N EPR, as well as for the Rainbow and Broken Spur hydrothermal fields (Fig. 4.26).

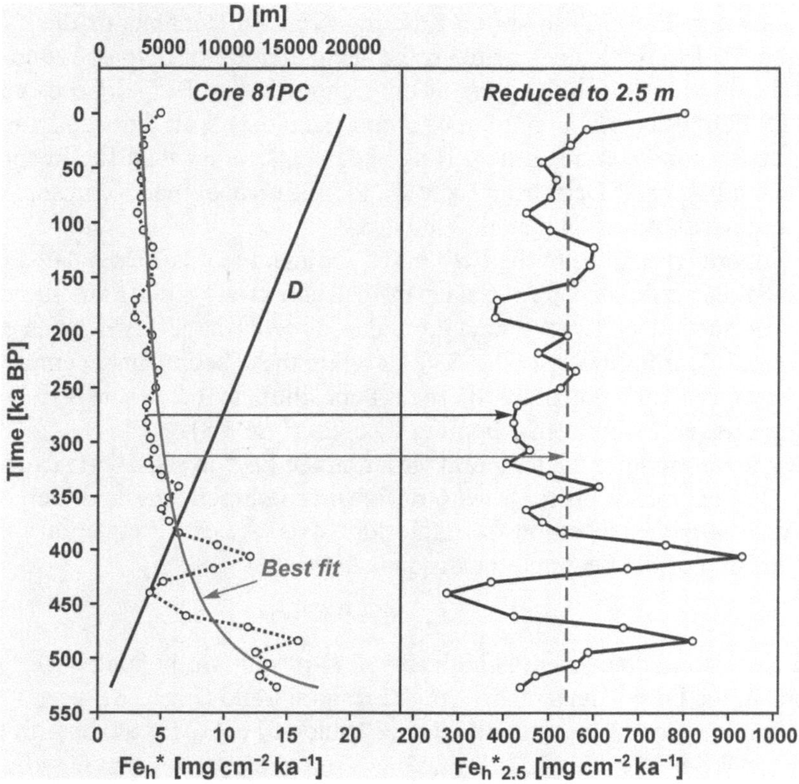

Fig. 5.8. Temporal variations in the distance of the deposition (D) from the axis of the Southeast Indian Ridge for sediments in Core 81 PC, and of Fe_h* estimated from the data of Kuhn et al. (2000) in comparison with variation in Fe_h* calculated from the best fit equation $\lg Fe_h^* = 2.96 - 0.57 \lg D$, as well as calculated values of Fe_h* reduced to the distance of 2.5 m from hydrothermal vents.

Temporal variations in the $Fe_h^*{}_{2.5}$ values show that in the periods of time 480 to 490 and 400 to 410 ka BP and in recent time there were stages of increased hydrothermal activity when the values of the $Fe_h^*{}_{2.5}$ exceeded 800 mg cm^{-2} ka^{-1} and the lowest value, <300 mg cm^{-2} ka^{-1}, was ~440 ka BP. During most of the time recorded in Core 81 PC the $Fe_h^*{}_{2.5}$ value varied from ~400 to ~600 mg cm^{-2} ka^{-1}. The variations in the $Fe_h^*{}_{2.5}$ value are quasicyclic with the duration of most of the cycles from 60 to 80 ka. Presumably, cycles with shorter periods also exist, like those on the EPR (Fig. 1.16), at Axial Seamount (Fig. 5.4) or in the TAG hydrothermal field (Fig. 5.2), but shorter intervals between samples are needed for their detection.

The estimations of the $Fe_h*_{2.5}$ values have been based on the assumption that the distances from the points of sediment deposition to the ridge axes are equal to the distances from the hydrothermal vents. More often the distances of the deposition points of the sediments in the cores to the axes are shorter than the distances to the vents, especially where the distances are not known for the ancient or proposed hydrothermal vents or fields that we are dealing with. In this case apparently some of the results obtained are underestimated somewhat including those based on data from Cores KC-1 and 81 PC. The use of two or more cores of sediments containing hydrothermal matter derived from one source should increase the accuracy of the estimated values.

The history of hydrothermal activity and mineral formation along **the East Pacific Rise axis between 20°30'S and 22°00'S** in the last 50 ka has been reconstructed using data of the study of 10 sediment cores that were collected 20 km to the west of the EPR axis.

This part of the EPR axis is one of the most hydrothermally active areas in the World Ocean. The temporal cyclicity of hydrothermal activity in this area was mentioned in Sect. 1.1.4. and was recognized by Dekov (1994) in the composition of the metalliferous sediments in three sections 10, 20, and 40 km west of the EPR axis. The amplitude of the variation in the contents of Fe_{CFB} is somewhat low, on the average about 10 percent relative to the contents. Obviously the low amplitude is caused by the high hydrothermal background related to the permanent existence of hydrothermal activity at some place or other, or simultaneously at places in the part of the EPR axis being considered or in adjacent parts. The contribution of hydrothermal matter from the adjacent neighboring sites that are in different stages of hydrothermal activity may result in the interference of their hydrothermal signals. Nevertheless, Fourier analysis carried out on materials in the study of 1524 samples have made it possible to demonstrate that more than half of the cycles detected had periods of 10±5 ka (Dekov 1994). This has not only provided evidence of the existence of cyclicity in the hydrothermal activity in individual hydrothermal fields but also has shown that information for individual hydrothermal fields can be extracted from integral indicators of the hydrothermal activity in the metalliferous sediments that accumulated at least as far as 40 km from the EPR axis.

Studies of metalliferous and transitional sediments collected along sections parallel or subparallel to a spreading axis enable the reconstruction of the history of hydrothermal activity within separate segments of a spreading ridge. The analysis of isopleths of the accumulation rate of hydrothermal Fe in spatiotemporal dissection is a very productive approach to these reconstructions.

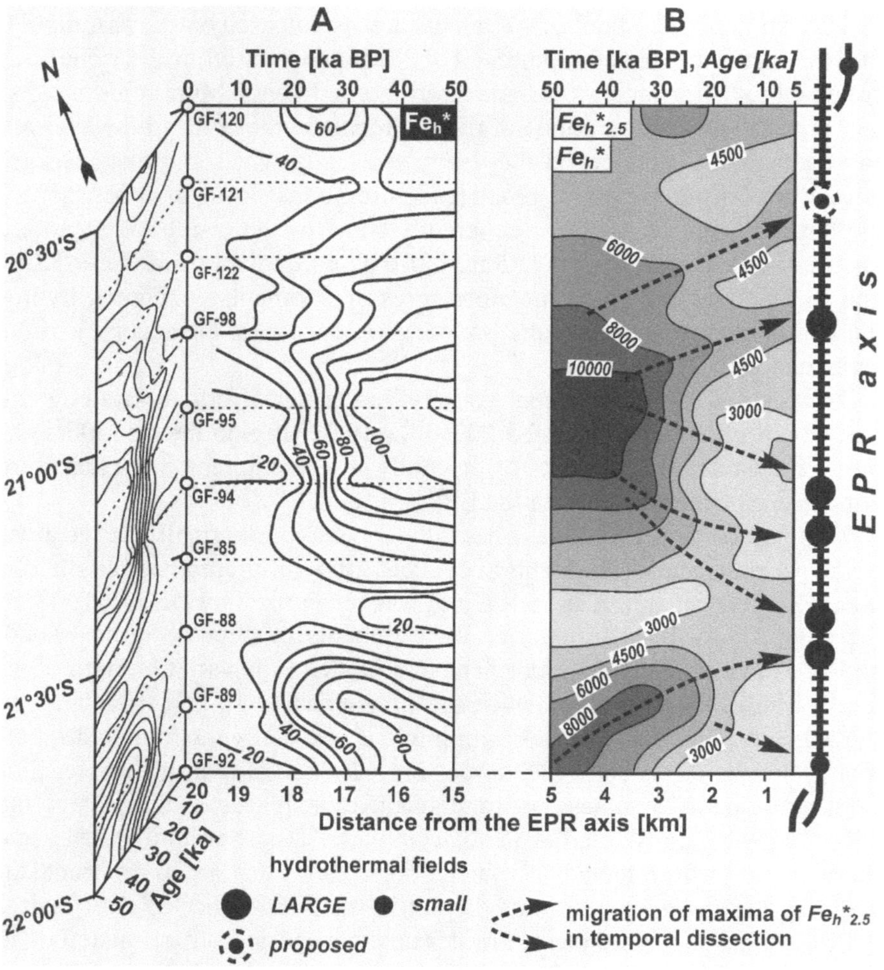

Fig. 5.9. A – Temporal evolution of Fe_h* [mg cm^{-2} ka^{-1}] in surface metalliferous sediments in the last 50 ka along a section parallel to the EPR axis and gradually moving to the west from the axis with the half-rate of spreading of 10 cm a^{-1}. After Dekov (1994) with changes and additions. B – Reconstruction of $Fe_h*_{2.5}$ [mg cm^{-2} ka^{-1}] for the last 50 ka and the distribution of Fe_h* [mg cm^{-2} ka^{-1}] in the lowermost sediments of the metalliferous stratum within 5 km to the west of the EPR axis.

Following my advice and recommendations, Dekov (1994) plotted his data for the accumulation rates of hydrothermal Fe in sediments of the section mentioned that is 20 km west of the EPR axis in spatiotemporal dissection and obtained a scheme of the evolution of Fe_h* in the surface metalliferous sediments in the section during the last 50 ka (Fig. 5.9A). It was based on the assumption that the half-rate of spreading of the EPR in this area was equal to that at 20°S, which is about 10 cm a^{-1} (Rea 1978), and 5 km of the Pacific plate motion corresponds to a time interval of 50 ka at this half-rate of spreading.

Based on these data for the distribution of Fe_h* in the surface metalliferous sediments of this moving section, the values of $Fe_h*_{2.5}$ have been reconstructed (Fig. 5.9B) by using the equation (4.1), $lg(Fe_h*/Fe_h*_{2.5}) = 0.23 - 0.56lgD$, that shows the relative decrease of Fe_h* with increasing distance from hydrothermal vents in the Pacific Ocean (Sect. 4.4). Corrections for the westerly direction of bottom currents in this area were calculated and applied from the data of Walter and Stoffers (1985) and Gurvich (1998). In this reconstruction as well as in previous ones (Fig. 5.6, 5.8) it was assumed that: a) the hydrothermal Fe has been contributed to the sediments only from hydrothermal plumes or their fragments; and b) the accumulation rate of the hydrothermal Fe at the time of deposition of the sediments was equal to its settling flux.

At present the sediments that accumulated at the EPR axis 50 ka BP are located 5 km to the west and east of the axis because of the spreading; the half-rate of spreading is 10 cm a^{-1} (or 0.1 km ka^{-1}). Therefore the plot in Fig. 5.9B shows not only the evolution of the settling flux of the hydrothermal Fe at distance of 2.5 m from the hydrothermal vents during the last 50 ka, but it is also a reconstruction of the distribution of Fe_h* in the lowermost part of the metalliferous stratum within 5 km to the west of the EPR axis.

During the last 50 ka the reconstructed settling flux of the hydrothermal Fe along the EPR axis between 20°30'S and 22°00'S decreased to approximately half of the initial value (Fig. 5.9B). From 50 to 40 ka BP the average value of $Fe_h*_{2.5}$ was ~7000 mg cm^{-2} ka^{-1}, from 40 to 30 ka BP it was ~6900 mg cm^{-2} ka^{-1}, from 30 to 20 ka BP it was 5200 mg cm^{-2} ka^{-1}, from 20 to 10 ka BP it was ~3700 mg cm^{-2} ka^{-1}, and from 10 to 5 ka BP it was ~3600 mg cm^{-2} ka^{-1}. Apparently the general intensities of hydrothermal activity and mineral formation along the EPR axis between 20°30'S and 22°00'S followed a similar trend.

The reconstructed values of $Fe_h*_{2.5}$ are similar or higher than settling fluxes of the hydrothermal Fe reduced to 2.5 m from hydrothermal vents for the hydrothermal fields of the Endeavor Ridge and 13°N EPR, and lower than those for the Rainbow and Broken Spur hydrothermal fields (Fig. 4.26). The maximal reconstructed values of $Fe_h*_{2.5}$, ~15000 mg cm^{-2} ka^{-1}, are similar to the settling fluxes of the hydrothermal Fe reduced to 2.5 m from the vents of the Rainbow and Broken Spur hydrothermal fields.

High values of the settling fluxes of the hydrothermal Fe in the mentioned hydrothermal fields are of very local distribution. And the reconstructed average values of $Fe_h*_{2.5}$ characterize the segment ~160 km long. The settling flux of the hydrothermal Fe near individual vents can be lower, similar or higher than near the vents in the Rainbow and Broken Spur fields. But the reconstructed values of $Fe_h*_{2.5}$ for the segment ~160 km long can be explained only by a great number of hydrothermal vents. This agrees well with the conclusions from studies of the non-buoyant hydrothermal plumes (Sect. 4.3.2).

Fig. 5.9B enables the reconstruction of the position and migration of the sites of hydrothermal activity on the EPR axis between 20°30'S and 22°S for the last 50 ka. Temporal migration of the sites is reflected by the temporal migration of the maxima of the $Fe_h*_{2.5}$ values.

By using the temporal dissection method these can be shown as trajectories. Reconstruction of the recent sites of hydrothermal activity is the least accurate because of the accumulation of the youngest sediments in the cores at a maximum distance from the EPR axis (Fig. 5.9A). Nevertheless this reconstruction shows that the reconstructed positions of the recent sites agree well with the locations of the existing hydrothermal fields (Fig. 5.9B). This evidence shows that the reconstructions of the earlier positions are reasonably accurate.

According to the reconstruction of the history for the period from 50 to 30 ka BP, when hydrothermal activity had more than twice the intensity than in recent time, two main hydrothermal centers existed in the central and southern parts of the segment of the EPR axis that was studied. During this time the southern center was migrating to the north at the rate of ~1 m a^{-1} and in the period from 35 to 30 ka BP a gradual decrease in the hydrothermal activity in the central part of the segment and the splitting of this hydrothermal center began to take place. In the period from 30 to 20 ka BP the decrease in hydrothermal activity continued in the central and southern parts of the segment. The splitting of the hydrothermal center in the central part of the segment may have resulted in the existence of separate large hydrothermal fields; two fields of this group were migrating to the north at the rate of ~0.5 m a^{-1}, and three other fields were migrating to the south at the rates of 0.6–0.9 m a^{-1}. The migration of the southern center to the north continued at the rate of ~0.6 cm a^{-1}. From 20 ka BP until recent time the intensity of the hydrothermal activity was much lower than earlier and did not vary significantly. During this time the migration of the fields in the central group to the north continued at the same rate, and the rate of migration of the southern fields in the group decreased. The migration of the southern center to the north continued at a lower rate of ~0.3 m a^{-1}. Presumably, a new hydrothermal field that resulted from the splitting of the southern center appeared in the southern part of the segment. This field migrated to the south.

The pattern of migration of the maximum values for $Fe_h*_{2.5}$ that were used for the reconstruction of hydrothermal activity cannot be explained by the movement of the ocean crust or by bottom currents. It is next to impossible for: a) close parts of a lithosphere plate to move at the rate of n·10 cm a^{-1} in the opposite directions; and b) during many thousands of years the directions of the bottom currents in adjacent areas of a deep ocean would differ as much as the directions of the migration of the maximum values for $Fe_h*_{2.5}$, or even cross each other.

High values of the settling flux and the accumulation rate of the hydrothermal Fe in sediments usually accompany the formation of large or numerous small massive sulfide bodies at the hydrothermal orifices (Lisitzin et al. 1990). It is certainly within reason to suggest that, at the active stage in the life of high-temperature hydrothermal vents that are accompanied by the formation of metalliferous sediments, the settling flux and the accumulation rate of the hydrothermal Fe in the vicinity of the hydrothermal vents on an axis of a spreading ridge reflect the rate of formation of massive sulfides. If this is the case the diagram in Fig. 5.9B also reflects the present occurrence of hydrothermal sulfide mineral masses (if they are preserved) that formed on the EPR axis between 20°30'S and 22°00'S during the last 50 ka and located to the west of it at distances up to 5 km. The characteristics of the relationship of the settling flux or the accumulation rate of the hydrothermal Fe on an axis of a spreading ridge near hydrothermal vents and the occurrence of masses of hydrothermal sulfide minerals probably vary along the rift systems in the World Ocean. It is expected that this relationship will be better defined in the future.

5.2. Reconstruction of the history of hydrothermal activity and mineral formation based on the studies of deep-sea drilling cores

The accumulation rate of the hydrothermal metal-bearing matter in sediments is a very useful parameter for reconstructing the intensity of hydrothermal activity and mineral formation in the geological past. The values of this parameter in sediments collected during the deep-sea drilling program vary from place to place and along the cores. Usually the sediments in the lower parts of the cores accumulated earlier and closer to the spreading axes than the overlying sediments. Therefore on comparisons of the values measured even in one core and especially in many cores, techniques of systematizing and consideration of these data are of great importance. Certainly the best way to analyze and study the temporal variations in the accumulation rates of the hydrothermal metal-bearing matter in sediments is to do detailed palinspastic mapping for the chosen time intervals both in separate areas and in the whole ocean. However, such studies of the ocean floor require detailed information and data that are available only from the analyses of many samples from many drill cores. Obtaining this kind of data is laborious and expensive, and is not likely to be possible now or in the near future. Data and work of this kind has been made possible for the study of recent stages in geologic history in only one area of metalliferous sediments in the Southeast Pacific.

The following kinds of studies are possible at the present time: 1) studies of variations in the accumulation rates of the hydrothermal metal-bearing matter in sediments from sections or narrow strip-sections along spreading tracks or along perpendiculars to the axes of spreading ridges, and 2) studies of variations in the accumulation rates of the hydrothermal metal-bearing matter in sediments from sections parallel or subparallel to the axes of spreading ridges.

The first kind of studies was attempted earlier by Boström (1973), Gurvich et al. (1984), Lyle et al. (1986). However, the analysis of the evolution of the hydrothermal activity was complicated in these studies because temporal variations in the distances of the points of sediment deposition to the spreading axes were not taken into account, and these distances for the sediments could vary in each core from zero to many hundreds of kilometers. This omission can be avoided by systematizing the data in relation to temporal coordinates and by plotting the age of the samples on the ordinate axes and the time since the beginning of the formation of strata on the abscissa axes. In this case, if the scales of both axes are equal, the sequence of the data for each core falls on a straight line that is at an angle of 45° to the ordinate axis. The point where this line crosses the ordinate axis corresponds to the age of the sediments that lie directly on the basalt basement and the point where this line crosses the abscissa axis shows the duration of time for the accumulation of the whole sedimentary series in the place where the core was collected. The line is not continuous if there are hiatuses in the strata and the breaks correspond to the times and duration of the hiatuses. Use of the temporal coordinates is convenient for comparing data from different sediment cores as: a) for the ordinate axis there is no need to consider differences in sedimentation rates between coeval sediments from different cores and differences in sedimentation rates in sediments accumulated at different times, and b) for the abscissa axis there is no need to consider differences in spreading rates at different times, if it is assumed here that in each moment of geologic time the spreading rate within a part of a spreading ridge crossed by a strip-section was identical.

For demonstrating this method strip-sections (or sections), usually not more than 10° to 15° in width, have been chosen in the southern and northern parts of the Pacific and Atlantic Oceans and the locations of the drill holes from these sections are shown in Fig. 3.3. Data for the accumulation rate of the hydrothermal Fe in the metalliferous and non-metalliferous sediments that have an admixture of the hydrothermal metal-bearing matter have been used.

The southern section in the Pacific Ocean is located west of the EPR axis between 19°S and ~6°S (Fig. 3.3) and in general its direction corresponds to the direction of the Pacific plate motion. The distribution of the values of Fe_h* in the sediments of this section in the temporal coordinates is shown in Fig. 5.10A. It covers the time period from 40 Ma BP to 0 Ma BP. During this period the values of Fe_h* varied over a wide range and maximum values occurred about 38–36, 25–24, 18, 14, 9–8, and 5 Ma BP.

Variations in the accumulation rates of hydrothermal Fe and Mn do not correlate with the variations in the spreading rate (Fig. 5.10A, 5.11). Only the maximum values for Fe_h* indicated at ~5 Ma BP and 38–36 Ma BP may have resulted from increases in the spreading rate. The increase at ~5 Ma BP can be seen in Fig. 5.11. As for the increase at 38–36 Ma BP, according to Zonenshain and Khain (1989), the average rate of crust accretion in Eocene time for the Pacific Ocean was almost one and a half times higher than at present. The period of time from 33–32.5 Ma BP to 46–44.5 Ma BP is characterized by increased values of a relative momentary spreading rate[41] in the Pacific Ocean (Schreider 1982).

The maxima of values for Fe_h* in sediments accumulated ~27–25, 18, 14, and 9–8 Ma BP (Fig. 5.10A) are synchronous with the earlier measured maxima of values for the accumulation rate of Mn (Lyle et al. 1986, 1987). They correlate with events of tectonic reorganizations in the Southeast Pacific (Lyle et al. 1986, 1987; Rea and Leinen 1986). The maximum at ~25 Ma BP apparently is synchronous with the formation of the rift between the Cocos and Nazca plates, 25–23 Ma BP, as well as with the jump of the spreading center 175 km to the west, from the Selkirk Ridge segment of an ancient spreading ridge to the EPR in the period from 24 to 22 Ma BP (Mammerickx et al. 1980). The maximum at ~18 Ma BP apparently is synchronous with the spreading center jump 500 km to the west, from the Mendoza Rise segment of an ancient spreading ridge to the EPR, that took place 20–18.5 Ma BP. At approximately the same time the jumps

[41] Ratio of the linear spreading rate measured between two adjacent magnetic anomalies in a profile to the spreading rate measured between two foregoing magnetic anomalies at the same profile.

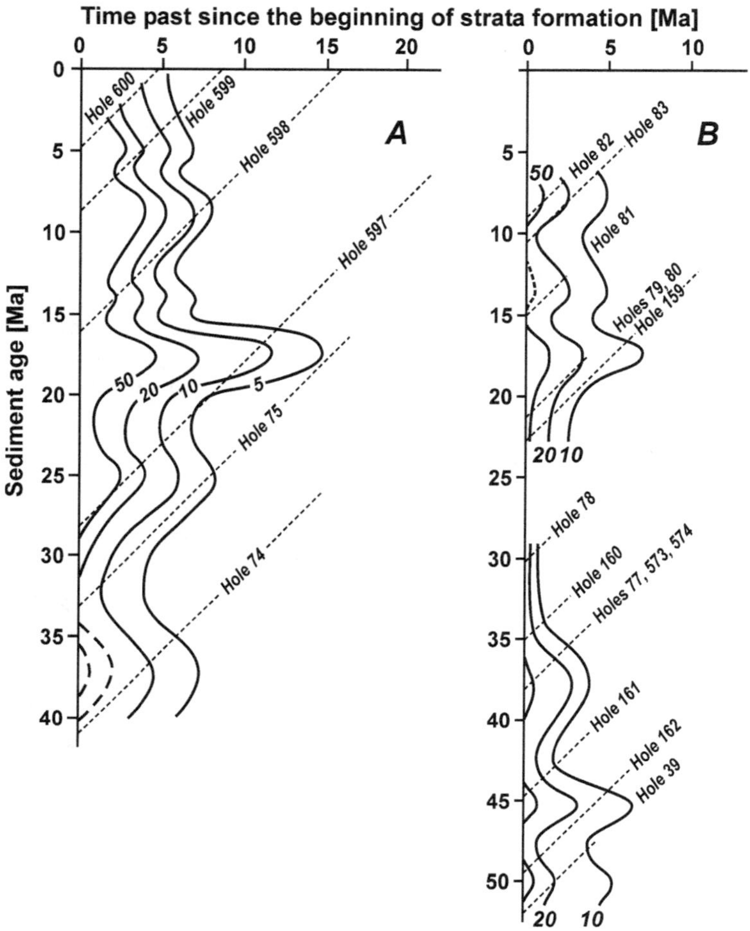

Fig. 5.10. Distribution of Fe_h* [mg cm^{-2} ka^{-1}] in bottom sediments of the southern (A) and northern (B) sections in the Pacific Ocean.

westward took place from the Gallego Rise and Roggeveen Rise, the segments of ancient spreading ridges, to the EPR (Mammerickx et al. 1980). The maximum at ~14 Ma BP apparently is synchronous with the major southward jump of the Cocos–Nazca spreading center, that took place ~14.5 Ma BP (Werner et al. 2003) and with the existence at this time of two spreading centers, between the Galapagos and unnamed faults; they were the EPR and the parallel ridge about ~400 km to the west of it. About 11 Ma BP the parallel center became extinct (Mammerickx and Klitgord 1982). Apparently the maximum at 9–8 Ma BP resulted from the jump of the spreading center from the ancient Galapagos Rise to the EPR. Both ridges were active in the time interval from 8.2 to 6.5 Ma BP (Mammerickx et al. 1980), and this is reflected clearly in the increase of the accumulation rate of the hydrothermal metal-bearing matter in the metalliferous sediments of DSDP Site 319 in the Bauer Depression (Dymond et al. 1977). And furthermore, between 10.5 and 8 Ma BP the Pacific-Riviera Ridge changed in strike direction from north-northwest to northeast (Lyle et al. 1986; Rea and Leinen 1986).

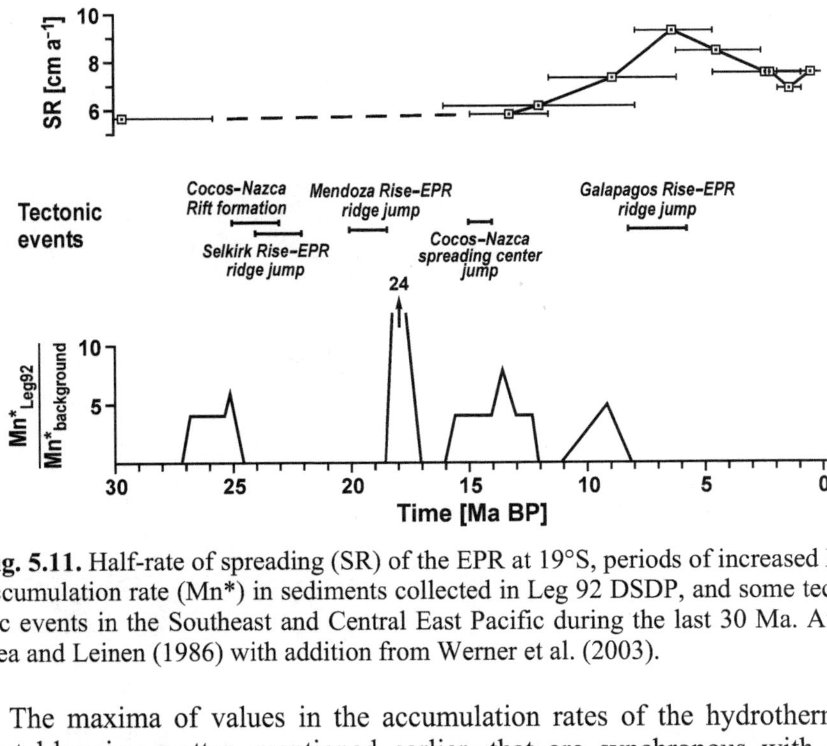

Fig. 5.11. Half-rate of spreading (SR) of the EPR at 19°S, periods of increased Mn accumulation rate (Mn*) in sediments collected in Leg 92 DSDP, and some tectonic events in the Southeast and Central East Pacific during the last 30 Ma. After Rea and Leinen (1986) with addition from Werner et al. (2003).

The maxima of values in the accumulation rates of the hydrothermal metal-bearing matter, mentioned earlier, that are synchronous with the events of tectonic reorganizations and especially the maximum ~18 Ma BP are significantly high (Fig. 5.10A). They cannot be explained by the decreasing distances from the sites of sediment deposition to spreading centers alone. Lyle et al. (1986, 1987) have assumed that breakings and changes in the strike of the ridges associated with the events of tectonic reorganizations could have increased the intensity of the hydrothermal activity substantially because of the fracturing of the ocean crust that enabled more active interaction of ocean water with hot rocks.

Based on the values of Fe_h* the temporal variations in the total accumulation of the hydrothermal Fe have been estimated in the sediments in the whole section within a band 1 km wide (Fe_{1km}). That is to say, estimates have been made of the temporal evolution in the contribution of hydrothermal Fe at ~19°S from 1 km of the rift length to the sediments located west of the EPR axis. These estimated values reflect the evolution in the intensity of the hydrothermal activity and the mineral formation related to it. The resulting estimates are shown in Fig. 5.12A.

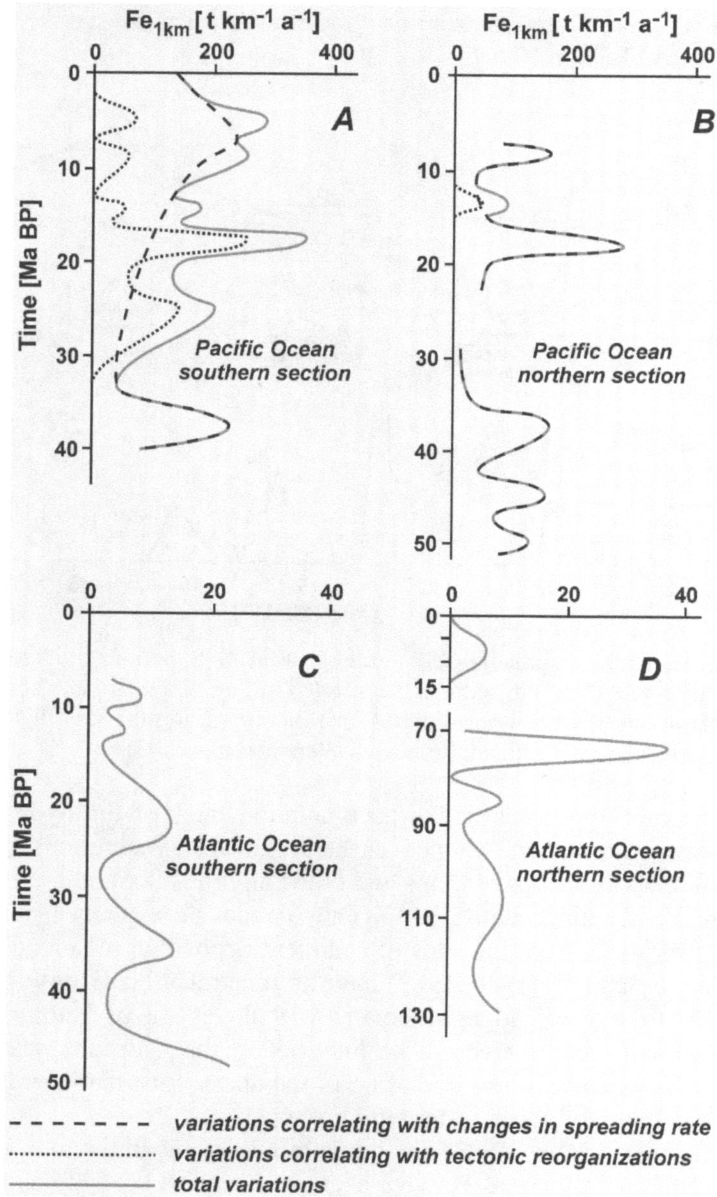

Fig. 5.12. Temporal variations of accumulation of hydrothermal Fe in sediments of the sections within bands 1 km wide along tracks of lithosphere plates.

The southern section is located to the west of the EPR axis within the field of recent metalliferous sediments in the Southeast Pacific. In the area around 19°S the value of Fe_{1km} for the recent metalliferous sediments is about 150 t km^{-1} a^{-1} (Fig. 5.12) which is comparable to the average value for the last 40 Ma. Within the field of the recent metalliferous sediments about 550000 tons of hydrothermal Fe accumulates annually (Table 1.2); and if the length of the rift within the field is ~4000 km the average value of Fe_{1km} for the two flanks of the EPR is about 140 t km^{-1} a^{-1}; for one flank it is about 70 t km^{-1} a^{-1}, which is less than twice that for the western flank of the ridge in the area around 19°S.

Most of the holes in **the northern section in the Pacific Ocean** have been located between 0° and 16°N (Fig. 3.3). Distribution of the values of Fe_h* in the basal sediments of this section in the temporal coordinates is shown in Fig. 5.10B. It covers the period of time from 52 to 6 Ma BP. Because of the lack of data there is a void from 29 to 23 Ma BP. In general the variations of Fe_h* are close to those for the section in the southern part of the Pacific Ocean. The maxima of values for Fe_h* occurred about ~50, 45, 39–37, 18–17, 14–13, and 8–7 Ma BP and all of the maxima or the grouping of them are synchronous with the periods of time and episodes when there was an increase in the average spreading rate of the North Pacific (Fig. 5.13). A long period in Eocene time, when there was an increase in the spreading rate, coincided with a period of time when there was an increase in the average spreading rate for the whole Pacific Ocean. In Eocene time the average rate of accretion of the Pacific crust was at a maximum for all of Cenozoic time (Zonenshain and Khain 1989). Presumably, the general increase in Fe_h* in the Eocene sediments of the section corresponds to this increase in the spreading rate, and there are maxima of Fe_h* at ~50, 45, and 39–37 Ma BP above the background of this increase. The maximum at ~45 Ma BP is synchronous with the increase in the spreading rate of 45–42 Ma BP (Schwan 1985). The maxima at 18–17 and 8–7 Ma BP are synchronous with episodes of sharp increases in the average spreading rate in the North Pacific dated about 18 and 8 Ma BP. Apparently the maximum at ~14–13 Ma BP is synchronous with the major southward jump of the Cocos–Nazca spreading center that occurred about 14.5 Ma BP (Werner et al. 2003). The synchronism of the maxima of Fe_h* at ~18–17 and 8–7 Ma BP in the sediments of the northern section with the maxima at ~18 and 9–8 Ma BP in the sediments of the southern section is worthy of notice (Fig. 5.10). Probably the tectonic reorganizations at or about this time that have been mentioned when considering the southern section could also have influenced the Fe_h* values in the sediments of the northern section.

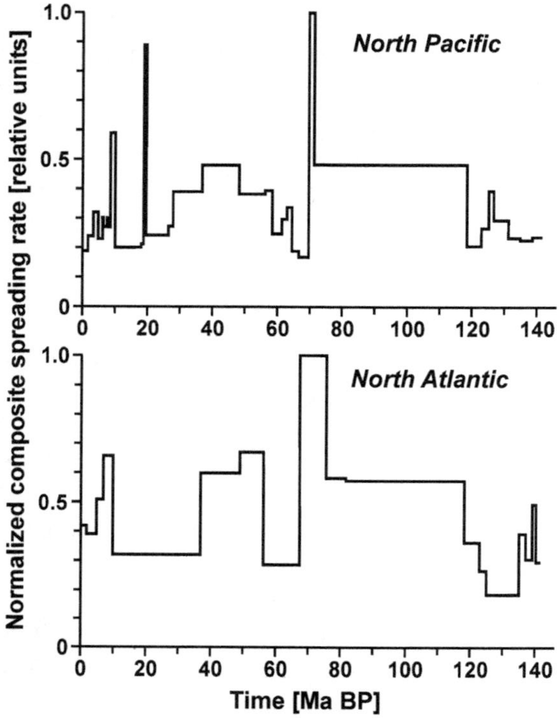

Fig. 5.13. Evolution of average normalized spreading rates for the North Pacific and the North Atlantic. Composed from Rich et al. (1986).

The values of Fe_{1km} in the sediments of the northern section to the west of the EPR axis have been estimated for the last 52 Ma (Fig. 5.12B). On the average these values are twice as low as those in the sediments of the southern section. Probably this is caused by a lower influence of tectonic reorganizations and slower spreading rates, at least in Neogene time. The general directions of the bottom currents may be another reason. At present, in the northern section area, an easterly direction of transport of the particulate hydrothermal matter is prevalent, and a westerly direction prevails in the southern section area (Lonsdale 1986; Lupton et al. 2004). Easterly and northeasterly directions of bottom currents have been prevalent in the northern section area at least since Eocene time (Lisitzin et al. 1980a).

An empirical relationship exists between the values of Fe_{1km} for the sections in the Pacific Ocean and the values of $Fe_h{}^*{}_{2.5}$ on the axis of the EPR:

$$Fe_h{}^*{}_{2.5} = 61 Fe_{1km}, \quad (5.1)$$

where $Fe_h{}^*{}_{2.5}$ is expressed as mg cm^{-2} ka^{-1}, and Fe_{1km} as t km^{-1} a^{-1}.

For the area crossed by the southern section near 19°S the value of Fe_{1km} for recent time is 140 t km^{-1} a^{-1} (Fig. 5.12). With this value of Fe_{1km} the value of $Fe_h*_{2.5}$ is ~8500 mg cm^{-2} ka^{-1}; with correction for the westerly bottom currents (Fig. 5.14) ~6000 mg cm^{-2} ka^{-1}. This value is close to the values of $Fe_h*_{2.5}$ as reconstructed between 20°30'S and 22°00'S in the last 50 ka (Fig. 5.9).

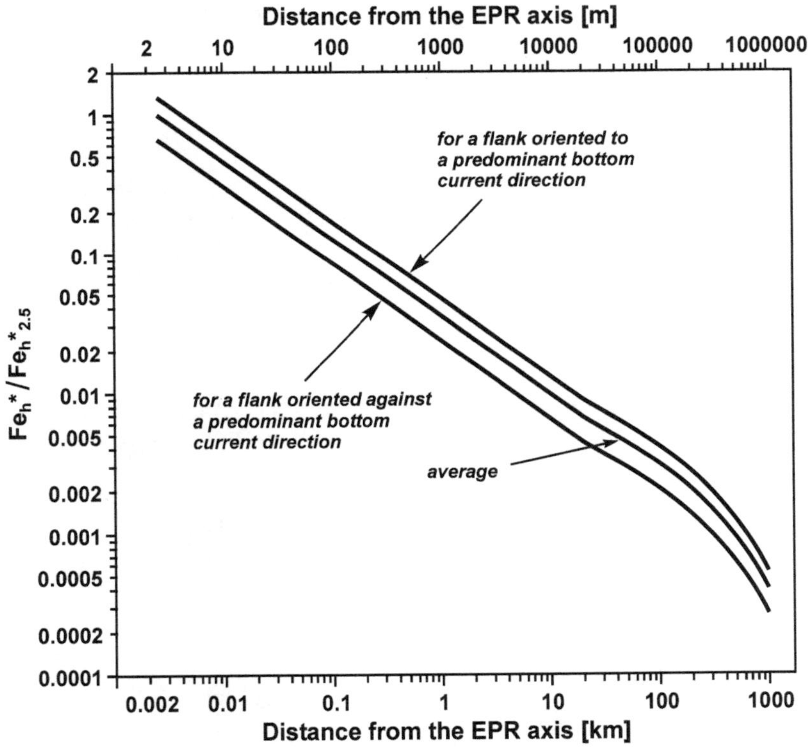

Fig. 5.14. Relative decrease in Fe_h* with distance in comparison with Fe_h* at 2.5 m from hydrothermal vents at different flank expositions and directions of bottom currents. Average for the field of recent metalliferous sediments in the Southeast Pacific and the East Pacific Rise between 5°S and 12°N.

The positions of holes of **the southern section in the Atlantic Ocean** are shown in Fig. 3.3. The section is located on the east and on the west sides of the Mid Atlantic Ridge axis between 30°S and 25°S. The distribution of the values of Fe_h* in the basal sediments of the section in the temporal coordinates is shown in Fig. 5.15A. It covers the time period from ~50 to ~6 Ma BP, and there are data that indicate that from ~66–65 Ma BP accumulation of the hydrothermal Fe in the basal sediments of the section was almost absent (Lisitzin et al. 1990). In general the values of Fe_h* in the

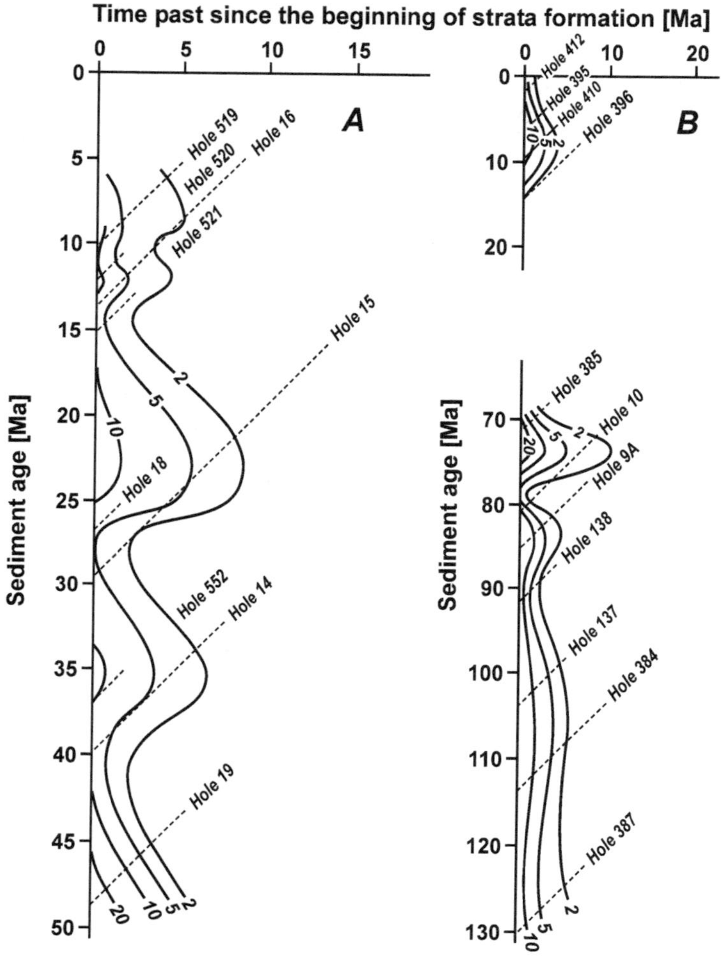

Fig. 5.15. Distribution of Fe_h* [mg cm^{-2} ka^{-1}] in bottom sediments of the southern (A) and northern (B) sections in the Atlantic Ocean.

sediments of the section are much lower than those in the sediments of the sections in the Pacific Ocean. During the time periods of ≥49, ~36–35, 24–20, ~12, and ~9–8 Ma BP there were maxima of the values of Fe_h*. The first and the third maxima are synchronous with the maxima of values of the accumulation rate of Mn in the basal sediments in the western part of the same section ~50–46 and 25–21 Ma BP (Boström 1973; Gurvich et al. 1984). Some of the maxima of the values of Fe_h* are synchronous with the maxima of the spreading rate. According to the data from Schreider (1985), the spreading rate was increasing during the time period from 59 to 52 Ma BP. Apparently the maximum for Fe_h* ≥49 Ma BP was synchronous with the maximum in the spreading rate ~52 Ma BP. The spreading rate was also increasing during the time period from 42 to 38–37 Ma BP (Schreider 1985) and 38–35 Ma BP it began to decrease (Schwan 1985). The latter time interval was a period with a maximum spreading rate, and the maximum in the Fe_h* ~36–35 Ma BP apparently is synchronous with this maximum.

The positions of holes of **the northern section in the Atlantic Ocean,** located between 25°N and 45°N, are shown in Fig. 3.3. The distribution of the values of Fe_h* in the basal sediments of this section in the temporal coordinates is shown in Fig. 5.15B. It covers the periods of time from ~130 to ~70 Ma BP and from ~15 Ma BP to 0 Ma BP. In general the values of Fe_h* in the sediments of the section are close to the values in the southern section in the Atlantic Ocean. In the time period from 130 to ~70 Ma BP two maxima of the values of Fe_h* existed at ~85 Ma BP and 75–73 Ma BP. During the time interval from 130 to 100 Ma BP the values of Fe_h* did not vary significantly, but there are few data for this interval. Within the time period 15 to 0 Ma BP only one maximum occurs at ~8 Ma BP. Appreciable minima of the values of Fe_h* occurred at about 80–78, 70, and 15 Ma BP. These periods of time are in good agreement with those of decreasing spreading rates in the North Atlantic (Fig. 5.13). The highest values of Fe_h* were at ~75–73 Ma BP. This maximum was synchronous with the maximum in the spreading rate in the North Atlantic. The other maximum of Fe_h* at ~8 Ma BP was synchronous with another maximum in the spreading rate (Fig. 5.13). Only the average spreading rate is known for the whole period of time from 120 to 80 Ma BP and that is why comparisons for all parts of the period are not possible.

Variations in the values of Fe_{1km} have been estimated for the time period from ~50 to ~6 Ma within the southern section (Fig. 5.12C) and for the time periods from ~130 to ~70 Ma BP and from ~15 to 0 Ma BP within the northern section in the Atlantic Ocean (Fig. 5.12D). The positions of these sections both to the east and to the west of the Mid-Atlantic Ridge axis were taken into account, and the values of Fe_{1km} have been calculated as half-sums of the values for the eastern and western flanks of the ridge when obtaining data and results for comparison.

On the average the values of Fe_{1km} for the sections in the Atlantic Ocean are 10 to 20 times lower than those for the sections in the Pacific Ocean. Obviously major differences are caused mainly by the much slower spreading rates of the Mid-Atlantic Ridge in comparison with the East Pacific Rise. Presumably the direction of the bottom currents in the section areas is another reason. In the Pacific Ocean the particulate hydrothermal matter has been transported mainly in a latitudinal direction and this has favored its accumulation on the flanks of the EPR. In the Atlantic Ocean the particulate hydrothermal matter has been transported mainly in longitudinal directions (Lisitzin et al. 1980a). This would account for a reduction in the amount of particulate hydrothermal matter deposited on the flanks of the MAR in the sections sampled, and the morphology of the MAR could be another reason for the deposition of a reduced amount of hydrothermal matter. With low spreading rates, practically no hydrothermal matter leaves the rift valley and it does not reach the flank areas (Fig. 4.5). With an increase in the spreading rate, which could be up to threefold (see Fig. 5.13), not only hydrothermal activity and the incidence of hydrothermal plumes increased. The morphology of the rift probably also changed and particulate hydrothermal matter was released to the flanks of the ridge (Fig. 4.5). During the Cretaceous time, because of the lower vertical gradients of temperature and salinity in the ocean water, hydrothermal plumes could rise much higher above the ocean bottom than at present (Vogt 1989) and particulate hydrothermal matter also could leave the rift valley and reach the flank areas.

Section subparallel to the EPR axis between 5°S and 12°N. Reconstruction of the history of hydrothermal activity and mineral formation along the EPR axis between 5°S and 12°N in the last 18 Ma has been carried out on materials obtained during the study of sediment samples from the cores of the ODP Leg 138 (the collection of M. Levitan). Seven stations of this Leg have been located along a section that is more than 1600 km long on the western flank of the EPR and subparallel to the EPR axis (Fig. 5.16). The half-rate of spreading in this segment of the EPR is 6 to 8 cm a^{-1}.

Both metalliferous and non-metalliferous sediments with admixture of the hydrothermal metal-bearing matter were collected in these cores (Mayer et al. 1992; Gurvich et al. 1995b). The hydrothermal metal-bearing matter was contributed to these sediments from the axial part of the EPR. Chemical composition of the sediments at ODP Sites 845, 846, 848–854 and DSDP Site 83 has been studied by Gurvich et al. (1995b). Data for these sediments together with data for the sediment age and physical properties from Hays (1972), Mayer et al. (1992), and Shackleton et al. (1995), have been used for calculating the rates of accumulation of the hydrothermal Fe. The values of Fe_h* have been averaged for every 0.5 Ma period and plotted along the backtracked paths of the sites. For Sites 848 to 854 isopleths of the values of Fe_h* in spatiotemporal dissection have been drawn, and for Sites 845, 846, and 83 graphs of the values of Fe_h* along the backtracked paths have been drawn (Fig. 5.16).

It is clearly shown in Fig. 5.16 that maximum values of Fe_h* occur in the basal sediments that accumulated in the vicinity of the EPR axis. They are similar to the values that occur in the vicinity of the EPR axis within the recent field of metalliferous sediments in the Southeast Pacific (Fig.

1.24), and in basal metalliferous sediments collected during DSDP Legs 9, 16, and 85 (Fig. 5.10B; Leinen and Stakes 1979; Jarvis 1985), but lower than in recent and ancient metalliferous sediments that accumulated in the axial part of the EPR at 19°S to 22°S (Fig. 5.9A, 5.10A).

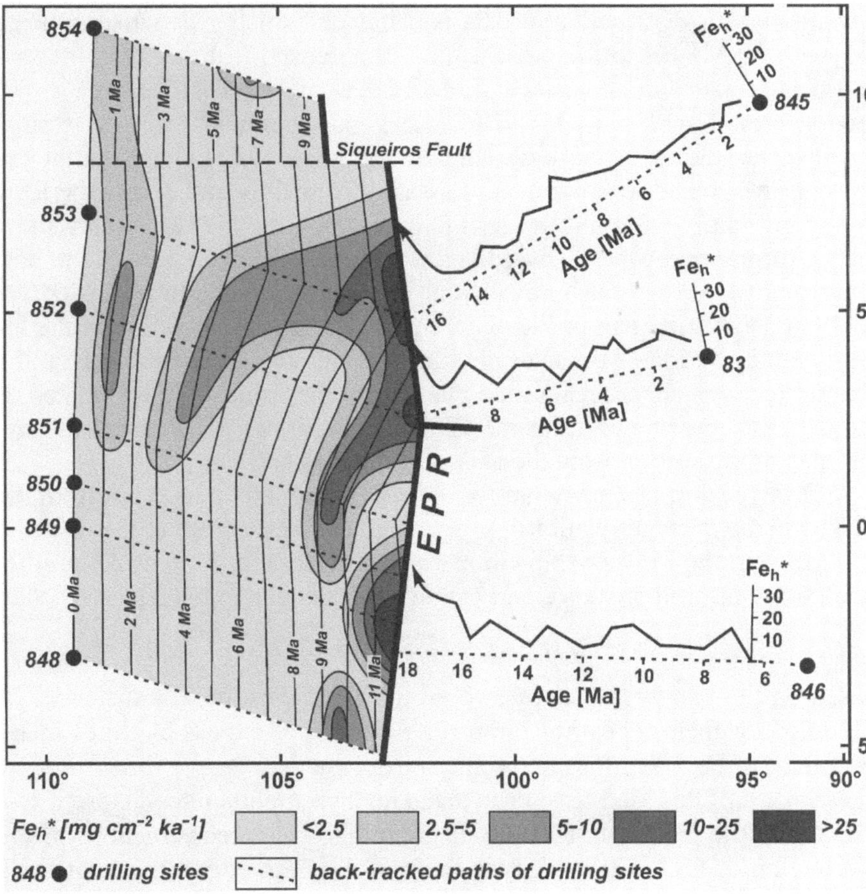

Fig. 5.16. Accumulation rates of hydrothermal Fe in bottom sediments of the section on the western flank of the EPR between 5°S and 12°N subparallel to the EPR axis in spatiotemporal dissection, and in bottom sediments along the backtracked paths of Sites 845, 83, and 846.

The isopleth plot demonstrates the temporal evolution of Fe_h* in surface sediments in a section subparallel to the EPR axis that gradually diverges from the axis along with the Pacific plate with the half-rate of spreading. The graphs showing data plotted for Sites 845, 846, and 83 demonstrate temporal evolution of Fe_h* in surface sediments along the backtracked paths of these sites (Fig. 5.16).

The values of Fe_h* in the sediments from the eastern flank of the EPR are approximately twice as high as those in the sediments from the western flank located at the same latitude at similar distances from the EPR axis. Apparently this distribution of data is evidence, which shows that, at least since 12–10 Ma BP to the present time, the transport of the hydrothermal particulate matter in an easterly direction has prevailed in the area of the section between 5°S and 12°N. It is in good agreement with the results obtained in studies of the hydrology in the area, which show that at the present time the bottom water moves eastward mainly and across the EPR axis (Lonsdale 1976), and the results of paleooceanological studies show that in former times the eastward and northeastward movement of the bottom water in this area has prevailed at least since Eocene time (Lisitzin et al. 1980a). The diagram in Fig. 5.16 enables the characterization of the intensity and position of the hydrothermal activity along the EPR axis in the general features only, because the changes in the values of Fe_h* related to the increase in distance from the EPR axis were not considered, and the distance of the section from the axis is as great as 850 km.

In Sect. 4.7 it was shown that for distances from 20 to 2000 km from the EPR axis the average relationship of a decrease in the Fe_h* at the distance D (km) from the EPR axis, in comparison with that for Fe_h* at 20 km distance from the EPR axis, is clearly shown by the equation (4.3):

$$Fe_h^*/Fe_{h20}^* = (1.58–0.446\lg D)^2.$$

Based on the known values of Fe_h* in the sediments and the distances of the points of their deposition from the EPR axis it is possible to estimate the values of Fe_h* at different distances from the axis.

Estimates of this kind have been made for 11 temporal intervals from 12–10.5 Ma BP to 0.5–0 Ma BP, and diagrams showing reconstructions of the distribution of Fe_h* on the western flank of the EPR during these intervals have been drawn (Fig. 5.17). In these diagrams the near-axial values of Fe_h* correspond to a distance of 20 km from the EPR axis, the boundary value for the equation (4.3). Areas of intensive and very intensive hydrothermal activity and mineral formation on the EPR axis that occurred in the geological past are clearly visible on these diagrams. In sediments adjacent to these areas the reconstructed values of Fe_{h20}* are 10–25 and >25 mg cm^{-2} ka^{-1} respectively. They are the values determined for the western flank of the EPR oriented against the predominant direction of bottom currents. Actually the sediments of the section have accumulated on the western flank of the EPR and a correction for the bottom currents has not been made in the reconstructed values of Fe_h* at a distance of 20 km to the west of the EPR axis.

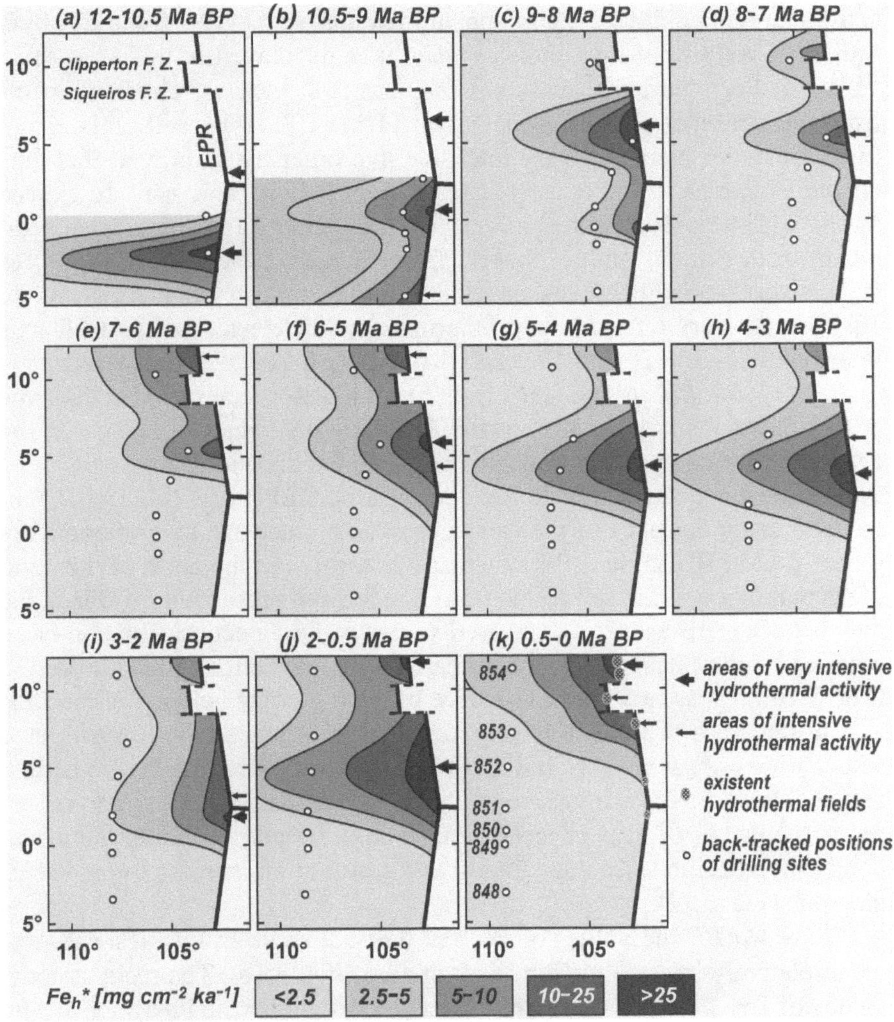

Fig. 5.17. Diagrams of paleodistribution of Fe_h* in bottom sediments to the west of the EPR axis and reconstructed positions and relative intensities of hydrothermal activity on the EPR axis for the last 12 Ma. The positions of the hydrothermal fields that existed are from Hekinian et al. (1983, 1985), Hekinian and Fouquet (1985), Gente et al. (1986), McConachy et al. (1986), Fouquet et al. (1988, 1996), Haymon et al. (1991), Poroshina et al. (1992), Baker et al. (1994), Von Damm (1995), Shank et al. (2000), Hannington et al. (2002).

12 to 10.5 Ma BP. Sediments accumulated during this time have been collected at Sites 848–850, 852, 845, 846, and 83 (Fig. 5.16). Low values of Fe_h* (<2.5 mg cm^{-2} ka^{-1}) in the basal sediments from Site 848, accumulated >11 Ma BP (Fig. 5.17a), are evidence of the absence of hydrothermal activity in the vicinity of their deposition points. Later, 11–10 Ma BP, hydrothermal activity of low intensity may have developed in this area. High values of Fe_h*, up to 60 mg cm^{-2} ka^{-1}, occur in the near-axial sediments from Site 849 that accumulated 11.5 to 11 Ma BP. Apparently this is evidence of the existence of very intensive hydrothermal activity at this time on the EPR axis in the area of 2°–3°S (Fig. 5.17a). This activity existed during ~1.5 Ma. High values of Fe_h* (>25 mg cm^{-2} ka^{-1}) in the basal sediments from Site 83 and, probably, from Site 852 (there are no data for these sediments) (Fig. 5.16) are evidence that indicates that from ~11 to 10.5 Ma BP very intensive hydrothermal activity existed on the EPR axis in the area of ~3°N (Fig. 5.17a) and that about 10 Ma BP it decreased.

10.5 to 9 Ma BP. In sediments from Site 845 that accumulated at this time at a distance about 400 km from the EPR axis the values of Fe_h*, ~10 mg cm^{-2} ka^{-1}, are comparable to those for Fe_h* in sediments from Site 83 that accumulated approximately 40 to 50 km from the EPR axis (Fig. 5.16). Obviously this is evidence of the existence of very intensive hydrothermal activity ~10 Ma BP on the EPR axis that was located to the north of the point of accumulation of the sediments from Site 83, and presumably it was in the area between 5°S and 7°S. Increased values of Fe_h* occur in the non-basal sediments from Sites 848, 850, 851 that accumulated during this time period (Fig. 5.16). Apparently this is evidence of hydrothermal activity on the EPR axis in the areas 5°S and 1°N (Fig. 5.17b). According to the reconstructed values of Fe_{h20}*, at 5°–7°N and 1°N there was very intensive hydrothermal activity and it was less intensive at 5°S. About 9 Ma BP an increase occurred in the values of Fe_h* in the sediments in the southern and northern sections in the Pacific Ocean apparently as a result of an increase in the spreading rate (Fig. 5.12).

9 to 8 Ma BP the values of Fe_h* in the sediments from Sites 848–852 were relatively low, <5 mg cm^{-2} ka^{-1} or zero (Fig. 5.16). The reconstructed values of Fe_{h20}* (Fig. 5.17c) are evidence of hydrothermal activity of low intensity or its absence on the EPR axis in the southern part of the area in question. Only in the area 0° to 1°S near the axis of the EPR are the reconstructed values of Fe_{h20}* from 10 to 25 mg cm^{-2} ka^{-1} (Fig. 5.17c), and the reconstructed values of Fe_{h20}* in the sediments from Sites 845, 853, and 83 are evidence of the existence of hydrothermal activity in the central part of the area. Apparently the highest intensity was located between 5°N and 7°N (Fig. 5.17c). Hydrothermal activity was low in intensity or absent in the northern part of the area.

8 to 7 Ma BP the values of Fe_h* in the sediments from Sites 848 to 852 were low, <2.5 mg cm^{-2} ka^{-1} or zero (Fig. 5.16). The reconstructed values near the axis (Fig. 5.17d) are evidence of hydrothermal activity of low intensity or its absence on the EPR axis in the southern half of the area. In the sediments from other sites and especially from Sites 853 and 845 the values of Fe_h* were much higher (Fig. 5.16). The reconstructed values of Fe_{h20}*, 10 to 25 mg cm^{-2} ka^{-1}, in the sediments to the west of the EPR (Fig. 5.17d) are evidence that 8 to 7 Ma BP intensive hydrothermal activity existed only in the area of 5°–6°N. At this time hydrothermal activity of low intensity probably appeared to the north of the Clipperton Fracture Zone.

7 to 6 Ma BP the pattern in general was similar. Hydrothermal activity persisted in the area of 5°–6°N. However, the increase in Fe_h* in the sediments from Site 854 (Fig. 5.16) is evidence that hydrothermal activity north of the Clipperton Fracture Zone increased at this time and the reconstructed values of Fe_{h20}* are 10 to 25 mg cm^{-2} ka^{-1} (Fig. 5.17e).

The increase in Fe_h* in the sediments from Site 846 accumulated ~8–6.5 Ma BP (Fig. 5.16), presumably, can be attributed to the deposition point of these sediments. They accumulated directly north of the ancient Galapagos Rise at the period of time from 8.2 to 6.5 Ma BP when both the EPR and the ancient Galapagos Rise were active (Mammerickx et al. 1980).

6 to 5 Ma BP the values of Fe_h* in the sediments from Sites 848 to 851 and 846 were low, <2.5 mg cm^{-2} ka^{-1} or zero. The value of Fe_h* in the sediments from Site 852 increased in spite of the increase in the distance from the place of their deposition to the EPR axis. In sediments from Site 853 the value of Fe_h* persisted (Fig. 5.16). The reconstructed value of Fe_{h20}* at ~5°–6°N exceeds 25 mg cm^{-2} ka^{-1} (Fig. 5.17f). This is evidence of very intensive hydrothermal activity on the EPR axis. Hydrothermal activity persisted to the north of the Clipperton Fracture Zone, however a decrease in the Fe_h* values in the sediments from Site 854 (Fig. 5.16) indicates that it became less intensive.

5 to 3 Ma BP the values of Fe_h* in the sediments from Sites 848 to 850, and 846 were low, <2,5 mg cm^{-2} ka^{-1} or zero. The increase in the Fe_h* values in the sediments from Site 852 persisted and reached a maximum, >5 mg cm^{-2} ka^{-1}, 5–4 Ma BP. About 4–3 Ma BP the value of Fe_h* in the sediments from Site 851 increased to 2.5–5 mg cm^{-2} ka^{-1}. The value of Fe_h* also increased in the sediments from Site 83 and was >10 mg cm^{-2} ka^{-1} in the time period from 4 to 3 Ma BP. In the sediments from Sites 853, 854, and 845 the values of Fe_h* gradually decreased (Fig. 5.16). The reconstructed values of Fe_{h20}* (Fig. 5.17g,h) indicate that in the southern part of the area in question from 5 to 3 Ma BP the hydrothermal activity on the EPR axis was low in intensity or absent. In the northern part of the area, from 2°–3°N to 5°–6°N, it was more intensive. Hydrothermal activity decreased to the north of the Clipperton Fracture Zone.

3 to 2 Ma BP the values of Fe_h* in the sediments of all of the sites, except for Sites 851 and 83, were very low or zero (Fig. 5.16). The reconstructed values of Fe_{h20}* are evidence that at this time very intensive hydrothermal activity occurred in the area at ~2°N (Fig. 5.17i).

From 6–5 Ma BP to 3–2 Ma BP the area of very intensive hydrothermal activity moved along the EPR axis over a distance of 400–450 km (from 5°–6°N to ~2°N). The average rate of this movement was about 15 cm a^{-1}.

2 to 0.5 Ma BP the values of Fe_h* in the sediments from Sites 848 to 850, and 846 were very low or zero. In the sediments from Sites 851 to 853 accumulated ~700 km to the west of the EPR axis the values of Fe_h* increased to >2.5 mg cm^{-2} ka^{-1}. A maximum in the Fe_h* values was reached ~1 Ma BP. The value of Fe_h* also increased and exceeded 2.5 mg cm^{-2} ka^{-1} in the sediments from Site 854 in the time period from ~2–1.5 Ma BP to at least 0.3 Ma BP (when the youngest sediments studied from Site 854 accumulated). On the eastern flank of the EPR, in the sediments from Sites 845 and 83, the value of the Fe_h* increased, reached a maximum about 1.5 Ma BP, and then decreased (Fig. 5.16). Apparently the distribution of the reconstructed values of Fe_{h20}* (Fig. 5.17j) indicates the existence of very intensive hydrothermal activity between the Siqueiros Fracture Zone and the area of the Galapagos Triple Junction from 2 to 0.5 Ma BP, on the one hand, and on the other hand they may indicate the appearance of very intensive hydrothermal activity north of the Clipperton Fracture Zone. There are other data that are evidence of the latter. Basal metalliferous sediments that accumulated from 2.2 to 1.1 Ma BP were sampled in the vicinity of the Clipperton Fracture Zone during DSDP Leg 54 (Rosendahl et al. 1980). According to the estimations that were based on the results from this Leg (Rosendahl et al. 1980; Donnelly 1980a,b; Schrader et al. 1980), the values of Fe_h* in these sediments range from tens to hundreds of mg cm^{-2} ka^{-1}.

0.5 to 0 Ma BP the values of Fe_h* in the sediments from all the sites to the west of the EPR axis, except for Sites 854 and 853, were zero. In the sediments from Site 854 the value of Fe_h* exceeded 2.5 mg cm^{-2} ka^{-1} (Fig. 5.16), in the sediments from Site 853 it was about 2 mg cm^{-2} ka^{-1}. The distribution of the reconstructed values of Fe_{h20}* (Fig. 5.17k) is an evidence of the existence of very intensive hydrothermal activity north of 11°N, and of intensive hydrothermal activity between 11° and 7°N. Apparently the very intensive hydrothermal activity that existed from 2 to 0.5 Ma BP in the area between the Siqueiros Fracture Zone and the area of the Galapagos Triple Junction became extinct.

The reconstructed values of Fe_{h20}* between 11°N and 12°N, >25 mg cm^{-2} ka^{-1}, and between 7°N and 11°N, 10–25 mg cm^{-2} ka^{-1}, are close to the average values of Fe_h* in recent metalliferous sediments in the vicinity of the EPR axis in the area between 11°N and 14°N where the values, estimated from data of Lisitzin (1978) and Cherkashev (1990, 1992), range from 20 to 50 mg cm^{-2} ka^{-1}. Agreement in the data such as this encourages and allows one to consider the diagrams in Fig. 5.17a-k as reasonably accurate, especially since the accuracy of the reconstructed values for the recent sediments, accumulated at the longest distance (~600 km) from the EPR axis, is the lowest.

The reconstructed locations of recent (0.5–0 Ma BP) areas of hydrothermal activity coincide with the positions of the existing hydrothermal fields along the EPR axis (Fig. 5.17k). At present hydrothermal activity of greatest intensity and associated mineral formation on the EPR axis within the area in question exists to the north of the Clipperton Fracture Zone between 11°N and 13°N; the areas of intensive hydrothermal activity and mineral formation occur between the Clipperton and Siqueiros Fracture Zones; the areas with hydrothermal activity of lower intensity and mineral formation are at ~7°N, ~4°N, and ~2°N (Fig. 5.17k). If the reconstruction for the time period from 2 to 0.5 Ma BP is correct, it is evident that very intensive hydrothermal activity developed at this time on the EPR axis between 11°N and 13°N.

During the last 12 Ma massive accumulations of hydrothermal sulfide minerals formed in the hydrothermal fields along the EPR between 5°S and 12°N. The approximate place and time of their formation can be determined from the locations of the reconstructed areas of hydrothermal activity (Fig. 5.17a-k). Because of the movement of the Pacific lithosphere plate at the present time these accumulations or their remnants are located at different distances from the EPR axis. If these accumulations have been preserved and covered by sediments or lava flows, their present locations apparently coincide with the areas that have high accumulation rates of hydrothermal Fe in the lowermost basal sediments. The locations of these sediments and the accumulation rates of hydrothermal Fe on the EPR axis, reduced to the distance of 2.5 m from hydrothermal vents ($Fe_h{}^*{}_{2.5}$), can be estimated from the data for the sediments from the drilling cores of the section explored to the west of the EPR axis (Fig. 5.16). In this reconstruction as well as in the reconstructions based on the studies of sediment cores (Sect. 5.1) it is assumed that: a) the hydrothermal Fe has been contributed to the sediments only from hydrothermal plumes or their fragments; and b) the accumulation rate of the hydrothermal Fe at the time of deposition of the sediments was equal to its settling flux. Estimations have been made by using a combination of the two equations (4.1) and (4.3) and with the application of corrections for bottom currents. The complex graph used in obtaining the reconstructed values of $Fe_h{}^*{}_{2.5}$ from the values of $Fe_h{}^*$ and the distance from the EPR axis is shown in Fig. 5.14. In the case of the sediments from the western flank of the EPR and the predominant easterly direction of the bottom currents the line "for a flank oriented against a predominant bottom current direction" can be used. The difference between figures on this line and the line "for a flank oriented to a predominant bottom current direction"[42] is a factor of ~0.5, which has been estimated by comparing the values for Fe_h* in sediments from the eastern and western flanks of the EPR that accumulated at the same latitude and at similar distances from the EPR axis.

The calculated reconstructed values of $Fe_h*_{2.5}$ have been plotted along the backtracked paths of the drilling sites with reference to the age of the sediments from the lowermost part of the basal layer or the age of the crust (Fig. 5.18B). The range of variations of the values of $Fe_h*_{2.5}$ on the EPR axis between 5°S and 12°N during the last 12 Ma is from <1000 to 7500 mg cm^{-2} ka^{-1}. The values of $Fe_h*_{2.5}$ corresponding to the more intensive, >4000 mg cm^{-2} ka^{-1}, and less intensive, 2000–4000 mg cm^{-2} ka^{-1}, hydrothermal activity are approximately twice as low as that on the EPR axis between 20°30'S and 22°00'S during the last 50 ka (Fig. 5.9). This is in good agreement with the general relationship of the accumulation rate of the hydrothermal Fe with the spreading rate, other factors being consistent. Between 5°S and 12°N the half-rate of spreading is 6 to 8 cm a^{-1}, and between 20°30'S and 22°00'S it is ~10 cm a^{-1}. According to the general relationships shown in Figure 4.35, when the half-rate of spreading is 6 to 8 cm a^{-1}, the Fe_h* value is approximately twice as low as it is when the spreading rate is 10 cm a^{-1}.

High settling fluxes and accumulation rates of the hydrothermal Fe in sediments near hydrothermal vents usually accompany the formation and accumulation of sulfide minerals within hydrothermal fields. That is why Fig. 5.18B shows not only the reconstructed distribution of the near-vent accumulation rates of the hydrothermal Fe and the values of Fe_h* in the lowermost part of the basal sediment layer, but also the recent locations of the areas where there are buried accumulations of hydrothermal sulfide minerals (if they have been preserved) that formed during the last 12 Ma along the EPR axis and moved to the west-northwest with the Pacific plate. The higher the values of Fe_h* in the lowermost part of the basal sediment layer the greater the probability of occurrence of buried accumulations of hydrothermal sulfide minerals. Obviously the quantitative aspects of this relationship will be defined in the future.

It must be noted that apparently huge masses of hydrothermal polymetallic mineral matter that formed during the last ~150 Ma are buried in the present ocean. This matter is rich in Fe, Cu, Zn, and other metals. Estimates for the accumulation of the hydrothermal polymetallic mineral matter formed on the axis of the East Pacific Rise and buried at present in

[42] This line can be used for the sediments from the section located to the west of the EPR axis between 20°30'S and 22°00'S (Fig. 5.9) where the westerly direction of the bottom currents prevails.

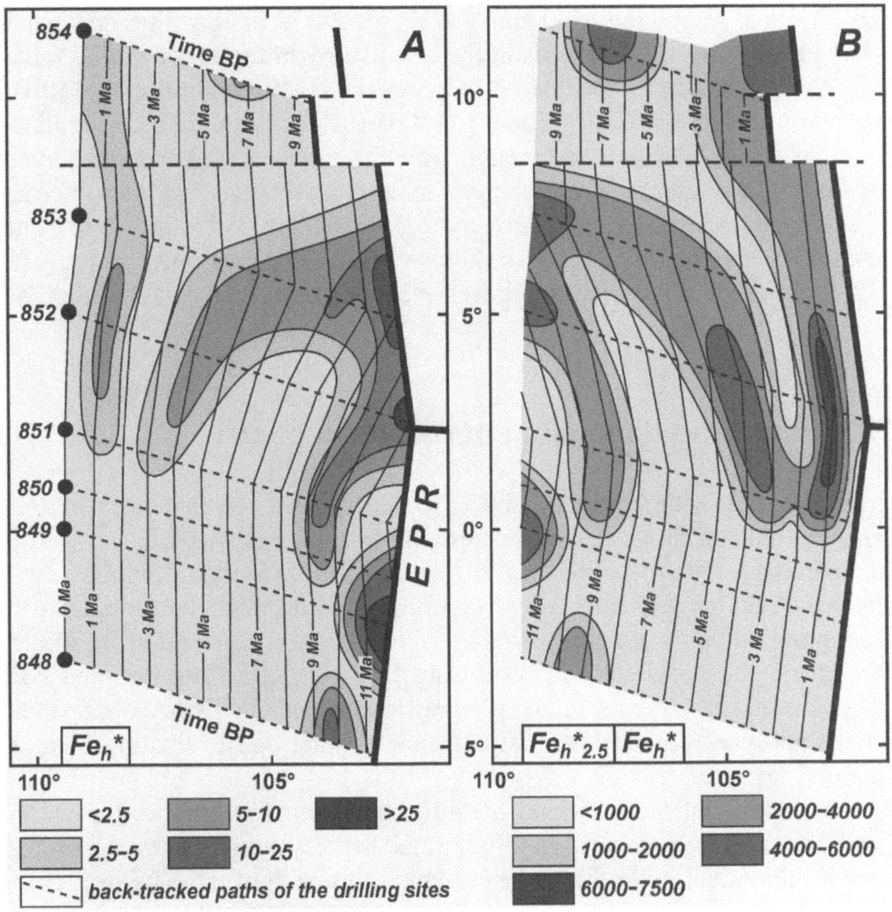

Fig. 5.18. A – Temporal evolution of Fe_h* [mg cm^{-2} ka^{-1}] in surface sediments in the last 12 Ma along a section subparallel to the EPR axis and gradually moving to the west-northwest from the axis with the half-rate of spreading. B – Reconstruction of $Fe_h*_{2.5}$ [mg cm^{-2} ka^{-1}] for the last 12 Ma as well as the distribution of Fe_h* [mg cm^{-2} ka^{-1}] in the lowermost sediments of the basal sediment layer within the area from the section to the EPR axis. Isochrones show sediment and crust age. The distribution reflects the probability of present occurrence of buried accumulations of hydrothermal sulfide minerals.

the Pacific can be made for the benefit of discussion and argument. In the metalliferous sediments in the Southeast Pacific $550 \cdot 10^3$ tons of hydrothermal Fe accumulates annually (Table 1.2). If this amount is considered to be dispersed hydrothermal Fe (actually the amount is greater because not all of the dispersed hydrothermal Fe accumulates in the metalliferous sediments), but it comprises 95–98% of the total hydrothermal contribution of

Fe, which is estimated $(560 \div 580) \cdot 10^3$ t a^{-1}. 2–5% of this amount, that is $(10 \div 30) \cdot 10^3$ t a^{-1}, accumulates in the massive hydrothermal metallic mineral material. If this is considered to be an average amount, during $150 \cdot 10^6$ years along the East Pacific Rise $(1.5 \div 4.5) \cdot 10^{12}$ tons of the Fe accumulated in the massive hydrothermal metallic mineral matter. If the average content of Fe in this matter is ~30%, then an amount of the massive hydrothermal metallic mineral matter for recovery and use is ca. $(5 \div 15) \cdot 10^{12}$ tons. This matter (if it is preserved) is covered by sediments or lava flows and it is located in the Pacific Ocean to the west-northwest and to the east of the East Pacific Rise axis.

5.3. Scale and dimension of reconstructions

In previous sections of this book it has been shown that metalliferous and low-metalliferous sediments provide information that can be used in the reconstruction of the history of hydrothermal activity and in the search for accumulated and buried hydrothermal mineral deposits of significant economic value as mineral resources. The data and concepts from the research on the proximal and distal metalliferous sediments reported here can be used for different kinds of reconstruction of the history of these deposits and can provide essential data for locating and evaluating them as mineral resources.

The studies of the proximal metalliferous sediments enable the reconstruction of the history of hydrothermal activity and mineral formation within individual hydrothermal fields and the search of large hydrothermal mineral accumulations and answer questions related to their position and composition when exploring and evaluating them as resources. The parameters that can be reconstructed and that have been considered are shown in Table 5.1.

The properties of the proximal metalliferous sediments that accumulated within hydrothermal fields of fast, medium, and slow-spreading rifts are similar. But the occurrence and intensity of high-temperature hydrothermal activity on slow-spreading ridges is much lower than on medium- and fast-spreading ridges. High-temperature hydrothermal fields are far apart on the slow-spreading ridges (German and Parson 1998), and fields of distal metalliferous sediments are small or absent. If they do form they usually occur within deep rift valleys, because the nonbuoyant hydrothermal plumes rarely extend beyond such valleys (Sect. 4.3). Fields of distal metalliferous sediments of considerable size can form in the vicinity of groups or numerous groups of hydrothermal fields, and only their combined contribution

Table 5.1. Scales of reconstruction of the history of hydrothermal activity and mineral formation and of search for hydrothermal accumulations in the deep ocean by using materials from studies of metalliferous and low-metalliferous sediments.

Scales of studies / *studied sediments*	Spreading rates	Objects of reconstruction and search	Dimensions of objects, km	Distances to objects, km	Reconstructed and searched parameters of objects	Temporal intervals of reconstruction
Local / *proximal*	slow ÷ medium	large hydrothermal accumulations (active and relict) within hydrothermal fields	$n \cdot 10^{-1}$	$n \cdot 10^{-2} \div (1 \div 2)$	position of large hydrothermal accumulations, intensity of hydrothermal activity and mineral formation on them, metallogenic specialization	$0 \div n \cdot 10^{1}$ ka BP
Areal / *distal*	medium ÷ fast	hydrothermal fields (active and relict) and/or their groups, areas of occurrence of hydrothermal accumulations	$n \cdot 10^{0} \div n \cdot 10^{1}$	$n \cdot 10^{0} \div n \cdot 10^{1}$	position of hydrothermal fields and/or their groups, intensity of hydrothermal activity and mineral formation, present position of areas and probability of occurrence of hydrothermal accumulations	$0 \div n \cdot 10^{2}$ ka BP
Regional / *distal*	medium ÷ fast	areas of hydrothermal activity and mineral formation (recent and ancient) within large rift segments, areas of occurrence of hydrothermal accumulations	$n \cdot 10^{1} \div n \cdot 10^{2}$	$n \cdot 10^{1} \div n \cdot 10^{2}$	position of areas of recent and ancient hydrothermal activity and mineral formation, present position of areas and probability of occurrence of hydrothermal accumulations	$0 \div n \cdot 10^{1}$ Ma BP
Global / *distal*	slow ÷ fast	oceans and their parts	$>n \cdot 10^{3}$		intensity of hydrothermal activity and mineral formation	$0 \div 1.5 \cdot 10^{2}$ Ma BP

of hydrothermal matter can provide sufficient amounts of material for the formation of these sediments. For this reason fields of distal metalliferous sediments of considerable size can form only in and near medium-spreading and mainly fast-spreading ridges. The existence of such fields is evidence of the medium and fast spreading rates of ridges that are or were contributors of the metal-bearing matter to these sediments. It was shown in Sects. 5.1 and 5.2 how the results of studies of the distal metalliferous sediments enable the solving of problems in the reconstruction of their history, in the search and exploration for them and in determining their area and regional distribution. The parameters that can be reconstructed are shown in Table 5.1.

The manner of investigation and approach in the exploration and research of sections parallel or subparallel to the axes of spreading ridges can vary depending on the direction and velocities of bottom currents. The best areas for locating sections for investigation are those where the bottom currents are slow and their predominant directions of flow are across the spreading axes. The section to the west of the EPR axis from 5°S to 12°N is a good example (Fig. 5.16–5.18), where it has been shown that the distance of the sections from the axes of the spreading ridges can be hundreds of kilometers. Sections of minimum distance from the axes must enable the collection of cores that are of sufficient length for the studies. It is not necessary to use continuous core sections for the study and reconstruction of the sequence of historical events. Fragments of core sections can be used for obtaining information and data that are needed: the age of the sediments, the accumulation rate of the hydrothermal Fe, and the distance from the axis of the spreading ridge that contributed the hydrothermal matter to these sediments. However, in such cases the accuracy of the reconstructed history is reduced. In general, the closer the sediment strata studied are to the basement rocks the greater the accuracy. The equations 4.1 and 4.3, which have been used for the reconstructions, are not universal, and they describe the average relationships in the present Pacific and recent Southeast Pacific Ocean. In other areas and at other times in the geological history the character of such relationships could have been different. Corrections can be obtained from materials located at sections or strip-sections along spreading tracks or perpendicular to the axes of spreading ridges, especially if they are located on both flanks of ridges. They can give information on variations in the accumulation rates of the hydrothermal Fe with increasing distances from the axes of spreading ridges both at present and in the geological past.

It has been shown above that at the present time there are approaches and methods for the study of the geology of metalliferous sediments, and for the features or parameters that provide valuable information on the intensity of hydrothermal activity that produced mineral formation, on their locations during geologic time, on the present position of areas and sites where buried hydrothermal accumulations of sulfide minerals can occur, and, in the case of proximal metalliferous sediments, on the locations and special metallogenic features of the accumulations of hydrothermal sulfide minerals that are formed simultaneously with these sediments.

Undoubtedly, with further development of research methods the opportunities for exploration and the possibilities of reconstructing the history and genesis more specifically will be extended greatly in breadth and in accuracy. Nevertheless, the data available at the present time, obtained in the study of the metalliferous sediments, enable the solving of exploration and research questions and the reconstruction of geological events on a local and regional scale. A list of the geological features and parameters considered today in research and in the reconstruction of geological events, the dimensions of features in different scales of studies and distances to them, as well as intervals of time for different scales of studies, are shown in Table 5.1.

Some notice is given in Table 5.1 to studies of a global scale. These can provide new and very important perspectives in the study of hydrothermal activity and mineral formation that have not as yet been considered.

There exists a number of approaches to the study of the global history of hydrothermal activity in the ocean. It is considered important to know the global evolution of hydrothermal mineral formation, climate of the Earth, composition of ocean water, etc. In approaches of this kind the evolution in the contents and ratios of chemical elements and/or their isotopes, which are proxies indicating the hydrothermal contribution to the ocean, are studied. But most of these elements have residence times that exceed the mixing time of the ocean water and cannot characterize particular oceans, let alone any of their parts. Iron, which is used as the main proxy for the content of hydrothermal material in metalliferous sediments, is of special and critical interest. According to my estimates, the hydrothermal contribution of Fe to the Pacific Ocean is about 40–50% of the total contribution of the labile Fe (the dissolved plus that precipitated from solution) to the deep ocean, and its residence time here is estimated about 10 years. From a geological point of view, the response of ocean sediments, and most of all the metalliferous sediments, to variations in the contributions of the hydrothermal Fe is practically instantaneous.

The physical-historical approach to studies of the accumulation of both the proximal and distal metalliferous sediments, as well as knowledge of the factors, which determine variations in the composition and accumulation rates of the metalliferous sediments, enable the reconstruction of the history of the hydrothermal activity and mineral formation within hydrothermal fields and within particular segments of the spreading ridges. In the latter case the present positions of buried massive hydrothermal accumulations can be determined with some probability. This has important significance, especially for the East Pacific Rise, a part of the World Ocean rift system where the most intensive hydrothermal activity and mineral forming processes take place. On the EPR, at least during the last 150 million years, from 5 to 15 trillion (!) tons of polymetallic hydrothermal metal-bearing material have formed. This material has been covered by sediments and lava flows and is located in the Pacific Ocean to the east and to the west-northwest of the EPR axis. At the present time this enormous mineral resource that is rich in such elements as Fe, Cu, Zn, etc. is not recoverable. However, as engineering and technological methods are developed, specific attention will be paid to this obvious and enormous source of metals and minerals. Research and the results of studies of the metalliferous sediments will provide essential information for use in the prospecting and exploration for them.

REFERENCES

Adachi M, Yamamoto K, Sugisaki R (1986) Hydrothermal chert and associated siliceous rocks from the Northern Pacific: their geological significance as indication of ocean ridge activity. Sedimentary Geology 46(1/2):125-148

Al-Karghuli A (1979) Röntgenographische Untersuchungen an Sedimenten aus dem Atlantis II Tief, Rotes Meer. Heidelberg

Almukhamedov AI, Matveenkov VV, Kashintsev GL (1981) Chemical composition of basalts from the axial zone of the Red Sea rift at 18°N. Doklady Akademii Nauk SSSR (Trans USSR Acad Sci) 260(3):744-747

Almukhamedov AI, Juteau T, Matveenkov VV, Eissen JF, Kashintsev GL (1983) Geochemistry of Red Sea low-potash tholeiites. Geochemistry International 20(5):56-70

Almukhamedov AI, Kashintsev GL, Matveenkov VV (1985) Evolution of basalt volcanism in the Red Sea region. Nauka, Novosibirsk

Andreeva IA, Stepanova TV (1992) Minerals-indicators of ore in metalliferous sediments from MAR. In: Gramberg IS, Ainemer AI (eds) Hydrothermal sulfide ores and metalliferous sediments of the ocean. Nedra, Sankt-Petersburg, pp 175-177

Anschutz P, Blanc G (1995) Chemical mass balances in metalliferous deposits from the Atlantis II Deep, Red Sea. Geochim Cosmochim Acta 59(20):4205-4218

Anschutz P, Blanc G (1996) Heat and salt fluxes in the Atlantis II Deep (Red Sea). Earth Planet Sci Lett 142(1-2):147-159

Anschutz P, Turner JS, Blanc G (1998) The development of layering, fluxes through double-diffusive interfaces, and location of hydrothermal sources of brines in the Atlantis II Deep: Red Sea. Journ Geophys Res 103(C12):27809-27819

Anschutz P, Blanc G, Chatin F, Geiller M, Pierret M-C (1999) Hydrographic changes during 20 years in the brine-filled basins of the Red Sea. Deep-Sea Res, Part I 46:1779-1792

ARCYANA scientific team (1975) Transform fault and rift valley from bathyscaph and diving saucer. Science 190(4210):108-116

Arrhenius G, Bonatti E (1965) Neptunism and volcanism in the ocean. In: Progress in Oceanography. Pergamon Press, Oxford, pp 7-22

Bäcker H (1976) Fazies chemische Zusammensetzung rezenten Ausfällungen aus Mineralquellen im Roten Meer. Geologisches Jahrbuch D17:151-172

Bäcker H (1982) Metalliferous sediments of hydrothermal origin from the Red Sea. In: Halbach P, Winter P (eds) Marine Mineral Deposits. Glückauf, Essen, pp 102-136

Bäcker H, Schoell M (1972) New deeps with brines and metalliferous sediments in the Red Sea. Nature Physical Sciences 240:153-158

Bäcker H, Richter H (1973) Die rezente hydrothermalsedimentäre Lagerstätte Atlantis-II Tief im Roten Meer. Geologische Rundschau 62(3):697-741

Bäcker H, Lange J, Richter H (1975) Morphology of the Red Sea central graben between Subair Islands and Abul Kizaan. Geologisches Jahrbuch D13:79-135

Bäcker H, Lange J, Marchig V (1985) Hydrothermal activity and sulfide formation in axial valleys of the East Pacific Rise crest between 18 and 22°S. Earth Planet Sci Lett 72(1):9-22

Bach W, Banrjee NR, Dick HJB, Baker ET (2002) Discovery of ancient and active hydrothermal systems along the ultra-slow spreading Southwest Indian Ridge 10°-16°E. Paper 10.1029/2001GC000279. Geochem Geophys Geosyst 3(7):1-15.

Bagin VI, Pechersky DM (1977) Magneto-mineralogical studies of metalliferous sediments and ferromanganese concretions from the East Pacific Rise. Physics of the Solid Earth 13(9):659-664

Bagin VI, Bagina OL, Bogdanov YuA, Gendler TS, Lebedev AI, Lisitzin AP, Pechersky DM (1975) Iron in the metalliferous sediments of the Bauer Deep and East Pacific Rise. Geochemistry International 12(2):105-125

Bagin VI, Bagina OL, Lebedev AI, Rzaev AG (1979) The magnetic properties and the origin of forms of iron in sediments of the East Pacific Rise. Physics of the Solid Earth 15(1):49-55

Bagin VI, Gendler TS, Dekov VM, Polishchuk OB (1993a) The Mössbauer investigations of metal bearing sediments in the area of the East Pacific Rise (20°30'-22°00'S). Oceanology 33(5):674-681

Bagin VI, Gendler TS, Polishchuk OB, Dekov VM (1993b) Magnetic properties of metalliferous sediments from the East Pacific Rise axial zone. In: LP Zonenshain Memorial Conference on Plate Tectonics. IORAS, Moscow and GEOMAR, Kiel, pp 37-38

Bagin VI, Gendler TS, Dekov VM, Polishchuk OB, Gurvich EG (1996) Magnetic characteristics of surface layer metalliferous sediments from the near-axial zone of the East Pacific Rise between 20°30' and 22°00'S. Oceanology 36(2): 305-313

Baker ET (1994) A 6-year time series of hydrothermal plumes over the Cleft segment of the Juan de Fuca Ridge. Journ Geophys Res 99(B3):4889-4904

Baker ET, Hammond SR (1992) Hydrothermal venting and the apparent magmatic budget of the Juan de Fuca Ridge. Journ Geophys Res 97(B3):3443-3456

Baker ET, Massoth GJ (1987) Characteristics of hydrothermal plumes from two vent fields on the Juan de Fuca Ridge, northeast Pacific Ocean. Earth Planet Sci Lett 85(1/3):59-73

Baker ET, McDuff RE, Massoth GJ (1990) Hydrothermal venting from the summit of a ridge axis seamount: Axial Volcano, Juan de Fuca Ridge. Journ Geophys Res 95:12843-12854

Baker ET, Feely RA, Mottl MJ, Sansone FT, Wheat CG, Resing JA, Lupton JE (1994) Hydrothermal plumes along the East Pacific Rise, 8°40'N to 11°50'N; plume distribution and relationship to the apparent magmatic budget. Earth Planet Sci Lett 128(1-2):1-17

Baker ET, Chen YJ, Morgan JP (1996) The relationship between near-axis hydrothermal cooling and the spreading rate of mid-ocean ridges. Earth Planet Sci Lett 142:137-145

Baker ET, Edmonds HN, Michael PJ, Bach W, Dick HJB, Snow JE, Walker SL, Banerjee NR, Langmuir CH (2004) Hydrothermal venting in magma deserts: The ultraslow-spreading Gakkel and Southwest Indian Ridges. Geochem Geophys Geosyst 5(8):1-29, doi:10.1029/2001GC000712

Baker ET, Lowell RP, Reising JA, Feely RA, Embley RW, Massoth GJ, Walker SL (2004a) Decay of hydrothermal output following the 1998 seafloor eruption at Axial Volcano: Observations and models. Journ Geophys Res 109(B01205):1-14

Balashov YuA (1976) Geochemistry of the rare earth elements. Nauka, Moscow

Balashov YuA, Turanskaya NV, Bratishko RH (1974) Zonality of distribution of the rare earth elements in oceans. Geokhimiya (Geochemistry) 5:751-762

Barrett TJ, Jarvis I (1988) Rare earth element geochemistry of metalliferous sediments from DSDP Leg 92, the East Pacific Rise transect. Chem Geol 67:243-259

Barrett TJ, Taylor PN, Lugowsky J (1987) Metalliferous sediments from DSDP leg 92: The East Pacific Rise transect. Geochim Cosmochim Acta 51(9):2241-2253

Barriga FFA, Costa IMA, Relvas JMR, Ribeiro A, Fouquet Y, Ondreas H, Parson L (1997) The Rainbow serpentinites and serpentinite-sulfide stockwork (Mid-Atlantic Ridge, AMAR segment): a preliminary report of the FLORES results. EOS 78(46):F832

Baturin GN (1971) Deep-sea metal-bearing sediments of hydrothermal genesis. In: Zenkevich LA (ed) History of the World Ocean. Nauka, Moscow, pp 259-277

Baturin GN (1993) Ocean ores. Nauka, Moscow

Baturin GN, Kochenov AV, Trimonis ES (1969) Composition and origin of iron-ore sediments and hot brines in the Red Sea. Okeanologiya (Oceanology) 9(3):442-451

Batuyev BN, Krotov AG, Markov VF, Cherkashev GA, Krasnov SG, Lisitzin ED (1994) Massive sulfide deposits discovered and sampled at 14°45'N, Mid-Atlantic Ridge. BRIDGE Newsletter 6:6-10

Baumann A, Richter H, Schoell M (1973) Suakin deep: Brines and hydrothermal sediments in the deepest part of the Red Sea. Geologische Rundschau 62(3): 684-697

Bemis KG, Von Herzen RP, Mottl MG (1993) Geothermal heat flux from hydrothermal plumes on the Juan de Fuca Ridge. Journ Geophys Res 98:6351-6366

Bendel V, Fouquet Y, Auzende J-M, Lagabrielle Y, Grimaud D, Urabe T (1993) The White Lady hydrothermal field, North Fiji back-arc basin, Southwest Pacific. Econ Geol 88(8):2237-2249

Bender M, Broecker W, Gornitz V, Middel U, Kay K, Sun SS, Biscaye P (1971) Geochemistry of three cores from the East Pacific Rise. Earth Planet Sci Lett 12(4):425-433

Berner RA (1973) Phosphate removal from seawater by adsorption of volcanogenic ferric oxides. Earth Planet Sci Lett 18(1):77-88

Bertine KK, Lee DS (1983) Antimony content and speciation in the water column and interstitial waters of Saanich Inlet. In: Trace metals in seawater. Plenum Press, New York, London, pp 21-38

Bignell RD, Ali SS (1976) Geochemistry and stratigraphy of Nereus Deep, Red Sea. Geologisches Jahrbuch D17:173-186

Bignell RD, Cronan DS, Tooms JS (1976a) Red Sea metalliferous brine precipitates. Geol Assoc Canada, Spec Pap 14:147-184

Bignell RD, Cronan DS, Tooms JS (1976b) Metal dispersion in the Red Sea as an aid to marine geochemical exploration. In: Applied Earth Science 85. Austin and Sons, Hertford, pp B274-B278

Binns RA, Scott SD (1993) Actively forming polymetallic sulfide deposits associated with felsic volcanic rocks in the Eastern Manus back-arc basin, Papua New Guinea. Econ Geol 88(8):2226-2236

Bischoff JL (1969) Red Sea geothermal brine deposits: their mineralogy, chemistry, and genesis. In: Degens ET, Ross DA (eds) Hot brines and recent heavy metal deposits in the Red Sea. Springer, New York, pp 368-401

Bischoff JL, Sayles FL (1972) Pore fluid and mineralogical studies of recent marine sediments: Bauer Depression region of the East Pacific Rise. Journ Sed Petrol 42(3):711-721

Blaise B, Bornhold BD (1987) Geochemistry of northern Juan de Fuca Ridge sediments, northeast Pacific. Geol Surv Canada Paper 87-1A/1. Canadian Gov Publ Centre, Ottawa, pp 127-142

Bloch S (1978) Phosphorus distribution in smectite-bearing basal metalliferous mud. Chem Geol 22:353-359

Blum N, Puchelt H (1991) Sedimentary-hosted polymetallic massive sulfide deposits of the Kebrit and Shaban Deeps, Red Sea. Mineralum Deposita 26:917

Bogdanov YuA (1980) Pelagic sedimentary process in the Pacific Ocean. IOAN, Moscow

Bogdanov YuA (1990a) Sedimentation and history of hydrothermal activity. In: Lisitzin AP (ed) Geological structure and hydrothermal formations of the Juan de Fuca Ridge. Nauka, Moscow, pp 140-146

Bogdanov YuA (1990b) Report of Lithology Group. In: Report of Cruise 21 of R/V Akademik Mstislav Keldysh. Vol. 2. IO RAS, Moscow, pp 3-230

Bogdanov YuA (1992) History and cyclic development of hydrothermal ore formation in the ocean. In: Mirlin EG (ed) Metallogeny of present and ancient oceans. Geoexpert, Moscow, pp 45-51

Bogdanov YuA (1996) Hydrothermal sulfide mineralization in ocean rifts. Okeanologiya (Oceanology) 36(2):277-287

Bogdanov YuA (1997) Hydrothermal deposits of the Mid-Atlantic Ridge. Scientific World, Moscow

Bogdanov YuA, Chekhovskikh EM (1979) Sedimentation rates and accumulation rates. In: Smirnov VI (ed) Metalliferous sediments of the Southeast Pacific. Nauka, Moscow, pp 110-121

Bogdanov YuA, Gurvich EG (1986a) Hydrothermal manifestations at 18°N. In: Lisitzin AP, Bogdanov YuA (eds) Metalliferous sediments of the Red Sea. Nauka, Moscow, pp 54-59

Bogdanov YuA, Gurvich EG (1986b) Spatial and temporal evolution of lithologic and mineral composition. In: Lisitzin AP, Bogdanov YuA (eds) Metalliferous sediments of the Red Sea. Nauka, Moscow, pp 119-126

Bogdanov YuA, Gurvich EG (1986c) The source of ore matter and evolution of the hydrothermal system of ocean rifts. In: Lisitzin AP, Bogdanov YuA (eds) Metalliferous sediments of the Red Sea. Nauka, Moscow, pp 218-230

Bogdanov YuA, Lisitzin AP (1990) Composition of hydrothermal deposits. Reasons for differentiation of hydrothermal matter. In: Lisitzin AP (ed) Geology and hydrothermal formations of the Juan de Fuca Ridge. Nauka, Moscow, pp 80-86

Bogdanov YuA, Sagalevich AM (2002) Geological investigations with the Mir deep-sea manned submersibles. Scientific World, Moscow

Bogdanov YuA, Grigor'ev VN, Muraviov VI, Serova VV, Trimonis ES (1979a) Distribution of metalliferous sediments in the Southeast Pacific. In: Smirnov VI (ed) Metalliferous sediments of the Southeast Pacific. Nauka, Moscow, pp 72-95

Bogdanov YuA, Gurvich EG, Lisitzin AP (1979b) A model of organic carbon accumulation in bottom sediments of the Pacific Ocean. Geochemistry International 16(3):151-159

Bogdanov YuA, Lisitzin AP, Migdisov AA, Smirnov VI, Starostin VI (1979c) On genesis of metalliferous sediments. In: Smirnov VI (ed) Metalliferous sediments of the Southeast Pacific. Nauka, Moscow, pp 249-275

Bogdanov YuA, Gurvich EG, Lisitzin AP (1983a) Mechanism of ocean sedimentation and differentiation of chemical elements in the ocean. In: Lisitzin AP (ed) Biogeochemistry of the ocean. Nauka, Moscow, pp 165-200

Bogdanov YuA, Zhuleva EV, Zonenshain LP (1983b) Atlas of underwater photos from the Red Sea rift. Nauka, Moscow

Bogdanov YuA, Gurvich EG, Butuzova GYu, Lisitzina NA, and Scientific team (1986) Metalliferous sediments of the Red Sea. Nauka, Moscow

Bogdanov YuA, Gordeev VV, Levitan MA (1987) Localization and lithology of metalliferous sediments in the Indian Ocean. In: Lisitzin AP, Gurvich EG (eds) Metalliferous sediments of the Indian Ocean. Nauka, Moscow, pp 20-27

Bogdanov YuA, Khvorova IV, Serova VV, Gorbunova ZN (1989) Sedimentation in the rift zone of the Juan de Fuca Ridge. International Geology Review 31(8):753-760

Bogdanov YuA, Serova VV, Khvorova IV (1990) Lithology of sediments. In: Lisitzin AP (ed) Geological structure and hydrothermal formations of the Juan de Fuca Ridge. Nauka, Moscow, pp 87-94

Bogdanov YuA, Lisitzin AP, Muraviov KG (1992) General regularities of formation of hydrothermal deposits. In: Lisitzin AP (ed) Hydrothermal activity of the Mid-Atlantic Ridge (TAG field); geology, geochemistry, and ore formation. Nauka, Moscow, pp 172-189

Bogdanov YuA, Gorbunova ZN, Serova VV (1993a) Lithologic and mineralogical characteristics of sediments. In: Lisitzin AP (ed) Hydrothermal systems and sedimentary formations of the Mid-Atlantic Ridge. Nauka, Moscow, pp 16-39

Bogdanov YuA, Lisitzin AP, Serova VV (1993b) General characteristics of the sedimentary formation. In: Lisitzin AP (ed) Hydrothermal systems and sedimentary formations of the Mid-Atlantic Ridge. Nauka, Moscow, pp 9-16

Bogdanov YuA, Rona PA, Gurvich EG, Kuptsov VM, Rimsky-Korsakov NA, Sagalevich AM, Hannington MD (1994) Relict sulfide mounds of the TAG hydrothermal field, Mid-Atlantic Ridge (26°N, 45°W). Okeanologiya (Oceanology) 34:590-599

Bogdanov YuA, Gurvich EG, Lisitzin AP, Isaeva AB, Ivanov GV, Krasnyuk AD (1995a) New cores of metalliferous sediments from the TAG hydrothermal field (Mid-Atlantic Ridge 26°N, 45°W) collected in BRAVEX/94 expedition. Doklady Rossiiskoi Akademii Nauk (Trans Russian Acad Sci) 345(2):219-222

Bogdanov YuA, Gurvich EG, Lisitzin AP, Muraviov KG, Isaeva AB, Ivanov GV, Krasnyuk AD (1995b) A newly discovered relict hydrothermal mound in the TAG hydrothermal field, Mid-Atlantic Ridge, 26°N. BRIDGE Newsletter 8: 16-19

Bogdanov YuA, Sagalevich AM, Ashadze AM, Chernyaev ES, Gurvich EG, Lukashin VN, Ivanov GV, Peresypkin VI (1995c) The 14°45'N hydrothermal field of the Mid-Atlantic Ridge. Doklady Rossiiskoi Akademii Nauk (Trans Russian Acad Sci) 343(3):353-357

Bogdanov YuA, Gurvich EG, Lisitzin AP, Isaeva AB, Ivanov GV, Krasnyuk AD (1995d) Metalliferous sediment cores collected during BRAVEX-94 expedition. BRIDGE Newsletter 8:20-24

Bogdanov YuA, Sagalevich AM, Chernyaev ES, Ashadze AM, Gurvich EG, Lukashin VN, Ivanov GV, Peresypkin VI (1995e) A study of the hydrothermal field at 14°45'N on the Mid-Atlantic Ridge using the "Mir" submersibles. BRIDGE Newsletter 9:9-13

Bogdanov YuA, Gurvich EG, Lisitzin AP, Serova VV, Gorbunova ZN (1997) Sediments of the active rift zone in the western Woodlark Basin and the development of hydrothermal and volcanic activity. Mar Geol 142:143-170

Bogdanov YuA, Bortnikov NS, Vikent'ev IV, Gurvich EG, Sagalevich AM (1997a) A new type of modern mineral forming systems: black smokers of the 14°45'N hydrothermal field, Mid-Atlantic Ridge. Geology of Ore Deposits 39(1):68-90

Bogdanov YuA, Gurvich EG, Lukashin VN (1997b) Geochemistry of ore occurrences in the 14°45'N hydrothermal field, Mid-Atlantic Ridge. Geokhimiya (Geochemistry) 4:413-422

Bogdanov YuA, Sagalevich AM, Galkin SV, Chernyaev ES, Lukashin VN (1997c) Geological and biological investigations in the north part of the Mohns Ridge (Norwegian Sea). Okeanologiya (Oceanology) 37(4):609-616

Bogdanov YuA, Gurvich EG, Kashintsev GL, Gorbunova ZN, Vlasova IE (1998) Bottom sediment core from the hydrothermal field of the Mid-Atlantic Ridge at 14°45'N. Oceanology 38(2):221-227

Bogdanov YuA, Sagalevich AM, Gurvich EG, Vikent'ev IV, Lein AYu, Pimenov NV, Peresypkin VI, Gordeev VYu, Voitov DM (1999) Underwater geological research of the Rainbow hydrothermal field (Mid-Atlantic Ridge). Trans Russian Acad Sci 365(5):657-662

Bogdanov YuA, Gurvich EG, Lein AYu, Sagalevich AM, Novikov GV, Peresypkin VI, Bortnikov NS, Vikent'ev IV (2000) Hydrothermal manifestations from the Logachev and Rainbow fields (Mid-Atlantic Ridge) – a new type of hydrothermal deposits of ocean rifts. Russian Journal of Earth Sciences 2(4). http://eos.wdcb.rssi.ru/rjes/v02/rej0004.htm

Bogdanov YuA, Bortnikov NS, Vikent'ev IV, Lein AYu, Gurvich EG, Sagalevich AM, Simonov VA, Ikorsky SV, Stavrova OO, Apollonov VN (2002) Mineralogical and geochemical features of hydrothermal sulfide ores and fluids from the Rainbow field associated with serpentinites, Mid-Atlantic Ridge (36°14'N). Geology of Ore Deposits 44(6):510-542

Bogdanov YuA, Lein AYu, Vikent'ev IV, Bogdanova OYu, Sivtsov AV, Gorshkov AI, Isaeva AB (2004) Low-temperature hydrothermal accumulations from ocean rifts connected genetically with serpentinites. In: Vinogradov ME and Lappo SS (eds) New ideas in oceanology. Vol. 2. Nauka, Moscow, pp 372-402

Bogdanov YuA, Lisitzin AP, Sagalevich AM, Gurvich EG (2005) Hydrothermal ore-genesis of the ocean floor. Nauka, Moscow

Bolger GW, Betzer PR, Gordeev VV (1978) Hydrothermally derived manganese suspended over the Galapagos Spreading Center. Deep-Sea Res 25:721-733

Bonatti E (1967) Mechanism of deep-sea volcanism in the South Pacific. In: Abelson PH (ed) Researches in Geochemistry. Vol. 2. John Wiley and Sons, New York, London, pp 453-491

Bonatti E, Zerbi M, Kay R, Rydell H (1976) Metalliferous deposits from the Apennine ophiolites: Mesozoic equivalents of modern deposits from oceanic spreading centers. Geol Soc Amer Bull 87(1):83-94

Bornhold BD, Tiffin DL, Currie RG (1981) Trace metal geochemistry of sediments, Northeast Pacific ocean. Geol Surv Canada Pap 80-25. Canadian Gov Publ Centre, Ottawa

Bortnikov NS, Vikent'ev IV, Apollonov VN (2001) Rainbow serpentinite-related hydrothermal field, Atlantic mid-ocean ridge, 36°14'N. In: Mineralogical and geochemical features Mineral Deposits at the Beginning of the 21-st Century. Balkema, Rotterdam, pp 265-268

Boström K (1973) The origin and fate of ferromanganese active ridge sediments. Stockholm Contributions in Geology 27(2):148-243

Boström K, Peterson MNA (1966) Precipitates from hydrothermal exhalations on the East Pacific Rise. Econ Geol 61(7):1258-1265

Boström K, Peterson MNA (1969) The origin of aluminum-poor ferromanganese sediments in areas of high heat flow on the East Pacific Rise. Mar Geol 7(5): 427-448

Boström K, Valdes S (1969) Arsenic in ocean floors. Lithos 2:351-360
Boström K, Joensuu O, Moore C, Boström B, Dalziel M, Horowitz A (1973a) Geochemistry of barium in pelagic sediments. Lithos 6(2):159-174
Boström K, Kraemer T, Gartner S (1973b) Provenance and accumulation rates of opaline silica, Al, Ti, Fe, Mn, Cu, Ni, and Co in Pacific pelagic sediments. Chem Geol 11(2):123-148
Boström K, Joensuu O, Valdes S, Riera M (1974) Geochemical history of South Atlantic Ocean sediments since Late Cretaceous. Mar Geol 12(2):85-121
Boström K, Joensuu O, Valdes S, Charm W, Glaccum R (1976) Geochemistry and origin of East Pacific sediments sampled during DSDP Leg 34. In: Init Reports DSDP. Vol. 34. US Govt Print-Office, Wash (DC), pp 559-574
Boström K, Lysén L, Moore C (1978) Biological matter as a source of authigenic matter in pelagic sediments. Chem Geol 23(1-4):11-20
Bowers TS, Campbell AC, Measures CI, Spivack AJ, Khadem M, Edmond JM (1988) Chemical controls on the composition of vent fluids at 11-13°N and 21°N, East Pacific Rise. Journ Geophys Res 93(B5):4522-4536
Brewer PG and Spencer DW (1969) A note on the chemical composition of the Red Sea brines. In: Degens ET, Ross DA (eds) Hot brines and recent heavy metal deposits in the Red Sea. New York, Springer, pp 174-179
Briais A (1995) Structural analysis of the segmentation of the Central Indian Ridge between 20°30'S and 25°30'S (Rodriguez Triple Junction). Marine Geophysical Research 17:431-467
Brockamp O, Goulart E, Harder H, Heydemann A (1978) Amorphous copper and zinc sulfides in the metalliferous sediments of the Red Sea. Contrib Miner Petrol 68(1):85-88
Bromfield SM, David DJ (1976) Sorption and oxidation of manganous ions and reduction of manganese oxide by cell suspensions of a manganese oxidizing bacterium. Soil Biology and Biochemistry 8:37-43
Brooks RR, Kaplan IR and Peterson MNA (1969) Trace metal composition of the Red Sea geothermal brines and interstitial water. In: Degens ET, Ross DA (eds) Hot brines and recent heavy metal deposits in the Red Sea. New York, Springer, pp 180-203
Butterfield DA, Massoth GJ (1994) Geochemistry of North Cleft segment vent fluids: Temporal changes in chlorinity and their possible relation to recent volcanism. Journ Geophys Res 99(B3):4951-4968
Butuzova GYu (1984) Mineralogy and certain aspects of ore-bearing sediment genesis in the Red Sea; Part II, General processes of mineralization and ore-formation in the Atlantis II Deep. Lithology and Mineral Resources 19(4): 293-311
Butuzova GYu (1989) Types of modern hydrothermal and hydrothermal-sedimentary formations in the active zones of the World Ocean. Lithology and Mineral Resources 24(5):409-426
Butuzova GYu (1998) Hydrothermal-sedimentary ore forming processes in the Red Sea rift zone. GEOS, Moscow
Butuzova GYu, Lisitzina NA (1986a) Sediments of the Atlantis II Deep. General characteristics of mineral composition of ore-bearing matter of metalliferous

sediments from the Red Sea. In: Lisitzin AP, Bogdanov YuA (eds) Metalliferous sediments of the Red Sea. Nauka, Moscow, pp 200-209

Butuzova GYu, Lisitzina NA (1986b) Sediments of the Atlantis II Deep. Lithology and mineral composition. In: Lisitzin AP, Bogdanov YuA (eds) Metalliferous sediments of the Red Sea. Nauka, Moscow, pp 105-118

Butuzova GYu, Lyapunov SM (1995) Rare earth elements in hydrothermal-sedimentary deposits from the Red Sea. Lithology and Mineral Resources 30(1): 13-26

Butuzova GYu, Dritz VA, Lisitzina NA (1979) Dynamics of clay mineral formation in ore-bearing sediments from the Atlantis II Deep (Red Sea). Litologiya i Poleznyye Iskopayemyye (Lithology and Mineral Resources) 1:30-42

Butuzova GYu, Lisitzina NA, Alexandrova VA, Shurygina EV (1980) Structure and composition of ore-bearing stratum from the Atlantis II Deep (Red Sea). Litologiya i Poleznyye Iskopayemyye (Lithology and Mineral Resources) 3: 26-41

Butuzova GYu, Lisitzina NA, Lubchenko IYu (1981) Geochemical features of metalliferous sediments of the Atlantis-II Deep (Red Sea). Lithology and Mineral Resources 16(3):35-46

Butuzova GYu, Dritz VA, Lisitzina NA, Tsipursky SI (1983) New data of authigenic layer silicates in metalliferous sediments of the Atlantis-II Deep (Red Sea). Lithology and Mineral Resources 18(5):492-497

Butuzova GYu, Lisitzina NA, Gurvich EG (1986) Sediments of the Thetis Deep. Lithology and mineral composition. In: Lisitzin AP, Bogdanov YuA (eds) Metalliferous sediments of the Red Sea. Nauka, Moscow, pp 164-167

Butuzova GYu, Dritz VA, Morozov AA, Gorshkov AI, Sivtsov AV (1988a) Mineralogy and formation processes of oxidized Fe and Mn substances in the Red Sea Rift zone; Part I, Ore mineralogy. Lithology and Mineral Resources 23(4):318-329

Butuzova GYu, Morozov AA, Dritz VA, Gorshkov AI(1988b) Mineralogy and formation processes of oxidized Fe and Mn substances in the Red Sea Rift zone; Part 2, Main processes of mineral formation. Litologiya i Poleznyye Iskopayemyye (Lithology and Mineral Resources) 6:17-30

Campbell AC, Edmond JM (1989) Halide systematics of submarine hydrothermal vents. Nature 342:168-170

Campbell AC, Edmond JM, Colodner D, Palmer MR, Falkner KK (1987) Chemistry of hydrothermal fluids from the Mariana Trough back arc basin in comparison to mid-ocean ridge fluids. EOS 68(44):1531

Campbell AC, Gieskes JM, Lupton JE, Lonsdale PF (1988a) Manganese geochemistry in the Guaymas Basin, Gulf of California. Geochim Cosmochim Acta 52:345-357

Campbell AC, Palmer MR, Klinkhammer GP, Bowers TS, Edmond JM, Lawrence JR, Casey JF, Thompson G, Humphris S, Rona PA, Karson JA (1988b) Chemistry of hot springs on the Mid-Atlantic Ridge. Nature 335(6190):514-519

Cave RR, German CR, Thomson J, Nesbitt RW (2002) Fluxes to sediments underlying the Rainbow hydrothermal plume at 36°14'N on the Mid-Atlantic Ridge. Geochim Cosmochim Acta 66:1905-1923

Cave RR, Ravizza GE, German CG, Thomson J, Nesbitt RW (2003) Deposition of osmium and other platinum-group elements beneath the ultramafic-hosted Rainbow hydrothermal plume Earth Planet Sci Lett 212:65-79

Chan L-H, Edmond JM, Thompson GA, Gillis K (1992) Lithium isotopic composition of submarine basalts: implications for the lithium cycle in the oceans. Earth Planet Sci Lett 108:151-160

Chan L-H, Edmond JM, Thompson GA (1993) A lithium isotope study of hot springs and metabasalts from mid-ocean ridge hydrothermal systems. Journ Geophys Res 98(B6):9653-9659

Charlou J-L, Fouquet Y, Donval JP, Auzende J-M, Jean-Baptiste P, Stievenard M, Michel S (1996) Mineral and gas chemistry of hydrothermal fluids on an ultrafast spreading ridge: East Pacific Rise, 17° to 19° S (Naudur cruise, 1993) – phase separation processes controlled by volcanic and tectonic activity. Journ Geophys Res 101(B7):15899-15919

Charlou J-L, Donval JP, Fouquet Y, Jean-Baptiste P, Holm (2002) Geochemistry of high H_2 and CH_4 vent fluids issuing from ultramafic rocks at the Rainbow hydrothermal field (36°14'N, MAR). Chem Geol 191:345-359

Chelishchev NF, Gribanova NK, Novikov GV (1992) Sorption characteristics of ocean iron-manganese nodules and crusts. Nedra, Moscow

Cherkashev GA (1990) Metalliferous sediments from areas of ocean sulfide ore formation (on an example of the northern part of the East Pacific Rise). VNIIOkeangeologiya, Leningrad

Cherkashev GA (1992) Geochemistry of metalliferous sediments from areas of ore formation in the ocean. In: Gramberg IS, Ainemer AI (eds) Hydrothermal sulfide ores and metalliferous sediments of the ocean. Nedra, Sankt-Petersburg, pp 138-152

Cherkashev GA (1995) Hydrothermal input into sediments of the Mid-Atlantic Ridge. In: Parson LM, Walker CL, Dixon DR (eds) Hydrothermal vents and processes. Geol Soc London, Spec Publ 87. Geol Soc London, London, pp 223-229

Cherkashev GA (2004) Hydrothermal sulfide ore formation in the northern part of the Mid-Atlantic Ridge. VNIIOkeangeologiya, Sankt-Petersburg

Cherkashev GA, Krasnov SG, Egiazarov BKh, Ainemer AI (1985) Hydrothermal-sedimentary and hydrothermal ore formations in the World Ocean. VIEMS, Moscow

Chester R, Hughes MJ (1967) A chemical technique for the separation of ferromanganese minerals, carbonate minerals and adsorbed trace elements from pelagic sediments. Chem Geol 2(3):249-265

Chin CS, Coale KH, Elrod VA, Johnson KS, Massoth GJ, Baker ET (1994) In situ observations of dissolved iron and manganese in hydrothermal vent plumes, Juan de Fuca Ridge. Journ Geophys Res 99(B3):4969-4984

Choukroune P, Francheteau J, Hekinian R (1984) Tectonics of the East Pacific Rise near 12°50'N: submersible study. Earth Planet Sci Lett 68(2):115-127

Cochran JR, Martinez F (1988) Evidence from the northern Red Sea on the transition from continental to oceanic rifting. Tectonophysics 153(1/4):25-53

Cole TG (1985) The nature, origin and temperature of formation of smectites in the Bauer Deep, southeast Pacific. Geochim Cosmochim Acta 87:1512-1540
Cole TG (1988) The nature and origin of smectite in the Atlantis II Deep, Red Sea. Can Miner 26:755-765
Converse DR, Holland HD, Edmond JM (1984) Flow rates in the axial hot springs of the East Pacific Rise (21°N): Implications for the heat budget and the formation of massive sulfide deposits. Earth Planet Sci Lett 69(2):159-175
Corliss JB, Dymond J, Gordon LI, Edmond J, Von Herzen RP, Ballard RD, Green K, Williams D, Bainbridge A, Crane K, Van Andel TH (1979) Submarine thermal springs on the Galapagos rift. Science 203:1073-1083
Courtois C, Treuil M (1977) Distribution des terres rares et de quelques elements en trace dans les sediments recents des fosses de la Mer Rouge. Chem Geol 20(1):57-72
Cowen JP, Li Y-H (1991) The influence of a changing bacterial community on trace metal scavenging in a deep-sea particle plume. Journ Mar Res 49:517-542
Cowen JP, Massoth GJ, Baker ET (1986) Bacterial scavenging of Mn and Fe in a mid- to far-field hydrothermal particle plume. Nature 322(6075):169-171
Cowen JP, Massoth GJ, Feely RA (1990) Scavenging rates of dissolved manganese in a hydrothermal vent plume. Deep-Sea Res 37:1619-1637
Cox ME, McMurtry GM (1981) Vertical distribution of mercury in sediments from the East Pacific Rise. Nature 289:789-792
Craig H (1969) Geochemistry and origin of the Red Sea brines. In: Degens ET, Ross DA (eds) Hot brines and recent heavy metal deposits in the Red Sea. Springer, New York, pp 208-242
Crocket JH, Macdougall JD, Harris RC (1973) Gold, palladium and iridium in marine sediments. Geochim Cosmochim Acta 37:2547-2556
Cronan DS (1972) The Mid-Atlantic Ridge near 45°N. Al, As, Hg and Mn in ferruginous sediments from the median valley. Canadian Journ Earth Sci 9(3): 319-323
Cronan DS (1976) Basal metalliferous sediments from the Eastern Pacific. Geol Soc Amer Bull 87(6):928-934
Cronan DS (1980) Underwater Minerals. Academic Press, London
Cronan DS (1983) Metalliferous sediments in the CCOP/SOPAC region of the Southwest Pacific, with particular reference to geochemical exploration for the deposits. CCOP/SOPAC Technical Bulletin 4. Suva Stationery Manufacturing Co, Suva
Cronan DS, Hodkinson RA (1997) Geochemistry of hydrothermal sediments from ODP Sites 834 and 835 in the Lau Basin, Southwest Pacific. Mar Geol 141: 237-268
Cronan DS, Moorby SA, Glasby GP, Thomson J, Hodkinson R (1984) Hydrothermal and volcanoclastic sedimentation on the Tonga-Kermadec Ridge and its adjacent marginal basins. Marginal Basin Geology: Volcanic and associated sedimentary processes in modern and ancient marginal basins. Geol Soc London, Spec Publ 16. Blackwells, Oxford, pp 137-149

Cronan DS, Hodkinson R, Harkness DD, Moorby SA, Glasby GP (1986) Accumulation rates of hydrothermal metalliferous sediments in the Lau Basin, SW Pacific. Geo-Marine Letters 6:51-56

Daesslé LW, Cronan DS (2002) Late Quaternary hydrothermal sedimentation adjacent to the Central Lau Spreading Center. Mar Geol 182:389-404

Daesslé LW, Cronan DS, Marchig V, Wiedicke M (2000) Hydrothermal sedimentation adjacent to the propagating Valu Fa Ridge, Lau Basin, SW Pacific. Mar Geol 162:479-500

Danielsson L-G, Dyrsson D, Granéli A (1980) Chemical investigations of Atlantis II and Discovery brines in the Red Sea. Geochim Cosmochim Acta 44:2051-2066

Davydov MP (1992) Mineral indicators of ore formation in bottom sediments from the EPR. In: Gramberg IS, Ainemer AI (eds) Hydrothermal sulfide ores and metalliferous sediments of the ocean. Nedra, Sankt-Petersburg, pp 165-175

Davydov MP, Sudarikov SM, Dragan-Sushchova OI (1997) Distribution of iron and manganese in the bottom metalliferous sediments of the 21°-23°S region, East-Pacific Rise. Doklady Rossiiskoi Akademii Nauk (Trans Russian Acad Sci) 352(1):141-144

Davydov MP, Sudarikov SM, Alexandrov PA, Ashadze AM (2002) Geochemistry of metalliferous sediments from hydrothermal fields of the East Pacific Rise (11°30'-13°N). I. Geochemistry of Holocene sediments. Geokhimiya (Geochemistry) 3:319-339

Dekov VM (1994) Hydrothermal sedimentation in the Pacific Ocean. Nauka, Moscow

Dekov VM, Gurvich EG (1991) Relationship between grain size and chemical compositions of metal-bearing sediments in the axial zone of the East Pacific Rise (20°30'-22°N). Oceanology 31(5):620-624

Dekov VM, Gurvich EG (1993) Rare-earth element composition of metalliferous sediments from the near-axial zone of the East Pacific Rise (20°30'-22°N). Spisanie na Bulgarskogo Geologichesko Druzhestvo 54(2):838-845

Dekov VM, Marchig V, Rajta I, Uzonyi I (2003) Fe-Mn micronodules born in the metalliferous sediments of two spreading centers: East Pacific Rise and Mid-Atlantic Ridge. Mar Geol 199:10-21

Delaney JR, Robigou V, McDuff RE, Tivey MK (1992) Geology of a vigorous hydrothermal system on the Endeavor segment, Juan de Fuca Ridge. Journ Geophys Res 97:19663-19682

Demina LL, Serova VV, Tambiev SB (1992) Particle fluxes. The study using sediment traps. In: Lisitzin AP (ed) Hydrothermal accumulations of the Mid-Atlantic Ridge. Nauka, Moscow, pp 68-76

Derkachev AN, Likht FR, Nechayev VP, Nikolaeva NA, Grigor'eva TN (1989) Hydrothermal manifestations and metalliferous sediments in the area of the Juan de Fuca Ridge. Doklady Akademii Nauk SSSR (Trans USSR Acad Sci) 305(5):1203-1207

Desonie DL, Duncan RA (1990) The Cobb-Eickelberg Seamount chain: Hotspot volcanism with mid-ocean ridge basalt affinity. Journ Geophys Res 95:12697-12711

Detrick RS, Fox PJ, Shulz N, Pockalny R, Kong 1, Mayer L, Ryan WBF (1988) Geologic and tectonic setting of the MARK area. In: Littleton RM, Stewart NJ (eds) Proceedings of the Ocean Drilling Program covering legs 106 and 109 of the cruises of the drilling vessel JOIDES Resolution. Texas A & M University, Ocean Drilling Program, College Station, pp 15-22

Donnelly TW (1980a) Appendix. Chemical composition of deep-sea sediments – sites 9 through 425, legs 2 through 54. In: Powell R (ed) Initial Reports DSDP. Vol. 54. Texas A&M University, Ocean Drilling Program, College Station, pp 899-949

Donnelly TW (1980b) Secondarily modified sediments of the eastern Pacific: major element chemistry of Sites 420, 424, and 425, Deep Sea Drilling Project Leg 54. In: Powell R (ed) Initial Reports DSDP. Vol. 54. Texas A & M University, Ocean Drilling Program, College Station, pp 329-338

Douville E, Bienvenu P, Charlou J-L, Donval JP, Fouquet Y, Appriou P, Gamo T (1999a) Yttrium and rare earth elements in fluids from various deep-sea hydrothermal systems. Geochim Cosmochim Acta 63(5):627-643

Douville E, Charlou J-L, Donval JP, Hureau D, Appriou P (1999b) Le comportement de l'arsenic (As) et de l'antimoine (Sb) dans les fluids provenant de différents systèmes hydrothermaux océaniques. Compt Rend Acad Sci Paris, Sciences de la terre et des planets 328:97-104

Douville E, Charlou J-L, Oelkers EH, Bienvenu P, Jove Colon CF, Donval JP, Fouquet Y, Prieur D, Appriou P (2002) The Rainbow vent fluids (36°14'N, MAR): the influence of ultramafic rocks and phase separation on trace metal content in Mid-Atlantic Ridge hydrothermal fluids. Chem Geol 184:37-48

Dubinin AV (2004) Geochemistry of the rare earth elements in the ocean. Geokhimiya (Geochemistry) 4:339-358

Dubinin AV, Volkov II (1986) The rare-earth elements in the East Pacific Rise metalliferous sediments. Geochemistry International 23(9):62-76

Dubinin AV, Volkov II (1988) A mechanism of rare-earth accumulation in metalliferous sediments of the East Pacific Rise. Geokhimiya (Geochemistry) 4: 557-566

Dubinin AV, Volkov II (1992) Geochemistry of bottom sediments from the East Pacific Rise: general characteristics of metal-bearing matter. Litologiya i Poleznyye Iskopayemyye (Lithology and Mineral Resources) 6:3-24

Durant P, Prieur D, Jeanthon C, Jacq E (1990) Occurrence and activity of heterotrophic manganese oxidizing bacteria associated with alvinellids (Polychaetous annelids) from a deep hydrothermal vent site on the East Pacific Rise. Compt Rend Acad Sci Serie III - Sciences de la Vie 310:273-278

Dymond J (1981) Geochemistry of Nazca plate surface sediments: An evaluation of hydrothermal, biogenic, detrital, and hydrogenous sources. Geol Soc Amer Mem 154:133-173

Dymond J, Roth S (1988) Plume dispersed hydrothermal particles: A time-series record of settling flux from the Endeavor Ridge using moored sensors. Geochim Cosmochim Acta 52:2525-2536

Dymond J, Veeh HH (1975) Metal accumulation rates in the southeast Pacific and the origin of metalliferous sediments. Earth Planet Sci Lett 28(1):13-22

Dymond J, Corliss JB, Heath GR, Field CW, Dasch EJ, Veeh HH (1973) Origin of metalliferous sediments from the Pacific Ocean. Geol Soc Amer Bull 84(10): 3355-3372

Dymond J, Corliss JB, Stillinger R (1976) Chemical composition and metal accumulation rates of metalliferous sediments from Sites 319, 320, and 321. Initial Reports DSDP. Vol. 34. US Govt Print-Office, Wash(DC), pp 575-588

Dymond J, Corliss JB, Heath GR (1977) History of metalliferous sedimentation at Deep Sea Drilling Site 319 in the South Eastern Pacific. Geochim Cosmochim Acta 41(6):741-753

Dziak RP, and Fox CG (1999) The January 1998 earthquake swarm at Axial Volcano, Juan de Fuca Ridge: Hydroacoustic evidence of seafloor volcanic activity. Geophys Res Lett 26:3429-3432

Edenborn HM, Paquin Y, Chateauneuf G (1985) Bacterial contribution to manganese oxidation in a deep coastal sediment. Estuarine and Coastal Marine Science 21:801-815

Edmond JM, Von Damm KL (1985) Chemistry of ridge crest hot springs. Biol Soc Wash Bull 6:43-47

Edmond JM, Campbell AC, Palmer MP, Klinkhammer GP (1986) Preliminary report on the chemistry of hydrothermal fluids from the Mid-Atlantic Ridge. EOS 67(44):1021

Edmond JM, Campbell AC, Palmer MP, German CR (1990) Geochemistry of hydrothermal fluids from the Mid-Atlantic Ridge: TAG and MARK. EOS 71: 1650-1651

Edmond JM, Campbell AC, Palmer MP, German CR, Klinkhammer GP, Edmonds HN, Elderfield H, Thompson G, Rona P (1995) Time series studies of vent fluids from the TAG and MARK sites (1986, 1990) Mid-Atlantic Ridge: a new solution chemistry model and a mechanism for Cu/Zn zonation in massive sulfide ore bodies. In: Parson LM, Walker CL, Dixon DR (eds) Hydrothermal vents and processes. Geol Soc London, Spec Publ 87. Geol Soc London, London, pp 77-86

Edmonds HN, German CR (2004) Particle geochemistry in the Rainbow hydrothermal plume, Mid-Atlantic ridge. Geochim Cosmochim Acta 68(4):759-772

Edmonds HN, Michael PJ, Baker ET, Connelly DP, Snow JE, Langmuir CH, Dick HJB, Mühe R, German CR, Graham DW (2003) Discovery of abundant hydrothermal venting on the ultraslow-spreading Gakkel Ridge, Arctic Ocean. Nature 421:252-256

Ehrlich HL (1983) Manganese oxidizing bacteria from a hydrothermally active area on the Galapagos Rift. Ecological Bulletin (Stockholm) 35:357-366

Ehrlich HL (1985) Mesophilic manganese-oxidizing bacteria from a hydrothermal discharge area at 21° North on the East Pacific Rise. In: Planetary Ecology. Van Norstrand, Reinhold, pp 186-194

Elderfield H, Schultz A (1996) Mid-ocean ridge hydrothermal fluxes and the chemical composition of the ocean. Annu Rev Earth Sci 24:191-224

Elderfield H, Greaves MJ, Rudnicki MD (1993a) Aluminum reactivity in hydrothermal plumes at the Mid-Atlantic Ridge. Journ Geophys Res 98(B6):9667-9670

Elderfield H, Mills RA, Rudnicki MD (1993b) Geochemical and thermal fluxes, high temperature venting and diffuse flow from mid-ocean ridge hydrothermal systems: the TAG hydrothermal field, Mid-Atlantic Ridge 26°N. In: Magmatic Processes and Plate Tectonics. Geol Soc London, Spec Publ 76. Geol Soc London, London, pp 295-307

Embley RW, Ganse R, Malahoff A, Hammond SR (1982) Tectonics of the Blanco Trough and Blanco – Juan de Fuca intersection. EOS 63(45):1146

Embley RW, Chadwick WW Jr, Clague D, Stakes D (1999) 1998 eruption of Axial Volcano: Multibeam anomalies and seafloor observations. Geophys Res Lett 26:3425-3428

Emel'yanov EM (1975) Fe, Mn and Ti in bottom sediments of the Atlantic Ocean. Litologiya i Poleznyye Iskopayemyye (Lithology and Mineral Resources) 3:3-19

Emel'yanov EM (1982) Sedimentogenesis in the basin of the Atlantic ocean. Nauka, Moscow

Emerson S, Cranston RE, Liss PS (1979) Redox species in a reducing fjord: Equilibrium and kinetic considerations. Deep-Sea Res 26A:859-873

Emerson S, Kalhorn S, Jacobs L, Tebo BM, Nealson KH, Rosson RA (1982) Environmental oxidation rate of manganese(II): bacterial catalysis. Geochim Cosmochim Acta 46:1073-1079

Feely RA, Lewison M, Massoth GJ, Robert-Baldo G, Lavelle JW, Byrne RH, Von Damm KL, Curl HC Jr (1987) Composition and dissolution of black smoker particulates from active vents on the Juan de Fuca Ridge. Journ Geophys Res 92(B11):11347-11363

Feely RA, Geiselman TL, Baker ET, Massoth GJ, Hammond SR (1990) Distribution and composition of buoyant and non-buoyant hydrothermal plume particles from the ASHES vent at Axial Volcano, Juan de Fuca Ridge. Journ Geophys Res 95(B8):12855-12874

Feely RA, Massoth GJ, Trefry JH (1991a) Scavenging of oxyanions in buoyant hydrothermal plumes on the Southern Juan de Fuca Ridge. EOS 72:235

Feely RA, Trefry JH, Massoth GJ, Metz S (1991b) A comparison of the scavenging of phosphorus and arsenic from seawater by hydrothermal iron oxyhydroxides in the Atlantic and Pacific Oceans. Deep-Sea Res 36(6):617-623

Feely RA, Massoth GJ, Baker ET, Lebon GT, Geiselmam TL (1992) Tracking the dispersal of hydrothermal plumes from the Juan de Fuca Ridge using suspended matter compositions. Journ Geophys Res 97(B3):3457-3468

Feely RA, Massoth GJ, Trefry JH, Backer ET, Paulson AJ, Lebon GT (1994) Composition and sedimentation of hydrothermal plume particles from North Cleft segment, Juan de Fuca Ridge. Journ Geophys Res 99(B3):4985-5006

Feely RA, Baker ET, Marumo K, Urabe T, Ishibashi J, Gendron J, Lebon GT, Okamura K (1996) Hydrothermal plume particles and dissolved phosphate over superfast-spreading southern East Pacific Rise. Geochim Cosmochim Acta 60(13): 2297-2323

Feely RA, Trefry JH, Lebon GT, German CR (1998) The relationship between P/Fe and V/Fe ratios in hydrothermal precipitates and dissolved phosphate in seawater. Geophys Res Lett 25(13):2253-2256

Field MP, Sherrell RM (2000) Dissolved and particulate Fe in a hydrothermal plume at 9°45'N, East Pacific Rise: Slow Fe(II) oxidation kinetics in Pacific plumes. Geochim Cosmochim Acta 64(4):619-628

Fisher K, Puchelt H (1972) Barium 56. Handbook of Geochemistry. Vol. II-3. Springer, Berlin, pp 56-A-1–56-O-22

Fleet AJ, Robertson AHF (1980) Ocean-ridge metalliferous and pelagic sediments of the Semail Nappe, Oman. Journ Geol Soc London 37:403-422

Forsyth FW (1972) Mechanism of earthquakes and plate motions in the East Pacific. Earth Planet Sci Lett 17:189-193

Fouquet Y, Auclair G, Cambon P, Etoubleau J (1988) Geological setting, mineralogical and geochemical investigations on sulfide deposits near 13 degrees N on the East Pacific Rise. Marine Geology 84:145-178

Fouquet Y, Von Stackelberg U, Charlou J-L, Donval JP, Erzinger J, Foucher JP, Herzig PM, Mühe R, Soakai S, Wiedicke M, Whitechurch H (1991) Hydrothermal activity and metallogenesis in the Lau back-arc basin. Nature 349(6312):778-781

Fouquet Y, Von Stackelberg U, Charlou J-L, Erzinger J, Herzig PM, Mühe R, Wiedicke M (1993) Metallogenesis in back-arc environments: the Lau Basin example. Econ Geol 88(8):2150-2177

Fouquet Y, Knott R, Cambon P, Fallick A, Rickard D, Desbruyeres D (1996) Formation of large sulfide mineral deposits along fast spreading ridges: Example from off-axis deposits at 12°43' on the East Pacific Rise. Earth Planet Sci Lett 144:147-162

Fouquet Y, Charlou J-L, Ondréas H, Radford-Knoery J, Donval JP, Douville E, Apprioual R, Cambon P, Pellé H, Landuré JY, Normand A, Ponsevera E, German C, Parson L, Barriga F, Costa I, Relvas J, Ribeiro (1997) A Discovery and first submersible investigations on the Rainbow Hydrothermal Field on the MAR (36°14'N). EOS 78(46):832

Fouquet Y, Barriga F, Charlou J-L, Elderfield H, German C, Ondréas H, Parson L, Radford-Knoery J, Relvas J, Ribeiro A, Schultz A, Apprioual R, Cambon P, Costa I, Donval JP, Douville E, Landuré JY, Normand A, Pellé H, Ponsevera E, Riches S, Santana H, Stephen M (1998) FLORES diving cruise with the Nautile near the Azores – First dives on the Rainbow Field: hydrothermal seawater/mantle interaction. InterRidge News 7:24-28

Francheteau J, Ballard RD (1983) The East Pacific Rise near 21°N, 13°N and 21°S: inferences for along strike variability of axial processes of the Mid-Ocean Ridge. Earth Planet Sci Lett 64(1):93-116

Francheteau J, Needham HD, Chourkroune P and CYAMEX Scientific Team (1979) Massive deep sea sulfide ore deposits discovered on the East Pacific Rise. Nature 277:523-528

Froelich PN, Bender ML, Heath GR (1977) Phosphorus accumulation rates in metalliferous sediments on the East Pacific Rise. Earth Planet Sci Lett 34(3): 351-359

Gamo T, Nakayama E, Shitashima K, Isshiki K, Obata H, Okamura K, Kanayama S, Oomori T, Koizumi T, Matsumoto S, Hasumoto H (1996) Hydrothermal plumes at the Rodriguez triple junction, Indian ridge. Earth Planet Sci Lett 142(1-2):261-270

Gamo T, Chiba H, Yamanaka T, Okudaira T, Hashimoto J, Tsuchda S, Ishibashi J, Kataoka S, Tsunogai U, Okamura K, Sano Y, Shinjo R (2001) Chemical characteristics of newly discovered black smoker fluids and associated hydrothermal plumes at the Rodriguez Triple Junction, Central Indian Ridge. Earth Planet Sci Lett 193:371-379

Gendler TS, Polishchuk OB, Dekov VM, Novakova AA (1993) Mössbauer spectra and magnetic properties of metalliferous sediments of the EPR (21°S). In: ICAME 93 Book of Abstracts. Vancouver, p 3

Gente P, Auzende J-M, Renard V, Fouquet Y, Bideau D (1986) Detailed geological mapping by submersible of the East Pacific Rise axial graben near 13°N. Earth Planet Sci Lett 78(2/3):224-236

Gente P, Mevel C, Auzende J-M, Karson JA, Fouquet Y (1991) An example of recent accretion on the Mid-Atlantic Ridge: the Snake Pit neovolcanic ridge (MARK area, 23°22'N). Tectonophysics 190(1):1-29

Geologic Research in the Central Atlantic (1991) Sharapov VN (ed). Nauka, Novosibirsk

German CR, Angel MV (1995) Hydrothermal fluxes of metals to the oceans: a comparison with anthropogenic discharge. In: Parson LM, Walker CL, Dixon DR (eds) Hydrothermal vents and processes. Geol Soc London Spec Publ 87. Geol Soc London, London, pp 365-372

German CR, Parson LM (1998) Distributions of hydrothermal activity along the Mid-Atlantic Ridge: interplay of magmatic and tectonic controls. Earth Planet Sci Lett 160(14):327-341.

German CR, Von Damm K (2004) Hydrothermal processes In: Elderfield H (ed) Treatise on geochemistry. Vol. 6 The oceans and marine geochemistry (eds Holland HD, Turekian KK). Elsevier-Pergamon, Oxford, pp 181-222

German CR, Klinkhammer GP, Edmond JM, Mitra A, Elderfield H (1990) Hydrothermal scavenging of rare-earth elements in the ocean. Nature 345(6275): 516-518

German CR, Campbell AC, Edmond JM (1991) Hydrothermal scavenging at the Mid-Atlantic Ridge: Modification of trace element dissolved fluxes. Earth Planet Sci Lett 107:101-114

German CR, Higgs NC, Thomson J, Mills R, Elderfield H, Blusztajn J, Fleer AP, Bacon MP (1993) A geochemical study of metalliferous sediments from the TAG Hydrothermal Mound, 26°08'N, Mid-Atlantic Ridge. Journ Geophys Res 98(B6):9683-9692

German CR, Klinkhammer G, Rudnicki MD (1996a) The Rainbow hydrothermal plume, 36°15'N, MAR. Geophys Res Lett 23:2979-2982

German CR, Parson LM and the HEAT Scientific Party (1996b) Hydrothermal exploration near the Azores Triple Junction: Tectonic control of venting at slow-spreading ridges? Earth Planet Sci Lett 138:105-119

German CR, Bourlès DL, Brown ET, Hergt J, Colley S, Higgs NC, Ludford EM, Nelsen TA, Feely RA, Raisbeck G, Yiou F (1997) Hydrothermal scavenging on the Juan de Fuca Ridge: $^{230}Th_{xs}$, ^{10}Be, and REEs in ridge-flank sediments. Geochim Cosmochim Acta 61(19):4067-4078

German CR, Baker ET, Mevel C, Tamaki K, the FUJI Science Team (1998) Hydrothermal activity along the Southwest Indian Ridge. Nature 395:400-493

German CR, Connely DP, Evans AJ, Murton BJ, Curewitz D, Okino K, Parson LM (2001) Hydrothermal activity along the Central Indian Ridge: Ridges, hot spots and philately. EOS Fall Meeting Suppl 82(47):OS42E-06

German CR, Colley S, Palmer MR, Khripounoff A, Klinkhammer GP (2002) Hydrothermal plume-particle fluxes at 13°N on the East Pacific Rise. Deep-Sea Res I 49:1921-1940

Geyh M, Höhndorf A (1976) The contribution of complementary ^{14}C and Th/U analyses to the stratigraphy of the Red Sea sediments. Geologisches Jahrbuch D17(1):79-91

Girdler RW (1985) The evolution of the Gulf of Aden in space and time. In: School of physics. University Newcastle-upon-Tyne, pp 747-762.

Girin YuP, Gordeev VV, Gurvich EG, Lukashin VN, Migdisov AA, Barskaya NV (1979) Chemical composition of Fe-Mn nodules from metalliferous and normal pelagic sediments of the Southeast Pacific Ocean. Geochemistry International 16(1):45-61

Godfrey LV, Mills RA, Elderfield H, Gurvich EG (1994) Lead behavior at the TAG hydrothermal vent field, 26°N, Mid-Atlantic Ridge. Mar Chem 46:237-254

Goodwin AM (1964) Geochemical studies at the Heleniro range. Econ Geol 59: 684-718

Goodwin AM (1973) Archaean iron-formations and tectonic basins of the Canadian Shield. Econ Geol 68:915-933

Gorbunova ZN (1981) High-dispersion minerals in the sediments of the East Pacific Rise and adjacent regions. Lithology and Mineral Resources 16(3):246-256

Gorbunova ZN (1982) High-dispersion minerals in sediment cores from the Southeast Pacific. Okeanologiya (Oceanology) 22(3):454-459

Gorbunova ZN (1987) High-dispersion minerals in metalliferous sediments. In: Lisitzin AP, Gurvich EG (eds) Metalliferous sediments of the Indian Ocean. Nauka, Moscow, pp 76-86

Gorbunova ZN (1990) High-dispersion minerals in sediments of the Axial Seamount and adjacent areas. In: Lisitzin AP (ed) Geological structure and hydrothermal formations of the Juan de Fuca Ridge. Nauka, Moscow, pp 94-99

Gorbunova ZN, Lisitzin AP (1981) Main regularities of clay mineral distribution in surface sediments of the Southeast Pacific. Doklady Akademii Nauk SSSR (Trans USSR Acad Sci) 261(6):1402-1407

Gordeev VV (1986a) Geochemistry of sediments. In: Lisitzin AP (ed) Transform faults of the Indian Ocean. Nauka, Moscow, pp 159-167

Gordeev VV (1986b) Geochemistry of sediments. In: Lisitzin AP (ed) Transform faults of the Indian Ocean. Nauka, Moscow, pp 190-196

Gordeev VV (1988) Report of Geochemistry Group. In: Report of Cruise 15 of R/V Akademik Mstislav Keldysh. Vol. 2(2). IO RAS, Moscow, pp 3-100

Gordeev VV, Demina LL (1979) Direct observations of thermal springs on the bottom of the Pacific Ocean (Galapagos active zone, Hess Depression). Geokhimiya (Geochemistry) 6: 902-917

Gordeev VV, Lisitzin AP (1979) Trace elements. In: Volkov II (ed) Oceanology. Ocean chemistry. Vol. 1, Chemistry of ocean water. Nauka, Moscow, pp 337-375

Gordeev VV, Demina LL, Konkin AM (1979) Iron and manganese of hydrothermal origin in deep water of the Pacific Ocean (Galapagos active zone, Hess Depression). In: Smirnov VI (ed) Metalliferous sediments of the Southeast Pacific. Nauka, Moscow, pp 224-236

Grill EV, Chase RL, Macdonald RD, Murray JW (1981) A hydrothermal deposit from Explorer Ridge in the Northeast Pacific Ocean. Earth Planet Sci Lett 52:142-150

Grinenko VA, Ustinov VI (1982) Isotope equilibriums of hydrothermal systems. In: Galimov EM (ed) IX All-union symposium on stable isotopes in geochemistry. Vol. 1. GEOKHI, Moscow, pp 92-94

Gross GA (1984) The metallogenic significance of metalliferous sediments, Precambrian to Recent. Geol Surv Canada Paper 84-8. Canadian Gov Publ Centre, Ottawa

Gross GA (1987) Mineral deposits on the seabed. Marine Mining 6:109-119

Gross GA (1988) A comparison of metalliferous sediments, Precambrian to Recent. Krystalinikum 19:59-74

Gross GA (1991) Genetic concepts for iron-formation and associated metalliferous sediments. Econ Geol Monographs 8:51-81

Gross GA, McLeod CR (1987) Metallic minerals on the seabed. Geol Surv Canada Paper 86-21. Canadian Gov Publ Centre, Ottawa

Guney M, Nawab Z, Marhoun MA (1984) Atlantis-II-Deep's mineral reserves and their evaluation. In: Offshore Technology Conference. Vol. 3. Houston, p 33

Gurvich EG (1977) Trace elements in bottom sediments of the Southeast Pacific. IOAN, Moscow

Gurvich EG (1987) Geochemistry of antimony in the Pacific Ocean. In: The third congress of soviet oceanologists. Gidrometeoizdat, Leningrad, pp 79-80

Gurvich EG (1988) Report on geochemical investigations in Cruise 4 of R/V Geolog Fersman. IOAN, Moscow

Gurvich EG (1990) Report of Geochemistry Group. In: Report of Cruise 21 of R/V Akademik Mstislav Keldysh. Vol. 3. IOAN, Moscow

Gurvich EG (1992a) Metalliferous sediments of the World ocean: recent and ancient. In: 29-th Intern Geol Congress. Vol. 3. Kyoto, p 756

Gurvich EG (1992b) Recent and ancient ocean metalliferous sediments. In: Mirlin EG (ed) Metallogeny of modern and ancient oceans. Geoexpert, Moscow, pp 183-186

Gurvich EG (1994) Report of the group of geochemistry of sediments and ores. In: Report of Cruise 34 of R/V Akademik Mstislav Keldysh. IO RAS, Moscow

Gurvich EG (1998) Metalliferous sediments of the World Ocean. Scientific World, Moscow

Gurvich EG (2004) Pelagic metalliferous sediments, new approaches. In: Vinogradov ME and Lappo SS (eds) New ideas in oceanology. Vol. 2. Nauka, Moscow, pp 351-371

Gurvich EG, Bogdanov YuA (1986a) Characteristics of brines. In: Lisitzin AP, Bogdanov YuA (eds) Metalliferous sediments of the Red Sea. Nauka, Moscow, pp 100-105

Gurvich EG, Bogdanov YuA (1986b) Comparison of hydrothermal accumulations from the Red Sea rift zone. In: Lisitzin AP, Bogdanov YuA (eds) Metalliferous sediments of the Red Sea. Nauka, Moscow, pp 209-218

Gurvich EG, Bogdanov YuA (1986c) Sedimentation of the metal-bearing matter and facies of metalliferous sediments. In: Lisitzin AP, Bogdanov YuA (eds) Metalliferous sediments of the Red Sea. Nauka, Moscow, pp 232-243

Gurvich EG, Bogdanov YuA (1986d) Sediments of the Atlantis II Deep. Accumulation rates of the main components of the metal-bearing matter. Relationship to intensity of hydrothermal activity. In: Lisitzin AP, Bogdanov YuA (eds) Metalliferous sediments of the Red Sea. Nauka, Moscow, pp 138-145

Gurvich EG, Bogdanov YuA (1986e) Temporal and spatial variations of chemical composition of sediments. In: Lisitzin AP, Bogdanov YuA (eds) Metalliferous sediments of the Red Sea. Nauka, Moscow, pp 126-138

Gurvich EG, Bogdanov YuA (1986f) Sediments of the Gypsum Deep. Sediment chemistry. In: Lisitzin AP, Bogdanov YuA (eds) Metalliferous sediments of the Red Sea. Nauka, Moscow, pp 192-195

Gurvich EG, Bogdanov YuA (1986g) Sediments of the Thetis Deep. Sediment chemistry. In: Lisitzin AP, Bogdanov YuA (eds) Metalliferous sediments of the Red Sea. Nauka, Moscow, pp 167-175

Gurvich EG, Bogdanov YuA (1986h) Sediments of the Vema Deep. Sediment chemistry. In: Lisitzin AP, Bogdanov YuA (eds) Metalliferous sediments of the Red Sea. Nauka, Moscow, pp 187-189

Gurvich EG, Bogdanov YuA (1991) Report of Geology Group. In: Report of Cruise 23 of R/V Akademik Mstislav Keldysh. Vol. 2. IO RAS, Moscow

Gurvich EG, Lisitzin AP (1980a) Hafnium in bottom sediments of the Pacific and Indian Oceans. Doklady Akademii Nauk SSSR (Trans USSR Acad Sci) 250(3):722-726

Gurvich EG, Lisitzin AP (1980b) Lanthanum in bottom sediments of the Pacific and Indian Oceans. Doklady Akademii Nauk SSSR (Trans USSR Acad Sci) 250(5):1242-1244

Gurvich EG, Lisitzin AP (1980c) Hafnium accumulation rates on the bottom of the Pacific and Indian Oceans. Doklady Akademii Nauk SSSR (Trans USSR Acad Sci) 250(4):953-956

Gurvich EG, Lisitzin AP (1980d) Lanthanum accumulation rates on the bottom of the Pacific and Indian Oceans. Doklady Akademii Nauk SSSR (Trans USSR Acad Sci) 253(1):233-235

Gurvich EG, Lisitzin AP (1980e) Distribution of lanthanum in bottom sediments of the Pacific and Indian Oceans. Doklady Akademii Nauk SSSR (Trans USSR Acad Sci) 250(6):1447-1449

Gurvich EG, Lisitzin AP (1980f) Thorium contents in bottom sediments of the Pacific and Indian Oceans. Doklady Akademii Nauk SSSR (Trans USSR Acad Sci) 252(6):233-235

Gurvich EG, Zaikov VV (1992) Metalliferous siliceous sediments of Mugojarian paleoceanic basin. In: Popov VE (ed) Volcanogenic sedimentary ore formation. VSEGEI, Sankt-Petersburg, pp 37-39

Gurvich EG, Bogdanov YuA, Kurinov AD, Katargin NV (1976) Antimony in the metalliferous sediments of the Pacific Ocean. Oceanology 16(3):279-283

Gurvich EG, Bogdanov YuA, Kurinov AD, Katargin NV (1977) Scandium in bottom sediments of the Southeast Pacific. Geokhimiya (Geochemistry) 10:1582-1597

Gurvich EG, Bogdanov YuA, Lisitzin AP (1978) Behavior of barium in recent sedimentation in the Pacific. Geochemistry International 15(2):28-43

Gurvich EG, Bogdanov YuA, Lisitzin AP (1979) Behavior of barium in present sedimentation during formation of metalliferous sediments of the Pacific Ocean. Geochemistry International 16(1):62-79

Gurvich EG, Lisitzin AP, Kurinov AD (1980a) Hafnium. In: Lisitzin AP (ed) Geochemistry of elements hydrolysates. Nauka, Moscow, pp 181-200

Gurvich EG, Lisitzin AP, Kurinov AD (1980b) Thorium. In: Lisitzin AP (ed) Geochemistry of elements hydrolysates. Nauka, Moscow, pp 201-218

Gurvich EG, Lukashin VN, Lisitzin AP, Kurinov AD (1980c) Rare earth elements and yttrium. In: Lisitzin AP (ed) Geochemistry of elements hydrolysates. Nauka, Moscow, pp 71-116

Gurvich EG, Bogdanov YuA, Lisitzin AP (1984) Types of hydrothermal accumulations in the ocean and their evolution. In: Kholodov VN (ed) Evolution of sedimentary ore formation in the Earth history. Nauka, Moscow, pp 41-65

Gurvich EG, Bogdanov YuA, Shimkus KM, Sokolova EA (1986a) Sediments of the Suakin Deep. In: Lisitzin AP, Bogdanov YuA (eds) Metalliferous sediments of the Red Sea. Nauka, Moscow, pp 64-71

Gurvich EG, Bogdanov YuA, Sval'nov VN, Shimkus KM (1986b) Chemical composition of background sediments from the Red Sea rift. In: Lisitzin AP, Bogdanov YuA (eds) Metalliferous sediments of the Red Sea. Nauka, Moscow, pp 49-54

Gurvich EG, Levitan MA, Lisitzin AP (1988) Ancient metal-bearing formations of the Indian Ocean. Doklady Akademii Nauk SSSR (Trans USSR Acad Sci) 302(4):915-919

Gurvich EG, Bogdanov YuA, Zaikov VV, Zonenshain LP (1989) Middle Devonian metalliferous sediments in the South Urals. In: 28-th Intern Geol Congress. Vol. 1. Wash (DC), p 600

Gurvich EG, Bogdanov YuA, Lukashin VN, Rudenko MV (1995a) Report of the group of geology and geomorphology. In: Report of Cruise 35 of R/V Akademik Mstislav Keldysh. IO RAS, Moscow
Gurvich EG, Levitan MA, Kuz'mina TG (1995b) Chemical composition of Leg 138 sediments and history of hydrothermal activity. In: Proceedings of the Ocean Drilling Program, Scientific Results. Vol. 138. Texas A & M University, Ocean Drilling Program, College Station, pp 769-778
Gurvich EG, Bogdanov YuA, Vikent'ev IV, Gordeev VYu, Lein AYu, Peresypkin VI, Pimenov NV, Rudenko MV (1998) Report of the Geology Group. In: Report of Cruise 42 of the R/V Akademik Mstislav Keldysh. IO RAS, Moscow
Hackett JP, Bischoff JL (1973) New data on the stratigraphy, extent and geologic history of the Red Sea geothermal deposits. Econ Geol 68:553-564
Halliday AN, Davidson JP, Holden P, Owen RM, Olivarez AM (1992) Metalliferous sediments and the scavenging residence time of Nd near hydrothermal vents. Geophys Res Lett 19:761-764
Handschumacher DW (1976) Post-Eocene plate tectonics of the Eastern Pacific. The Geophysics of the Pacific Ocean Basin and its margin. Geophys Monogr 19. Amer Geophys Union, 1976
Hannington MD, Herzig PM, Thompson G, Rona PA (1990) Metalliferous sulfide-oxide sediments from the TAG hydrothermal field (26°N), Mid-Atlantic Ridge. EOS 71:1653
Hannington MD, Petersen S, Herzig PM, Jonasson IR (2002) A Global database of seafloor hydrothermal systems, including a digital database of geochemical analyses of seafloor polymetallic sulfides: prepared for the International Seabed Authority. Version 1.0. Central Data Repository. http://triton.ori.u-tokyo.ac.jp/~intridge/
Hartmann M (1973) Untersuchung von suspendiertem Material in den Hydrothermallaugen des Atlantis-II-Tiefs. Geologische Rundschau 62(3):742-754
Hartmann M (1980) Atlantis II Deep geothermal brine system: Hydrographic situation in 1977 and changes since 1965. Deep-Sea Res 27(2A):161-171
Hartmann M (1985) Atlantis-II-Deep geothermal brine system. Chemical processes between hydrothermal brines and Red Sea deep water. Mar Geol 64:157-177
Hartmann M, Scholten JC, Stoffers P, Wehner F (1998) Hydrographic structure of brine-filled deeps in the Red Sea – new results from Shaban, Kebrit, Atlantis II, and Discovery Deeps. Mar Geol 144:311-330
Hashimoto J, Ohta S, Gamo T, Chiba H, Yamaguchi T, Tsuchida S, Okidaira T, Watabe H, Yamanaka T, Kitazawa M (2001) First hydrothermal vent communities from the Indian Ocean discovered. Zool Sci 18:717-721
Hauschild J, Grevemeyer I, Kaul N, Villinger R (2003) Asymmetric sedimentation on young ocean floor at the East Pacific Rise, 15°S. Mar Geol 193:49-59
Haymon R, Kastner M (1981) Hot spring deposits on the East Pacific Rise at 21°N: Preliminary description of mineralogy and genesis. Earth Planet Sci Lett 53:363-381
Haymon RM, Fornari DJ, Edwards MH, Carbotte SM, Wright D, McDonald KC (1991) Hydrothermal vent distribution along the East Pacific Rise crest (9

degrees 09' – 54'N) and its relationship to magmatic and tectonic processes on fast-spreading mid-ocean ridges. Earth Planet Sci Lett 104(2-4):513-534

Hays JD, Cook HE, Jenkins G, Cook FM, Fuller J, Goll R, Orr W (1972) Site 83. In: Init Reports DSDP. Vol. 9. US Govt Print-Office, Wash (DC), pp 537-614

Heath GR, Dymond J (1977) Genesis and transformation of metalliferous sediments from the East Pacific Rise, Bauer Deep, and Central Basin, northwest Nazca plate. Geol Soc Amer Bull 88:723-733

Heath GR, Dymond J (1981) Metalliferous sediment deposition in time and space: East Pacific Rise and Bauer Basin, Northern Nazca Plate. Geol Soc Amer Mem 154:175-197

Hein JP, Ross CR, Yeh H (1979) Mineralogy and diagenesis of surface sediments from Domes area A, B and C. In: Marine geology and oceanography of the pacific manganese nodule province. Plenum Press, New York, pp 365-396

Heinze PM (1985) Vergleich zweier quartärzeitlicher Kastenlotkerne aus dem Pacific in Bezug auf ihre geochemische und mineralogische Zusammensetzung und Bestimmung der jeweiligen Sedimentationsrate mittels biostratigraphischer Bearbeitung der planktonischen Foraminifiren (westlich des EPR bei 17°S/113°W und 13°N/104°W). Univ Hannover, Hannover

Hekinian R, Fevrier M (1979) Comparison between deep-sea hydrothermal deposits recovered from recent spreading ridges. In: La Genese des Nodules de Manganese. Colloques Internationaux du Centre National de la Recherch Scientifique 289. Centre National de la Recherche Scientifique, Paris, pp 167-178

Hekinian R, Fouquet Y (1985) Volcanism and metallogenesis of axial and off-axial structures on the East Pacific Rise near 13 degrees N. Econ Geol 80(2): 221-249

Hekinian R, Francheteau J, Renard V, Ballard R, Choukroune P, Cheminee JL, Albarede F, Minster JF, Charlou J-L, Marty JC, Boulegue J (1983) Intense hydrothermal activity at the rise axis of the East Pacific Rise near 13°N: Submersible witness the growth of a sulfide chimney. Mar Geophys Res 6(1):1-14

Hekinian R, Francheteau J, Ballard RD (1985) Morphology and evolution of hydrothermal deposits at the axis of the East Pacific Rise. Oceanologica Acta 8(2):221-249

Hembelen Ch, Roether W, Stoffers P (1996) Östliches Mittelmeer, Rotes Meer, Arabisches Meer, Cruise No 31. Meteor Berichte 96-4. Univ Hamburg, Hamburg

Hendricks RL, Reisbick FB, Mahaffey EJ, Roberts DB, Peterson MNA (1969) Chemical composition of sediments and interstitial brines from the Atlantis II, Discovery and Chain deeps In: Degens ET, Ross DA (eds) Hot brines and recent heavy metal deposits in the Red Sea. Springer Verlag, New York, pp

Herman YP (1965) Etudes des sediments Quaternaires de la Mer Rouge. PhD Thesis. Universite de Paris, Paris, pp 341-415

Herron EM (1972) Sea-floor spreading and the Cenozoic history of the East Central Pacific. Geol Soc Amer Bull 83(6):1671-1692

Herron EM, Cande SC, Hall BR (1981) An active spreading center collides with a subduction zone: A geophysical survey of the Chile margin triple junction. Geol Soc Amer Mem 154:683

Herzig PM, Plüger WL (1988) Exploration for hydrothermal activity near the Rodriguez Triple Junction, Indian Ocean. Canadian Mineralogist 26:721-736

Holland H (1987) The Chemical Evolution of the Atmosphere and Oceans. Princeton University Press, Princeton

Holm NG, Charlou J-L (2001) Initial indications of abiotic formation of hydrocarbons in the Rainbow ultramafic hydrothermal system, Mid-Atlantic Ridge. Earth Planet Sci Lett 191(1-2):1-8

Horowitz A (1970) The distribution of Pb, Ag, Sn, Tl, and Zn in sediments on active oceanic ridges. Mar Geol 9:241-259

Horowitz A, Cronan DS (1976) The geochemistry of basal sediments from the North Atlantic Ocean. Mar Geol 20:205-228

Hot brines and recent heavy metal deposits in the Red Sea (1969). Degens ET, Ross DA (eds). Springer Verlag, New York

Hutchinson RW, Engels GG (1970) Tectonic significance of regional geology and evaporite lithofacies in North-Eastern Ethiopia. Royal Soc London Phil Trans 267:313-320

Hydrothermal systems and sedimentary formations of the Mid-Atlantic Ridge (1993). Lisitzin AP (ed). Nauka, Moscow

Initial Reports of the Deep-Sea Drilling Project (1970-1987). Vol. 2-96. US Govt Print-Office, Wash (DC)

Ito E, Harris DM, Anderson AT (1983) Alteration of oceanic crust and geologic cycling of chlorine and seawater. Geochim Cosmochim Acta 47(9):1613-1624

Iyer SD, Shyam Prasad M, Gupta SM, Nirmal Charan S (1997) Evidence for recent hydrothermal activity in the Central Indian Basin. Deep-Sea Res I 44(7):1167-1184

James RH, Elderfield H (1996) Dissolved and particulate trace metals in hydrothermal plumes at the Mid-Atlantic Ridge. Geophys Res Lett 23(23):3499-3502

James RH, Elderfield H, Palmer MR (1995a) The chemistry of hydrothermal fluids from the Broken Spur site, 29°N Mid-Atlantic Ridge. Geochim Cosmochim Acta 59(4):651-659

James RH, Elderfield H, Rudnicki MD, German CR, Palmer MR, Chin C, Greaves MJ, Gurvich EG, Klinkhammer GP, Ludfort E, Mills RA, Thomson J, Williams A (1995b) Hydrothermal plumes at Broken Spur, 29°N Mid-Atlantic Ridge: Chemical and physical characteristics. Hydrothermal vents and processes. Geol Soc London, Spec Publ 87. Geol Soc London, London, pp 97-110

Jarvis J (1985) Geochemistry and origin of Eocene-Oligocene metalliferous sediments from the Central Equatorial Pacific: Deep Sea Drilling Project Sites 573 and 574. In: Init Reports DSDP. Vol. 85. US Govt Print-Office, Wash (DC), pp 781-804

Jean-Baptiste P, Mantisi F, Pauwells H, Grimaud D, Patriat P (1992) Hydrothermal ^{3}He and manganese plumes at 19°29'N on the Central Indian Ridge. Geophys Res Lett 19(17):1787-1790

Jedwab J, Boulegue J (1984) Graphite crystals in hydrothermal vents. Nature 310(5972):41-43

Johnson HP, Pruis MJ (2003) Fluxes of fluid and heat from the oceanic crustal reservoir. Earth Planet Sci Lett 216:565-574

Kaplan IR, Sweeney RE, Nissenbaum A (1969) Sulfur isotope studies on Red Sea geothermal brines and sediments. In: Degens ET and Ross DA (eds) Hot brines and recent heavy metal deposits in the Red Sea. Springer Verlag, New York, pp 474-498

Karl SM (1984) Sedimentologic, diagenetic, and geochemical analysis of upper Mesozoic ribbon cherts from the Franciscan assemblage at the Marine Headlands, California. In: Franciscan Geology of Northern California. SEPM, Pacific Section, Los Angeles, pp 71-88

Karpoff AM, Walter AV, Pflumio C (1988) Metalliferous sediments within lava sequences of the Sumail ophiolite (Oman); mineralogical and geochemical characterization, origin and evolution. Tectonophysics 151:223-245

Karson JA, Thompson G, Humphris SE, Edmond JM, Bryan WB, Brown JR, Winters AT, Pockalny RA, Casey JF, Campbell AC, Klinkhammer G, Palmer MR, Kinzler RJ, Sulanowska MM (1987) Along axis variations in sea floor spreading in the MARK area. Nature 328:681-685

Kennett JP (1982) Marine Geology. Engelwood, Prentice-Hall

Khripunoff A, Alberic P (1991) Settling of particles in a hydrothermal vent field (East Pacific Rise 13°N) measured with sediment traps. Deep-Sea Res 38(6):729-744

Khripunoff A, Vangriesheim A, Crassous P, Segonzac M, Colaço A, Dasbruyères D, Barthelemy R (2001) Particle flux in the Rainbow hydrothermal field (Mid-Atlantic Ridge): Dynamics, mineral and biological composition. Journ Marine Res 59:633-656

Kishida K, Sohrin Y, Okamura K, Ishibashi J (2004) Tungsten enriched in submarine hydrothermal fluids. Earth Planet Sci Lett 222(3-4):819-827

Klinkhammer GP, Bender ML (1980) The distribution of manganese in the Pacific Ocean. Earth Planet Sci Lett 46:361-384

Klinkhammer GP, Hudson A (1986) Dispersal pattern for hydrothermal plumes in the South Pacific using manganese as a tracer. Earth Planet Sci Lett 79:241-249

Klinkhammer GP, Elderfield H, Greaves MJ, Rona PA, Nelsen T (1986) Manganese geochemistry near high-temperature vents in the Mid-Atlantic rift valley. Earth Planet Sci Lett 80(3-4):230-240

Klinkhammer GP, German CR, Elderfield H, Greaves MJ, Mitra A (1994) Rare earth elements in hydrothermal fluids and plume particulates by inductively coupled plasma mass spectrometry. Mar Chem 45:179-186

Klinkhammer GP, Chin CS, Wilson C, German CR (1995) Venting from the Mid-Atlantic Ridge at 37°17'N: the Lucky Strike hydrothermal site. In: Parson LM, Walker CL, Dixon DR (eds) Hydrothermal vents and processes. Geol Soc London Spec Publ 87. Geol Soc London, London, pp 87-96

Klitgord KD, Mudie JD, Larson PA, Grow JA (1973) Fast seafloor spreading on the Chile Ridge. Earth Planet Sci Lett 20:93-99

Koblents-Mishke OI (1977) Primary production. In: Vinogradov ME (ed) Oceanology. Ocean biology. Vol. 1, Biological structure of the ocean. Nauka, Moscow, pp 62-65

Krasnov SG, Maslov MN, Andreev NM (1988) Hydrothermal mineralization in the southern part of the East Pacific Rise. Doklady Akademii Nauk SSSR (Trans USSR Acad Sci) 302(1):161-164

Krasnov SG, German NE, Cherkashev GA (1992a) Distribution and the main factors of formation of metalliferous sediment composition. In: Gramberg IS, Ainemer AI (eds) Hydrothermal sulfide ores and metalliferous sediments of the ocean. Nedra, Sankt-Petersburg, pp 129-138

Krasnov SG, Kreiter II, Poroshina IM (1992b) The distribution of hydrothermal vents on the East Pacific Rise (21°20' – 22°40'S) based on a study of dispersion patterns of hydrothermal plumes. Oceanology 32(3):375-381

Krasnov SG, Cherkashev GA, Stepanova TV, Batuyev BN, Krotov AG, Malin BV, Maslov MN, Markov VF, Poroshina IM, Samovarov MS, Ashadze AM, Lazareva LI, Ermolaev IK (1995a) Detailed geological studies of hydrothermal fields in the North Atlantic. In: Parson LM, Walker CL, Dixon DR (eds) Hydrothermal vents and processes. Geol Soc London Spec Publ 87. Geol Soc London, London, pp 43-64

Krasnov SG, Poroshina IM, Cherkashev GA (1995b) Geological settings of high-temperature hydrothermal activity and massive sulfide formation on fast- and slow-spreading ridges. In: Parson LM, Walker CL, Dixon DR (eds) Hydrothermal vents and processes. Geol Soc London Spec Publ 87. Geol Soc London, London, pp 17-32

Krauskopf, K B (1956) Factors controlling concentrations of thirteen rare metals in sea water. Geochim Cosmochim Acta 9(1):1-32

Krishna KS, Gopala Rao D, Ramana MV, Subrahmanyan V, Sarma K, Pilipenko AI, Shcherbakov VS, Radhakrishna Murthy IV (1995) Tectonic model for the evolution of oceanic crust in the northeastern Indian Ocean from the Late Cretaceous to the early Tertiary. Journ Geophys Res 100(B10):20011-20024

Krishnaswami S (1976) Authigenic transition elements in Pacific pelagic clays. Geochim Cosmochim Acta 40(4):425-434

Ku TL (1969) Uranium series isotopes from the Red Sea hot brine area. In: Degens ET and Ross DA (eds) Hot brines and recent heavy metal deposits in the Red Sea. Springer Verlag, New York, pp 512-524

Ku TL, Thurber DL, Mathieu GG (1969) Radiocarbon chronology of Red Sea sediments. In: Degens ET and Ross DA (eds) Hot brines and recent heavy metal deposits in the Red Sea. Springer Verlag, New York, pp 348-359

Kuhn T, Burger H, Castradori D, Halbach P (2000) Volcanic and hydrothermal history of the ridge segments near the Rodriguez Triple Junction (Central Indian Ocean) deduced from sediment geochemistry. Mar Geol 169:391-409

Kunzendorf H, Walter P, Stoffers P, Gwozdz R (1984/1985) Metal variations in divergent plate-boundary sediments from the Pacific. Chem Geol (1/2):113-133

Kunzendorf H, Stoffers P, Gwozdz R (1988) Regional variations of REE patterns in sediments from active plate boundaries. Mar Geol 84(3/4):191-199

Kuptsov VM (1989) Methods of chronology of Quaternary sediments from oceans and seas. Nauka, Moscow

Kuptsov VM (1993) Geochronological studies of sediment cores by isotope methods. In: Lisitzin AP (ed) Hydrothermal systems and sedimentary formations of the Mid-Atlantic Ridge. Nauka, Moscow, pp 50-54

Kuptsov VM, Palkina AM (1986) Radiocarbon dating of sediments. In: Lisitzin AP, Bogdanov YuA (eds) Metalliferous sediments of the Red Sea. Nauka, Moscow, pp 24-36

Lalou C, Thompson G, Arnold M, Brichet E, Druffel ERM, Rona PA (1990) Geochronology of TAG and Snake Pit hydrothermal fields, Mid-Atlantic Ridge: witness to a long and complex hydrothermal history. Earth Planet Sci Lett 97(1-2):113-128

Lalou C, Reyss J-L, Brichet E, Arnold M, Thompson G, Fouquet Y, Rona P (1993) New age data for Mid-Atlantic Ridge hydrothermal sites TAG and Snake Pit chronology revisited. Journ Geophys Res 98(B6):9705-9713

Lalou C, Reyss J-L, Brichet E, Rona P, Thompson G (1995) Hydrothermal activity on a 10^5-year scale at a slow-spreading ridge, TAG hydrothermal field, Mid-Atlantic Ridge 26°N. Journ Geophys Res 100(9B):17855-17862

Lalou C, Reyss J-L, Brichet E, Krasnov S, Stepanova T, Cherkashev G, Markov V (1996) Initial chronology of a recently discovered hydrothermal field at 14°45'N, Mid-Atlantic Ridge. Earth Planet Sci Lett 144:483-490

Lange J (1974) Geochemische Untersuchungen an pelagischen Sedimenten des Atlantischen und Pacifischen Ozeans (DSDP, Leg I-VII). Georg-August-Univ, Göttingen

Lebedev LM, Tsepin AI (1990) Mineralogy of hydrothermal bodies. In: Lisitzin AP (ed) Geological structure and hydrothermal formations of the Juan de Fuca Ridge. Nauka, Moscow, pp. 69-77

Lein AYu, Grichuk DV, Gurvich EG, Bogdanov YuA (2000) A new type of hydrothermal fluids enriched in hydrogen and methane in the Mid-Atlantic Ridge rift zone. Trans Russian Acad Sci 375(3):380-383

Leinen M, Stakes D (1979) Metal accumulation rates in the central equatorial Pacific during Cenozoic time. Geol Soc Amer Bull 90(1):357-375

Leinen M, Pisias P (1984) An objective technique for determining endmember composition and for partitioning sediments according to their sources. Geochim Cosmochim Acta 48:47-62

Levin LE, Baskakova DK, Virta AN (1985) Buried metal-bearing accumulations within the sedimentary cover of oceans and seas (Pacific segment). VIEMS, Moscow

Levin LE, Varentsov IM, Baskakova DK, Virta AN (1987) Presence of metal-bearing accumulations in the Pacific sedimentary cover and its relationship to spreading and volcanism. Bulleten' Moskovskogo Obshchestva Ispytateley Prirody (Bull Moscow Soc Naturalists). Geol Sect 62(4):3-17

Lewison MA, Feely RA, Curl HC (1986) Composition of smoker particulates from active vents within the caldera of Axial volcano on the Juan de Fuca Ridge. EOS 67(44):1027

Lilley MD, Feely RA, Trefry JH (1995) Chemical and biochemical transformations in hydrothermal plumes. In: Humphris SE, Zierenberg RA, Mullineaux LS, Thomson RE (eds) Seafloor hydrothermal systems: Physical, chemical, biological, and geological interactions. Geophysical Monograph 91. AGU, pp 369-391

Lisitzin AP (1974) Ocean sedimentation. Nauka, Moscow

Lisitzin AP (1978) Processes of ocean sedimentation. Nauka, Moscow

Lisitzin AP (1980) Distribution of terrigenous material. In: Lisitzin AP (ed) Oceanology. Ocean geology. Geological history of the ocean. Nauka, Moscow, pp 172-192

Lisitzin AP (1984) Sediment body of the ocean. In: Geology of the ocean bottom based on deep-sea drilling data. Nauka, Moscow, pp 12-62

Lisitzin AP (1993) Hydrothermal systems of the World Ocean – contribution of endogenic material. In: Lisitzin AP (ed) Hydrothermal systems and sedimentary formations of the Mid-Atlantic Ridge. Nauka, Moscow, pp 147-245

Lisitzin AP (2001) Lithology of lithosphere plates. Geology and Geophysics 42(4):522-559

Lisitzin AP (2004) Geology of the World Ocean in the third millennium – new approaches, achievements, and outlook. In: Vinogradov ME and Lappo SS (eds) New ideas in oceanology. Vol. 2. Nauka, Moscow, pp 7-66

Lisitzin AP, Fisher A, Heezen B (1973) Geochemistry and lithology of drill cores from the Pacific Ocean. In: I-st Intern Geochem Congress. Vol IV-1 – Sedimentary processes. GEOKHI, Moscow, pp 405-416

Lisitzin AP, Bogdanov YuA, Mudmaa IO, Serova VV, Zverinskaya IB, Lebedev AI, Lukashin VN, Gordeev VV (1976) Metalliferous sediments and their genesis. In: Lisitzin AP (ed) Geological and geophysical research in the Southeast Pacific. Nauka, Moscow, pp 289-379

Lisitzin AP, Bogdanov YuA, Levitan MA, Nikolaev SD, Chekhovskikh EM (1980a) History of Mesozoic-Cenozoic sedimentation in the World Ocean. In: Lisitzin AP (ed) Oceanology. Ocean geology. Geological history of the ocean. Nauka, Moscow, pp 407-427

Lisitzin AP, Lukashin VN, Emel'yanov EM, Zverinskaya IB (1980b) Aluminum. In: Lisitzin AP (ed) Geochemistry of elements hydrolysates. Nauka, Moscow, pp 18-49

Lisitzin AP, Lukashin VN, Emel'yanov EM, Zverinskaya IB (1980c) Titanium. In: Lisitzin AP (ed) Geochemistry of elements hydrolysates. Nauka, Moscow, pp 117-149

Lisitzin AP, Gordeev VV, Bogdanov YuA (1987) Geochemistry of metalliferous sediments from the Indian Ocean. In: Lisitzin AP, Gurvich EG (eds) Metalliferous sediments of the Indian Ocean. Nauka, Moscow, pp 100-127

Lisitzin AP, Bogdanov YuA, Zonenshain LP, Kuz'min MI, Sagalevich AM (1989) Hydrothermal manifestations of the Mid-Atlantic Ridge at 26°N (TAG hydrothermal field). Izvestiya Akademii Nauk SSSR (News USSR Acad Sci), Geology Series 12:3-20

Lisitzin AP, Bogdanov YuA, Gurvich EG (1990) Hydrothermal accumulations of ocean rift zones. Nauka, Moscow

Lisitzin AP, Bogdanov YuA, Gurvich EG (1992a) Hydrothermal springs and ores on the ocean floor. In: Mirlin EG (ed) Metallogeny of present and ancient oceans. Geoexpert, Moscow, pp 14-39

Lisitzin AP, Crook K, Bogdanov YuA, Gurvich EG (1992b) Hydrothermal field of the Manus Basin rift zone. Izvestiya Akademii Nauk SSSR (News of the USSR Acad Sci). Geology Series 10:34-55

Lisitzin AP, Malahoff OR, Bogdanov YuA., Gurvich EG (1992c) Hydrothermal accumulations from the northern part of the Lau Basin (Pacific ocean). Izvestiya Akademii Nauk SSSR (News USSR Acad Sci). Geology Series 4:5-24

Lisitzin AP, Lukashin VN, Gordeev VV, McConachy TF, Scott SD, Shevchenko VP (1997) Hydrological and geochemical anomalies associated with hydrothermal activity in SW Pacific marginal and back-arc basins. Mar Geol 142(1-4):7-45

Little SA, Stolzenbach KD, Von Herzen RP (1987) Measurements of plume flow from a hydrothermal vent field. Journ Geophys Res 92:2587-2596

Lonsdale P (1976) Abyssal circulation in the Southeast Pacific and some geological implications. Journ Geophys Res 81(6):1163-1176

Ludford EM, Palmer MR, German CR, Klinkhammer GP (1996) The geochemistry of Atlantic hydrothermal particles. Geophys Res Lett 23(23):3503-3506

Lukashin VN (1981) Geochemistry of trace elements in processes of sedimentation in the Indian Ocean. Nauka, Moscow

Lukashin VN (1983) Forms of elements in sediments. In: Lisitzin AP (ed) Biogeochemistry of the ocean. Nauka, Moscow, pp 312-344

Lukashin VN (1987) Forms of chemical elements in metalliferous sediments of the Indian Ocean. In: Lisitzin AP, Gurvich EG (eds) Metalliferous sediments of the Indian Ocean. Nauka, Moscow, pp 128-150

Lukashin VN, Lisitzin AP (1980) Gallium. In: Lisitzin AP (ed) Geochemistry of elements hydrolysates. Nauka, Moscow, pp 50-70

Lukashin VN, Lisitzin AP, Emel'yanov EM (1980) Zirconium. In: Lisitzin AP (ed) Geochemistry of elements hydrolysates. Nauka, Moscow, pp 150-180

Lukashin VN, Cherkashev GA, Isaeva AB (1990) Chemical composition of bottom sediments. In: Lisitzin AP (ed) Geological structure and hydrothermal formations of the Juan de Fuca Ridge. Nauka, Moscow, pp 128-140

Lukashin VN, Rusakov VYu, Lisitzin AP, Lein AYu, Isaeva AB, Serova VV, Karpenko AA (1999) Study of particle fluxes in the Broken Spur hydrothermal vent field (29°N, Mid-Atlantic Ridge). Explor Mining Geol 8(3-4):341-353

Lukashina NP (1993) Stratigraphy of Late Quaternary sediments and conditions of sedimentation within the rift valley of the Mid-Atlantic Ridge between 23° and 26°N. In: Lisitzin AP (ed) Hydrothermal systems and sedimentary formations of the Mid-Atlantic Ridge. Nauka, Moscow, pp 39-50

Lupton JE (1995) Hydrothermal plumes: near and far field. In: Humphris SE, Zierenberg RA, Mullineaux LS, Thomson RE (eds) Seafloor hydrothermal systems: Physical, chemical, biological, and geological interactions. Geophysical Monograph. Vol. 91. AGU, Wash (DC), pp 317-346

Lupton JE, Craig H (1981) A major ^3He source on the East Pacific Rise. Science 214:13-18

Lupton JE, Pyle DG, Jenkins WJ, Green R, Evans L (2004) Evidence for an extensive hydrothermal plume in the Tonga-Fiji region of the South Pacific. Geochem Geophys Geosyst 5(1):1-18, doi:10.1029/2003GC000607

Lyle M, Owen RM, Leinen M (1986) History of hydrothermal sedimentation at the East Pacific Rise, 19°S. In: Init Reports DSDP. Vol. 92. US Govt Print-Office, Wash (DC), pp 585-596

Lyle M, Leinen M, Owen RM, Rea DK (1987) Late Tertiary history of hydrothermal deposition at the East Pacific Rise, 19°S: correlation to volcano-tectonic events. Geophys Res Lett 14(6):595-598

Malahoff A, McMurtry G, Hammond S, Embley R (1984) High temperature fields – Juan de Fuca Ridge Axial Volcano. EOS 65(45):1112

Mammerickx J, Herron E, Dorman L (1980) Evidence for two fossil spreading ridges in the South-East Pacific. Geol Soc Amer Bull 91(5/1):263-271

Mammerickx J, Klitgord KD (1982) Northern East Pacific Rise: evolution from 25 my BP to the present. Journ Geophys Res 87:6751-6759

Mandernack KW, Tebo BM (1993) Manganese scavenging and oxidation at hydrothermal vents and in vent plumes. Geochim Cosmochim Acta 57(16): 3907-3923

Marchig V, Erzinger J (1986) Chemical composition of Pacific sediments near 20°S: changes with increasing distance from the East Pacific Rise; and Appendix. In: Init Reports DSDP. Vol. 92. US Govt Print-Office, Wash (DC), pp 371-381

Marchig V, Gundlach H (1982) Iron-rich metalliferous sediments on the East Pacific Rise: prototype of undifferentiated metalliferous sediments on divergent plate boundaries. Earth Planet Sci Lett 58:361-382

Marchig V, Erzinger J, Heinze P-M (1986) Sediment in the black smoker area of the East Pacific Rise (18.5°S). Earth Planet Sci Lett 76(1/2):93-106

Marchig V, Gundlach H, Buchholz M (1987) Ore formation at rapidly diverging plate margins. In: Results of Cruise Geometer 4. BRG Circular. Vol. 4, pp 3-14

Marchig V, Gundlach H, Holler G, Wilke M (1988) New discoveries of massive sulfides on the East Pacific Rise. Mar Geol 84(3):179-190

Marienfeld P, Marchig V (1992) Indications of hydrothermal activity at the Chile Ridge spreading center. Mar Geol 105:241-252

Massoth GJ, Baker ET, Feely RA, Curl HC Jr (1984) Hydrothermal signals away from the southern Juan de Fuca Ridge. EOS 65:1112

Massoth GJ, Baker ET, Lupton JE, Feely RA, Butterfield DA, Von Damm KL, Roe KL, Lebon GT (1994) Temporal and spatial variability of hydrothermal manganese and iron at Cleft segment, Juan de Fuca Ridge. Journ Geophys Res 99(B3):4905-4923

Mayer LA, Pisias N, Janecek T and the Shipboard Scientific Party (1992) Proceedings of the Ocean Drilling Program, Initial Reports. Vol. 138 (Parts 1 and 2). Texas A & M University, Ocean Drilling Program, College Station

McConachy TF, Ballard R, Mottl MJ, Von Herzen RP (1986) Geologic form and settling of a hydrothermal vent field at lat 10°56'N, East Pacific Rise: A detailed study using Angus and Alvin. Geology 14:295-298

McMurtry GM, Yeh H-W (1981) Hydrothermal clay mineral formation of East Pacific Rise and Bauer basin sediments. Chem Geol 32:189-205

McMurtry GM, Malahoff A, Feely RA, Massoth GJ (1984) Geology and chemistry of hydrothermal nontronite deposits from the Juan de Fuca Ridge. EOS 65(45):1112

McMurtry GM, De Carlo EH, Kim KH (1991) Accumulation rates, chemical partitioning, and Q-mode factor analysis of metalliferous sediments from the North Fiji Basin. Mar Geol 98:271-295

Mendel V (1997) Processus d'accretion au niveau des dorsales ultra-lentes: L'example de la Dorsale Sud-Ouest Indienne. PhD thesis. Universite Louis Pasteur, Strasbourg

Metalliferous sediments of the Indian Ocean (1987) Lisitzin AP, Gurvich EG (eds). Nauka, Moscow

Metalliferous sediments of the Southeast Pacific (1979) Smirnov VI (ed). Nauka, Moscow

Metz S, Trefry JH (2000) Chemical and mineralogical influences on concentrations of trace metals in hydrothermal fluids. Geochim Cosmochim Acta 64(13):2267-2279

Metz S, Trefry JH, Nelsen T (1988) History and geochemistry of a metalliferous sediment core from the Mid-Atlantic Ridge. Geochim Cosmochim Acta 52(10):2369-2378

Mevel C, Auzende J-M, Cannat M, Donval DP, Dubois J, Fouquet Y, Gente P, Grimaud D., Karson JA, Segonzag M, Stievenard M (1989) La ride du Snake Pit (dorsale Medio-Atlantique, 23°22'N): resultats preliminaries de la campagne HYDROSNAKE. Compt Rend Acad Sci Paris 308(6):545-552

Michard A, Albarede F (1985) Hydrothermal uranium uptake at ridge crests. Nature 317(6034):244-246

Michard A, Albarede F (1986) The REE content of some hydrothermal fluids. Chem Geol 55(1/2):51-60

Michard A, Albarede F, Michard G, Minster JF, Charlou J-L (1983) Rare-earth elements and uranium in high-temperature solutions from the East Pacific Rise vent field (13°N). Nature 303:795-797

Michard G, Albarede F, Michard A, Minster JF, Charlou J-L, Tan N (1984) Chemistry of solutions from the 13°N East Pacific Rise hydrothermal site. Earth Planet Sci Lett 7(3):297-307

Migdisov AA, Bogdanov YuA, Lisitzin AP, Gurvich EG, Lebedev AI, Lukashin VN, Gordeev VV, Girin YuP, Sokolova EG (1979) Geochemistry of metalliferous sediments from the Southeast Pacific. In: Smirnov VI (ed) Metalliferous sediments of the Southeast Pacific. Nauka, Moscow, pp 122-200

Miller AR, Densmore CD, Degens ET, Hathaway JC, Manheim FT, Mcfarlin PF, Pocklington R, Jokela A (1966) Hot brines and recent iron deposits of the Red Sea. Geochim Cosmochim Acta 30(3):341-359

Mills RA (1992) A geochemical and isotopic study of hydrothermal sediments from the Mid-Atlantic Ridge, 26°N. PhD diss. Darwin College, Cambridge

Mills RA, Elderfield H (1995) Hydrothermal activity and the geochemistry of metalliferous sediments. In: Humphris SE, Zierenberg RA, Mullineaux LS, Thomson RE (eds) Seafloor hydrothermal systems: Physical, chemical, biological, and geological interactions. Geophysical Monograph. Vol. 91. AGU, Wash (DC), pp 392-407

Mills RA, Elderfield H, Thomson J (1993) A dual origin for the hydrothermal component in a metalliferous sediment core from the Mid-Atlantic Ridge. Journ Geophys Res 98(B6):9671-9681

Mills RA, Thomson J, Elderfield H, Hinton RW, Hyslop E (1994) Uranium enrichment in metalliferous sediments from the Mid-Atlantic Ridge. Earth Planet Sci Lett 124(1-4):35-47

Minster J, Jordan T (1978) Present day plate motion. Journ Geophys Res 83:5331-5354

Missack EA (1988) Mineralogy and phase relations of the massive sulfides and metalliferous sediments of the axial rift valley, Red Sea. Heidelberger Geowissenschaftliche Abhandlungen 23:1-213

Missack EA, Stoffers P, and El Goresy A (1989) Mineralogy, paragenesis and phase relations of copper-iron sulfides in the Atlantis II Deep, Red Sea. Mineralum Deposita 24:82

Mitra A, Elderfield H, Greaves MJ (1994) Rare earth elements in submarine hydrothermal fluids and plumes from the Mid-Atlantic Ridge. Mar Chem 46:217-235

Monin AS, Plakhin EA, Stunzhas PA (1980a) About stratification of hot brines in the Atlantis II Deep. Doklady Akademii Nauk SSSR (Trans USSR Acad Sci) 255(2):458-462

Monin AS, Zonenshain LP, Litvin VM, Sorokhtin OG (1980b) Structure of the Red Sea rift. Doklady Akademii Nauk SSSR (Trans USSR Acad Sci) 254(5): 1198-1202

Monin AS, Bogdanov YuA, Zonenshain L.P. and Scientific team (1985) Underwater geologic investigations from manned submersibles. Nauka, Moscow

Moorby SA, Knedler KE, Glasby GP, Hodkinson R, Cronan DS (1986) Lithology, color, mineralogy, and geochemistry of marine sediments from the Lau Basin, Havre Trough, and Tonga-Kermadec Ridge. NZOI Oceanographic Field Report 27. Ward, Wellington

Mortlock RA, Froelich PN, Feely RA, Massoth GJ, Walker SL, Butterfield DA, Lupton JL (1993) Silica and germanium in Pacific Ocean hydrothermal vents and plumes. Earth Planet Sci Lett 119:365-378

Mossman DG, Hefferson KG (1978) On the possible primary precipitation of atacamite and other metal chlorides in certain stratabound deposits. Chem Geol 21(1/2):151-159

Mottl MJ (2003) Partitioning of energy and mass fluxes between mid-Ocean ridge axes and flanks at high and low temperature. In: Halbach P, Tunnicliffe V, Hein J (eds) Energy and mass transfer in marine hydrothermal systems. DUP, Berlin, pp 271–286

Mottl MJ, McConachy TF (1990) Chemical processes in buoyant hydrothermal plumes on the East Pacific Rise near 21°N. Geochim Cosmochim Acta 54: 1911-1927

Müller A, Schneider W, Zachmann D (1985) Structure-related sediments on the East Pacific Rise at 13° to 21°S. Neues Jahrbuch Geol Paläontol 4:203-224

Murdmaa IO (1979) Ocean sediments and sedimentary rocks In: Bezrukov PL (ed) Oceanology. Ocean geology. Ocean sedimentation and magmatism. Nauka, Moscow, pp 104-162

Murray J, Renard AF (1891) Deep-sea deposits. Report "Challenger" expedition (1873-1876). Eyre & Spottiswoode, London

Murray JW (1979) Iron oxide. In: Barns RG (ed) Marine minerals. Min Soc Am. Short Course Notes. Vol. 6. Wash (DC), pp 47-98

Mustafa Z, Nawab Z, Horn R, Le Lann F (1984) Economic interest of hydrothermal deposits. The Atlantis II Project. In: Second International Seminar on the Offshore Mineral Resources Brest: GERMINAL. Brest, pp 507-539

Nealson KH, Tebo BM, Rosson RA (1988) Occurrence and mechanisms of microbial oxidation of manganese. Advances Applied Microbiology 33:279-318

Nelsen TA, Metz S, Trefry J (1986) Sedimentology and composition of sediments near a black smoker field on the Mid-Atlantic Ridge. EOS 67(44):1022

Nelsen TA, Klinkhammer GP, Trefry JH, Trocine RP (1986/1987) Real-time observation of dispersed hydrothermal plumes using nephelometry: examples from the Mid-Atlantic Ridge. Earth Planet Sci Lett 81:245-252

Novikov GV (1994) Ion-exchange mobility of metals from ocean nodules and crusts. In: Lisitzin AP (ed) Geology of oceans and seas. Vol. 2. IO RAN, Moscow, pp 236-237

Olivarez AM, Owen RM (1989) REE/Fe variations in hydrothermal sediments: Implications for the REE content of seawater. Geochim Cosmochim Acta 53: 757-762

Onishi H (1969) Antimony 51. In: Handbook of Geochemistry. Vol. II-1. Springer, Berlin, pp 51-B-51-O

Oreshkin VN (1977) Study of distribution and migration of cadmium in the Pacific Ocean with use of methods of atomic absorption and atomic fluorescent analysis with non-flame atomization of powdered samples. IOAN, Moscow

Oudin E, Cocherie A (1988) Fish debris record the hydrothermal activity in the Atlantis II Deep sediments (Red Sea). Geochim Cosmochim Acta 52:177-184

Oudin E, Thisse Y, Ramboz C (1984) Fluid inclusion and mineralogical evidence for high-temperature saline hydrothermal circulation in the Red Sea metalliferous sediments: Preliminary results. Marine Mining 5:3-31

Ovchinnikov LN and Masalovich AM (1981) Experimental study of hydrothermal ore formation. Nauka, Moscow

Owen RM, Olivarez AM (1988) Geochemistry of rare earth elements in Pacific hydrothermal sediments. Mar Chem 25:183-196

Palmer MR, Edmond JM (1989) Cesium and rubidium in submarine hydrothermal fluids: Evidence from Sr:Ca and $^{87}Sr/^{86}Sr$ ratios. Earth Planet Sci Lett 95(1):8-14

Parrot J-F, Delaune-Mayere M (1974) Les terres d'ombre du Bassit (nord-ouest Syrien). Comparaison avec les termes similaires du Troodos (Chypre). Cahiers de l'Office de la Recherche Scientifique et Technique Outre-Mer, Paris, Serie Geologique 6:147-160

Parson LM, Fouquet Y, Ondreas H, Barriga F, Relvas J, Riberio A, Charlou J-L, German CR (1997) Non-transform discontinuity settings for contrasting hydrothermal systems on the MAR: Rainbow and FAMOUS at 36°14' and 36°34'N. EOS 78(46):F832

Parson L, Gracia E, Coller D, German C, Needham D (2000) Second-order segmentation; the relationship between volcanism and tectonism at the MAR, 38°N – 35°40'N. Earth Planet Sci Lett 17:213-251

Pelayo AM, Stein S, Stein CA (1994) Estimation of oceanic hydrothermal heat flux from heath flow and depths of mid-ocean ridge seismicity and magma chambers. Geophys Res Lett 21:713-716

Peters Tj (1988) Geochemistry of manganese-bearing cherts associated with Alpine ophiolites and the Hawasina formations of Oman. Mar Geol 84:229-238

Peterson MNA, Edgar NT, Von der Borch CC, Rex RW (1970) Cruise leg summary and discussion. In: Init Reports DSDP. Vol. 2. US Govt Print-Office, Wash (DC)

Pfeifer H-R, Oberhänsli H, Epprecht W (1988) Geochemical evidence for a synsedimentary hydrothermal origin of Jurassic iron-manganese deposits at Gonzen (Sargans, Helvetic Alps, Switzerland). Mar Geol 84:257-272

Pierret M-C, Clauer N, Blanc G (2000) Hydrothermal impact on sediments of Red-Sea deeps by the study of some trace metals. Oceanologica Acta 23(7): 783-792

Piper DZ (1974) Rare earth elements in the sedimentary cycle: a summary. Chem Geol 14(4):285-304

Piper DZ, Graef PA (1974) Gold and rare-earth elements in sediments from the East Pacific Rise. Mar Geol 17:287-297

Plüger WL, Herzig PM, Becker KP, Deissman G, Schops D, Lange J, Jenisch A, Ladage S, Richnow HH, Schulze T, Michaelis W (1990) Discovery of hydrothermal fields at the Central Indian Ridge. Marine Mining 9:73-86

Poroshina IM, Krasnov SG, Dubinin EP, Gurevich NI (1992) Morphotectonic criteria of ore presence at the EPR. In: Gramberg IS, Ainemer AI (eds) Hydrothermal sulfide ores and metalliferous sediments of the ocean. Nedra, Sankt-Petersburg, pp 9-24

Pottorf RJ, Barnes HL (1983) Mineralogy, geochemistry, and ore genesis of hydrothermal sediments from the Atlantis II Deep, Red Sea. Econ Geol 5:198-223

Price RC, Kennedy AK, Riggs-Sneeringer M, Frey FA (1986) Geochemistry of basalts from the Indian Ocean triple junction: implications for generation and evolution of Indian Ocean ridge basalts. Earth Planet Sci Lett 78:379-396

Proceedings of the Ocean Drilling Program (1988–2004) Scientific Results. Vol. 101-194. Texas A&M University, Ocean Drilling Program, College Station

Puchelt H, Laschek D (1984) Marine Erzvorkommen im Roten Meer. Fridericiana Zeitschrift der Universität Karlsruhe 34:3-17

Puchelt H, Stoffers P (1997) Massive sulfide ores from the Red Sea (Cruises Sonne 29 and Meteor 31). In: Vokes Symposium, NTNU

Ramboz C and Danis M (1990) Superheating in the Red Sea? The heat mass balance of the Atlantis II deep revisited. Earth Planet Sci Lett 97:190

Ramboz C, Oudin E and Thisse Y (1988) Geyser-type discharge in Atlantis II Deep, Red Sea: evidence of boiling from fluid inclusions in epigenetic anhydrite. Canadian Mineralogist 26:765

Ravizza G, McMurtry GM (1993) Osmium isotopic variations in metalliferous sediments from the East Pacific Rise and the Bauer Basin. Geochim Cosmochim Acta 57:4301-4310

Rea DK (1978) Asymmetric sea-floor spreading and a nontransform axis offset: the East Pacific Rise 20°S survey area. Geol Soc Amer Bull 89:836-844

Rea DK, Blakely RJ (1975) Short-wavelength magnetic anomalies in a region of rapid sea-floor spreading. Nature 255:126-128

Rea DK, Leinen M (1986) Neogene controls on hydrothermal activity and paleoceanography of the Southeast Pacific Ocean. In: Init Reports DSDP. Vol. 92. US Govt Print-Office, Wash (DC), pp 597-617

Rehkämper M, Nielsen SG (2004) The mass balance of dissolved thallium in the ocean. Mar Chem 85:125-139

Reid JL (1981) On the mid-depth circulation of the World ocean. In: Evolution of physical oceanography. MIT Press, Cambridge (MA), pp 70-111

Renard V, Hekinian R, Francheteau J, Ballard RD, Backer H (1985) Submersible observations at the axis of the ultrafast-spreading East Pacific Rise (17°30' to 21°30'S). Earth Planet Sci Lett 75(4):339-353

Renner RM, Glasby GP, Walter (1997) Endmember analysis of metalliferous sediments from the Galapagos Rift and East Pacific Rise between 2°N and 42°S. Applied Geochemistry 12:383-395

Revelle RR (1944) Marine bottom samples collected in the Pacific Ocean by the "Carnegie" on her seventh cruise. Carnegie Inst Publ 556. Carnegie Inst, Wash (DC)

Rich J, Johnson GL, Jones J, Campsie J (1986) A significant correlation between fluctuations in seafloor spreading rates and evolutionary pulsation. Paleoceanography 1(1):85-95

Riddihough RP (1977) A model for recent plate interactions off Canada's West Coast. Canadian Journ Earth Sci 14:384-396

Riech V (1990) Calcareous ooze, volcanic ash, and metalliferous sediments in the Quaternary of the Lau and North Fiji Basins. Geologisches Jahrbuch D92:109-162

Riech V, Marchig V, Sunkel G, Weiss W (1990) Hydrothermal and volcanic input in sediments of the Lau Back-Arc Basin, SW Pacific. Marine Mining 9:183-203

Rimskaya-Korsakova MN, Dubinin AV (2003) Rare earth elements in sulfides of submarine hydrothermal vents from the Atlantic Ocean. Doklady Rossiiskoi Akademii Nauk (Trans Russian Acad Sci) 389(5):672-676

Robertson AFH, Boyle JF (1983) Tectonic setting and origin of metalliferous sediments in the Mesozoic Tethys Ocean In: Rona PA, Boström K, Laubier L,

Smith KL Jr (eds) Hydrothermal processes at seafloor spreading centers. Plenum Press, New York, pp 595-663

Robertson AHF, Hudson JD (1973) Cyprus umbers: chemical precipitates on a Tethyan ocean ridge. Earth Planet Sci Lett 28(1):93-101

Robertson AHF, Varnavas SP, Panagos AG (1987) Ocean ridge origin and tectonic setting of Mesozoic sulfide and oxide deposits of the Argolis peninsula of the Peloponnesus, Greece. Sedimentary Geology 53(1):1-32

Rona PA (1984) Hydrothermal mineralization at seafloor spreading centers. Earth Science Review 20(1):1-104

Rona PA, Scott SD (1993) A special issue on sea-floor hydrothermal mineralization: new perspectives Preface. Econ Geol 88(8):1935-1975

Rona PA, Trivett DA (1992) Discrete and diffuse heat transfer at ASHES vent field, Axial Volcano, Juan de Fuca Ridge. Earth Planet Sci Lett 109(1):57-71

Rona PA, Klinkhammer G, Nelson TA, Trefry JH, Elderfield H (1986) Black smokers, massive sulfides and vent biota at the Mid-Atlantic Ridge. Nature 321(6065):33-37

Rona PA, Widenfalk L, Boström K (1987) Serpentinized ultramafics and hydrothermal activity at the Mid-Atlantic Ridge crest. Journ Geophys Res 91(B2): 1417-1427

Rona PA, Hannington MD, Thompson G (1990) Evolution of hydrothermal mounds, TAG hydrothermal field, Mid-Atlantic Ridge 26°N, 45°W. EOS 71: 1650

Rona PA, Bogdanov YuA, Gurvich EG, Rimski-Korsakov NA, Sagalevich AM, Hannigton MD, Thompson G (1993a) Relict hydrothermal zones in the TAG hydrothermal field, Mid-Atlantic Ridge 26°N, 45°W. Journ Geophys Res 98(B6):9715-9730

Rona PA, Hannington MD, Raman CV (1993b) Major active and relict seafloor hydrothermal mineralization: TAG hydrothermal field, Mid-Atlantic Ridge 26°N, 45°W. Econ Geol 88:1989-2017

Ronov AB, Yaroshevsky AA, Migdisov AA (1990) Chemical composition of the Earth crust and geochemical balance of the major elements. Nauka, Moscow

Rosendahl BR, Hekinian R, Briqueu L and Shipboard Scientific Party (1980) Initial Reports of the Deep-Sea Drilling Project. Vol. 54. US Govt Print-Office, Wash (DC)

Rosson RA, Nealson KH (1982) Manganese binding and oxidation by spores of a marine bacillus. Journ Bacteriol 151:1027-1034

Roth SE, Dymond J (1989) Transport and settling of organic material in a deep-sea hydrothermal plume: evidence from particle flux measurements. Deep-Sea Res 36(8):1237-1254

Rudnicki MD (1990) Hydrothermal plumes at the Mid-Atlantic Ridge. PhD Diss. St Edmund's College, Cambridge

Rudnicki MD, Elderfield H (1992) Theory applied to the Mid-Atlantic Ridge hydrothermal plumes: the finite-difference approach. Journ Volcanol Geotherm Res 50:161- 172

Rudnicki MD, Elderfield H (1993) A chemical model of the buoyant and neutrally buoyant plume above the TAG vent field, 26 degrees N, Mid-Atlantic Ridge. Geochim Cosmochim Acta 57:2939-2957

Rudnicki MD, German CR (2002) Temporal variability of the hydrothermal plume above the Kairei vent field, 25°S, Central Indian Ridge. Geochem Geophys Geosyst 3(2):1010 doi:101029/2001GC000240

Ruhlin DE, Owen RM (1986) The rare earth geochemistry of hydrothermal sediments from the East Pacific Rise: Examination of a seawater scavenging mechanism. Geochim Cosmochim Acta 50(3):393-400

Rydell H, Kraemer T, Boström K, Joensuu O (1974) Postdepositional injection of uranium rich solutions into the East Pacific Rise. Mar Geol 17:154-164

Sakai H, Osaki S and Tsukagishi M (1970) Sulfur and oxygen isotopic geochemistry of sulfate in black ore deposits of Japan. Geochem Journ 4:27

Satian MA (1985) Metalliferous silicites of the Late Mesozoic ophiolite series from the Lesser Caucasus. Izvestiya AN Armyanskoy SSR (News of the Acad Sci of the Armenian SSR). Earth Sciences XXXVIII(5):19-32

Savenko AV (2000) On the mechanism of boron accumulation in oceanic metalliferous sediments Oceanology 40(2):201-203

Sayles FL, Bischoff JL (1973) Ferromanganoan sediments in the equatorial East Pacific. Earth Planet Sci Lett 19(3):331-336

Sayles FL, Ku T-L, Bowker PC (1975) Chemistry of ferromanganoan sediments of the Bauer Deep. Geol Soc Amer Bull 1975 86:1423-1431

Sborshchikov IM, Almukhamedov AI, Matveenkov VV (1981) Geologic structure of the Red Sea axial zone. Vulkanologiya i Seismologiya (Volcanology and Seismology) 2:49-59

Scheirer DS, Baker ET, Johnson KTM (1998) Detection of hydrothermal plumes along the Southern Indian Ridge near the Amsterdam-St Paul Plateau. Geophys Res Lett 25:97-100

Schlich R (1982) The Indian Ocean: aseismic ridges, spreading centers and ocean basins. In: The ocean basins and margins. Vol. 6. Plenum Press, New York, pp 51-147

Schlich R, Munschy M, Royer JY (1986) Proposal for oceanic drilling at the Rodriguez Triple Junction (Indian Ocean). Ocean Drilling Program, Texas A&M University, College Station

Schmidt M, Botz R, Faber E, Schmitt M, Poggenburg J, Garbe-Schönberg D, Stoffers P (2003) High-resolution methane profiles across anoxic brine-seawater boundaries in the Atlantis II, Discovery, and Kebrit Deeps (Red Sea). Chem Geol 200(3-4):359-375

Schneider W, Schumann D (1979) Tonminerale in Normalsedimenten, hydrothermal beinflussten Sedimenten und Erzschlammen des Roten Meeres. Geologische Rundschau 68:631-648

Schoell M (1980) The hydrogen and carbon isotopic composition of methane from natural gases of various origins. Geochim Cosmochim Acta 44:649

Scholten JC, Stoffers P, Walter P, Plüger W (1991) Evidence for episodic hydrothermal activity in the Red Sea from the composition and formation of hydrothermal sediments, Thetis Deep. Tectonophysics 190:109-117

Scholten JC, Stoffers P, Garbe-Schönberg D, Moammar M (2000) Hydrothermal Mineralization in the Red Sea. In: Cronan DS (ed) Handbook of Marine Mineral Deposits. CRC Press, Boca Raton, pp 369-395

Schrader EL, Furbish WJ, Mattey D, May, JA (1980) Geochemistry and carbonate petrology of selected sediment samples from Deep Sea Drilling Project Leg 54, Eastern Pacific. In: Powell R (ed) Initial Reports DSDP. Vol. 54. Texas A & M University, Ocean Drilling Program, College Station, pp 319-328

Schreider AA (1982) Distribution of relative momentary spreading rates in the Pacific Ocean. In: Lisitzin AP (ed) Geology of oceans and seas. Vol. 2. IOAN, Moscow, pp 81-82

Schreider AA (1985) Variations of relative momentary spreading rates in the Atlantic Ocean. In: Yastrebov VS (ed) Technical means and methods of ocean and sea exploration. IOAN, Moscow, pp 60-61

Schultz A, Delaney JR, McDuff RE (1992) On the partitioning of heat flux between diffuse and point source seafloor venting. Journ Geophys Res 97(B12):12299-12314

Schwan W (1985) The worldwide active Middle/Late Eocene geodynamic episode with peaks at ±45 and ±37 my BP, and implications on problems of orogeny and seafloor spreading. Tectonophysics 115:197-234

Schwertmann U, Murad E (1983) Effect of pH on the formation of goethite and hematite from ferrihydrite. Clays and Clay Minerals 31(4):277-284

Schwertmann U, Thalmann H (1976) The influence of Fe(II), (Si), and pH on the formation of lepidocrocite and ferrihydrite during oxidation of aqueous $FeCl_2$ solutions. Clay Mineral 11:189-200

Sclater JG, Fisher RL, Patriat P, Tapscott C, Parsons B (1981) Eocene to recent development of the Southwest Indian Ridge, a consequence of the evolution of the Indian Triple Junction. Geophys Journ Roy Astron Soc 64(3):587-604

Scott MR, Scott RB, Morse JW, Betzer PR, Butler LW, Rona PA (1978) Metal enriched sediments from the TAG Hydrothermal Field. Nature 276(5690): 811-813

Selk BW (1982) Sedimentary evidence for hydrothermal activity in the Blanco Fracture Zone. EOS 63(45):1146

Shackleton NJ, Crowhurst S, Hagelberg T, Pisias NG, Schneider DA (1995) A new Neogene time scale: Application to Leg 138 Sites. In: Proceedings of the Ocean Drilling Program, Scientific Results. Vol. 138. Texas A & M University, Ocean Drilling Program, College Station, pp 73-101

Shadlun TN, Bortnikov NS, Bogdanov YuA, Gurvich EG (1992) Mineral composition, textures and conditions of formation of modern sulfide ores in the Manus Basin rift zone. Geology of Ore Deposits 5:3-21

Shank T, Fornari DI, Wigham B, Scheirer D, Getsiv J, Perfit M, Tolstoy M (2000) Biology of newly-discovered AHA hydrothermal vent field near 1°44'N on the East Pacific Rise axis. EOS Trans Amer Geophys Union 81(48):T51D-15

Shanks WC, Bischoff JL (1977) Ore transport and depositional system: A geochemical model. Geochim Cosmochim Acta 41(10):1507-1519

Shanks WC, Bischoff JL (1980) Geochemistry, sulfur isotope composition and accumulation rates of Red Sea geothermal deposits. Econ Geol 75(3):445-459

Shearme S, Cronan DS, Rona PA (1983) Geochemistry of sediments from the TAG hydrothermal field, Mid-Atlantic Ridge at latitude 26°N. Mar Geol 51(3/4):269-291

Sherrell RM, Field MP, Ravizza G (2000) Uptake and fractionation of rare earth elements on hydrothermal plume particles at 9°45'N, East Pacific Rise. Geochim Cosmochim Acta 63(11/12):1709-1722

Shimmield GB, Price NB (1988) The scavenging of U, ^{230}Th and ^{231}Pa during pulsed hydrothermal activity at 20°S, East Pacific Rise. Geochim Cosmochim Acta 52(3):669-677

Shterenberg LE, Vasil'eva GL, Voronin BI, Korina EA (1981) Gold- and silver-minerals in metalliferous sediments of the Pacific ocean. Izvestiya Akademii Nauk SSSR (News of the USSR Acad Sci). Geology Series 7:151-154

Singer A, Stoffers P (1981) Hydrothermal vermiculite from the Atlantis II Deep, Red Sea. Clays and Clay Minerals 29:454-458

Singer A, Stoffers P (1987) Mineralogy of a hydrothermal sequence in a core from the Atlantis-II Deep, Red Sea. Clay Minerals 22(3):251-267

Sirocko F, Garbe-Schonberg Dieter, Devey C (2000) Processes controlling trace element geochemistry of Arabian Sea sediments during the last 25,000 years. Glob Planet Change 26: 217–303

Skornyakova NS (1964) Dispersed iron and manganese in sediments of the Pacific ocean. Litologiya i Poleznyye Iskopayemyye (Lithology and Mineral Resources) 5:3-20

Skornyakova NS (1970) Dispersed iron and manganese in sediments of the Pacific ocean. In: Bezrukov PL (ed) Pacific Ocean. Vol. 6. Book 1. Sedimentation in the Pacific Ocean. Nauka, Moscow, pp 159-202

Sohrin Y, Gamo T, and INDOYO Shipboard Party (1999) CTD observations to search for hydrothermal activity on the Southwest Indian Ridge and the Central Indian Ridge just north of the Rodriguez Triple Junction: The Yokosuka/Shinkai MODE'98 Leg 3 INDOYO cruise. Deep-Sea Res I 15:7-11

Sorokin YuI (1977) Production of microflora. In: Vinogradov ME (ed) Oceanology. Ocean biology. Vol. 2, Biological productivity of the ocean. Nauka, Moscow, pp 209-233

Spiess FN, MacDonald KC, Atwater T, Ballard R, Carranza A, Cordoba D, Cox C, Diaz Garcia VM, Francheteau J, Guerrero J, Hawkins J, Haymon R, Hessler R, Juteau T, Kastner M, Larson R, Luyendyk B, Macdougall JD, Miller S, Normark W, Orcutt J, Rangin C (1980) East Pacific Rise: hot springs and geophysical experiments. Science 207(4438):1421-1433

Spivack AJ, Edmond JM (1987) Boron isotope exchange between seawater and the oceanic crust. Geochim Cosmochim Acta 51(5):1033-1043

Spivack AJ, Palmer MR, Edmond JM (1987) The sedimentary cycle of boron. Geochim Cosmochim Acta 51(7):1939-1950

Staritsyna GN, Krasnov SG, Poroshina IM, Tabunov SM (1989) Magmatism and tectonics of the axial zone of the East Pacific Rise in the area of 13°N in relation to hydrothermal activity. Doklady Akademii Nauk SSSR (Trans USSR Acad Sci) 308(2):432-435

Stepanova TV, Krasnov SG (1992) Chemical composition of ores. In: Gramberg IS, Ainemer AI (eds) Hydrothermal sulfide ores and metalliferous sediments of the ocean. Nedra, Sankt-Petersburg, pp 79-88

Stoffers P, Moammar M and Scientific party (1998) Cruise Report Sonne 121 Red Sea. Reports 88. Geol-Paläont Inst Univ Kiel, Kiel

Stoffers P, Schmitz W, Glasby GP (1983) Geochemistry of deep sea sediments from the Indian Ocean collected during DSDP legs 22, 24, 26 and 27. Chemie der Erde 42(1):15-30

Strakhov NM (1976) Problems of geochemistry of recent ocean lithogenesis. Nauka, Moscow

Strekopytov SV (1997) Geochemistry of the rare earth elements in sediments and nodules of the Pacific Ocean. IO RAN, Moscow

Sudarikov SM (1992) Characteristics of submarine thermal springs. In: Gramberg IS, Ainemer AI (eds) Hydrothermal sulfide ores and metalliferous sediments of the ocean. Nedra, Sankt-Petersburg, pp 39-57

Sudarikov SM, Krasnov SG, Kreiter II (1992) Dispersion halos of thermal springs in ocean water. In: Gramberg IS, Ainemer AI (eds) Hydrothermal sulfide ores and metalliferous sediments of the ocean. Nedra, Sankt-Petersburg, pp 107-128

Sudarikov SM, Davidov MP, Bazelyan VL, Tarasov VG (1995) Distribution and transformation of Fe and Mn in hydrothermal plumes and sediments and the potential function of microbiocenoses. In: Parson LM, Walker CL, Dixon DR (eds) Hydrothermal vents and processes. Geol Soc London Spec Publ 87. Geol Soc London, London, pp 249-255

Sudarikov SM, Roumiantsev AB (2000) Structure of hydrothermal plumes at the Logachev vent field, 14°45'N, Mid-Atlantic Ridge: evidence from geochemical and geophysical data. Journ Volcanol Geotherm Res 101:245-252

Sval'nov VN (1986) Calcareous biogenic-chemogenic accumulations. In: Lisitzin AP, Bogdanov YuA (eds) Metalliferous sediments of the Red Sea. Nauka, Moscow, pp 20-21

Sval'nov VN (1991) Dynamics of pelagic lithogenesis. Nauka, Moscow

Sval'nov VN, Shevchenko AYa, Uspenskaya TYu, Gurvich EG, Zavadskaya NN (1985) Composition of hydrothermal crusts from the Red Sea. Litologiya i Poleznyye Iskopayemyye (Lithology and Mineral Resources) 3:40-52

Tambiev SB (1988) Biogenic sedimentation of manganese from hydrothermal solutions of submarine high-temperature vents. In: Lisitzin AP (ed) Geology of oceans and seas. Vol. 3. IOAN, pp 123-124

Tambiev SB (1989) Biogenic sedimentation of manganese in areas of submarine high-temperature vents (on a study of material from sediment traps). Doklady Akademii Nauk SSSR (Trans USSR Acad Sci) 307(2):457-461

Tambiev SB, Demina LL (1992) Biogeochemistry and fluxes of manganese and some other metals in region of hydrothermal activities (Axial Mountain, Juan de Fuca Ridge and Guaymas Basin, Gulf of California). Deep-Sea Res 39(3/4):687-703

Tapscott GR, Patriat P, Fisher RL, Sclater JG, Hoskins H, Parsons B (1980) The Indian Ocean triple junction. Journ Geophys Res 85(B9):4723-4739

Taylor SR (1968) Geochemistry of andesites. In: Ahrens LH (ed) Origin and distribution of the elements. Pergamon Press, Oxford, New York, pp 559-583

Tebo BM, Emerson S (1986) Microbial manganese(II) oxidation in the marine environment: A quantitative study. Biogeochemistry 2:149-161

Thisse Y, Guennoc P, Pouit G, Nawab Z (1983) The Red Sea: a natural geodynamic and metallogenic laboratory. Episodes 1(3):3-9

Thompson G (1983) Hydrothermal fluxes in the ocean. In: Riley JP, Chester R (eds) Chemical oceanography. Vol. 8. Academic Press, London, pp 271-337

Thompson RE, Delaney GR, McDuff RE, Janecky DR, McClain JS (1992) Physical characteristics of the Endeavor Ridge hydrothermal plume during July 1988. Earth Planet Sci Lett 111(1):141-154

Thurnherr AM, Richards KJ (2001) Hyrdographic and high-temperature heat flux of the Rainbow hydrothermal site 36°14'N, Mid-Atlantic Ridge. Journ Geophys Res 106(C5):9411-9426

Tikhomirov VN (1987) Helium in bottom water of the triple junction area. In: Lisitzin AP, Gurvich EG (eds) Metalliferous sediments of the Indian Ocean. Nauka, Moscow, pp 97-100

Trefry JH, Metz S (1989) Role of hydrothermal precipitates in the geochemical cycling of vanadium. Nature 342:531-533

Trefry JH, Metz S, Trocine RP, Nelsen TA (1986) Geochemistry, and history of hydrothermal precipitates in a Mid-Atlantic vent field. EOS 67(44):1022

Trefry JH, Butterfield DB, Metz S, Massoth GJ, Trocine RP, Feely RA (1994) Trace metals in hydrothermal solutions from Cleft segment on the southern Juan de Fuca Ridge. Journ Geophys Res 99(B3):4925-4935

Trocine RP, Trefry JH (1988) Distribution and chemistry of suspended particles from an active hydrothermal vent site on the Mid-Atlantic Ridge at 26°N. Earth Planet Sci Lett 88(1/2):1-15

Truesdell AH (1975) Summary of Section II: geochemical techniques in exploration. In: United Nations Symposium on the development and use of geothermal resources. Proceedings. Vol. 1. San Francisco, p Iiii

Tunnicliffe V, Botros M, de Burgh ME, Dinet A, Johnson HP, Juniper SK, McDuff RE (1986) Hydrothermal vents of Explorer Ridge, North-East Pacific. Deep-Sea Res 33(3):401-412

Turekian KK, Bertine KK (1971) Deposition of molybdenum and uranium along the major ocean ridge systems. Nature 229:250-251

Turekian KK, Wedepohl KH (1961) Distribution of the elements in some major units of the Earth's crust. Geol Soc Amer Bull 72(2):175-192

Varnavas SP (1988) Hydrothermal metallogenesis at the Wilkes Fracture Zone - East Pacific Rise intersection. Mar Geol 79:77-104

Varnavas SP, Panagos AG (1984) Mesozoic metalliferous sediments from the ophiolites of Ermolini, Greece; analogue to mid-ocean ridge ferromanganese deposits. Chem Geol 42:227-242

Vinogradov AP (1962) Average contents of chemical elements in the major types of igneous rocks of the Earth's crust. Geokhimiya (Geochemistry) 7:555-571

Vinogradov AP (1967) Introduction to ocean geochemistry. Nauka, Moscow

Vinogradov ME, Shushkina EA, Kopelevich OV, Sherbastov SV (1996) Photosynthetic production of the World Ocean on satellite and expedition data. Okeanologiya (Oceanology) 36(4):566-575

Vogt PR (1989) Volcanogenic upwelling of anoxic, nutrient-rich water: A possible factor in carbonate-bank/reef demise and benthic faunal extinctions? Geol Soc Amer Bull 101:1225-1245

Von Damm KL (1990) Seafloor hydrothermal activity: Black smoker chemistry and chimneys. Ann Rev Earth Planet Sci 18:173-204

Von Damm KL (1995) Controls on the chemistry and temporal variability of seafloor hydrothermal fluids. In: Humphris SE, Zierenberg RA, Mullineaux LS, Thomson RE (eds) Seafloor hydrothermal systems: Physical, chemical, biological, and geological interactions. Geophysical Monograph 91. AGU, pp 222-247

Von Damm KL, Bischoff JL (1987) Chemistry of hydrothermal solutions from the southern Juan de Fuca Ridge. Journ Geophys Res 92(B11):11334-11346

Von Damm KL, Edmond JM, Grant B, Measures CI, Walden B, Weiss RF (1985a) Chemistry of submarine hydrothermal solutions at 21°N, East Pacific Rise. Geochim Cosmochim Acta 49(11):2197-2220

Von Damm KL, Edmond JM, Measures CI, Grant B (1985b) Chemistry of submarine hydrothermal solutions at Guaymas Basin, Gulf of California. Geochim Cosmochim Acta 49(11):2221-2237

Von Damm KL, Lilley MD, Shanks III WC, Brockington M, Bray AM, O'Grady KM, Olson E, Graham A, Proskurowski G, the SouEPR Science Party (2003) Extraordinary phase separation and segregation in vent fluids from the southern East Pacific Rise. Earth Planet Sci Lett 206:365-378

Von der Borch CC, Rex RW (1970) Amorphous iron oxide precipitates in sediments cored during Leg 5, Deep Sea Drilling Project. In: Init Reports DSDP. Vol. 5. US Govt Print-Office, Wash (DC), pp 541-544

Von der Borch CC, Nesteroff WD, Galehouse J (1971) Iron-rich sediments cored during Leg 8 of the Deep Sea Drilling Project. In: Init Reports DSDP. Vol. 8. US Govt Print-Office, Wash (DC), pp 829-836

Walker SL, Baker ET (1988) Particle-size distributions within hydrothermal plumes over the Juan de Fuca Ridge. Mar Geol 78:217-226

Walter P, Stoffers P (1985) Chemical characteristics of metalliferous sediments from eight areas of Galapagos Rift and East Pacific Rise between 2°N and 42°S. Mar Geol 65(3/4):271-287

Walter P, Schwarz B, Scholten J, Stoffers P (1986) Geochemie und Mineralogie der Sedimente des Mittelindischen Rückens zwischen 21°S und der Rodriguez Triple Junction (SO 43). In: GEMINO 2: Geothermale Metallogenese Indischer Ozean. RWTH, Aachen, pp 1-38

Walter P, Stoffers P, Glasby GP, Marchig V (1990) Major and trace element geochemistry of Lau Basin sediments. Geologisches Jahrbuch D92:163-188

Weiss RF (1977) Hydrothermal manganese in the deep sea: scavenging residence time and $Mn/^3He$ relationships. Earth Planet Sci Lett 37:257-262

Werner R, Hoernle K, Barckhausen U, Hauff F (2003) Geodynamic evolution of the Galapagos hot spot system (Central East Pacific) over the past 20 m.y.:

Constraints from morphology, geochemistry, and magmatic anomalies. Geochem Geophys Geosyst 4(12):1-28, doi:10.1029/2003GC000576
Whitmarsh RB, Weser OE, Ross DA and Shipboard Scientific Party (1974) Initial Reports of the Deep-Sea Drilling Project. Vol. 23. US Govt Print-Office, Wash (DC)
Wilson DS, Hey RN, Nishimura CE (1984) Propagation as a mechanism of ridge reorientation of the Juan de Fuca Ridge. Journ Geophys Res 89:9215-9226
Winterer EL (1973) Sedimentary facies and plate tectonics of equatorial Pacific. Amer Assoc Petrol Geol Bull 57(2):265-282
Wolery TJ, Sleep NH (1976) Hydrothermal circulation and geochemical flux at mid-ocean ridges. Journ Geol 84:249-275
Wong H, Degens E (1984) The crust beneath the Red Sea – Gulf of Aden – East African system: a review. Mitt Geol Paleontol 56:53-94
Yamamoto K (1987) Geochemical characteristics and depositional environments of cherts and associated rocks in the Franciscan and Shimato terranes. Sedimentary Geology 52(1/2):65-108
Zaikov VV, Zaikova EV (1986) Metalliferous sediments in Devonian basalts from the Mugojary. In: Ore-bearing, ore and nonmetallic formations of the Ural. UNC AN SSSR, Sverdlovsk, pp 23-25
Zaikov VV, Zaikova EV, Korinevsky VG (1984) Possible analogues of ocean metalliferous sediments and Fe-Mn nodules in Silurian of the Southern Ural. In: Kholodov VN (ed) Evolution of sedimentary ore formation in the Earth history. Nauka, Moscow, pp 178-189
Zaikova EV (1985) Genetic heterogeneity of Paleozoic siliceous rocks from the North Mugojary. Doklady Akademii Nauk SSSR (Trans USSR Acad Sci) 282(5):1206-1209
Zaikova EV (1991) Siliceous rocks of ophiolite associations (on example of Mugojary). Nauka, Moscow
Zelenov KK (1964) Iron and manganese in exhalations of the Banu-Vuhu underwater volcano (Indonesia). Doklady Akademii Nauk SSSR (Trans USSR Acad Sci) 155(6):1317-1320
Zhabina NN, Sokolov VS (1982) Compounds of sulfur in ore-bearing sediments from the Atlantis II Deep (Red Sea). Geologicheskij Zhurnal (Geological Journal) 2:64-72
Zhivago AV (1987) Geomorphology and tectonics in the area of triple junction of mid-ocean ridges in the western part of the Indian Ocean. In: Lisitzin AP, Gurvich EG (eds) Metalliferous sediments of the Indian Ocean. Nauka, Moscow, pp 7-20
Zierenberg RA, Shanks WC (1983) Mineralogy and geochemistry of epigenetic features in metalliferous sediments, Atlantis-II Deep, Red Sea. Econ Geol 78(1):57-73
Zonenshain LP, Khain VV (1989) Variations in tectonic activity of the Earth during the past 150 million years. Doklady Akademii Nauk SSSR (Trans USSR Acad Sci) 305(2):402-405
Zonenshain LP, Monin AS, Sorokhtin OG (1981) Tectonics of the Red Sea rift at 18°. Geotektonika (Geotectonics) 2:3-22

Zonenshain LP, Kuz'min MA, Lisitzin AP, Bogdanov YuA, Baranov BV (1989) Tectonics of the Mid-Atlantic rift valley between the TAG and MARK areas (24-26°N): evidence for vertical tectonism. Tectonophysics 159(1-2):1-23

Zonenshain LP, Kuz'min MI, Baranov BV (1992) Relief, tectonics and magmatism. In: Lisitzin AP (ed) Hydrothermal formations of the Mid-Atlantic Ridge. Nauka, Moscow, pp 12-44

INDEX

Kolibri 360 GmbH
Ferdinand-Jühlke-Straße 7
99095 Erfurt, Deutschland
produktsicherheit@kolibri360.de